PROSTAGLANDIN ABSTRACTS

A Guide to the Literature
Volume 1: 1906-1970

42,

PROSTAGLANDIN ABSTRACTS

A Guide to the Literature
Volume 1: 1906-1970

(ONLY)

Richard M. Sparks

Population Information Program

Department of Medical and Public Affairs
Science Communication Division
The George Washington University Medical Center
Washington, D. C.

IFI/PLENUM · NEW YORK-WASHINGTON-LONDON

Library of Congress Cataloging in Publication Data

Sparks, Richard M
  Prostaglandin abstracts.

  Abstracts prepared by authors of articles or members of the Science Communication
Division staff, George Washington University Medical Center.
  CONTENTS: v. 1. 1906-1970.
  1. Prostaglandin—Abstracts. 2. Prostaglandin—Bibliography. I. George Washington
University, Washington, D.C. Medical Center. Science Communication Division. II.
Title. [DNLM: 1. Prostaglandins—Abstracts. ZQU90 S736p]
QP801.P68S65   574.1'9247   73-21780
ISBN 0-306-67011-9

The preparation of this volume was supported by the United States Agency for
International Development through contracts with the Worcester Foundation
for Experimental Biology and the George Washington University.

IFI/Plenum Data Corporation is a subsidiary of
Plenum Publishing Corporation
227 West 17th Street, New York, N. Y. 10011

United Kingdom edition published by Plenum Press, London
A Division of Plenum Publishing Company, Ltd.
Davis House (4th Floor), 8 Scrubs Lane, Harlesden, London, NW10 6SE, England

CONTENTS

ACKNOWLEDGEMENTS

The editor wishes to thank the many organizations and individuals who have contributed to the success of this project, especially the Upjohn Company of Kalamazoo, Michigan, Drs. J. R. Weeks, J. C. Babcock, G. E. Underwood and the late Dr. L. E. Rhuland; the Worcester Foundation for Experimental Biology of Shrewsbury, Massachusetts, Ms. Julia Lobotsky and Ms. Gerry Seward; and staff personnel who prepared and edited much of the material in this volume, James R. Heath III and Nancy E. Stillerman.

FOREWORD

This volume contains abstracts of most of the significant scientific literature dealing with prostaglandins published between 1906—when certain biologically active tissue extracts first stimulated the speculation of researchers—and 1970—when the use of prostaglandins for experimental control of fertility and induction of labor had been reported from six countries. Of the more than 4000 articles now identified which were published between 1906 and 1972, approximately half had appeared in print by the end of 1970. A second volume will cover the material printed in 1971 and 1972 with special emphasis on the role of prostaglandins in reproductive physiology.

The reasons for publishing this considerable compilation of data are twofold. On the one hand, prostaglandins, a family of fatty acid derivatives first identified in human semen, have been shown to have great and varied effects on all aspects of human physiology due to activity at the cellular level as mediators in the formation of cyclic AMP. The study of prostaglandins is considered today one of the most promising fields in the biological sciences.

Secondly, as is now recognized, prostaglandins play an important role in reproduction, influencing both male and female fertility. It is hoped that this volume will make the results of early prostaglandin research available to investigators throughout the world, including especially those in developing countries who may not have easy access to such material.

At a time of increasing world concern over rapid population growth and excess or unwanted pregnancies, researchers throughout the world are seeking new or improved methods for the voluntary control of fertility. Prostaglandins hold great promise in this field, and have already been used clinically in 15 countries to induce menstruation, or terminate first and second trimester pregnancies. So far, side effects such as nausea, vomiting, and diarrhea have limited the usefulness of the presently available prostaglandin compounds and/or delivery systems under most circumstances.

But the search for better means of birth control continues. It seems likely that prostaglandins, which are the natural mediators of reproductive functions, or some new prostaglandin analogs administered in the form of a medicated vaginal tampon, intrauterine injection, or possibly in combination with other drugs will eventually play an important role in fertility control.

At the time of this writing, the United States Food and Drug Administration is expected to license prostaglandin $F_{2\alpha}$ for sale in the United States within the next two months for use in the termination of second trimester pregnancy. The Upjohn Company of Kalamazoo, Michigan hopes to begin marketing immediately.

The Office of Population, United States Agency for International Development, has supported applied research in prostaglandins with special reference to use in developing countries. USAID seeks specifically to identify or improve present birth control technologies so that they can be effectively and conveniently used by men and women in developing countries. This volume is designed above all to assist researchers in the developed and developing countries who share this objective. We hope that this increase in the availability of scientific information about prostaglandins will hasten the day when what the United Nations has called "the basic right of every person to determine the number and spacing of their offspring" is not only a basic human right but also a genuine reality throughout the world.

November 1, 1973

R. T. Ravenholt, M.D., M.P.H.
Director

J. Joseph Speidel, M.D., M.P.H.
Chief, Research Division

Office of Population
United States Agency for International Development

EXPLANATORY NOTE

Every citation in this bibliography was verified with a copy on file in the library of the Population Information Program of the George Washington University Medical Center. If a satisfactory author's or prepared abstract accompanied the article, it was used and credit was given. Otherwise, abstracts were prepared by members of the Science Communication Division staff. Those individuals, identified by their initials at the end of the abstracts are: Frank D. Bradley, Mary Ellen Hashmall, James R. Health III, John W. Johnston, Irvin C. Mohler, R. Allen Pierce, Paul Richmond, Charles W. Shilling, Richard M. Sparks, Nancy E. Stillerman, George Wolfhard, Michael Towers, and Arthur R. Turner. If a prepared abstract was used, but the format was changed without alteration of the content regarding prostaglandins, it has been identified as "Author modified." If no informative abstract accompanied the article, and none could be prepared, as, for instance, when a translation could not be readily obtained, the citation has been included without an abstract.

Since this bibliography is a guide to the literature, citations and abstracts have been prepared for all the types of literature—journal articles, reviews, books, book chapters, newspaper articles, abstracts, symposia—in which prostaglandins have been discussed. In the preparation of abstracts, however, primary consideration has been given to the research significance of the article, that is to the amount of new information contained in it. For example, most editorials and reviews are not abstracted in detail because the scientific information upon which they were based is not new or fully provided in the article. Furthermore, the original research is also included in this bibliography and can be readily identified through the subject and author indexes.

Citations are grouped chronologically and alphabetically by the surname of the senior author. Editorials, notes, and materials with no author indicated are listed under Anonymous, alphabetically by title.

0001
JAPPELLI, G. and G.M. SCAFA
  Sur les effets des injections intraveineuses d'extrait prostatique du chien. [On the effects of intra-
    venous injections of dog prostate extract.]
  Archivio Italiano di Biologia 45: 165-189. 1906.

An extract of dog prostate was injected intravenously into dogs and rabbits, and some general effects
were noted. When injected intravascularly into the dog, the extract showed a high level of toxicity;
there was a paralysing effect on the central respiratory system and a definite influence on heart beat
and amplitude. Effects on the blood were also noted; in the dog there was an anticoagulative effect
and a coagulative one in the rabbit. It is suggested that all of these are due to a nucleo-protein present
in the prostate extract (NES) 0015

0002
THAON, P.
  Toxicite des extraits de prostate; leur action sur la pression arterielle et le rythme cardiaque. [Toxic-
    ity of prostate extracts; their action on arterial pressure and cardiac rhythm.]
  Comptes Rendus des Seances de la Societe de Biologie et de ses Filiales. 63: 111-112. 1907.

Rabbits were injected intravenously with a saline preparation of bull prostate gland. 5-10 sec after
administration, arterial pressure rose steadily 2 to 3 centimeters for 30 to 40 sec. This was followed
by a severe toxic effect which caused a fall in pressure below the original level and eventual death of
the animal. No intracardial coagulations or pulmonary infarctions were found during autopsy. The
fall in pressure was not thought to be a hypotensive effect, but due to the extreme toxicity of the
substance. It is questioned whether the release of the substance in the blood could cause some of the
effects associated with prostate problems. Experiments using distilled water preparations of bull
prostates are also mentioned. These seemed to have more effect on arterial pressure than the saline
preparations. (NES) 0019

0003
DUBOIS, C. and L. BOULET
  Action des extraits de prostate sur les mouvements de l'intestin. [Action of prostate extracts on
    intestinal movement.]
  Comptes Rendus des Seances de la Societe de Biologie et de ses Filiales. 71: 536-537. 1911.

In vivo and in vitro experiments were conducted to test the reaction of dog, cat, rabbit and sheep
intestinal strips (duodenum, jejumum, ileum or rectum) to extracts from the prostate gland. Aqueous
and glycerine preparations were made using tissue from dog, bull and ram prostates. The intestinal
strips were dipped into 2 test tubes, one containing 200 cc of a normal nutritive serum and another
containing 200 c.c. of the serum plus the prostate extract (the quantity of the extract varied from
0.25g-5g). The 2 test tubes were kept at a constant temperature of 38-39°C. The strips dipped in the
prostate preparations were partially or completely inhibited (6 cases). There was also a lessening in
tonus. Experiments were also carried out on the intestine in situ. 5g were injected into the saphenous
vein, a reduction was noted in 3 out of 4 cases. Similar in vitro experiments were performed using
extracts (aqueous and glycerine) from testicle, spleen, muscle and liver; however, the results were
contradictory. It is suggested that the inhibiting effect of the prostate extract is due to an internal
secretion of the gland. (NES) 0020

0004
BATTEZ, G., and L. BOULET
   Action de l'extrait de prostate humaine sur la vessie et sur la pression arterielle. [Action of human prostate extract on the vascular system and arterial pressure.]
   Comptes Rendus des Seances de la Societe de Biologie et de ses Filiales. 44: 8-9. 1913.

   The action of human prostate extract on the dog vascular system was compared to prior experiments using dog prostate extract. A dog was injected in the saphenous vein with 12.5 cg/kg of the extract. Violent contractions of the vessels lasting 25 sec were seen 26 sec after the injection. A definite hypotensive effect was also noted. However, a second injection had less effect. It is concluded that human and dog prostate have the same mode of action. (NES) 0023

0005
DUBOIS, C. and L. BOULET
   Action des extraits de prostate hypertrophiee sur la vessie. [Actions of extracts from hypertrophied prostate on the vascular system.]
   Comptes Rendus des Seances de la Societe de Biologie et de ses Filiales. 82: 1054-1055. 1919.

   8 dogs, anesthetized with either curarine or chloralose, were injected intravenously with extract (20cg - 1 g/kg/dog) from a hypertrophied human prostate gland. In 3 cases, there was a small to large constriction of the vessels and a fall in arterial pressure. In 5, the pressure was equally lowered, but there was less contraction. It is also noted that after removal of tumor from the prostate, that the function of the prostate returned to normal and that the unknown vascular hormone is present which maintains contractility. (NES) 0048

0006
KURZROK, R. and C.C. LIEB
   Biochemical studies of human semen. II. The action of semen on the human uterus.
   Proceedings of the Society for Experimental Biology and Medicine. 28: 268-272. 1930.

   Human seminal fluid was added to organ baths containing strips of human uterine muscle. A given uterine strip would relax in response to one semen sample and contract in response to a sample from a different man. However, a semen sample could cause relaxation in a strip from one uterus and simultaneously contract a strip from a different uterus. The uteri from fertile women responded to semen by relaxing, while uteri from women with a history of sterility responded by contracting. The active principle in semen was dializable and did not require the presence of sperm to be active. (JRH) 0065

0007
von EULER, U.S.
   An adrenaline-like action in extracts from the prostatic and related glands.
   Journal of Physiology. 81: 102-112. 1934.

   Two biologically active substances were isolated from the prostate gland and seminal vesicles of several mammalian species. One of these resembled epinephrine in biological activity and chemical properties. The other showed vasodilator and smooth muscle stimulating properties. (JRH) 0001

0008
COCKRILL, J.R., E.G. MILLER, Jr., and R. KURZROK
The substance in human seminal fluid affecting uterine muscle.
American Journal of Physiology. 122: 577-580. 1935.

Semen samples collected from 75 men were tested on human myometrial strips from over 400 uteri. Samples from 65 of the men caused relaxation to the uterus while the others caused contractions. Mild alkali treatment of contracting semen caused it to relax the uterus. Strong alkali (pH 11 or higher) inactivated it completely as did boiling. Treatment of boiled semen with acetyl chloride restored its activity. Esterase diminished the activity of the semen. Physostigmine enhanced the effect of semen on uterine muscle. Pretreatment of the muscle with atropine completely blocked the effect of the semen on the uterine strips. Because of these properties and because they could be duplicated by acetylcholine, the authors conclude that the active principle in seminal fluid is acetylcholine or something closely related to it. (JRH) 0052

0009
von EULER, U.S.
A depressor substance in the vesicular gland.
Journal of Physiology. 84: 21P-22P. 1935.

The relationship of vesiglandin, isolated from monkey seminal vesicles; substance P, isolated from the brain and small intestine; and the vasodepressor extracted from human seminal fluid and seminal vesicle secretion is discussed. It is suggested that the human seminal fluid depressor might be a mixture of vesiglandin and substance P. (JRH) 0003

0010
GOLDBLATT, M.W.
Properties of human seminal plasma.
Journal of Physiology. 84: 208-218. 1935.

Both human seminal plasma and absolute alcohol (or acetone) extracts of seminal plasma i v caused a strong reduction in blood pressure in anesthetized (ether or urethane) or pithed cats and rabbits both with and without atropine. The active principle was destroyed by boiling with alkali but not nitrous acid. By other assays and tests, it was shown that the active principle was not histamine or choline. It was not possible to determine if the active substance came from the seminal vesicle or prostate gland. (JRH) 0002

0011
von EULER, U.S.
On the specific vaso-dilating and plain muscle stimulating substances from accessory genital glands in man and certain animals (prostaglandin and vesiglandin).
Journal of Physiology. 88: 213-234. 1936.

The chemical and biological properties of prostaglandin were investigated. Prostaglandin is soluble in water, alcohol, acetone, and under certain conditions, ether and chloroform. It is destroyed by very low pH (less than pH 1) and was unstable at pH values greater than 7. Biologically, it is a powerful vasodilator and hypotensive agent and a stimulator of intestinal muscle. PG generally enhances uterine activity and its biological effects are not blocked by atropine. (JRH) 0053

0012
von EULER, U.S.
Action of adrenaline, acetylcholine and other substances on nerve-free vessels (human placenta).
Journal of Physiology. 93: 129-143. 1938.

Prostaglandin obtained from seminal fluid produced a sustained vasoconstriction when added to the fluid perfusing isolated human placentas. This type of effect had not been previously observed. It was attributed to direct action on the vascular muscle cells. (JRH) 0004

0013
ASPLUND, J.
A quantative determination of the content of contractive substances in human sperm and their significance for the motility and vitality of the spermatozoa.
Acta Physiologica Scandinavica. 13: 103-108. 1947.

The amount of contractive substance in human seminal plasma was bioassayed on rabbit small intestine as prostaglandin equivalents using a PG extract as a standard. The seminal plasma contained not only prostaglandin and choline but at least one more unidentified contractive agent which is not sensitive to atropine. (JRH) 0031

0014
ASPLUND, J.
Some preliminary experiments in connection with the effect of prostaglandin on the uterus and tubae in vivo.
Acta Physiologica Scandinavica. 13: 109-114. 1947.

Intravenous or intraperitoneal administration of prostaglandin extracted from human seminal fluid caused an increase in uterine motility and a lowering of blood pressure in urethane anesthetized rabbits. A comparison of routes of administration showed the following order of effectiveness on rabbit blood pressure: intravenous > intraperitoneal > intrauterine > intravaginal. A modified Rubin's test for tubal patency showed that the PG increased patency of the tubal ostium when adminstered by all of the above routes. These results suggest that PG may be important in sperm transport. (JRH) 0028

0015
BERGSTROM, S.
Prostaglandinets kemi. [Chemistry of prostaglandins.]
Nordisk Medicin. 42: 1465-1466. 1949.

Abstract only. Prostaglandin has been concentrated about 500,000 times from ovine prostate glands. The active principle contains an acidic group and the methyl ester is slightly more active than the free acid. The biological activity ceases on acylation or hydrogenation. In countercurrent distribution of the purest samples the activity parallels the U.V. absorption maximum at 280 mμ. No nitrogen could be demonstrated. (Author) 0007

0016
von EULER, U.S.
Prostaglandinets autofarmakologiska verkningar. [Pharmacodynamic effects of prostaglandin].
Nordisk Medicin. 42: 1463-1465. 1949.

A brief review is given of the pharmacodynamic actions of a specific constituent of the secretions from accessory genital glands of the human male and of sheep. The active principle, called prostaglandin, is a weak unsaturated fatty acid forming water-soluble salts with barium and alkalis. Prostaglandin exerts a strong stimulating action on smooth muscle, such as the intestine, the bladder and the uterus. It lowers the blood pressure chiefly by contracting the vessels of the liver. The vessels of the lungs and placenta are likewise constricted. In cases of sterility the prostaglandin content of human semen was often found to be low. (Author) 0008

0017
HAWKINS, D.F. and A.H. LABRUM
Function of the prostate.
British Medical Journal. 2: 1236. 1956.

This letter to the editor is in reference to an earlier letter which suggested that the prostate gland is functionless. It is pointed out that large quantities of prostaglandin have been reported in prostatic secretions and that these PG's may have an important physiological role in reproductive physiology. It was found that the PG content of human seminal fluid is directly related to the maintenance of spermatozoal activity and inversely related to the percentage of abnormal spermatozoa in semen samples. (JRH) 0011

0018
BERGSTROM, S. and J. SJOVALL
The isolation of prostaglandin.
Acta Chemica Scandinavica. 11: 1086. 1957.

Abstract only. The authors report the isolation of PGF in pure crystalline form. The compound (102°-103°) is an unsaturated hydroxy acid which does not contain nitrogen. It causes contraction of the rabbit duodenum at ~5×10^{-9} g/ml. There is at least one other active acidic factor in the sheep prostate gland. (JRH) 0014

0019
ELIASSON, R.
A comparative study of prostaglandin from human seminal fluid and from prostate gland of sheep.
Acta Physiologica Scandinavica. 39: 141-146. 1957.

Prostaglandins extracted from human seminal fluid and from prostate gland of sheep have been compared chemically and biologically. It has not been possible to distinguish the two substances by paper chromatography, thermostability or biological tests, including rabbit jejunum, rat jejunum, guinea-pig ileum, guinea-pig uterus, rabbit blood pressure and cat blood pressure. (Author) 0013

0020
PICKLES, V.R.
A plain-muscle stimulant in the menstruum.
Nature. 180. 1198-1199. 1957.

Acetone extracts of menstrual fluid were dissolved in saline. The extract had smooth muscle stimulating properties which were not abolished by atropine in several bioassay systems. Doses of extract
too small to cause stimulation alone potentiated the response to other smooth muscle stimulants
(histamine, acetlycholine or epinephrine). On the basis of pharmacological and solubility characteristics, the active principle appears to be a lipoid-soluble acid of the general type described by Vogt.
(JRH) 0029

0021
VOGT, W.
Pharmacologically active lipid-soluble acids of natural occurrence.
Nature. 179: 300-304. 1957.

Prostaglandins are briefly discussed in this review as being lipid soluble acids with smooth muscle
stimulating and hypotensive properties. (JRH) 0012

0022
ELIASSON, R.
Formation of prostaglandin in vitro.
Nature. 182: 256-257. 1958.

Prostaglandin was extracted from seminal fluid and sheep seminal vesicles and the PG content bioassayed. The seminal fluid contained about 10 times more PG per ml than the vesicular tissue did.
When seminal fluid was incubated for varying periods, no change in PG content was found. However,
incubation of homogenized seminal vesicles showed the production of up to 50 units of PG/g tissue in
160 min. Addition of lecithinase A (from cobra venom) to the homogenates increased PG production,
but compound 48/80 (a releasor of histamine and slow reacting substance) inhibited biosynthesis of
PG in vitro. (JRH) 0016

0023
ELIASSON, R.
The spasmolytic effect of patulin.
Experentia. 14: 460-461. 1958.

Patulin 3-10µg/ml (a compound isolated from *Penicillium patulum* culture filtrates) inhibited the
contraction of the isolated guinea pig ileum induced by a variety of stimulators including prostaglandin. (JRH) 0018

0024
TOH, C.C. and A. MOHIUDDIN
Vasoactive substances in the nasal mucosa.
British Journal of Pharmacology and Chemotherapy. 13: 113-117. 1958.

Extracts of the nasal mucosa of sheep and dogs contained two intestinal smooth muscle stimulating
substances. One was shown to be histamine. The other was not identified. However, on the basis of

physical and chemical properties, it appears to be a lipid soluble, unsaturated acid. It is apparently not darmstoff, irin, the unsaturated G acid of human plasma or the unsaturated fatty acid of envenomed egg yolk, all of which stimulate smooth muscle. (JRH) 0022

0025
VOGT, W.
Naturally occurring lipid-soluble acids of pharmacological interest.
Pharmacological Reviews. 10: 407-435. 1958.

Prostaglandin literature is comprehensively reviewed with a detailed discussion of pharmacological properties. 12 prostaglandin references are cited. (JRH) 0017

0026
AMBACHE, N.
Further studies on the preparation, purification and nature of irin.
Journal of Physiology. 146: 255-294. 1959.

An extensive chemical and biological study of irin was performed. It is considered unlikely that irin and prostaglandin are identical since they have different heat stabilities and R_F values. (JRH) 0024

0027
BERGSTROM, S., H. DUNER, U.S. von EULER, B. PERNOW and J. SJOVALL
Observations on the effects of infusion of prostaglandin E in man.
Acta Physiologica Scandinavica. 45: 145-151. 1959.

Chemically pure PGE was infused i v in doses of 0.2-0.7μg/kg/min in two healthy male subjects over periods of 4-10 min. Tachycardia, reddening of the face, headache and an oppressive feeling in the chest were noted. Systemic arterial blood pressure and cardiac output fell moderately. (Authors modified) 0026

0028
BERGSTROM, S., R. ELIASSON, U.S. von EULER and J. SJOVALL
Some biological effects of two crystalline prostaglandin factors.
Acta Physiologica Scandinavica. 45: 133-144. 1959.

The effects of two crystalline fractions of PGE and PGF were tested on a variety of isolated organs and on the blood pressure of the rabbit. Characteristic differences in the activity ratio for PGE and PGF were noted for different organs. The biological actions of partially purified preparations of prostaglandin could be largely but not wholly explained by assuming that they contained a mixture of PGF and PGE. (Authors modified) 0025

0029
ELIASSON, R.
Studies on prostaglandin. Occurrence, formation and biological actions.
Acta Physiologica Scandinavica. 46 Supp. 158: 1-73. 1959.

This extensive early review of PG literature contains sections on: general technique for isolation and assay; results of assays of biological substances for PG's; and the relationship of PG's to each other and other compounds with similar properties. There are 137 references cited. (JRH) 0021

0030
KARLSON, S.
The influence of seminal fluid on the motility of the non-pregnant human uterus.
Acta Obstetrica et Gynecologica Scandinavica. 38: 503-521. 1959.

Human seminal fluid introduced into the vagina or uterus generally caused contraction of the corpus and relaxation of the cervical canal. It is mentioned that seminal fluid contains prostaglandins. (JRH) 0027

0031
PICKLES, V.R.
Myometrial responses to the menstrual plain-muscle stimulant.
Journal of Endocrinology. 19: 150-157. 1959.

Human menstrual stimulant was extracted and examined on a variety of bioassay tissues. The menstrual stimulant was a lipid which caused contraction of human and guinea pig uteri in vitro and the guinea pig uterus in vivo. The response of both types of uterine tissue was similar, but there was more variability with the human uterus. (JRH) 0030

0032
BERGSTROM, S. and J. SJOVALL
The isolation of prostaglandin E from sheep prostate glands.
Acta Chemica Scandinavica. 14: 1701-1705. 1960.

Crystalline PGE was isolated from frozen sheep prostate glands. This PG had smooth muscle stimulating and vasodepressor properties. The most likely empirical formula is $C_{20}H_{34}O_5$. (JRH) 0034

0033
BERGSTROM, S. and J. SJOVALL
The isolation of prostaglandin F from sheep prostate glands.
Acta Chemica Scandinavica. 14: 1693-1700. 1960.

Sheep prostate glands were extracted for prostaglandin. Pure crystalline PGF was obtained. On the basis of various chemical analyses, the formula $C_{20}H_{36}O_5$ has been suggested. (JRH) 0033

0034
BERGSTROM, S., L. KRABISCH and J. SJOVALL
Smooth muscle stimulating factors in ram semen.
Acta Chemica Scandinavica 14: 1706-1710. 1960.

Extraction and chromatographic separation of lipid soluble, acidic material in ram seminal plasma has shown the presence of at least two different factors with smooth muscle stimulating and blood pressure lowering effects. One of the factors shows certain properties similar to those of the PGE previously isolated from sheep prostate glands. (Authors) 0032

0035
CHAKRAVARTY, N.
The occurrence of a lipid-soluble smooth-muscle stimulating principle ('SRS') in anaphylactic re-
action.
Acta Physiologica Scandinavica. 48: 167-177. 1960.

Both histamine and slow reacting substance (SRS) were released from guinea pig lungs which had
been sensitized to albumin when they were challenged with antigen (anaphylactic reaction). There
was a strong correlation between the amount of histamine and SRS released in the different animals.
Chemically, SRS was found to be a lipid soluble acid. Differential sensitivity tests of prostaglandin
and SRS on the guinea pig ileum and rabbit duodenum indicated that they were different substances.
(JRH) 0035

0036
ELIASSON, R. and N. POSSE
The effect of prostaglandin on the nonpregnant human uterus in vivo.
Acta Obstetrica et Gynecologica Scandinavica. 39: 112-126. 1960.

Prostaglandin extracted from human seminal plasma was placed in the posterior fornix of 8 normal
nonpregnant human volunteers of proven fertility at different phases of the menstrual cycle. Recordings of uterine activity showed that during the menstrual, proliferative and secretory phases, the
uterus was almost insensitive to PG, but at the time of ovulation PG caused an increase in activity of
the corpus followed in some cases by an inhibition. However, PG caused an inhibition when oxytocin
was infused simultaneously. It seems likely that PG in human seminal fluid could facilitate the
migration of sperm from the vagina into the uterus. (JRH) 0036

0037
PICKLES, V.R. and H.J. CLITHEROE
Further studies of the menstrual stimulant.
Lancet. 2: 959-960. 1960.

The menstrual stimulant has been separated into 3 active components. The most potent of the 3
resembles irin and prostaglandin. (JRH) 0037

0038
HAWKINS, D.F. and A.H. LABRUM
Semen prostaglandin levels in fifty patients attending a fertility clinic.
Journal of Reproduction and Fertility. 2: 1-10. 1961.

The levels of the smooth muscle stimulating substance prostaglandin in the semen of patients attending a fertility clinic were estimated. The data obtained were analysed in relation to the clinical and
laboratory findings on these patients. While the results of this preliminary survey were not conclusive,
they did suggest that the semen prostaglandin level is related in some positive way to the processes
leading to conception. Further investigation of a larger group of subfertile couples where the wife is
normal and the husband oligospermic is particularly indicated. (Authors) 0039

0039
ABRAHAMSSON, S., S. BERGSTROM and B. SAMUELSSON
The absolute configuration of prostaglandin F_{2-1}.
Journal of the Chemical Society. 332. 1962.

A brief report on the absolute configuration of PGF_2 is made. A structural diagram is deduced from an electron density map. (JRH) 0046

0040
BERGSTROM, S. and B. SAMUELSSON
Isolation of prostaglandin E_1 from human seminal plasma.
Journal of Biological Chemistry. 237: PC3005-PC3006. 1962.

Prostaglandins were extracted from human seminal plasma. The prostaglandins were purified and separated by chromatography. The presence of PGE_1 was proven by chromatographic and mass spectra analysis and comparison to authentic PGE_1 (JRH) 0043

0041
BERGSTROM, S., L. KRABISCH, B. SAMULELSSON and J. SJOVALL
Preparation of prostaglandin F from prostaglandin E. Prostaglandins and related factors 6.
Acta Chemica Scandinavica. 16: 969-974. 1962.

Reduction of prostaglandin E(PGE) $(C_{20}H_{34}O_5)$ with sodium borohydride yielded two isomeric reduction products PGF_1 and PGF_2, $(C_{20}H_{36}O_5)$ of which PGF_1 was identical with a compound earlier isolated from sheep prostate glands and designated PGF. The data presented indicate that PGF is an unsaturated dihydroxyketocarboxylic acid and that PGF_1 and PGF_2 are stereoisomeric, unsaturated trihydroxycarboxylic acids. (Authors) 0010

0042
BERGSTROM, S., B. SAMUELSSON and J. SJOVALL
Structure, occurrence and properties of prostaglandins and SRS.
Federation Proceedings. 21: 281. 1962.

Abstract only. Borohydride reduction of PGE yielded PGF_1 and PGF_2. This indicated that PGE is an unsaturated monohydroxy-monoketo acid. It has been established unequivocally that PGE_1 is one of the smooth muscle active factors present in the seminal fluid from both sheep and man. (JRH) 0045

0043
BERGSTROM, S., R. RYHAGE, B. SAMUELSSON and J. SJOVALL
The structure of prostaglandin E, F_1 and F_2.
Acta Chemica Scandinavica. 16: 501-502. 1962.

The authors briefly describe the methods they used to determine the structure of PGE, PGF_1 and PGF_2. The determinations were based on mass spectral analysis of fractions isolated by gas chromatography and they were confirmed by synthesis. (JRH) 0041

0044

BYGDEMAN, M. and R. ELIASSON
The effect of prostaglandin on the motility of the non-pregnant human uterus in vitro.
International Journal of Fertility. 7: 354. 1962.

Abstract only. The effect of PG's extracted from human seminal plasma on human myometrium in vitro was studied. The myometrium was most sensitive to the PG's at or near the time of ovulation. The normal response to seminal PG's was relaxation; however, if the sensitivity of the myometrium was decreased experimentally or if very low doses of PG were used, the myometrium responded with contractions. (JRH) 0009

0045

DAKHIL, T. and W. VOGT
Hydroperoxyde als trager der darmerregenden wirkung hochungesattigter fettsauren. [Hydroperoxides as the carriers of intestinal effects of polyunsaturated fatty acids.]
Naunyn-Schmiedebergs Archiv fur Experimentelle Pathologie und Pharmakologie. 243: 174-186. 1962.

It is briefly mentioned that prostaglandins have smooth muscle stimulating properties that are not dependent on the formation of hydroperoxides as is the case with the polyunsaturated acids investigated in this paper. (JRH) 0047

0046

SANDBERG, F., A. INGELMAN-SUNDBERG, L. LINDGREN and G. RYDEN
In vitro effects of prostaglandin on different parts of the human Fallopian tube.
Nature. 193: 781-782. 1962.

Human Fallopian tubes from non-pregnant women were cut into 4 equal segments and suspended in an organ bath. Prostaglandin extracted from seminal plasma was added to the bath. The PG stimulated the proximal segment and inhibited the 3 distal segments. No differences were noted with tubes from women in the proliferatory or secretory phases. (JRH) 0044

0047

ABRAHAMSSON, S.
A direct determination of the molecular structure of prostaglandin F_{2-1}.
Acta Crystallographica. 16: 409-418. 1963.

The molecular structure and configuration of PGF_{2-1} was determined by X-ray diffraction and electron-density projection. The molecular structure was found to be the same as that previously proposed on the basis of other methods. Structural diagrams are given. (JRH) 0128

0048

ANGGARD, E. and S. BERGSTROM
Biological effects of an unsaturated trihydroxy acid ($PGF_{2\alpha}$) from normal swine lung. Prostaglandin and related factors 13.
Acta Physiologica Scandinavica. 58: 1-12. 1963.

The effects on various isolated smooth muscle organs and on the cardiovascular system of cats and rabbits were studied. The rabbit duodenum, guinea-pig ileum, rat duodenum and colon, hen rectal

12

caecum, guinea pig uterus and rat uterus were stimulated to a slow contraction. Rabbit duodenum and oestrogen treated rat uterus proved the most sensitive organs, with a threshold concentration as low as 0.001 μg/ml. I v injection in rabbits of 15-30 μg/kg caused a fall in blood pressure, with no changes in right ventricular pressure or heart rate. I v injection of 15-30 μg/kg in cats increased right ventricular pressure, with a concomitant depression of the systemic blood pressure. A bradycardia, which could be abolished by vagotomy or atropine, occurred 10-30 seconds later. It was concluded, that in cats the fall in blood pressure resulted partly from a reduced cardiac output due to constriction of the pulmonary vessels, partly from the bradycardia, and to a minor extent only from direct effects on muscle vessels. (Authors) 0105

0049
ANGGARD, E. and B. SAMUELSSON
Smooth muscle stimulating lipids in sheep lung.
Acta Physiologica Scandinavica. 59 Supp. 213: 170. 1963.

Abstract only. Smooth muscle stimulating lipids were isolated from sheep lung. The main active agent was $PGF_{2\alpha}$. On the basis of isotope dilution techniques, the $PGF_{2\alpha}$ content was about 0.5 μg/g lung. PGE_2 was also present. (JRH) 0104

0050
BERGSTROM, S. and U.S. von EULER
The biological activity of prostaglandin E_1, E_2 and E_3.
Acta Physiologica Scandinavica. 59: 493-494. 1963.

The relative activity of PGE_1, PGE_2 and PGE_3 on the rabbit jejunum, the guinea pig ileum, and on rabbit blood pressure was measured. On the rabbit jejunum, the order of reactivity was $PGE_2 > PGE_1 > PGE_3$; on the rabbits' blood pressure, the order was $PGE_1 > PGE_2 > PGE_3$. On the guinea pig ileum, the order was $PGE_1 > PGE_2 > PGE_3$. The differences in activity ratios of the 3 PG's on the rabbit jejunum and blood pressure can be used to estimate the amount of PGE_2 in a mixture of the 3 compounds. (JRH) 0107

0051
BERGSTROM, S., R. RYHAGE, B. SAMUELSSON and J. SJOVALL
Degradation studies on prostaglandins. Prostaglandins and related factors 10.
Acta Chemica Scandinavica. 17: 2271-2280. 1963.

PGE_1, $PGF_{1\alpha}$ and $PGF_{1\beta}$ were oxidatively degraded to determine their structural configuration. The side chains were found to be $-(CH_2)_6 COOH$ and $-CH=CH-CHOH-(CH_2)_4CH_3$. The technique and structures found are described. (JRH) 0112

0052
BERGSTROM, S., L.A. CARLSON and L. ORO.
Effects of prostaglandins on catecholamine-induced changes in the free fatty acids of plasma in the dog.
Biochemical Journal. 89: 27p-28p. 1963.

Abstract only. Intravenous injections of 5 μg of PGE_1, PGE_2 or PGE_3 during the constant infusion of norepinephrine (NE) greatly reduced the increase in plasma free fatty acids (FFA), glycerol and blood pressure normally caused by NE in anesthetized dogs. $PGF_{1\alpha}$ was without effect. Continuous intra-arterial infusion of PGE_1 had more effect on continuous NE infusion than intravenous infusion did (blocked FFA increased and caused a fall in blood pressure). However, PGE_1 only slightly reduced the glucose response to NE. (JRH) 0123

0053
BERGSTROM, S. and B. SAMUELSSON
    Isolation of prostaglandin E_1 from calf thymus. Prostaglandins and related factors 20.
    Acta Chemica Scandinavica 17: S282-S287. 1963

    Lipid soluble, acidic material with smooth muscle stimulating activity has been extracted from calf thymus and separated by different chromatographic techniques. PGE_1 was identified as the major biologically active component. The concentration of this factor, determined by isotope experiments was found to be 0.8 μg per g tissue. (Authors modified) 0113

0054
BERGSTROM, S.
    Prostaglandins—a group of hormonal compounds of widespread occurrence.
    Biochemical Pharmacology. 12: 413-414. 1963.

    The author briefly reviews prostaglandin literature dealing with the occurrence of different prosta-glandins from various animal sources. There are 9 references cited. (JRH) 0135

0055
BERGSTROM, S., R. RYHAGE, B. SAMUELSSON and J. SJOVALL
    Prostaglandins and related factors 15. The structures of prostaglandin E_1, $F_{1\alpha}$ and $F_{1\beta}$.
    Journal of Biological Chemistry. 238: 3555-3564. 1963.

    PGE_1 isolated from sheep vesicular glands has been shown to have the structure (2-(6-carboxyhexyl)-3-(3-hydroxyocten-1-yl)-4-hydroxycyclopentanone. Reduction of prostaglandin E_1 yields two epi-meric compounds, prostaglandin $F_{1\alpha}$ and prostaglandin $F_{1\beta}$, which differ in the steric position of the hydroxyl group formed by reduction of the keto group. (Authors modified) 0117

0056
BYDGEMAN, and R. ELIASSON
    A comparative study on the effect of different prostaglandin compounds on the motility of the
        isolated human myometrium.
    Medicina Experimentalis. 9: 409-415. 1963.

    PGE_1, PGE_2, PGE_3, $PGF_{1\alpha}$ $PGF_{2\alpha}$ or a PG extract of human seminal fluid were added to an organ bath containing strips of human myometrial muscle. The PGE's and seminal fluid extract all inhibited myometrial activity. The PGE's were all about equally potent in this test system. The PGF's (up to 0.5 μg/ml) only inhibited one out of 8 experiments. Half of the strips responded to the PGF's with a slight increase in tonus. The PGF's were also equipotent. (JRH) 0137

0057
BYGDEMAN, M. and R. ELIASSON
   The effect of prostaglandin from human seminal fluid on the motility of the non-pregnant human
      uterus in vitro.
   Acta Physiologica Scandinavica. 59: 43-51. 1963.

   Prostaglandin was extracted from pooled human seminal fluid and the concentration of PG/ml
   standardized by bioassay on rabbit jejunum. Human myometrial strips from 60 uteri were arranged
   for recording contractions and divided into groups according to the phase of the menstrual cycle. The
   PG inhibited spontaneous contractions in 90% of the uteri, but they were 3-5 times more sensitive to
   PG at the time of ovulation. Some uteri were stimulated by PG. The stimulated group could be
   divided into 2 subgroups: those which were stimulated at all dose levels, and those which were
   stimulated by low doses and inhibited by higher doses. (JRH) 0106

0058
BYGDEMAN, M. and R. ELIASSON
   Potassium and the reaction pattern of the human myometrium to prostaglandin in vitro.
   International Journal of Fertility. 8: 869. 1963.

   Abstract only. A total extract of prostaglandin from human seminal fluid (HSF-PG) normally produces inhibition of spontaneous motility in isolated myometrium from the human uterus. Occasionally, stimulation occurs instead. By means of variations in the extracellular potassium concentration, the sensitivity and the reactivity pattern to HSF-PG are influenced. The inhibitor action of
   HSF-PG is enhanced at low potassium concentrations ($<$5 mEquiv./1.) and is decreased at high
   potassium concentrations (10-16 mEquiv./1). Increasing the potassium in the bath fluid may also
   change the reactivity pattern from inhibition to stimulation. (Authors) 0101

0059
BYGDEMAN, M. and R. ELIASSON
   Potassium and the reactivity pattern of the isolated human myometrium to prostaglandin from
      human seminal fluid.
   Experientia. 19: 180-181. 1963.

   Strips of nonpregnant human myometrium were suspended in physiological solutions containing
   different concentrations of potassium. The strips were then exposed to prostaglandin extracted from
   human seminal plasma. Lowering the potassium level increased the inhibitory effect of the PG on the
   uterus. Increasing the potassium had the opposite effect, and in some cases, caused the PG to
   stimulate the uterine strips (13-16 m Eqv K^+/1). (JRH) 0124

0060
EGLINTON, G., R.A. RAPHAEL, G.N. SMITH, W.J. HALL and V.R. PICKLES
   Isolation and identification of two smooth muscle stimulants from menstrual fluid.
   Nature. 200: 960,993-995. 1963.

   Prostaglandins were extracted from 3,675 menstrual specimens. The prostaglandins were isolated and
   identified by a combination of chromatographic, mass spectral and bioassay tests. Two PG's were
   found; $PGF_{2\alpha}$ and PGE_2 with $PGF_{2\alpha}$ being most abundant. The possible physiological roles of these
   PG's in the menstrual cycle are discussed. (JRH) 0108

0061
ELIASSON, R.
Prostaglandin-properties, actions and significance.
Biochemical Pharmacology. 12: 405-412. 1963.

The author reviews the literature on prostaglandins. Special attention is given to the pharmacological properties of prostaglandins in reproductive physiology. There are 20 references cited. (JRH) 0136

0062
HALL, W.J. and V.R. PICKLES
The dual action of menstrual stimulant A2 (prostaglandin E_2).
Journal of Physiology. 169: 90P-91P. 1963.

Components A 1 and A 2 of the menstrual stimulant have been identified as $PGF_{2\alpha}$ and PGE_2, respectively. Component A 2 (PGE_2) has been shown to have 2 effects on the guinea pig myometrium: 1. a transient stimulation and 2. a long lasting potentiation of other stimulants such as vasopressin. (JRH) 0115

0063
HOLMES, S.W., E.W. HORTON and I.H.M. MAIN
The effect of prostaglandin E_1 on responses of smooth muscle to catechol amines, angiotensin and vasopressin.
British Journal of Pharmacology and Chemotherapy. 21: 538-543. 1963.

Intravenous injection of PGE_1 (10 μg/kg) into urethane anesthetized rabbits and rats and into pentobarbital anesthetized cats caused a long lasting reduction in the pressor response to epinephrine, norepinephrine, angiotensin or vasopressin. PGE_1 injected intraarterially reduced the pressor response on blood flow in the hind limb of anesthetized cats, but had no effect on the vasodilation caused by isoprenaline. PGE_1 (up to 20 μg/kg) reduced the duration, but not the force of cat nictitating membrane contractions in response to epinephrine or sympathetic preganglionic stimulation. PGE_1 had no effect on stimulation of rabbit isolated auricles by isoprenaline or epinephrine. Relaxations of the rabbit isolated duodenum induced by epinephrine were also unaltered. Contractions of the rabbit vas deferens in vivo and in vitro were reduced by PGE_1. The inhibitory effects of PGE_1 on catecholamines is non specific since it also inhibits other agents. (JRH) 0111

0064
HORTON, E.W., I.H.M. MAIN and C.J. THOMPSON
The action of intravaginal prostaglandin E_1 on the female reproductive tract.
Journal of Physiology. 168: 54P-55P. 1963.

PGE_1 (1 μg/kg) injected intravenously into urethane anesthetized rabbits caused a fall in blood pressure and reduced tone and peristalsis in the Fallopian tube. Acetylcholine or bradykinin produced a similar fall in blood pressure but did not alter tubal contractility. Intravaginal application of larger doses of PGE_1 caused similar effects after a variable delay. Tubal ligation or separation did not alter the effects of intravaginal PGE_1, indicating that the vaginally applied PGE_1 must be absorbed into the blood before causing its effect. PGE_1 also inhibited contractions of the cervix, but no effect was found on the uterine horns. (JRH) 0130

0065
HORTON, E.W.
  Action of prostaglandin E_1 on tissues which respond to bradykinin.
  Nature. 200: 892-893. 1963.

  The effects of PGE_1 were compared to the effects of bradykinin in a number of biological systems. Both PGE_1 and bradykinin caused vasodilation in the cat hind limb and both increased capillary permeability (Pontamine sky blue reaction). However, PGE_1, unlike bradykinin, did not: cause pain (human blister base); induce epinephrine release (cat adrenal medulla); or reduce air entry (guinea pig lung). PGE_1 caused contraction of the rat uterus, guinea pig ileum and rat duodenum, while brady-kinin relaxed them. (JRH) 0114

0066
HORTON, E.W. and I.H.M. MAIN
  A comparison of the biological activities of four prostaglandins.
  British Journal of Pharmacology and Chemotherapy. 21: 182-189. 1963.

  The relative activities of PGE_1, PGE_2, PGE_3 and $PGF_{1\alpha}$ were compared on the guinea pig ileum, rabbit jejunum, hamster colon, rat uterus, pithed rat blood pressure, cat gastrocnemius blood flow, cat hind limb skin blood flow and rabbit Fallopian tubal tone and peristalsis. The PGE's were qualitatively similar in effect in all systems with PGE_1 and PGE_2 being about equally effective, and PGE_3 always being less effective. $PGF_{1\alpha}$ was not as potent a vasodilator in the cat gastrocnemius or skin blood flow or as potent a depressor on the rabbit blood pressure. On the rabbit jejunum, $PGF_{1\alpha}$ was twice as potent as PGE_1, while PGE_1 was 40 times more active than $PGF_{1\alpha}$ on the guinea pig ileum. The only qualitative difference between the PG's was that the PGE's decreased tone and contractility in the rabbit Fallopian tube while $PGF_{1\alpha}$ increased tone. (JRH) 0110

0067
HORTON, E.W. and C.J. THOMPSON
  Thin layer chromatography and bio-assay of prostaglandins.
  Journal of Physiology. 167: 15P. 1963.

  A sensitive specific technique for assaying PGE_1 involving thin layer chromatography and parallel bioassay on atropinized rabbit duodenum and the hamster colon is described. The PG's from human seminal fluid were extracted with ether and dried. They were then dissolved in methanol-chloroform and chromatographed on silical gel plates. A marker plate with authentic PGE_1 was run simultaneously. The spot corresponding to PGE_1 was eluted and bioassayed. As little as 1 μg of PGE_1 can be detected. (JRH) 0103

0068
LEE, J.B., R.B. HICKLER, C.A. SARAVIS and G.W. THORN
  Sustained depressor effect of renal medullary extract in the normotensive rat.
  Circulation Research. 13: 359-366. 1963.

  Crude homogenates of rabbit medulla produced a sustained depression of blood pressure when injected intravenously into the normotensive, anesthetized, pentolinium-treated rat. Similar activity was present in rat medulla, but not in extracts of renin-free cortex or certain nonrenal tissues. The activity responsible for sustained depressor activity was a dialyzable, ethanol soluble, low molecular weight (<4500 mol wt) substance(s) which was resistant to a mixture of peptide hydrolases and destroyed by heating for one hour at 100°C. (Authors modified) 0129

0069
NUGTEREN, D.H.
  The enzymic conversion of γ-linolenic acid into homo-γ-linolenic acid.
  Biochemical Journal. 89: 28p-29p. 1963.

  Abstract only. The enzymatic conversion of γ-linolenic acid to homo-γ-linolenic acid by the microsomal or mitochondrial fraction of rat liver is described. (JRH) 0127

0070
PICKLES, V.R.
  Active lipids in menstrual fluid.
  Biochemical Pharmacology. 12: 429-430. 1963.

  A group of lipid smooth-muscle stimulants is liberated from the human endometrium during menstruation, and may in part be recovered from the menstrual fluid. These stimulants probably cause the strong rhythmical uterine contractions that expel the decidua menstrualis. (Author) 0126

0071
PICKLES, V.R. and W.J. HALL
  Some physiological properties of the "menstrual stimulant" substances A1 and A2.
  Journal of Reproduction and Fertility. 6: 315-317. 1963.

  The "menstrual stimulant Component A" substances, which have now been identified as certain prostaglandins, generally stimulate smooth-muscle preparations but may inhibit the human myometrium in certain circumstances. A tentative re-assessment of their physiological significance is suggested. (Authors) 0118

0072
RAMWELL, P.W. and J.E. SHAW
  The nature of non-cholinergic substances released from the cerebral cortex of cats on direct and
    indirect stimulation.
  Journal of Physiology. 169: 51P-52P. 1963.

  The nature of a substance released from the cat cerebral cortex after direct or indirect stimulation was studied. Based on electrophoretic and solubility characteristics, it was concluded that the substance is not acetylcholine, bradykinin, or serotonin and that is has an acidic nature. (JRH) 0131

0073
RAMWELL, P.W. and J.E. SHAW
  The spontaneous and evoked release of noncholinergic substances from the cerebral cortex of cats.
  Life Sciences. 6: 419-426. 1963.

  Superfusates of the somatosensory cortex of pentobarbital anesthetized cats contained a substance which stimulated the rat uterus in the presence of atropine, 2-Brom LSD, and which is not destroyed by incubation with chymotrypsin. Therefore, the active substance was not acetylcholine, serotonin or a kinin. The amount of active substance secreted could be increased by direct transcallosal stimulation, contralateral radial nerve stimulation or analeptic drugs. (JRH) 0120

18

0074
SAMUELSSON, B.
Isolation and identification of prostaglandins from human seminal plasma. 18. Prostaglandins and related factors.
Journal of Biological Chemistry. 238: 3229-3234. 1963.

The prostaglandins were extracted from human seminal plasma and the various PG's isolated by chromatographic means. The prostaglandins identified were: PGE_1, PGE_2, PGE_3, $PGF_{1\alpha}$ and $PGF_{2\alpha}$. Details of the techniques used are given. (JRH) 0116

0075
SAMUELSSON, B.
Prostaglandins and related factors. 17. The structure of prostaglandin E_3.
Journal of the American Chemical Society. 85: 1878-1879. 1963.

Experiments which elucidated the structure of PGE_3 are described. The structure was determined on the basis of mass spectrographic and nuclear magnetic resonance studies of derivatives of PGE_3 isolated from sheep seminal vesicles. A structural diagram is given. (JRH) 0125

0076
SAMUELSSON, B.
Prostaglandins of human seminal plasma.
Biochemical Journal. 89: 34p. 1963.

Abstract only. The prostaglandins were extracted from human seminal plasma. The PG's present were isolated, purified and identified by chromatography, mass spectrometry and infra red spectrophotometry. Seminal plasma contained PGE_1, PGE_2, PGE_3, $PGF_{1\alpha}$ and $PGE_{2\alpha}$. The concentration of PGE's was about 20 μg/ml and the concentration of PGF's was 3-5 μg/ml. (JRH) 0122

0077
SAMUELSSON, B. and G. STALLBERG
Structure and synthesis of a derivative of prostaglandin E_1. Prostaglandins and related factors 16.
Acta Chemica Scandinavica. 17: 810-816. 1963.

The synthesis of a derivative of PGE_1 is described. This compound was developed as an aid to the determination of prostaglandin structure. The techniques used in this synthesis are described. (JRH) 0109

0078
SANDBERG, F., A. INGELMAN-SUNDBERG and G. RYDEN
The effect of prostaglandin E_1 on the human uterus and the Fallopian tubes in vitro.
Acta Obstetrica et Gynecologica Scandinavica. 42: 269-278. 1963.

The effect of PGE_1 on the contractility of the human uterus and Fallopian tubes was measured in vitro. The Fallopian tubes were cut in 4 equal segments and grouped according to phase of the menstrual cycle (secretory, proliferative and post-menopausal). Uterine strips from the isthmus and corpus were grouped in the same way. PGE_1 stimulated the contractility of the proximal segment of the Fallopian tube and relaxed the distal 3 segments. PGE_1 inhibited the uterine strips, but the tonus of the isthmus was not affected. PGE_1 was more inhibitory on the tonus of the corpus during the proliferative phase of the cycle. In post-menopausal Fallopian tubes the first 2 segments showed the same type of response as the first segment in premenopausal tubes. A working hypothesis for the physiological role of PG's in the fertilization process is discussed. (JRH) 0121

0079
SANDBERG, F., A. INGELMAN-SUNDBERG and G. RYDEN
The effect of purified prostaglandin E_1 on the human uterus and tubes.
International Journal of Fertility. 8: 869. 1963.

Abstract only. Pure PGE_1 had the same biological effect as prostaglandin isolated from seminal plasma according to Eliasson. The effect on the tube consisted of contraction of the proximal quarter and relaxation of the remainder. The effect upon the uterus was relaxation. (Authors modified) 0102

0080
SANDBERG, F., A. INGELMAN-SUNDBERG and G. RYDEN
The specific effect of prostaglandin on different parts of the human Fallopian tube.
Journal of Obstetrics and Gynaecology of the British Commonwealth. 70: 130-134. 1963.

The Fallopian tubes from 64 women were cut into 4 segments of equal length and suspended in an organ bath. The tubes were divided into 4 groups according to the phase of the menstrual cycle (proliferative, ovulatory, secretory and post-menopausal). The prostaglandin was extracted from human seminal fluid. The proximal section of the tubes responded to the PG with an increase in amplitude of contraction and tonus, while the distal 3 segments relaxed. There was no difference in response between menstrual phases, but in post-menopausal tubes the proximal 2 segments showed responses similar to the proximal segment of the other tubes (i.e. increased contractility). The possible implications of PG actions on the Fallopian tubes in reproductive physiology are discussed. (JRH) 0119

0081
STEINBERG, D., M. VAUGHAN, P.J. NESTEL and S. BERGSTROM
Effects of prostaglandin E opposing those of catecholamines on blood pressure and on triglyceride breakdown in adipose tissue.
Biochemical Pharmacology. 12: 764-766. 1963.

PGE_1 (0.1 μg/ml) significantly reduced the lipolytic effects of epinephrine, norepinephrine, ACTH and glucagon on rat adipose tissue in vitro (measured as glycerol release). PGE_1 given as 2 doses 30 min apart also inhibited TSH stimulated lipolysis. PGE_2 produced similar results, but was apparently less potent, while PGE_3 had no effect in preliminary experiments. PGE_1 also produced some inhibition of basal lipolysis, but the results were variable. In pentobarbital anesthetized dogs, PGE_1 (12.5 μg/kg) alone produced a sharp fall in blood pressure. When PGE_1 was given with epinephrine or norepinephrine, it greatly reduced the normal rise in blood pressure due to the catecholamines. It is possible that the vasodepressor actions of PG's are due to inhibition of endogenous catecholamines. (JRH) 0132

0082
VOGT, W.
Pharmacologically active acidic phospholipids and glycolipids.
Biochemical Pharmacology. 12: 415-420. 1963.

It is mentioned that prostaglandins have been reported to form soluble salts with alkaline earth ions. This would indicate that acidic lipids could exert their effect by forming lipid soluble salts with Ca^{++} thus influencing transport of Ca^{++} through the cell membrane. (JRH) 0134

0083
ANGGARD, E. and B. SAMUELSSON
Prostaglandins and related factors. 28. Metabolism of prostaglandin E_1 in guinea pig lung: the structures of two metabolites.
Journal of Biological Chemistry. 239: 4097-4102. 1964.

Enzymes in the particle-free fraction of guinea pig lung homogenates convert PGE_1 (11α, 15-dihydroxy-9-ketoprost-13-enoic acid) into two less polar metabolites. The structures of these metabolites, 11α, 15-dihydroxy-9-ketoprostanoic acid and 11α-hydroxy-9, 15-diketoprostanoic acid have been determined. The syntheses of 11α-hydroxy-9, 15-diketoprost-13-enoic acid and of 11α-hydroxy-9, 15-diketoprostanoic acid are described. (Authors modified) 0215

0084
ANGGARD, E. and B. SAMUELSSON
Smooth muscle stimulating lipids in sheep iris. The identification of prostaglandin $F_{2\alpha}$. Prostaglandins and related factors 21.
Biochemical Pharmacology. 13: 281-283. 1964.

Irin was extracted from 6,878 sheep irides. Solvent partition and chromatographic techniques were used to separate the lipid soluble, smooth muscle stimulating components. One of the compounds was proven to be $PGF_{2\alpha}$. (JRH) 0231

0085
ANONYMOUS
Prostaglandin.
Lancet. 2: 134. 1964.

Prostaglandin research is reviewed with special emphasis given to its possible role in human reproductive physiology. (JRH) 0214

0086
BERGSTROM, S., L.A. CARLSON, L.G. EKELUND and L. ORO
Effect of prostaglandin E_1 on the metabolism of the free fatty acids of plasma in man.
Biochemical Journal. 92: 42P-43P. 1964.

Abstract only. Intravenous infusions of PGE_1 at a rate of 0.1-0.2 μg/kg/min increased the plasma levels of FFA and glycerol. At the same time the heart rate increased, while inconsistent changes were observed in the blood pressures. When norepinephrine and PGE_1 were infused together at rates of 0.2 μg/kg/min for 20 min, PGE_1 was found to inhibit or abolish the cardiovascular responses usually elicited by norepinephrine. The increase in the concentration of FFA and glycerol in plasma caused by norepinephrine was, however, only slightly reduced by PGE_1 in man. (Authors modified) 0219

0087
BERGSTROM, S., L.A. CARLSON and L. ORO
   Effect of prostaglandins on catecholamine induced changes in the free fatty acids of plasma and in
      blood pressure in the dog.
   Acta Physiologica Scandinavica. 60: 170-180. 1964.

   Dogs anesthetized with pentobarbital were constantly infused with norepinephrine (NE). PGE_1,
   PGE_2 and PGE_3 (5 μg i v) all reduced plasma free fatty acids (FFA) and blood pressure while $PGF_{1\alpha}$
   was ineffective. Constant infusion of PGE_1 intra-arterially was more effective than constant intravenous infusion in inhibiting NE-induced changes. Infusion of PGE_1 caused a fall in blood pressure
   below normal and almost completely blocked the normal NE-induced increase in plasma FFA and
   glycerol, but only slightly reduced the hyperglycemia induced by NE. FFA turnover studies revealed
   that PGE_1 completely blocked the NE-induced stimulation of lipolysis. (JRH) 0209

0088
BERGSTROM, S., H. DANIELSSON, D. KLENBERG and B. SAMUELSSON
   The enzymatic conversion of essential fatty acids into prostaglandins.
   Journal of Biological Chemistry. 239: PC4006-PC4009. 1964.

   Incubation of homo-γ-linolenic acid with homogenates of sheep seminal vesicles led to the production
   of PGE_1 and incubation with eicosa-5,8,11,14,17 pentaenoic acid produced PGE_3. When do-
   cosa-10,13,16-trienoic acid was incubated with the enzyme system, a compound resembling bis-
   homo-PGE_1 was formed. The dietary requirements of essential fatty acids could be related to their
   role as precursors of PG's. (JRH) 0217

0089
BERGSTROM, S., H. DANIELSSON and B. SAMUELSSON
   The enzymatic formation of prostaglandin E_2 from arachidonic acid Prostaglandins and related
      factors 32.
   Biochimica et Biophysica Acta. 90: 207-210. 1964.

   PGE_2 was biosynthesized by incubating tritium labelled arachidonic acid with a homogenate of sheep
   seminal vesicles under aerobic conditions for 1 hr. About 20% of the labelled arachidonic acid was
   converted to PGE_2. It is suggested that since essential fatty acids are precursors of PG's, some of the
   symptoms of essential fatty acid deficiency could be partly due to inadequate biosynthesis of PG's.
   (JRH) 0206

0090
BERRY, P.A. and H.O.J. COLLIER
   Bronchoconstrictor action and antagonism of a slow-reacting substance from anaphylaxis of guinea-
      pig isolated lung.
   British Journal of Pharmacology and Chemotherapy. 23: 202-216. 1964.

   Both slow-reacting substance A (SRS-A) and lung prostaglandin ($PGF_{2\alpha}$) caused an increase in
   resistance to inflation in the guinea pig lung in vivo. However, SRS-A and $PGF_{2\alpha}$ are different, since
   salicylate enhanced the response to $PGF_{2\alpha}$ (100 and 200 μg i v) while it completely blocked the
   response to twice the effective dose of SRS-A. (JRH) 0230

0091

BERTI, F., V. BORRONI, R. FUMAGALLI, P. MARCHI and G.P. ZOCCHE

Sull'attivita pressoria della prostaglandina. [On the pressure activity of prostaglandins.]

Atti della Accademia Medica Lombarda. 19: 397-403. 1964.

A few μg of PGE_1 i v caused a significant hypotension in cats, rats, rabbits and guinea pigs with rats being the least sensitive. The hypotensive effects are dose related and no tachyphylaxis was observed. In dogs, PGE_1 induced bradycardia, peripheral vasodilation and some pulmonary vasodilation. The hypotensive effects seem to be due to direct action on vascular smooth muscle. No significant species variations were observed. (JRH) 0203

0092

BYGDEMAN, M.

The effect of different prostaglandins on human myometrium in vitro.

Acta Physiologica Scandinavica 63 Supp. 242: 1-78. 1964.

The effect of prostaglandins on the contractility of the pregnant and non-pregnant human uterus was studied under in vivo conditions. Uterine contractility was recorded by measuring the amniotic pressure or by using the micro-balloon method and the tracings were analyzed by qualitative and quantitative methods. Prostaglandin E_1 (PGE_1), E_2 (PGE_2), $F_{1\alpha}$ ($PGF_{1\alpha}$) and $F_{2\alpha}$ ($PGF_{2\alpha}$) were administered as single intravenous injections or by continuous intravenous infusion. The intramuscular, intra-amniotic and vaginal routes of administration were applied in a limited number of cases. Both the PGE and PGF compounds had a stimulatory effect on uterine contractility. Single intravenous injections given during the first and second trimester of pregnancy elicited a rapid elevation of uterine tonus. The magnitude of this response as measured in mm Hg turned out to be dose dependent. A rough determination of the relative potency of the 4 compounds could be made by compiling the results obtained in the second trimester of gestation. Single intravenous injections given to non-pregnant women resulted as well in elevation of uterine tonus. However, any significant dose-response relationship could not be established in non-pregnant women, probably due to difficulties related to the method of recording. Uterine sensitivity was therefore investigated by determining the dose of prostaglandin that elicited a threshold response. This dose was approximately the same for the pregnant and non-pregnant uterus. The results did not indicate any major differences in sensitivity during the various phases of the menstrual cycle. Constant intravenous infusion of prostaglandin induced a gradual increment of uterine activity both in the second trimester of gestation and at term. Low infusion rates of PGE_1, PGE_2 and $PGF_{2\alpha}$ administered at or near term induced reasonably coordinated activity similar to that seen during spontaneous labor at term. The same low infusion rate of PGE_2 given at midpregnancy induced incoordinated activity with low intensity of the contractions. Administration of high doses of PGE_1 during the second trimester initially caused elevation of tonus superimposed on small frequent contractions. This state of uterine incoordination was followed by the development of more coordinated contractions of high intensity. The stimulatory response to single intravenous injections of PGE_1 and PGE_2 was compared in the first and second trimester of pregnancy with the corresponding effect of oxytocin and ergometrin. PGE_1 induced an elevation of uterine tonus of similar magnitude to that of oxytocin although the duration of the prostaglandin response was approximately three times longer. Injection of ergometrin under the same experimental conditions resulted in a less pronounced elevation of tonus but the response had considerably longer duration. Intravenous infusion of oxytocin during the second trimester resulted in more coordinated contractions than those obtained by PGE_1 and PGE_2. Intraamniotic and vaginal administration was not accompanied by increased uterine contractility. However, the doses applied were very low. Possible mechanisms of action of prostaglandin are discussed on the basis of these results. (Author modified). 0227

0093
BYGDEMAN, M. and B. SAMUELSSON
Quantitative determination of prostaglandins in human semen.
Clinica Chimica Acta. 10: 566-568. 1964.

A technique for the quantitative estimation of PGE_1, PGE_2, PGE_3, $PGF_{1\alpha}$ and $PGF_{2\alpha}$ in individual samples of human semen is described. Tritium labelled PGE_1 and $PGF_{1\alpha}$ (0.5 µg) were added to each sample and the PG's extracted with ether and separated by thin layer chromatography. The amount of individual PG's were then determined by spectrophotometric and gas chromatographic methods. Repetitive determination of the PG's in pooled human semen showed close agreement between each test and recovery of 65% of the labelled PGE_1 and 79% of the labelled $PGF_{1\alpha}$. The mean amounts of each PG in µg/ml was PGE_1 15.5, PGE_2 12.8, PGE_3 2.5, $PGF_{1\alpha}$ 2.0 and $PGF_{2\alpha}$ 2.2. (JRH) 0223

0094
COVINO, B.G. and J.B. LEE
The mechanism of action of renomedullary depressor activity in the intact normotensive dog.
Clinical Research. 12: 179. 1964.

Abstract only. The mechanism of vasodepressor action of rabbit renomedullary extract was studied in pentobarbitalized dogs. The results suggest that the mechanism of renomedullary vasodepression is by direct peripheral arteriolar dilation resulting in lowered peripheral resistance and fall in blood pressure with compensatory increase in heart rate and cardiac output. (JRH) 0226

0095
van DORP, D.A., R.K. BEERTHUIS, D.H. NUGTEREN and H. VONKEMAN
The biosynthesis of prostaglandins.
Biochimica Biophysica Acta. 90: 204-207. 1964.

Tritium labelled arachidonic acid was incubated with homogenates of sheep seminal vesicles. Chromatographic separation of the prostaglandins from the incubation mixture proved that the arachidonic acid had been converted to PGE_2 (more than 32% yield). It is suggested that the function of essential fatty acids may be to serve as precursors for prostaglandins. (JRH) 0224

0096
van DORP, D.A., R.K. BEERTHUIS, D.H. NUGTEREN and H. VONKEMAN
Enzymatic conversion of all-cis-polyunsaturated fatty acids into prostaglandins.
Nature. 203: 839-841. 1964.

[14]Carbon labelled homo-γ-linolenic acid was incubated with homogenates of sheep seminal vesicles. The reaction products contained radiolabelled PGE_1. The enzyme system was also able to convert other straight chain, all-cis-polyunsaturated (18 and 19 carbon) fatty acids into the corresponding nor-prostaglandins. The nor-prostaglandins had very little biological activity. (JRH) 0211

0097

FREUND, M., A. SAPHIER and J. WIEDERMAN
In-vitro studies of the effect of semen on the motility of the vagina, uterus, and uterine horns in the guinea pig.
Fertility and Sterility. 15: 188-201. 1964.

The presence of PG in human and ram semen and its effect on several types of smooth muscle are briefly mentioned in the introduction of this paper. Guinea pig semen reduced the frequency and increased the force of uterine contraction in the guinea pig uterus in vitro. (JRH) 0213

0098

GRAY, G.W.
Effect of drugs on intestinal release of stimulant acidic lipids in relation to simultaneous drug effect on intestinal mechanical activity in vitro.
Journal of Pharmacology and Experimental Therapeutics. 146: 215-224. 1964.

Prostaglandins are mentioned as being acidic lipids which have potent biological effects on smooth muscle. The effect of several drugs on the response of the rabbit intestine to an acidic lipid which was extracted from rabbit intestine was measured. The acidic lipid was assumed to be similar to darmstoff. (JRH) 0225

0099

GREEN, K. and B. SAMUELSSON
Prostaglandins and related factors: XIX. Thin-layer chromatography of prostaglandins.
Journal of Lipid Research. 5: 117-120. 1964.

Methods are described for separation of all known prostaglandins and some derivatives as free acids and methyl esters by thin-layer chromatography both on analytical and preparative scales. The use of silica gel containing silver nitrate was required for separations of compounds differing in the degree of unsaturation whereas ordinary silica gel was suitable for group separations of PGE, PGF_α, and PGF_β compounds. (Authors modified) 0208

0100

HICKLER, R.B., D.P. LAULER, C.A. SARAVIS, A.I. VAGNUCCI, G. STEINER and G.W. THORN
Vasodepressor lipid from the renal medulla.
Canadian Medical Association Journal. 90: 280-287. 1964.

A renomedullary vasodepressor lipid was extracted from rabbit kidney medullas. It was not a protein, peptide or phospholipid. It seemed to be a myotropic, unsaturated, hydroxy fatty acid like the prostaglandins. In rat isolated hind limbs, the renomedullary extract was a vasoconstrictor, while in the isolated rat kidney it was a vasodilator. In pentobarbital anesthetized dogs, the lipid produced a rapid sustained fall in blood pressure and an increase in cardiac output indicative of a decrease in peripheral vascular resistance. Significant amounts of the vasoactive lipid were not found in other tissue (JRH) 0222

0101

HORTON, E.W.
Actions of prostaglandins E_1, E_2 and E_3 on the central nervous system.
British Journal of Pharmacology and Chemotherapy. 22: 189-192. 1964.

PGE_1, PGE_2 and PGE_3 injected into the cerebral ventricles of unanaesthetized cats, produced sedation, stupor and signs of catatonia. The threshold dose was 3 μg/kg. Slight sedation was also observed following an intravenous injection, but a dose of 20 μg/kg was required. In chicks, intravenous injections of prostaglandins (10 to 400 μg/kg) caused respiratory depression, profound sedation, loss of normal posture and, with the higher doses, loss of the righting reflex. (Author modified) 0201

0102

HORTON, E.W. and C.J. THOMPSON
Thin-layer chromatography and bioassay of prostaglandins in extracts of semen and tissues of the male reproductive tract.
British Journal of Pharmacology and Chemotherapy. 22: 183-188. 1964.

The PG in human, ram and rabbit semen; ground rabbit, guinea pig, hamster, rat and mouse prostate and seminal vesicles; the bulb of the cat penis; and ferret testicles was extracted with ether and purified by thin layer chromatography. The PG concentration was then estimated by comparison to pure PGE_1 in a bioassay system (rabbit jejunum and hamster colon). Human semen contained 24-783 μg/ml, ram semen 7.3 μg/ml and rabbit semen less than 0.5 μg/ml PG (assayed as PGE_1). The various reproductive tissues from other animals did not contain any detectable PG (less than 0.5 μg/g). While this technique does not separate the individual PG's, it does exclude other non-lipid biologically active substances which could interfere with the bioassay. (JRH) 0210

0103

LEE, J.B., M.A. MAZZEO and B.H. TAKMAN
The acidic lipid characteristics of sustained renomedullary depressor activity.
Clinical Research. 12: 254. 1964.

Abstract only. The present studies were undertaken to elucidate the chemical characteristics of rabbit medullary depressor activity. The renomedullary depressor activity represents an unsaturated, acidic (-COOH) hydroxylated lipid, thus chemically, as well as physiologically, resembling the prostaglandin series of hormones. (Authors modified) 0228

0104

LEE, J.B., B.H. TAKMAN and B.G. COVINO
The isolation and chemical characteristics of renomedullary depressor substance.
Physiologist 7: 188. 1964.

Abstract only. The present studies suggest that the rabbit renomedullary depressor substance is an acidic lipid closely related to the prostaglandin E series of compounds. Important differences in its mechanism of vasodepression and thin layer mobilities suggest that it is not PGE_1, PGE_2, or PGE_3, but may represent an unknown derivative. (Authors modified) 0229

0105
MAIN, I.H.M.
The inhibitory actions of prostaglandins on respiratory smooth muscle.
British Journal of Pharmacology and Chemotherapy. 22: 511-519. 1964.

PGE_1 (1 ng/ml or more) relaxed the inherent tone of guinea pig and ferret tracheal muscle and reduced the response to acetylcholine applied before or after the PGE_1 in cat, monkey, rabbit, guinea pig and ferret tracheal muscle in vitro. When compared to PGE_1, the relative activities of PGE_2, PGE_3 and $PGF_{1\alpha}$ for relaxing acetylcholine stimulated cat tracheal muscle were 1.0, 0.2 and 0.002 respectively. In urethane anesthetized rabbit and guinea pig lungs, PGE_1 generally decreased the resistance to inflation and decreased the increase in resistance caused by vagal stimulation. In guinea pig lungs, PGE_1 also decreased the increase in resistance to inflation due to histamine. However, in the cat anesthetized with chloralose and pentobarbitone or urethane, PGE_1 (0.3 µg/kg) increased resistance to inflation and heart rate. The possible roles of PG's in lung physiology are discussed. (JRH) 0202

0106
SAMUELSSON, B.
Identification of a smooth muscle-stimulating factor in bovine brain. Prostaglandins and related factors 25.
Biochimica et Biophysica Acta. 84: 218-219. 1964.

A technique for identifying prostaglandins from tissue is described. The technique involves a series of chromatographic procedures and confirmation by mass spectrometry. Using this technique, ox brain was found to contain $PGF_{2\alpha}$ in concentrations of about 0.3 µg/g tissue wet weight. (JRH) 0204

0107
SAMUELSSON, B.
Identification of prostaglandin $F_{3\alpha}$ in bovine lung. Prostaglandins and related factors 26.
Biochimica et Biophysica Acta. 84: 707-713. 1964.

$PGF_{3\alpha}$ was isolated from bovine lung by reversed-phase partition chromatography, silicic acid chromatography and thin-layer chromatography. Final identification was achieved by gas-liquid chromatography of a trimethylsilyl ether derivative, infrared spectroscopy and mass spectrographic analysis. The stereo-chemistry at C-9 was determined by an isotope technique. $PGF_{2\alpha}$, earlier isolated from lung tissue of other species, was also identified. (Author modified) 0212

0108
SAMUELSSON, B.
Prostaglandins and related factors. 27. Synthesis of tritium-labeled prostaglandin E_1 and studies on its distribution and excretion in the rat.
Journal of Biological Chemistry. 239: 4091-4096. 1964.

Tritium labelled PGE_1 was synthesized and administered intravenously to rats. After 40 hr, 50% of the radioactivity was recovered in the urine and 10% in the feces as more polar metabolites of PGE_1. About the same amount of radioactivity was found in the bile as in the urine. Labelled PGE_1 was injected subcutaneously and the amount of tritium in various organs measured. The highest tritium concentration was in the kidneys and liver. Intermediate concentrations were found in the lungs, pituitary gland, adrenals, ovaries, uterus and heart. The brain, muscle, adipose tissue and thymus had the lowest levels. (JRH) 0216

0109
SANDBERG, F., A. INGELMAN-SUNDBERG and G. RYDEN
The effect of prostaglandin E_2 and E_3 on the human uterus and the Fallopian tubes in vitro.
Acta Obstetrica Gynecologica Scandinavica. 43: 95-102. 1964.

Using Magnus Kehrer's technique the effect of PGE_2 and PGE_3 (0.006-0.05γ/ml) has been investigated on different parts of human Fallopian tube from 41 women, and on corpus and isthmus of the uterus from 38 women in fertile age. PGE_2 exerts a specific action as demonstrated by an increase in tonus and amplitude maximum on the proximal part of the tube and an inhibitory action on the rest of the organ and on the uterus. PGE_3 produces an inhibitory effect throughout the Fallopian tube as well as on the uterus. (Authors modified) 0220

0110
SMITH, E.R., J.B. LEE and B.G. COVINO
Effect of renomedullary extract on vascular and nonvascular smooth muscle.
Journal of New Drugs. 4: 229-230. 1964.

Abstract only. Rabbit renomedullary homogenates were purified by chromatographic methods in an attempt to isolate a pure vasoactive substance. A chromatographically homogenous fraction finally was isolated by thin layer chromatography. Analysis revealed that this fraction represented a non-saturated acidic hydroxylated lipid. This pure fraction possessed potent vasodepressor activity but no nonvascular smooth-muscle stimulating activity. (Authors modified) 0232

0111
STEINBERG, D., M. VAUGHAN, P.J. NESTEL, O. STRAND and S. BERGSTROM
Effects of the prostaglandins on hormone-induced mobilization of free fatty acids.
Journal of Clinical Investigation. 43: 1533-1540. 1964.

PGE_1 (0.1 μg/kg) significantly reduced basal and epinephrine (E), norepinephrine, glucagon and ACTH stimulated lipolysis measured as glycerol release in rat adipose tissue in vitro. Twice as much PGE_1 (in 2 doses) also reduced the lipolysis caused by TSH. PGE_2 was slightly less effective than PGE_1 while PGE_3, $PGF_{1\alpha}$ and $PGF_{2\alpha}$ were much less effective as inhibitors of E-induced lipolysis. While PGE_1 inhibited hormone stimulated glycerol release and free fatty acid (FFA) release, it did not alter basal FFA release. PGE_1 also reduced E-stimulated phosphorylase activity. Injection of E into pentobarbital anesthetized dogs caused a rise in plasma FFA and glucose levels and an increase in blood pressure. PGE_1 (12.5 μg/kg i v) caused a slight fall in FFA, a pronounced decrease in blood pressure, but no significant change in glucose levels. When PGE_1 and E were given simultaneously, the E-induced change in FFA and blood pressure were blocked or reduced, but the glucose levels rose almost as much as with E alone. (JRH) 0205

0112
VAUGHAN, M.
Effect of pitressin lipolysis and on phosphorylase activity in rat adipose tissue.
American Journal of Physiology. 207: 1166-1168. 1964.

Pitressin stimulated glycerol release from rat epididymal adipose tissue in vitro was significantly reduced by addition of PGE_1 (0.1 μg/ml). (JRH) 0218

0113

WEEKS, J.R. and F. WINGERSON

Cardiovascular action of prostaglandin E_1 evaluated using unanesthetized relatively unrestrained rats. Federation Proceedings. 23: 327. 1964.

Abstract only. The cardiovascular actions of PGE_1 has been evaluated by some new techniques allowing use of unanesthetized relatively unrestrained rats. Threshold depressor action was about 1.8 $\mu g/kg$ by single i v injections, by continuous infusion 1.8 $\mu g/kg/min$. Repeated 10 $\mu g/kg$ i v doses showed no tachyphylaxis. Cardiac output increased when blood pressure was lowered by infusion of 10 $\mu g/kg/min$, indicating depressor action was due to lowered total peripheral resistance. Prostaglandin infusion blocked pressor responses to both angiotensin and epinephrine, indicating the mechanism of blockade to be reduced vascular reactivity rather than cardiovascular adrenergic blockade. (Authors modified) 0221

0114

ANGGARD, E. and B. SAMUELSSON

Biosynthesis of prostaglandins from arachidonic acid in guinea pig lung. Prostaglandins and related factors 38.

Journal of Biological Chemistry. 240: 3518-3521. 1965.

Homogenates of guinea pig lung transformed tritium-labeled arachidonic acid into $PGF_{2\alpha}$, PGE_2, 11α, 15-dihydroxy-9-ketoprost-5-enoic acid, and 11α-hydroxy-9, 15-diketoprost-5-enoic acid. (Authors modified) 0325

0115

ANGGARD, E.

The isolation and determination of prostaglandins in lungs of sheep, guinea pig, monkey and man. Biochemical Pharmacology. 14: 1507-1516. 1965.

Lipid extracts of lungs of sheep, guinea pig, monkey and man were analyzed for smooth muscle stimulating substances. Most of the spasmogenic activity was due to PG's. After preliminary purification, the PG's were isolated and identified chromatographically. $PGF_{2\alpha}$ was the main PG present and was identified in all lungs. Approximate concentrations of $PGF_{2\alpha}$/g wet weight calculated by means of isotope dilution were: sheep lung, 0.5 μg; guinea pig lung, 0.5 μg; human lung (autopsy material), 0.02 μg; and monkey lung (autopsy material), 0.2 μg. PGE_2 was identified in sheep lung. (Author modified) 0352

0116

ANGGARD, E. and B. SAMUELSSON

Prostaglandins and related factors. XLII. Metabolism of prostaglandin E_3 in guinea pig lung. Biochemistry. 4: 1864-1871. 1965.

$[^{14}C]PGE_3$ was prepared biosynthetically from 5,8,11,14,17-eicosa $[^{14}C]$ pentaenoic acid using homogenates of sheep vesicular glands. Enzymes in the supernatant fraction of a homogenate of guinea pig lung centrifuged 100,000 X g transformed the labeled PGE_3 into 2 less polar metabolites. These metabolites have been assigned the structures 11α, 15-dihydroxy-9-ketoprosta-5,17-dienoic and 11α-hydroxy-9,15-diketoprosta-5,17-dienoic acids. (Authors modified) 0337

0117
ANGGARD, E., K. GREEN and B. SAMUELSSON
Synthesis of tritium-labeled prostaglandin E_2 and studies on its metabolism in guinea pig lung. 37. Prostaglandins and related factors.
Journal of Biological Chemistry. 240: 1932-1940. 1965.

The synthesis of tritium-labeled PGE_2 is described. Enzymes in the particle-free fraction of guinea pig lung homogenates convert PGE_2 into two metabolites. These metabolites have been assigned the structures $11\alpha15$-dihydroxy-9-ketoprost-5-enoic acid and 11α-hydroxy-9,15-diketoprost-5-enoic acid. (Authors modified) 0331

0118
ANONYMOUS
Conversion of linoleic and linolenic acid metabolites to prostaglandins.
Nutrition Reviews. 23: 141-144. 1965.

The synthesis of PG's from metabolites of essential fatty acids is reviewed. It is suggested that their role as precursors of PG's may explain the biological significance of the essential fatty acids. (JRH) 0344

0119
AVANZINO, G.L., P.B. BRADLEY and J.H. WOLSTENCROFT
Actions of prostaglandins on brain stem neurones.
Journal of Physiology. 181: 34P. 1965.

Title only. 0321

0120
BABILLI, S. and W. VOGT
Nature of the fatty acids acting as 'slow reacting substance' (SRS-C).
Journal of Physiology. 177: 31P-32P. 1965.

Slow reacting substance C(SRS-C) was produced by injecting bee venom or phospholipase A into guinea pig lungs. SRS-C was shown to be a hydroxy fatty acid. Chromatographically and chemically it behaves like prostaglandins, especially PGF. It is suggested that SRS-C consists mainly of PG-like hydroxy acids which are preformed constituents of tissue phosphatides released by the action of phospholipase A. (JRH) 0314

0121
BERGSTROM, S., L.A. CARLSON, L-G. EKELUND and L. ORO
Cardiovascular and metabolic response to infusions of prostaglandin E_1 and to simultaneous infusions of noradrenaline and prostaglandin E_1 in man. Prostaglandin and related factors 35.
Acta Physiologica Scandinavica. 64: 332-339. 1965.

PGE_1 was infused at a rate of 0.1-0.2 μg/kg/min into 3 healthy, fasting male subjects for 20 min. No consistent changes in arterial pressures occurred. The heart rate increased about 20 beats/min. The concentration of free fatty acids (FFA) and of glycerol in plasma increased. A slight increase in oxygen consumption was observed. PGE_1 infused simultaneously with norepinephrine (NE) to 2 healthy, fasting male subjects reduced the increase in arterial pressures seen when only NE was infused to these subjects. Furthermore, PGE_1 completely inhibited the NE induced bradycardia. The increase in the concentration of FFA and glycerol in plasma caused by NE was only slightly reduced by PGE_1. PGE_1 had no effect on the calorigenic effect of NE. (Authors modified) 0313

0122
BERGSTROM, S., L.A. CARLSON, L-G. EKLUND and L. ORO
Effect of prostaglandin E_1 on blood pressure, heart rate and concentration of free fatty acids of plasma in man.
Proceedings of the Society for Experimental Biology and Medicine. 118: 110-112. 1965.

The infusion of PGE_1 (0.1-0.2 $\mu g/kg/min$) intravenously in 3 and intra-arterially in 1 human male subject caused an increase in heart rate (20 beats/min), blood free fatty acid (FFA), and blood glycerol, but little change in blood pressure. Infusion of norepinephrine (NE) caused a bradycardia, an increase in blood pressure, a rise in blood glycerol and FFA. When PGE_1 was infused simultaneously with the NE, the increase in blood pressure was reduced 50%, the bradycardia was abolished, but the increase in FFA and glycerol were only slightly reduced. In dogs, PGE_1 completely reversed the rise in FFA caused by NE infusion. (JRH) 0322

0123
BERGSTROM, S. and L.A. CARLSON
Influence of the nutritional state on the inhibition of lipolysis in adipose tissue by prostaglandin E_1 and nicotinic acid. Prostaglandin and related factors 46.
Acta Physiologica Scandinavica. 65: 383-384. 1965.

PGE_1 (0.01 $\mu g/ml$ and higher) inhibited the release of glycerol from rat epididymal fat pads of fed rats in vitro. However, PGE_1 (up to 1 $\mu g/ml$) had no effect on glycerol release from adipose tissue obtained from starved rats. Nicotinic acid inhibited lipolysis from adipose tissue from both fed and starved rats, but the effect was greatest in the fed rats. (JRH) 0311

0124
BERGSTROM, S. and L.A. CARLSON
Inhibitory action of prostaglandin E_1 on the mobilization of free fatty acids and glycerol from human adipose tissue in vitro. Prostaglandin and related factors.
Acta Physiologica Scandinavica. 63: 195-196. 1965.

Human subcutaneous adipose tissue was incubated in vitro. PGE_1 (1 $\mu g/ml$) significantly inhibited the basal release of glycerol and free fatty acid but did not alter glucose uptake. PGE_1 also inhibited epinephrine stimulated lipolysis. However, it was not possible to determine if the PG inhibited the hormone's activation of lipolysis or acted by an independent mechanism. (JRH) 0303

0125
BERGSTROM, S. and L.A. CARLSON
Lipid mobilization in essential fatty acid deficient rats. A preliminary note. Prostaglandin and related factors 39.
Acta Physiologica Scandinavica. 64: 479-480. 1965.

The effect of PGE_1 (1 $\mu g/ml$), nicotinic acid and norepinephrine (NE) on lipolysis in adipose tissue from rats fed normal and essential fatty acid (EFA) deficient diets was studied in vitro. PGE_1 and nicotinic acid were less effective in inhibiting glycerol and free fatty acid release from EFA-deficient adipose tissue than from control adipose tissue. NE stimulated glycerol and free fatty acid release equally in both types of adipose tissue. EFA-deficient adipose tissue had a faster rate of lipid mobilization than control adipose tissue. There may be a reduced synthesis of PGE_1 in EFA-deficient rats which could account for the increased lipolysis. (JRH) 0312

0126
BERGSTROM, S. and B. SAMUELSSON
Prostaglandins.
Annual Review of Biochemistry. 34: 101-108. 1965.

Prostaglandin literature is reviewed. There are sections on: isolation and structure determination; occurrence; biosynthesis; metabolism; and biological effects. There are 61 references cited. (JRH) 0333

0127
BERTI, F., R. LENTATI and M.M. USARDI
The species specificity of prostaglandin E_1 effects on isolated heart.
Medicina et Pharmacologia Experimentalis. 13: 233-240. 1965.

The effect of PGE_1 on the isolated cat, rabbit, rat, guinea pig and frog heart was studied in vitro. The cat, rabbit and rat hearts were relatively insensitive to PGE_1. Large doses (10-75 μg) caused moderate inotropic changes in these species but did not affect the response of the hearts to epinephrine. However, in the guinea pig, PGE_1 (1.0 μg) caused a long lasting (10 min) increase in heart rate, heart contraction and coronary blood flow. The effects of PGE_1 on the guinea pig heart were not prevented by β-adrenergic blockade. In the frog heart, PGE_1 produced long lasting positive inotropic effects which were similar to those of epinephrine. (JRH) 0326

0128
BRUNDIN, J.
Distribution and function of adrenergic nerves in the rabbit Fallopian tube.
Acta Physiologica Scandinavica. 66 Supp. 259: 5-57. 1965.

It is briefly mentioned that intravenous injection of PGE_1 into rabbits caused a relaxation of circular muscles of the isthmic region of the oviduct. Others have reported that PGE_1 causes contraction of longitudinal muscle in the human Fallopian tube in vitro. (JRH) 0309

0129
CARLSON, L.A. and P.R. BALLY
Inhibition of lipid mobilization.
In: Renold, A.E. and G.F. Cahill, Jr., eds., "Handbook of Physiology," Section 5. p. 557-574. Washington, American Physiological Society, 1965.

Prostaglandins are mentioned as being hormone substances which inhibit lipid mobilization from adipose tissue. The relative effectiveness and chemical structure of the prostaglandins is briefly reviewed. (JRH) 0327

0130
CARLSON, L.A.
Inhibition of the mobilization of free fatty acids from adipose tissue.
Annals of the New York Academy of Sciences. 131: 119-142. 1965.

A section on prostaglandins is included in this review of the inhibition of free fatty acids. PGE_1 appears to have antilipolytic properties which act directly on adipose tissue. (JRH) 0345

0131
COCEANI, F. and L.S. WOLFE
Prostaglandins in brain and the release of prostaglandin-like compounds from the cat cerebellar cortex.
Canadian Journal of Physiology and Pharmacology. 43: 445-450. 1965.

A water-soluble lipid material resistant to acid and alkali treatment was found in perfusates and extracts of brain tissue, which induced contraction of the 'slow type' on the isolated rat stomach fundus. The material isolated from brain tissue extracts behaved in solvent partition systems like the prostaglandin compounds and could be purified by silicic acid and thin-layer chromatography. The smooth-muscle-stimulating activity is likely due to the trihydroxyprostaglandin compounds. (Authors) 0340

0132
COLLINS, F.D. and J.F. CONNELLY
A fatty acid characteristic of a deficiency of linoleic acid in a case of hepatoma.
Lancet. 2: 883-885. 1965.

It is briefly mentioned that a deficiency in essential fatty acids could lead to a deficiency in prostaglandin synthesis causing a faster mobilization of fat from depots. (JRH) 0323

0133
CRUNDWELL, E., M.A. PINNEGAR and W. TEMPLETON
Synthesis of fatty acids with smooth muscle stimulant acitivity.
Journal of Medicinal Chemistry. 8: 41-45. 1965.

A series of fatty acids were synthesized and tested for smooth muscle stimulating activity. One of these, 12-hydroxyheptadec-*trans*-10-enoic, which represents a fragment of a prostaglandin, was 3 times more potent than ricinelaidic acid in stimulating the isolated hamster colon. (JRH) 0332

0134
DANIELS, E.G., J.W. HINMAN, B.A. JOHNSON, F.P. KUPIECKI, J.W. NELSON and J.E. PIKE
The isolation of an additional prostaglandin derivative from the enzymatic cyclization of homo-γ-linolenic acid.
Biochemical and Biophysical Research Communications. 21: 413-417. 1965.

The pooled mother liquors of several large scale biosynthetic runs, for the synthesis of PGE_1 by sheep seminal vesicle homogenates, were separated by column chromatography. One of the products was found to be identical with PGE_1-217 (PGA_1). Biologically PGA_1 has much less smooth muscle stimulating properties and antilipolytic properties than PGE_1, but it is about equipotent as a vasodepressor. (JRH) 0336

0135
DE PURY, G.G. and F.D. COLLINS
A raised level of free fatty acids in serum of rats deficient in essential fatty acids as a contributing cause of their fatty livers.
Biochimica et Biophysica Acta. 106: 213-214. 1965.

A hypothesis is developed to explain the fatty livers found in rats fed a diet deficient in essential fatty acids (EFA). Prostaglandins are known to inhibit catecholamine stimulated lipolysis and thus could be controllers of lipolysis. Since PG's are synthesized from EFA, EFA-deficient rats could also be PG-deficient rats. Therefore, they would have more lipase activity and consequently more free fatty acids in the serum which could contribute to the fatty livers. (JRH) 0306

0136
ELIASSON, R. and N. POSSE
Rubins's test before and after intravaginal application of prostaglandin.
International Journal of Fertility. 10: 373-377. 1965.

The effect of intravaginally applied human seminal PG's on the hindrance to a gas flow through the uterus and oviducts has been investigated in infertile women. All experiments have been performed at midcycle. In some patients, PG did not change the resistance, whereas in others a marked increase was recorded. This reactivity pattern is in agreement with the hypothesis that PG's may facilitate passive sperm migration from the vagina into the uterine cavity. (Authors modified) 0349

0137
GRANSTROM, E., U. INGER and B. SAMUELSSON
The structure of a urinary metabolite of prostaglandin $F_{1\alpha}$ in the rat. Prostaglandins and related factors 29.
Journal of Biological Chemistry. 240: 457-461. 1965.

The preparation of tritium-labeled $PGF_{1\alpha}$ is described. After administration to rats (60 μg subcutaneously) about 30% of the radioactivity was recovered in urine and 8% in feces. The labeled products in urine consisted almost exclusively of more polar metabolites. The main labeled urinary metabolite has been shown to be 2,3-dinor-$PGF_{1\alpha}$. (Authors modified) 0324

0138
HAMBERG, M. and B. SAMUELSSON
Isolation and structure of a new prostaglandin from human seminal plasma. Prostaglandins and related factors 43.
Biochimica et Biophysica Acta. 106: 215-217. 1965.

The separation and structural determination of 19 hydroxy-PGA_2 from human seminal plasma is described. Techniques involved solvent partition chromatography and identification by mass spectrometry, ultraviolet spectrometry and nuclear magnetic resonance. (JRH) 0304

0139

HANSSON, E. and B. SAMUELSSON

Autoradiographic distribution studies of ^{3}H-labeled prostaglandin E_1 in mice. Prostaglandins and related factors 31.

Biochimica et Biophysica Acta. 106: 379-385. 1965.

^{3}H-labeled prostaglandin E_1 (2-7 µg) was injected intravenously into mice and the distribution of radioactive material was studied by autoradiography. Autoradiograms obtained from sagittal sections of the whole body showed high concentration of ^{3}H in the kidney, liver and connective tissue. Lower but significant uptakes were also observed in the lungs and in the myometrium of uterus. The results were discussed in relation to current concepts of metabolism and biological effects of prostaglandin E_1 and also in relation to possible new sites of action. (Authors) 0305

0140

HOLMES, S.W.

Prostaglandin E_1, fatty acids and microbial growth.

Nature. 206: 405-406. 1965.

Experiments were performed to determine if PGE_1 had any antimicrobial powers. PGE_1 up to 1mg/ml had no effect on *Staphylococcus aureus, Streptococcus faecalis, Escherichia coli, Aspergillus niger*, or *Trichophyton mentagrophytes*. It did inhibit the growth of *Candida albicans*. PGE_1 apparently does not act as a naturally occurring antimicrobial agent. (JRH) 0310

0141

HORTON, E.W. and I.H.M. MAIN

Actions of prostaglandins on the spinal cord of the chick.

Journal of Physiology. 179: 18P-20P. 1965.

Earlier experiments by the authors indicated that prostaglandins produce their effects in chicks by acting on the spinal cord. Some experiments are described which will elucidate the actions of prostaglandins on the chick spinal cord. (JRH) 0355

0142

HORTON, E.W.

Biological activities of pure prostaglandins.

Experientia. 21: 113-118. 1965.

The literature on the use of pure PG's reviewed. There are sections on: actions on reproductive smooth muscle; actions on respiratory smooth muscle; actions on the cardio-vascular system; actions on intestinal smooth muscle, actions on the nervous system; actions on adipose tissue; actions on micro-organisms; metabolism of PGE_1; and miscellaneous observations (unpublished). There are 43 references cited. (JRH) 0342

0143

HORTON, E.W. and I.H.M. MAIN

Central nervous effects of prostaglandins.

British Journal of Pharmacology and Chemotherapy. 24: 305. 1965.

Title only. 0320

0144
HORTON, E.W. and I.H.M. MAIN
A comparison of the actions of prostaglandins $F_{2\alpha}$ and E_1 on smooth muscle.
British Journal of Pharmacology and Chemotherapy. 24: 470-476. 1965.

The biological activities of four prostaglandins have previously been reported; in the present paper the activity of a fifth prostaglandin, $F_{2\alpha}$, is compared with prostaglandin E_1 on smooth muscle preparations. Both prostaglandins contract intestinal smooth muscle of the rabbit, hamster and guinea-pig in vitro. Responses of the rabbit jejunum to prostaglandin $F_{2\alpha}$ are usually slower in onset and more prolonged than those of prostaglandin E_1. Both prostaglandins inhibit acetylcholine-induced contractions of the cat isolated tracheal chain, but $F_{2\alpha}$ has only one-thirtieth of the activity of E_1. The rat isolated uterus is contracted by both prostaglandins, but is more sensitive to $F_{2\alpha}$. On the other hand, the rabbit fallopian tube in vivo, which is relaxed by small doses of prostaglandin E_1, is contracted by $F_{2\alpha}$. Both prostaglandins increase blood flow through skeletal muscle in the cat and lower arterial blood pressure; $F_{2\alpha}$ is less potent in this respect than E_1. It is concluded that prostaglandin $F_{2\alpha}$ is a less potent inhibitor of smooth muscle than prostaglandin E_1, but that, on smooth muscle preparations which are contracted, $F_{2\alpha}$ is often more active than E_1. (Authors) 307

0145
HORTON, E.W. and I.H.M. MAIN
Differences in the effects of prostaglandin $F_{2\alpha}$, a constituent of cerebral tissue, and prostaglandin E_1
on conscious cats and chicks.
International Journal of Neuropharmacology. 4: 65-59. 1965.

$PGF_{2\alpha}$ produced no obvious effects when injected into the cerebral ventricles of 4 unanesthetized cats in doses of 15-100 μg/cats. In contrast, PGE_1 produced marked sedation and stupor in doses of 20 μg/cat. In 20 chicks, intravenous injection of $PGF_{2\alpha}$ (25-450 μg/kg) caused extension of the limbs and sometimes extension of the neck, followed by a tendency of the limbs to abduct. In lower doses (5-25 μg/kg), abduction of the limbs was the only change observed. Unlike PGE_1, $PGF_{2\alpha}$ caused no respiratory depression and little or no sedation in these doses. (Authors modified) 0343

0146
HORTON, E.W., I.H.M. MAIN and C.J. THOMPSON
Effects of prostaglandins on the oviduct, studied in rabbit and ewes.
Journal of Physiology. 180: 514-528. 1965.

Spontaneous contractions of the oviduct, uterine horn and cervix of the anaesthetized rabbit are inhibited by prostaglandin E_1 injected intravenously. The action on the oviduct is direct; it is not mediated via the central nervous system nor is it secondary to changes in blood pressure. When prostaglandin solutions in concentrations of 50 μg/ml. or more are administered intravaginally in rabbits, absorption into the circulation occurs in amounts sufficient to reduce the tone of the oviduct. Intra-aortic injections of prostaglandin lowered oviduct intra-luminal pressure in two ewes but raised it in three other ewes. In any one animal responses remained qualitatively similar throughout the experiment and a similar response was usually obtained subsequently with the isolated oviduct from that animal. Freshly collected ejaculates of ram semen injected intra-aortically had little or no effect upon the intra-luminal pressure of the oviduct. It is concluded that although prostaglandin can be absorbed from the vagina (of the rabbit) the amounts of prostaglandin in both rabbit and sheep semen are too small to have any significant effect upon oviduct smooth muscle even if complete and rapid vaginal absorption of seminal prostaglandin were to occur. (Authors) 0308

0147
HORTON, E.W. and I.H.M. MAIN
   Effects of prostaglandins on unanaesthetized animals.
   Journal of Physiology. 177: 34P. 1965.

   Title only. 0362

0148
INGELMAN-SUNDBERG, A., F. SANDBERG, G. RYDEN AND I. JOELSSON
   Les effets "in-vitro" des differents types de prostaglandines sur la motilite de l'uterus et des trompes
      non gravides. [The in vitro effects of different types of prostaglandins on the motility of non-
      pregnant uterus and tubes.]
   Bulletin de la Federation des Societes de Gynecologie et Obstetrique de Langue Francaise. 17:
      783-787. 1965.

Uterine and tubal sections from 147 women were used in a study of the effects of 7 prostaglandins on
the motility of reproductive tract in vitro using dosages of 0.0006 to 0.25 γ/ml. PGE_1 and PGE_2
caused contraction in the juxta-uterine region, but relaxation in other areas of the tube. $PGF_{1\alpha}$,
$PGF_{1\beta}$ and $PGF_{2\alpha}$ caused constriction of the tube, while $PGF_{2\beta}$ and PGE_3 relaxed it completely. All
PG's except $PGF_{1\alpha}$ and $PGF_{2\alpha}$ relaxed the uterus completely. $PGF_{1\alpha}$ caused light contraction
but not of the isthmus; $PGF_{2\alpha}$ caused constriction of all parts. $PGF_{1\beta}$ and $PGF_{2\beta}$ had the
least effect. The effects of PGE_1, PGE_2 and PGE_3 on the uterus in vivo are briefly mentioned
as a reduction in tonus and an increase in the flow of cervical mucous. (NES) 0368

0149
KISCHER, C.W.
   An organogenic block produced in culture.
   American Zoologist. 5: 710-711. 1965.

Abstract only. PGE_1 (2.8 X 10^{-4}M) completely blocks organogenesis of the feather in embryo chick
skin, while cytotoxic effects are nominal. Many of the epidermal cells appear vesiculated and nearly
adipic in structure, but they continue to proliferate and stratify as in normally developing skin. The
effect is not overcome by equivalent molar concentrations of norepinephrine or linoleic acid. It can
be reversed, however, by transfer to fresh normal media after one day, but not after two days.
Histochemical staining of treated cultures and controls reveals no obvious differences in distribution
of RNA, alkaline phosphatase, PAS material, or total proteins. (Authors modified) 0369

0150
KLENBERG, D. and B. SAMUELSSON
   The biosynthesis of prostaglandin E_1 studied with specifically ^{3}H-labelled 8,11,14-eicosatrienoic
      acids.
   Acta Chemica Scandinavica. 19: 534-535. 1965.

The fate of hydrogens at C-8, C-12 and C-11 of 8,11,14-eicosa-trienoic acid, during the biosynthesis
of PGE by homogenates of sheep seminal vesicles, was followed by using precursors labelled with
tritium at the appropriate carbons. The results showed that all the hydrogens were retained at their
original positions on C-8 and C-12 during the formation of the new bond between these carbons. Also
the hydrogen at C-11 was retained during the introduction of the hydroxyl group. (JRH) 0316

0151
KRNJEVIC, K.
Actions of drugs on single neurones in the cerebral cortex.
British Medical Bulletin. 21: 10-14. 1965.

PGE_1 is mentioned in a chart as having no effect when applied iontophoretically to neurones of the pentobarbital anesthetized cat cerebral cortex. (JRH) 0338

0152
KUPIECKI, F.P.
Conversion of homo-γ-linolenic acid to prostaglandin $F_{1\alpha}$ by ovine and bovine seminal vesicle extracts.
Life Sciences. 4: 1811-1815. 1965.

Radiolabelled homo-γ-linolenic acid was incubated with homogenates of sheep or bull seminal vesicles. The PG's were isolated and identified chromatographically from the incubation mixture. Both types of seminal vesicle produced $PGF_{1\alpha}$ and PGE_1. The sheep seminal vesicle homogenates produced more PGE_1 than $PGF_{1\alpha}$, while the bull seminal vesicle homogenates produced more $PGF_{1\alpha}$ than PGE_1. (JRH) 0330

1053
LEE, J.B., B.G. COVINO, B.H. TAKMAN and E.R. SMITH
Chemical and physiological properties of medullin.
Clinical Research. 13: 310. 1965.

Abstract only. Studies were undertaken to isolate and characterize the factor(s) in renal medulla extracts responsible for vasodepression and to determine the mechanism of hypotensive action. These studies revealed that rabbit renal medulla possesses a potent vasodepressor acidic lipid which lowers blood pressure by direct arteriolar dilation. This substance called medullin is not PGE_1, PGE_2 or PGE_3 but represents a more unsaturated derivative with less polar (-OH) groups. (Authors modified) 0365

0154
LEE, J.B., B.G. COVINO, B.H. TAKMAN and E.R. SMITH
Renomedullary vasodepressor substance, medullin: isolation, chemical characterization and physiological properties.
Circulation Research. 17: 57-77. 1965.

The biological, chemical and physical properties of medullin isolated from kidney medullas were compared to the properties of PGE_1. The results showed that medullin is closely related but not identical to PGE_1. Medullin was more mobile than PGE_1 chromatographically, stained more intensely with iodine vapors, showed absorption at higher wave lengths, and was 50 times weaker than PGE_1 in stimulating nonvascular smooth muscle. Medullin appears to be more unsaturated and less polar than PGE_1. (JRH) 0317

0155

LEE, J.B., R.B. HICKLER, C.A. SARAVIS and G.W. THORN
Sustained depressor effect of renal medullary extract in the the normotensive rat.
Nephron. 2: 117-118. 1965.

The authors briefly summarize experimental evidence on the nature of the renomedullary vasodepressor substance. In normotensive, anesthetized, pentolinium-treated rats, intravenous injection of crude homogenates of rabbit kidney medullas produce a sustained depression of blood pressure. Other studies have shown the active principle to be a low molecular weight, unsaturated, acidic hydroxylated lipid which resembles, chromatographically, the prostaglandins. (JRH) 0377

0156

LUNDBERG, W.O.
Some recent developments in fat nutrition.
Chemistry and Industry. April 3: 572-582. 1965.

In this review of the biochemistry of essential fatty acids (EFA), the author mentions that the recent discovery that EFA are precursors of prostaglandins provides the first evidence of a possible biological function for EFA. A review of the structure and synthesis of PG's from EFA is presented. (JRH) 0364

0157

MIYAZAKI, E.
[The mechanism of action of prostaglandin on smooth muscle contraction.]
Japanese Journal of Smooth Muscle Research. 1: 252. 1965.

Article in Japanese. Abstract not available at present. 0366

0158

NAIMZADA, M.K.
Attivita della prostaglandina E_1 sulla vescica seminale di cavia. [The activity of prostaglandin E_1 on the seminal vesicle of the guinea pig.]
Atti della Accademia Medica Lombarda. 20: 400-403. 1965.

PGE_1 sensitized the isolated guinea pig seminal vesicle to the stimulating effect of catecholamines (epinephrine and norepinephrine). It also increased markedly the contraction of the seminal vesicle induced by stimulation of the hypogastric nerve both in vivo and in vitro. The effects of PGE_1 are observed at very low doses (0.1-0.5 μg/ml in vitro and 5-20 μg/kg in vivo) and are long lasting. It is suggested that PGE_1 may have a specific action in the guinea pig, enhancing the sensitivity of the semen excreting apparatus to nervous stimuli. (Author modified) 0375

0159

NUGTEREN, D.H. and D.A. van DORP
The participation of molecular oxygen in the biosynthesis of prostaglandins.
Biochimica et Biophysica Acta. 98: 654-656. 1965.

Homo-γ-linolenic acid was incubated with homogenates of sheep seminal vesicles in an ^{18}O atmosphere to determine the origin of the 3 oxygen atoms that are incorporated into the PGE_1 moleculed during biosynthesis. The results of mass spectral analysis of the synthesized PGE_1 clearly showed that the 2 hydroxyl oxygens came from molecular oxygen. Since keto oxygens are known to exchange freely with the medium, no conclusions could be drawn as to the origin of the oxygen in the keto group. (JRH) 0354

0160
ORLOFF, J., J.S. HANDLER and S. BERGSTROM
    Effect of prostaglandin (PGE_1) on the permeability response of toad bladder to vasopressin, theo-
        phylline and adenosine 3',5'- monophosphate.
    Nature. 205: 397-398. 1965.

The effect of PGE_1 on the response of the toad bladder to vasopressin, theophylline and cAMP was examined. The latter three compounds increased the osmotic flow of water across the membrane when applied to the serosal surface. PGE_1 (1.7×10^{-6}M-1.7×10^{-11} M) significantly diminished the permeability response of the toad bladder to vasopressin and theophylline, but it had no effect on the response to cAMP. PGE_1 alone did not alter permeability. (JRH) 0315

0161
PICKLES, V.R. and P.F.V. WARD
    Menstrual stimulant component B, and possible prostaglandin precursors in the endometrium.
    Journal of Physiology. 178: 38P-39P. 1965.

A lipoid smooth-muscle stimulant called "the menstrual stimulant" has been previously isolated. This was subsequently divided into 3 components: A,B, and C. The A component was largely $PGF_{2\alpha}$. In this study, the B component was analyzed and found to contain eicosatrienoic and eicosatetraenoic acids which are prostaglandin precursors. It is concluded that the activity of the B component may be due to prostaglandin formed from these precursors. (JRH) 0363

0162
PICKLES, V.R., W.J. HALL, F.A. BEST and G.N. SMITH
    Prostaglandins in endometrium and menstrual fluid from normal and dysmenorrhoeic subjects.
    Journal of Obstetrics and Gynaecology of the British Commonwealth. 72: 185-192. 1965.

The endometrium, particularly during the secretory phase, contains a group of smooth-muscle stimu-
lants. The menstrual fluid contains the apparently identical substances, the principal one being a prostaglandin, $PGF_{2\alpha}$. The amount of $PGF_{2\alpha}$ recovered from the menstrual fluid is about ten times that recovered from curettings of secretory endometrium, which suggests that much of it is formed shortly before or during menstruation. In addition to $PGF_{2\alpha}$ and other myometrial spasmogens, the menstrual fluid contains smaller amounts of prostaglandins of the E series, which in large enough quantities may relax the human myometrium. A mixture of $PGF_{2\alpha}$ and PGE_2, in the proportion found in the menstrual fluid, is spasmogenic to the myometrium in vitro. A short series of patients referred to the hospital for treatment of severe primary dysmenorrhoea apparently showed a higher PGF/PGE ratio than did a much larger series of normal subjects. (Authors modified) 0302

0163
PINCUS, G.
    "The Control of Fertility," New York, Academic Press, p. 93-94. 1965.

Prostaglandins are briefly mentioned as being present in seminal and menstrual fluid. The PG's have potent effect on the contractility of female reproductive smooth muscle. (JRH) 0367

0164
RAMWELL, P.W., J.E. SHAW and J. KUCHARSKI
Prostaglandin: release from the rat phrenic nerve-diaphragm preparation.
Science. 149: 1390-1391. 1965.

Release of a substance that stimulates smooth muscle was detected from a phrenic nerve-diaphragm preparation from the rat upon direct and indirect stimulation. On thin-layer chromatography most of the active material behaved as a mixture of prostaglandins. The effect of electrical stimulation was mimicked by catecholamines but not by acetycholine or eserine. The effect of nerve stimulation was not antagonized by d-tubocurarine. (Authors) 0346

0165
RYHAGE, R. and B. SAMUELSSON
The origin of oxygen incorporated during the biosynthesis of prostaglandin E_1.
Biochemical and Biophysical Research Communications. 19: 279-282. 1965.

The origin of the oxygen incorporated during the formation of PGE_1 from dihomo-γ-linolenic acid was studied by incubating sheep seminal vesicle homogenates with arachidonic acid in an ^{18}O environment. The PGE_1 produced was analyzed by mass spectrometry in conjunction with gas chromatography. The results clearly show that the oxygens of the 2 hydroxyl groups come from oxygen gas, but this study does not allow conclusions to be drawn about the origin of the keto oxygen. (JRH) 0335

0166
SAMUELSSON, B.
On the chemistry, occurrence and metabolism of the prostaglandins.
Proceedings of the Second International Congress of Endocrinology, London, 1964. p. 847-856. Amsterdam, Excerpta Medica, 1965.

This extensive PG review contains sections on: chemistry, occurrence, metabolism, and biosynthesis of prostaglandins. There are 30 references cited. (JRH) 0378

0167
SAMUELSSON, B.
On the incorporation of oxygen in the conversion of 8,11,14.-eicosatrienoic acid to prostaglandin E_1.
Journal of the American Chemical Society. 87: 3011-3013. 1965.

The origin of the 3 oxygen atoms that are incorporated into PGE_1 during its biosynthesis from dihomo-γ-linolenic acid was studied by using ^{18}O atmosphere during the biosynthesis by sheep seminal vesicle homogenates. The results show that the oxygen in the keto group as well as the 2 hydroxyl oxygens come from molecular oxygen, and in fact, the keto oxygen is derived from the same molecule of oxygen as the oxygen in the hydroxyl group at C-11. The results suggest the biosynthesis involves 2 distinct reactions: a dioxygenase reaction and a monooxygenase reaction. Several schemes are described. (JRH) 0341

0168
SAMUELSSON, B.
  The prostaglandins.
  Angewandte Chemie: International Edition in English. 4: 410-416. 1965.

  Prostaglandins occur in many animal species and tissues. They are C_{20}-carboxylic acids containing one five-membered ring, one keto group, two or three hydroxy groups, and one to three C=C double bonds. Their structures were elucidated by chemical, radiochemical, and physical methods in combination with gas chromatography and mass spectrometry. The prostaglandins are biosynthesized from essential fatty acids. By their action smooth muscles are stimulated and lipid metabolism is affected. (Author) 0357

0169
SANDBERG, F., A. INGELMAN-SUNDBERG and G. RYDEN
  The effect of prostaglandin $F_{1\alpha}$, $F_{1\beta}$, $F_{2\alpha}$, and $F_{2\beta}$, on the human uterus and the Fallopian tubes in vitro.
  Acta Obstetrica et Gynecologica Scandinavica. 44: 585-594. 1965.

  The effect of $PGF_{1\alpha}$, $PGF_{1\beta}$, $PGF_{2\alpha}$ and $PGF_{2\beta}$ on human uterine tissue and Fallopian tubes in vitro was studied. There was no difference in response in specimens taken from different phases of the menstrual cycle; however, pieces of tissue from various parts of the uterus and Fallopian tubes showed considerable variability in response to the different prostaglandins. $PGF_{1\beta}$ produced a low incidence of response, and when there was a response, it was a weak increase in tonus in the tube and a decrease in tonus and amplitude in the uterus. $PGF_{2\alpha}$ produced a large increase in tonus in the tube and a lesser increase in the uterus. $PGF_{1\alpha}$ produced the greatest increase in tone in the proximal portion of the tube, but had little effect on the uterus. $PGF_{2\beta}$ produces a slight decrease in tone in the tube and a slight decrease in tone in the uterus. (JRH) 0356

0170
SASAMORI, S.
  Isolation of prostaglandins from human seminal fluid and studies on their stimulating effects on the intestinal smooth muscle.
  Sapporo Medical Journal. 28: 286-299. 1965.

  A mixture of PGE_1 and PGE_2 was isolated from human seminal fluid. The mixture was biologically active and injection of 50 μg/kg into a rabbit lowered blood pressure by 35%. Addition of 0.001 μg/ml stimulated rabbit duodenal strips. Removal of Ca^{++} or depolarization with K^+ blocked the effect of the PG's. The PG's did not affect the glycerol-treated intestinal smooth muscle-ATP system or the relaxing factor system under the present experimental conditions. (Authors modified) 0353

0171
SPERELAKIS, N. and D. LEHMKUHL
  Insensitivity of cultured chick heart cells to autonomic agents and tetrodotoxin.
  American Journal of Physiology. 209: 693-698. 1965.

  PGE_2 and $PGF_{2\alpha}$ in concentrations as high as 3.8×10^{-8} g/ml were ineffective in altering the frequency of firing of pacemaker or driven cells and initiating pacemaker activity in quiescent chick heart cells in vitro. (JRH) 0328

0172
SUZUKI, T. and W. VOGT
Prostaglandine in einem darmstoffpraparat aus froschdarm. (Prostaglandin in the darmstoff preparation of frog intestine.]
Naunyn Schmiedebergs Archiv fur Experimentelle Pathologie und Pharmakologie. 252: 68-78. 1965.

Darmstoff obtained from the bath fluid of the isolated gastro-intestinal tract of frogs was compared with prostaglandin by column and thin layer chromatography. The smooth muscle effect of this darmstoff preparation is mainly due to its PGE_1 and PGF_1 content. In addition another lipid component contributes weakly to the activity. Acidic phosphatides, which are major biologically active constituents of lipid extracts or hot water extracts of horse intestine, are not present or only to a small degree. Since they occur in a diffusible state, polar fatty acids, like prostaglandins, could be humoral transmitters. (Authors modified) 0361

0173
WALLACH, D.P.
The enzymic conversion of arachidonic acid to prostaglandin E_2 with acetone powder preparations of bovine seminal vesicles.
Life Sciences. 4: 361-364. 1965.

A technique for enzymatically converting arachidonic acid to PGE_1 with an acetone powder preparation of bull seminal vesicles is described. The method of preparing the acetone powder is given in detail. The synthesized PGE_2 was purified by chromatographic methods and bioassayed on guinea pig ileum. (JRH) 0329

0174
WALLER, G., H. THEORELL, and J. SJOVALL
Liver alcohol dehydrogenase as a 3β-hydroxy-5β-cholanic acid dehydrogenase.
Archives of Biochemistry and Biophysics. 111: 671-684. 1965.

Horse liver alcohol dehydrogenase was incubated for 30 min with 15 μM $PGF_{1\alpha}$, $PGF_{1\beta}$ or PGE_1. There was no evidence of enzymatic degradation of the PG's. (JRH) 0351

0175
WOLFE, L.S., F. COCEANI and M. SPENCE
Prostaglandins in brain.
Federation Proceedings. 24: 361. 1965.

Abstract only. A PG-like substance was found in perfusates of the cat cerebellum and in TCA acid ethanol or acetone extracts of fresh cat, rat and ox brains. PGF was the only PG found in extracts of 3 Kg of ox cerebrum. When assayed as PGE_1, the ox brain contained 0.8 μg PG/100g cerebrum. The perfusates contained PG (assayed as PGE_1) representing secretion of 1.6 ng/15 min/cm^2 cerebellar surface. (JRH) 0339

0176
WOLFE, L.S., F. COCEANI and M.W. SPENCE
Prostaglandins in brain.
Proceedings of the Eighth International Congress of Neurology, Vienna, 5-10 September, 1965. p. 159-164.

Perfusates of the surface of the cat cerebellum contained a slow acting, mild acid and base resistant, smooth muscle stimulating substance (rat fundus) which was not acetylcholine, serotonin, histamine or catecholamine. When the active substance was compared to PGE_1 on the rat fundus, the cerebellum secreted 0.4-1.6 μg/15 min/cm^2 surface area. A similar substance was extracted from ox and cat brains and the active substance isolated chromatographically. The active substance (95%) was found in the zone which corresponds to PGF. While not proven, the results suggest that the cerebellum secretes a PGF-like substance from its surface, probably PGF_2. (JRH) 0373

0177

AMBACHE, N., H.C. BRUMMER, J. WHITING and M. WOOD
Atropine-resistant substances in extracts of plexus-containing longitudinal muscle (PC-LM) from guinea-pig ileum.
Journal of Physiology. 186: 32P-33P. 1966.

Longitudinal muscles were separated from the guinea pig ileum. In some cases Auerbach's plexus adhered to the muscle and in others it did not. The longitudinal muscle was ground and extracted with distilled water. These extracts contained agents which contracted the histamine insensitive jird colon. A comparison of extracts from plexus containing preparations and plexus free preparations from the same animal were made (8 animals). The plexus containing preparations had 6-1.75 times the activity of the plexus free extracts. Ether extraction of the distilled water extracts showed that all of the activity was not in the acidic lipid fraction. However, chromatographic analysis of the acidic lipid fraction revealed the presence of PGE_2 and $PGF_{2\alpha}$. (JRH) 0531

0178

AMBACHE, N.
Biological characterization of, and structure-action studies on, smooth-muscle-contracting hydroxy-acids.
In: Pickles, V.R. and R.J. Fitzpatrick, eds., "Endogenous Substances Affecting the Myometrium. Proceedings of a Symposium held at the Department of Pharmacy," (Memoirs of the Society for Endocrinology No. 14) Univeristy of Bristol, 19-20 July, 1965. p. 19-28. New York, Cambridge University Press, 1966.

The author reviews the different bioassay systems used to study smooth muscle contracting hydroxy-acids obtained from extracts of biological tissues. The best bioassay systems for the identification of different types of lipid agents including prostaglandins are discussed along with some techniques that can be used to separate certain groups before bioassay. (JRH) 0549

0179

AMBACHE, N., H.C. BRUMMER, J.G. ROSE and J. WHITING
Thin-layer chromatography of spasmogenic unsaturated hydroxy-acids from various tissues.
Journal of Physiology. 185: 77P-78P. 1966.

Thin layer chromatography of spasmogenic unsaturated hydroxy-acids from rabbit ileum longitudinal muscle, rabbit cerebral hemispheres, and rabbit and cat iris revealed the presence of PGE and PGF compounds. The presence of PGE in rabbit-irin would explain its vasodilator properties and alkali-sensitivity. (JRH) 0509

0180

ANGGARD, E.
The biological activities of three metabolites of prostaglandin E_1.
Acta Physiologica Scandinavica. 66: 509-510. 1966.

Dihydroxy-PGE_1, 15-oxo-PGE_1 and 15-oxo-dihydroxy-PGE_1 were compared to PGE_1 for biological activity in several bioassay tissues (in vivo and in vitro). Dihydroxy-PGE_1 had from 14-41% of the activity of PGE_1 on rabbit ileum or guinea pig uterus in vitro. It was slightly less potent than PGE_1 in lowering the rabbit blood pressure, but more potent on guinea pig blood pressure. The 15-oxo metabolites of PGE_1 were much less potent than PGE_1 on all the test preparations. (JRH) 0514

0181
ANGGARD, E. and B. SAMUELSSON
  Metabolism of prostaglandins in the lung.
  Acta Physiologica Scandinavica. 68 Supp. 277: 232. 1966.

  Abstract only. In a homogenate of guinea-pig lung, prostaglandin E_1, E_2 and E_3 are metabolized by analogous reactions, e.g. saturation of the Δ^{13} double bond and oxidation of the secondary alcohol group at C-15. In swine lung only the latter transformation occurs. To study this reaction more closely, a purified enzyme was prepared from swine lung. It could be shown that the enzyme was specific for the hydroxyl group at C-15 in the prostaglandins and could thus be named 15-hydroxy-prostaglandin dehydrogenase. NAD^+ but not $NADP^+$ served as co-factor. The possible use of this enzyme in a sensitive and specific assay for prostaglandins is being investigated. (Authors modified) 0513

0182
ANGGARD, E. and B. SAMUELSSON
  Metabolites of prostaglandins and their biological properties.
  In: Pickles, V.R. and R.J. Fitzpatrick, eds., "Endogenous Substances Affecting the Myometrium.
    Proceedings of a Symposium held at the Department of Pharmacy," (Memoirs of the Society for
    Endocrinology No. 14) University of Bristol, 19-20 July, 1965. p. 107-117. New York, Cambridge
    University Press, 1966.

  The authors review a series of experiments on the metabolism of PGE_1 and PGE_2 and PGE_3. The biological activity of the metabolites of PGE_1 are also discussed. The 3 metabolites of PGE_1 were much less potent than the parent compound in vitro on isolated smooth muscle. However, when tested on blood pressure in vivo the 11α, 15-hydroxy-9-ketoprostanoic acid metabolite was more active in the guinea pig and only slightly less active in the rabbit than PGE_1 in lowering blood pressure. (JRH) 0544

0183
ANONYMOUS
  Closure of umbilical blood-vessels.
  Lancet. 2: 381. 1966.

  It is briefly mentioned that the lipid-soluble unsaturated hydroxy-acid, which has been isolated from umbilical cords and which causes contraction of umbilical arteries, may be allied to the prosta-glandins. (JRH) 0533

0184
ANONYMOUS
  Drugs affecting lipid metabolism.
  Science. 151: 1016-1017. 1966.

  Prostaglandins are briefly mentioned in this report of a symposium on drugs that affect lipid metab-olism. (JRH) 0590

0185
ANONYMOUS
Prostaglandins.
In: Damm, H.C., P.K. Besch and A.J. Goldwyn, eds., "The Handbook of Biochemistry and Biophysics," p. 243-247. Cleveland and New York, World Publishing Company, 1966.

The chemical structure and origin of the prostaglandins are briefly reviewed. Structural formulas for PG's are given. (JRH) 0505

0186
ANONYMOUS
Prostaglandin synthesized.
Science News. 89: 481. 1966.

This brief review reports the total synthesis of dihydro-PGE_1 by Upjohn chemists. The biological activity of PG's is briefly discussed. This synthetic method should make PG's more available for research. (JRH) 0529

0187
ANONYMOUS
Synthetic prostaglandin may be potent drug.
Chemical and Engineering News. July 4: 32-33. 1966.

In this brief review, the total synthesis of dihydro-PGE_1 is described in general terms. The biological activity of this compound is discussed and the possible uses of this and related compounds as drugs are mentioned. (JRH) 0500

0188
ASH, A.S.F.
Differential inhibition of bronchoconstriction in the guinea pig by extracts containing prostaglandins.
Proceedings of the 3rd International Pharmaceutical Meeting, Sao Paulo, 24-30 July, 1966. p. 167.

Abstract only. Partially purified PGE_2 was injected intravenously into guinea pigs in which bronchoconstriction had been induced by histamine, serotonin, bradykinin, acetylcholine or slow reacting substance A. PGE_1 inhibited the bronchoconstriction in all cases, but it was most effective against histamine and serotonin. Isolated tracheal chains constricted, by histamine or serotonin were relaxed by PGE_2 at concentrations that had no effect on acetylcholine contractions. 0597

0189
AVANZINO, G.L., P.B. BRADLEY and J.H. WOLSTENCROFT
Actions of prostaglandins E_1, E_2, and $F_{2\alpha}$ on brain stem neurones.
British Journal of Pharmacology and Chemotherapy. 27: 157-163. 1966.

PGE_1, PGE_2 $PGF_{2\alpha}$, linolenic and bis-homo-γ-linolenic acid were applied iontophoretically to brain stem neurones in decerebrate, unanesthetized cats. PGE_1 excited 18.5% and inhibited 4%, PGE_2 excited 27.5% and caused no inhibition. $PGF_{2\alpha}$ excited 26% and inhibited 10%. Some individual neurones were affected by one PG and unaffected by others, but never did one PG excite and the other inhibit. Tachyphylaxis to an individual PG was often observed, but never was a cross tachyphylaxis observed. Occasionally, neurones which were excited by PGE_1 were also excited by the PG precursors. There was no relationship to the response of a neurone to PG's and to acetylcholine. (JRH) 0558

0190
AVANZINO, G.L., P.B. BRADLEY and J.H. WOLSTENCROFT
Excitatory action of prostaglandin E_1 on brain-stem neurones.
Nature. 209: 87-88. 1966.

The effect of iontophoretic application of PGE_1 to individual neurones in the brain stem of anes-thetized, decerebrate cats was studied. PGE_1 stimulated 75 (28%) of the 265 neurones studied and had no effect on 184 (70%). Homo-gamma-linolenic acid excited about half of the neurones that were excited by PGE_1. Linolenic acid was also effective on some of the neurones excited by PGE_1. (JRH) 0510

0191
BEAL, P.F., III, J.C. BABCOCK and F.H. LINCOLN
A total synthesis of a natural prostaglandin.
Journal of the American Chemical Society. 88: 3131-3133. 1966.

A total synthesis of dl-dihydro-PGE_1 is described. The structure of the starting compounds and key intermediates are given. The synthesized PG showed similar biological activity in bioassay systems to natural dihydro-PGE_1. (JRH) 0564

0192
BECK, L., A.A. POLLARD, S.O. KAYAALP and L.M. WEINER
Sustained dilatation elicited by sympathetic nerve stimulation.
Federation Proceedings. 25: 1596-1606. 1966.

High frequency sympathetic nerve stimulation of dogs which had been adrenergically depleted re-sulted in long lasting vasodilation in a constantly perfused hind limb. PGE_1 (0.3-10 µg) injected intra-arterially into the perfused hind limb was the only vasodilator which mimicked the sustained dilation caused by high frequency sympathetic nerve stimulation in the depleted dog. (JRH) 0565

0193
BERGSTROM, S., L.A. CARLSON and L. ORO
Effect of different doses of prostaglandin E_1 on free fatty acids of plasma, blood glucose and heart rate in the nonanesthetized dog.
Acta Physiologica Scandinavica. 67: 185-193. 1966.

Intravenous infusion of PGE_1 (0.2, 0.4, 0.8 or 1.6 µg/kg/min) produced an increase in plasma free fatty acids (FFA) at the 0.2 µg/kg/min rate and a decrease in plasma FFA and glycerol at the higher doses. Blood glucose was increased at all infusion rates but was statistically significant only at 0.8 µg/kg/min. The heart rate increased with infusion rates of 0.2-0.8 µg/kg/min, but the highest infusion rate caused variable results. All the effects of the 0.2 µg/kg/min infusion rate were prevented by sympathetic ganglionic blockade. Thus PGE_1 may not only cause a direct inhibition of FFA mobiliza-tion in adipose tissue but at low doses may stimulate FFA mobilization of the sympathetic nervous system. (JRH) 0511

0194
BERGSTROM, S., L.A. CARLSON and L. ORO
Effect of prostaglandin E_1 on plasma free fatty acids and blood glucose in the dog.
Acta Physiologica Scandinavica. 67: 141-151. 1966.

Intravenous infusion of PGE_1 (0.2 μg/kg/min) into unanesthetized fasting dogs caused a rise in plasma free fatty acid (FFA) and glycerol during the infusion period. PGE_1 (0.12-1.0 μg/kg/min) infused into the aorta caused a rise in FFA in 9 conscious dogs and a fall in 3. FFA levels also increased in anesthetized dogs infused in the carotid artery or intravenously with PGE_1. Turnover rate studies showed the FFA changes to be due to changes in FFA mobilization rates. Glycerol level changes paralleled the changes in FFA. PGE_1 did not produce significant changes in blood glucose levels. Intra-aortic infusion of PGE_1 (0.3-0.8 μg/kg/min) inhibited the increase in plasma FFA induced norepinephrine, epinephrine, isoprenaline or dimethylphenylpiperazinium but did not alter the rise in blood glucose induced by the latter agent. (JRH) 0512

0195
BERGSTROM, S.
The prostaglandins.
In: Pincus, G. ed., "Recent Progress in Hormone Research, Proceedings of the Laurentian Hormone Conference," Vol. 22. p. 153-168. New York, Academic Press, 1966.

The author reviews the scientific literature on prostaglandins. There are sections on: isolation and structure; occurrence; biosynthesis; metabolism; and physiological action. There are 72 references cited. (JRH) 0571

0196
BOHLE, R., H. DITSCHUNEIT, J. AMMON and R. DOBERT
Uber die wirkung von prostaglandin E_1 (PGE_1) auf den fett-und kohlenhydratstoffwechsel des isolier-
ten rattenfettgewebes. [The effect of PGE_1 on fat and carbohydrate metabolism in isolated rat fatty tissue.]
Verhandlungen der Deutschen Gesellschaft fur Innere Medizin. 72: 465-469. 1966.

The effects of PGE_1 on the oxidation of labelled glucose to CO_2, on glucose uptake, on glycerine liberation, and on lipid and fatty-acid synthesis in rat epididymal fatty tissue were investigated. Epididymal fatty tissue of normally fed Wistar rats was placed in the incubation baths containing PGE_1, crystalline swine insulin, human growth hormone, or epinephrine shortly after the animals were killed. The anti-lipolytic action of PGE_1 was clearly demonstrated in the presence of glucose, and in the presence of increased lipolysis due to epinephrine and growth hormone. Glucose uptake and glucose oxidation were increased in the presence of epinephrine and growth hormone, but PGE_1 does not further intensify the effect of insulin on these processes. Basal lipid and fatty-acid synthesis was greatly increased by PGE_1, although the similar effect of insulin was much larger. The influence of PGE_1 on the metabolism of isolated fatty tissue is thus shown to be far more comprehensive than was previously believed. (MEMH) 0574

0197
BOHLE, E.
Uber wirkungen der prostaglandine auf den fett-und kohlenhydratstoffwechsel. [The influence of prostaglandins on fat and carbohydrate metabolism.]
Fette-Seifen-Anstrichmittel. 68: 938-939. 1966.

Abstract only. The results of in vivo and in vitro experimentation with the effects of PGE_1 on the metabolic performance of rat diaphragm and epididymal fatty tissue and on blood chemistry changes in conditions of increased lipolysis are summarized. PGE_1 stimulated CO_2 production, glucose up-take, the incorporation of glucose into triglycerides and, in certain conditions, fatty-acid synthesis from glucose. Lipolysis was inhibited by PGE_1. In increased lipolysis, PGE_1 influenced hyperlip-acidemia, hyperketonemia, and hypertriglyceridemia, but not the free glycerine content of the blood. The metabolic influence of PGE_1 is compared with that of insulin. (MEMH) 0503

0198
BOHLE, E., E. DOBERT, J. AMMON and H. DITSCHUNEIT
   Uber stoffwechselwirkungen von prostaglandinen. I. Der einfluss von prostaglandin E_1 auf den glucose- und fettstoffwechsel des epididymalen fettgewebes der ratte. [On the metabolic effects of prostaglandins. I. The influence of prostaglandin E_1 on carbohydrate and lipid metabolism of rat epididymal fatty tissue.]
   Diabetologia. 2: 162-168. 1966.

   PGE_1 (0.1 μg/ml) was incubated with rat adipose tissue in the presence or absence of insulin (500 μU), epinephrine (0.1 μg/ml) or human growth hormone (25 μg/ml). PGE_1 was more effective than insulin in inhibiting basal lipolysis and reduced the lipolytic effect of epinephrine and growth hor-mone. PGE_1 stimulated the oxidation of glucose to CO_2, but was only 1/10 as effective as insulin. Epinephrine and growth hormone increased the effect of PGE_1 on glucose metabolism. PGE_1 stim-ulated the glucose uptake caused by epinephrine, but did not alter the response to growth hormone. PGE_1 caused an 80% increase in the incorporation of glucose into lipids, but was less effective than the other agents. It is proposed that the stimulating effect of PGE_1 on glucose utilization is a consequence of regulatory changes in cellular metabolism and is closely related to antilipolytic activity. (JRH) 0594

0199
BOHLE, E., H. DITSCHUNEIT, F. MELANI, J. BEYER, K. SCHOFFLING and E.F. PFEIFFER
   Uber die regulation der unveresterten fettsauren. Unter besonderer berucksichtigung der prosta-glandine. [On the regulation of non-esterified fatty acids, with special attention to prosta-glandins.]
   Deutsche Medizinische Wochenschrift. 91: 1083-1090. 1966.

   Present knowledge of the endocrine regulation of the release of non-esterified fatty acids from fatty tissue, and their role in the metabolism, is reviewed. Primary disturbances of triglyceride regulation in fatty tissue can have a pathogenic influence on disturbances of carbohydrate metabolism. Insulin, and recently prostaglandins, have been found to exercise an antilipolytic effect in animal tissue. PGE_1 in particular has been found to inhibit the increased release of fatty acids in fasting rats, and to influence the heparin-induced activation of lipoprotein lipase. (MEMH) 0588

0200
BOHONOS, N. and H.D. PIERSMA
   Natural products in the pharmaceutical industry.
   Bioscience. 16: 706-714. 1966.

   Prostaglandins are briefly mentioned as being physiologically active agents that have been isolated from vertebrates. A brief history of research is given. (JRH) 0560

0201

BOYARSKY, S., P. LABAY and C. GERBER
  Prostaglandin inhibition of ureteral peristalsis.
  Investigative Urology. 4: 9-11. 1966.

  PGE_1 (1-30 μg/ml or 0.1-33.0 μg/kg i v) slows or stops ureteral peristalsis in vitro and in vivo in dogs.
  Action is dose dependent. Ureteral musculature has mechanisms in common with vascular and female
  genital musculature. (Authors modified) 0569

0202

BYGDEMAN, M. and B. SAMUELSSON
  Analyses of prostaglandins in human semen. Prostaglandins and related factors 44.
  Clinica Chemica Acta. 13: 465-474. 1966.

  A method for quantitative determination of PGE_1, PGE_2, PGE_3 $PGF_{1\alpha}$ and $PGF_{2\alpha}$ in human
  seminal fluid is described. The method involves group separation of PGE and PGF compounds by
  silicic acid chromatography and separation of the PGE compounds by thin layer chromatography.
  The PGE compounds are then determined by measuring the dienone chromophore (λmax 278 mμ)
  formed on treatement with alkali. The PGF compounds are separated and determined by gas-liquid
  chromatography. Semen from men with normal fertility contained: PGE_1, 25.0 μg/ml; PGE_2, 23.0
  μg/ml; PGE_3 5.5 μg/ml; $PGF_{1\alpha}$ 3.6 μg/ml and $PGF_{2\alpha}$, 4.4 μg/ml. In a group of men (10 patients)
  with suspected infertility of unknown origin were found 2 patients with very low concentrations of
  prostaglandins. (Authors) 0563

0203

BYGDEMAN, M., M. HAMBERG and B. SAMUELSSON
  The content of different prostaglandins in human seminal fluid and their threshold doses on the
    human myometrium.
  In: Pickles, V.R. and R.J. Fitzpatrick, eds., "Endogenous Substances Affecting the Myometrium.
    Proceedings of a Symposium held at the Department of Pharmacy," (Memoirs of the Society for
    Endocrinology No. 14) University of Bristol, 19-20 July, 1965. p. 49-64. New York, Cambridge
    University Press, 1966.

  A review of the literature on the quantitative and qualitative determination of the PG content of
  human seminal plasma is presented with an emphasis on technique; available data on the effects of 13
  prostaglandins on human myometrial strips is summarized. 21 references are cited. (RMS) 0554

0204

BYGDEMAN M. and O. HOLMBERG
  Isolation and identification of prostaglandins from ram seminal plasma. Prostaglandins and related
    factors 55.
  Acta Chemica Scandinavica. 20: 2308-2310. 1966.

  Ram seminal plasma was assayed quantitatively and qualitatively for prostaglandin content. The PG's
  found and their respective concentrations were: PGE_1 (28 μg/ml), PGE_2 (3.2 μg/ml), PGE_3 (0.2
  μg/ml), $PGF_{1\alpha}$ (5 μg/ml) and $PGF_{2\alpha}$ (2.3 μg/ml). No 19-hydroxy prostaglandins were found. (JRH)
  0527

0205
CARLSON, L.A. and L. ORO
Effect of prostaglandin E_1 on blood pressure and heart rate in the dog.
Acta Physiologica Scandinavica. 67: 89-99. 1966.

Infusion of PGE_1 either intravenously or intra-aortically into pentobarbital anesthetized dogs caused a fall in blood pressure and an increase in heart rate. The intra-aortic route was the most effective and the higher up the aorta the PG was infused, the more effective it was in lowering blood pressure. Infusion of PGE_1 into the carotid artery elevated blood pressure in 6 of 8 dogs, however, if the dogs were pretreated with a sympathetic blocking agent (Agentit®) or reserpine, PGE_1 then produced a fall in blood pressure. These blocking agents also prevented the increase in heart rate produced by PGE_1. Dogs infused with PGE_1 at rates sufficient to lower blood pressure still showed pressor responses to norepinephrine, epinephrine, dimethylphenylpiperazinium or electrical stimulation of the central ends of the cut vagus nerve. PGE_1 apparently lowers blood pressure by means independent of catecholamines. (JRH) 0516

0206
CLAYTON, D., H.C. HILL and R.I. REED
Mass spectrometry in natural product chemistry.
Advances in Mass Spectrometry. 3: 669-679. 1966.

Mass spectrometry was used to identify an unknown prostaglandin by comparing its mass spectrum to that of a known sample. The technique was sensitive enough to identify 5 μg of the compound from a thin layer chromatography plate. (JRH) 0584

0207
CLEGG, P.C., W.J. HALL and V.R. PICKLES
The action of ketonic prostaglandins on the guinea-pig myometrium.
Journal of Physiology. 183: 123-144. 1966.

The mechanism of the effects of PGE_1 or PGE_2 on the guinea pig uterus in vitro was examined. These PG's produce 2 effects: first, a direct stimulation of the uterus and second, a long lasting (20-80 min) enhancement in responsiveness of the uterus to other stimulants, vasopressin, electrical stimulation, oxytocin, $PGF_{2\alpha}$ or even PGE_1 and PGE_2 after the initial PGE_1 or PGE_2 had been washed out. The enhancement was not significantly altered by moderate changes of Mg^{++}, Ca^{++} or K^+ in the incubating medium. These two effects are postulated to involve two different receptor sites. The direct effect may be due to depolarization of the membrane, whereas enhancement may result from facilitation of excitation-contraction coupling. (JRH) 0508

0208
CLEGG, P.C.
Antagonism by protaglandins of the responses of various smooth muscle preparations to sympathomimetics.
Nature. 209: 1137-1139. 1966.

The effects of PGE_1, $PGF_{2\alpha}$, and $PGF_{1\alpha}$ on the response of rat was deferens, colon, fundic strip, and the guinea pig uterus and tracheal chain to the sympathomimetic agents phenylephrine, norepinephrine, epinephrine and isopropylnoradrenaline was investigated. It was found that irrespective of the nature of the direct effect of the prostaglandin (stimulation, inhibition, or no effect), and irrespective of the nature of the response of the preparation to sympathomimetics (stimulation or inhibition), the net effect of the prostaglandin was to depress the response of the tissue to sympathomimetics. However, often the first response of the tissue to sympathomimetics after exposure to the PG was potentiation, but this initial enhancement was followed by an irreversible suppression of response. Receptor protections studies revealed that the PG's do not compete for receptor sites but may act to bind the sympathomimetics to the receptor. (JRH) 0518

0209
CLEGG, P.C.
The effect of prostaglandins on the response of isolated smooth-muscle preparations to sympathomimetic substances.
In: Pickles, V.R. and R.J. Fitzpatrick, eds., "Endogenous Substances Affecting the Myometrium. Proceedings of a Symposium held at the Department of Pharmacy," (Memoirs of the Society for Endocrinology No. 14) University of Bristol, 19-20 July, 1965. p. 119-136. New York, Cambridge University Press, 1966.

The interaction of prostaglandins and sympathomimetics in a variety of smooth muscle preparations is reviewed. Both PGE's and PGF's antagonize the action of sympathomimetic agents (stimulation or inhibition). The inhibition is independent of the direct effect of the prostaglandin on the tissue (stimulation, inhibition or no effect). In some tissues, the inhibition of the effect of the sympathomimetic is potentiated for a short time by the PG before the long lasting inhibition starts. Receptor protection experiments show that PG's do not compete for the adrenergic receptor site. (JRH) 0556

0210
COCEANI, F. and L.S. WOLFE
On the action of prostaglandin E_1 and prostaglandins from brain on the isolated rat stomach.
Canadian Journal of Physiology and Pharmacology. 44: 933-950. 1966.

PGE_1 and PG extracted from ox brain (mainly $PGF_{2\alpha}$) caused contraction of the rat stomach fundus strips in concentrations as low as 10^{-10} g/ml. This contraction was oxygen dependent and was enhanced by sympathetic inhibitors and inhibited by either sympathetomimetic agents (epinephrine and NE) or reduction in temperature to $24°C$. Metabolic inhibitors such as cyanide, azide or CO inhibited PGE_1 while reduced glutathione and ascorbic acid enhanced PGE_1. It is suggested that PG's are converted to an active form by oxygen-requiring enzymes which either liberate bound Ca^{++} or enhance Ca^{++} influx. (JRH) 0572

0211
CROWSHAW, K.
Potential errors in the isolation and identification of millimicrogram quantities of prostaglandins.
Federation Proceedings. 25: 765. 1966.

Abstract only. A rapid technique involving solvent partition and thin layer chromatography for separation of PG's from biological tissues and fluids which contain PG's in concentrations of less than 0.1 μg is described. The purified PG's are then identified by microbioassay. This technique avoids the possible errors due to the presence of pharmacologically active lipid hydroperoxides in the biological sample. (JRH) 0567

0212
van DORP, D.A.
   The biosynthesis of prostaglandins.
   In: Pickles, V.R. and R.J. Fitzpatrick, eds., "Endogenous Substances Affecting the Myometrium.
     Proceedings of a Symposium held at the Department of Pharmacy," (Memoirs of the Society for
     Endocrinology No. 14) University of Bristol, 19-20 July, 1965. p. 39-47. New York, Cambridge
     University Press, 1966.

The author reviews the literature on the biosynthesis of prostaglandins. There are sections on:
synthesis by animal tissues; synthesis by human tissues; the mechanism of the conversion; and
prostaglandins and essential fatty acid deficiency. There are 25 references cited. (JRH) 0555

0213
van DORP, D.A.
   Biosynthese von prostaglandinen. [Biosynthesis of prostaglandins.]
   Fette-Seifen-Anstrichmittel. 68: 938. 1966.

Abstract only. It has been shown that essential fatty acids appear as "precursors" in the biosynthesis
of prostaglandins. The author devotes particular attention to the localization of the enzyme system
and to the question of the substrate-specificity of this system. (MEMH) 0504

0214
ELIASSON, R.
   Effect of bradykinin on the human uterus in vitro.
   Journal of Pharmacy and Pharmacology. 18: 396. 1966.

Bradykinin had no inhibitory effect on the human uterus in vitro. However, other strips from the
same uteri responded to PGE and other compounds in the usual manner. (JRH) 0540

0215
ELIASSON, R.
   The effect of different prostaglandins on the motility of the human myometrium.
   In: Pickles, V.R. and R.J. Fitzpatrick, eds., "Endogenous Substances Affecting the Myometrium.
     Proceedings of a Symposium held at the Department of Pharmacy," (Memoirs of the Society for
     Endocrinology No. 14) University of Bristol, 19-20 July, 1965. p. 77-88. New York, Cambridge
     University Press, 1966.

The author reviews the effect of partially purified prostaglandins extracted from human seminal fluid
on the human myometrium in vitro and in vivo. Also the effects of pure PGE_1, PGE_2, PGE_3, $PGF_{1\alpha}$
and $PGF_{2\alpha}$ on the myometrium are compared with the effects of the partially purified extracts in
vitro. (JRH) 0550

0216
ELIASSON, R.
   Effect of posterior pituitary hormones on the myometrial response to prostaglandin.
   Acta Physiologica Scandinavica. 66: 249-250. 1966.

Oxytocin and vasopressin were infused into 2 women at the estimated time of ovulation. Intravaginal
application of human seminal fluid prostaglandins caused an immediate decrease in uterine activity.

The PG's were generally ineffective during the proliferative phase, but in one woman there was a marked increase in uterine activity following intravaginal application of the PG's. No effect of PGE_1 on the responsiveness of human uterine strips treated with oxytocin and vasopressin was found in vitro. (JRH) 0525

0217
ELIASSON, R.
Mode of action of prostaglandins on the human non-pregnant myometrium.
Biochemical Pharmacology. 15: 755. 1966.

Addition of propranolol to an organ bath containing non-pregnant human myometrial strips completely blocked the effect of β-adrenergic agents but had no effect on the uterine responses to PGE_1 or PG extracts of human seminal plasma. Dehydroergotamine also failed to block PGE_1. Thus the response of the non-pregnant human uterus is not mediated by adrenergic receptors. (JRH) 0591

0218
ELIASSON, R. and P.L. RISLEY
Potentiated response of isolated seminal vesicles to catecholamines and acetylcholine in the presence of PGE_1.
Acta Physiologica Scandinavica. 67: 253-254. 1966.

The effect of subthreshold doses of PGE_1 on the response of isolated seminal vesicles from either normal adult or castrated (for 2 months) rats and guinea pigs to epinephrine, norepinephrine or acetylcholine was studied in vitro. The PGE_1 (12.5-25 ng/ml), in amounts 10-20 times lower than the amount needed to produce a contraction, increased the responsiveness of the vesicles to the other spasmogens and reduced the lag time in response to these agents. It also increased the steepness of the dose response curves. The PGE_1 effect persisted 16-36 min depending on the amount of PGE_1 used. (JRH) 0517

0219
von EULER, U.S.
Introductory survey: prostaglandin.
In: Pickles, V.R. and R.J. Fitzpatrick, eds., "Endogenous Substances Affecting the Myometrium. Proceedings of a Symposium held at the Department of Pharmacy," (Memoirs of the Society for Endocrinology No. 14) University of Bristol, 19-20 July, 1965. p. 3-18. New York, Cambridge University Press, 1966.

In this introductory survey, the history of prostaglandin research is reviewed. There are sections on earlier observations; actions of crude or semi-purified extracts; effects on the uterus; occurrence; isolation, chemistry; actions of pure compounds; effect of prostaglandins on fat mobilization; and the physiological significance of prostaglandins. There are 51 references cited. (JRH) 0552

0220
FELDBERG, W., and R.D. MYERS
Appearance of 5-hydroxytryptamine and an unidentified pharmacologically active lipid acid in effluent from perfused cerebral ventricles.
Journal of Physiology. 184: 837-855. 1966.

The brains of cats anesthetized with pentobarbital were perfused with an artificial cerebrospinal fluid. The perfusate contained at least 2 substances which contracted rat fundus strips. One of the substances was identified as serotonin. The other was an unknown hydroxy acid believed to be related to the prostaglandins or to irin. (JRH) 0524

0221

GUNSTONE, F.D.

Speculation on the role of epoxy acids as intermediates in the biosynthesis of polyunsaturated fatty acids.

Chemistry and Industry. Sept. 10: 1551-1554. 1966.

It is proposed that prostaglandin formation from arachidonic acid can be visualized as an epoxide rearrangement accompanied by a cyclization process. A diagrammatic scheme for this process is given. (JRH) 0593

0222

HALL, W.J.

Prostaglandins in human menstrual fluid and endometrial curettings.

In: Pickles, V.R. and R.J. Fitzpatrick, eds., "Endogenous Substances Affecting the Myometrium. Proceedings of a Symposium held at the Department of Pharmacy," (Memoirs of the Society for Endocrinology No. 14) University of Bristol, 19-20 July, 1965. p. 65-74. New York, Cambridge University Press, 1966.

The literature on the prostaglandin content of human menstrual fluid and endometrial curettings is reviewed. The human menstrual stimulant is composed of component A and component B. Component A was found to be mainly $PGF_{2\alpha}$ and component B was a mixture of hydroxy- and polyunsaturated fatty acids, some of which may be PG precursors. Curettings contain PGE and PGF compounds but in lower concentrations than found in the menstrual fluid. (JRH) 0551

0223

HALL, W.J.

Some effects of ketonic prostaglandins on guinea-pig myometrium.

Irish Journal of Medical Science. 483: 116. 1966.

Abstract only. Both PGE and PGF compounds stimulated the guinea pig myometrium in vitro. The PGE's were 100 times more potent than the PGF's. PGE produced contractions at concentrations as low as 10^{-9} M. They also potentiated the myometrial response to other spasmogens and electrical stimulation. This enhancement was long lasting and not affected by low Ca^{++} or high Mg^{++} in the incubating medium. (JRH) 0501

0224

HAESSLER, H.A. and J.D. CRAWFORD

Lipolysis in homogenates of adipose tissue: an inhibitor found in fat from obese rats.

Science. 154: 909-910. 1966.

Experiments with rats made obese by bilateral electrolytic lesions of the ventromedial nuclei of the hypothalmus suggest that the insensitivity of adipose tissue from these animals to the lipolytic effects of norepinephrine in vitro is due to the presence of an endogenous lipid-bound inhibitor of lipolysis. Solubility and biological characteristics of this inhibitor suggest that it may be similar to one of the prostaglandins. (JRH) 0589

0225
HAMBERG, M. and B. SAMUELSSON
  Novel biological transformations of 8,11,14-eicosatrienoic acid.
  Journal of the American Chemical Society. 88: 2349-2350. 1966.

Monohydroxy acids are formed from arachidonic acid by the same tissues that produce prostaglandins (sheep vesicular glands). (JRH) 0561

0226
HAMBERG, M. and B. SAMUELSSON
  Prostaglandins in human seminal plasma. Prostaglandins and related factors 46.
  Journal of Biological Chemistry. 241: 257-263. 1966.

Acidic lipids from 650 ml of human seminal plasma were extracted and assayed for prostaglandins.
By chromatography, spectroscopy, nuclear magnetic resonance, mass spectrometry and comparison
to known PG's, 13 prostaglandins were identified, 8 of which had not been reported previously. The
5 previously reported PG's were PGE_1, PGE_2, PGE_3, $PGF_{1\alpha}$ and $PGF_{2\alpha}$. The 8 new PG's were
PGA_1, PGA_2, PGB_1, PGB_2 and their 19-hydroxy derivatives. The 19-hydroxy PG's were present in
about 4 times higher concentrations than the PGE compounds. (JRH) 0535

0227
HAUGE, A., P.K.M. LUNDE and B.A. WAALER
  Effects of some vasoactive substances on vascular resistance and vascular capacity in isolated blood-
    perfused lungs.
  Acta Physiologica Scandinavica. 68 Supp. 277: 69. 1966.

PGE_1 perfused into isolated rabbit lungs caused a reduction in vascular resistance similar to that
caused by epinephrine but did not alter vascular capacity as does epinephrine. (JRH) 0523

0228
HERZOG, J., H. JOHNSTON and D.P. LAULER
  Natriuretic effect of prostaglandin E_1 in the dog kidney.
  Clinical Research. 14: 491. 1966.

Abstract only. PGE_1 (0.01-2.0 μg/min) was infused into the left renal artery of anesthetized hydropenic mongrel dogs with the right kidney used as a control. Within the dose range employed, PGE_1
exerted no significant depressor effect, and was not associated with a change in heart rate. At all
infusion rates, PGE_1 produced natriuresis from the infused kidney, and at higher doses minimal
natriuresis was also seen from the contralateral kidney. Glomerular filtration rate remained constant.
The natriuresis was associated with an increase in effective renal plasma flow, a parallel rise in urine
flow, and decreases in para amino hippuric acid extraction and in urine osmolality on the infused
side. In some instances, positive free water clearances were noted. The findings are consistent with
PGE_1 serving as an intrarenal natriuretic hormone. (Authors modified) 0579

0229
HICKLER, R.B., A.E. BIRBARI, E.U. QURESHI and M.L. KARNOVSKY
Purification and characterization of the vasodilator and antihypertensive lipid of rabbit renal medulla.
Association of American Physicians' Transactions. 79: 278-283. 1966.

The renomedullary vasodepressor lipid from rabbit kidneys was compared to PGA_1 and PGA_2. On the basis of spectrophotometric chromatographic and biological behavior, the vasodepressor lipid appears identical with PGA_2. (JRH) 0573

0230
HICKLER, R.B., A.E. BIRBARI, D.E. KAMM and G.W. THORN
Studies on a vasodepressor and antihypertensive lipid of rabbit renal medulla.
In: Milliez, P. and P. Tcherdakoff, eds., "L'Hypertension Arterielle," [Arterial Hypertension], International Club on Arterial Hypertension, First Meeting, 5-7 July, 1965. p. 188-202. Paris, L'Expansion Scientifique Francaise, 1966.

The long chain, fatty acid nature and vasodepressor properties of the vasodepressor lipid from rabbit kidney medullas suggested that it might be a prostaglandin. However, it differed from PGE_1 in that PGE_1 was more polar and the compounds had different chromatographic behaviors. It is still possible that the renal vasodepressor lipid is a prostaglandin, but it is not PGE_1. (JRH) 0570

0231
HORTON, E.W. and I.H.M. MAIN
The identification of prostaglandins in central nervous tissues of the cat and fowl.
Journal of Physiology. 185: 36P-37P. 1966.

Prostaglandins were extracted from cat cerebral tissue, whole chicken brains and spinal cords. Silicic acid column chromatography of the cat brain extracts produced 2 peaks corresponding to PGE and PGF. Thin layer chromatography of these peaks proved the presence of PGF, probably $PGF_{2\alpha}$. However, there was not enough PGE present for positive identification. Using similar techniques, chicken brains and spinal cords were shown to contain PGE_2 and $PGF_{2\alpha}$. (JRH) 0530

0232
HORTON, E.W. and I.H.M. MAIN
The relationship between the chemical structure of prostaglandins and their biological activity.
In: Pickles, V.R. and R.J. Fitzpatrick, eds., "Endogenous Substances Affecting the Myometrium. Proceedings of a Symposium held at the Department of Pharmacy," (Memoirs of the Society for Endocrinology No. 14) University of Bristol, 19-20 July, 1965. p. 29-37. New York, Cambridge University Press, 1966.

The relationship of the structure of 7 PG's with their biological activity in a series of bioassay systems is reviewed. The effects of alterations in structure on biological activity are also discussed. It is concluded that parallel bioassay on rabbit jejunum and cat trachea will distinguish between PGE's and PGF's but no bioassay system tested so far will identify individual prostaglandins in each series. (JRH) 0548

0233
KARIM, S.M.M.
   Identification of prostaglandins in human amniotic fluid.
   Journal of Obstetrics and Gynaecology of the British Commonwealth. 73: 903-908. 1966.

   Human amniotic fluid obtained after spontaneous rupture of amniotic membranes during labor was collected and extracted for prostaglandins. Chromatographic and bioassay analysis revealed the presence of PGE_1 (1 ng/ml), PGE_2 (0.5 ng/ml), $PGF_{1\alpha}$ (140 ng/ml) and $PGF_{2\alpha}$ (30 ng/ml). These prostaglandins may play a physiological role in parturition. (JRH) 0545

0234
KARIM, S.M.M.
   Isolation and identification of two smooth muscle stimulating substances in human umbilical cord extracts.
   British Journal of Pharmacology and Chemotherapy. 27: 445. 1966.

   Title only. 0559

0235
KARIM, S.M.M.
   A smooth muscle contracting substance in extracts of human umbilical cord.
   Journal of Pharmacy and Pharmacology. 18: 519-530. 1966.

   Aqueous extracts of human umbilical arteries and vein have been shown to contain a smooth muscle contracting substance. The active principle has been distinguished from smooth muscle contracting substances which are found in mammalian tissues, namely esters of choline, histamine, 5-hydroxy-tryptamine, bradykinin, angiotensin and darmstoff. Evidence indicates that the smooth muscle stimulating activity of umbilical blood vessel extracts is due to a lipid-soluble unsaturated hydroxy acid. The possible physiological function of the active substance is discussed. (Author) 0598

0236
KUPIECKI, F.P. and J.R. WEEKS
   Prolonged intravenous infusion of prostaglandin E_1 (PGE_1) on lipid metabolism in the rat.
   Federation Proceedings. 25: 719. 1966.

   Abstract only. PGE_1 given by continuous I.V. infusion to EFA deficient rats at 1 mg/kg/day for 8 weeks did not affect the scaly skin and tail. A similar 6 week infusion to normal rats lowered neutral liver lipids and plasma free fatty acids. Baseline lipolysis in the isolated fat pads from the PGE_1 infused rats did not differ significantly from controls but epinephrine stimulated lipolysis in fat pads from infused rats was double that of controls. (Authors modified) 0562

0237
LANDESMAN, R. and K. WILSON
   Smooth muscle active substances in the amniotic fluid.
   Clinical Obstetrics and Gynecology. 9: 554-564. 1966.

   It is briefly mentioned in this review that lipid smooth muscle stimulants found in human amniotic fluid resemble prostaglandin. (JRH) 0580

0238

LEE, J.B., J.Z. GOUGOUTAS, B.H. TAKMAN, E.G. DANIELS, M.F. GROSTIC, J.E. PIKE, J.W. HINMAN and E.E. MUIRHEAD
  Vasodepressor and antihypertensive prostaglandins of PGE type with emphasis on the identification of medullin as PGE_2-217.
  Clinical Research. 14: 231-232. 1966.

  Title only. 0577

0239

LEE, J.B., J.Z. GOUGOUTAS, B.H. TAKMAN, E.G. DANIELS, M.F. GROSTIC, J.E. PIKE, J.W. HINMAN and E.E. MUIRHEAD
  Vasodepressor and antihypertensive prostaglandins of PGE type with emphasis on the identification of medullin as PGE_2-217.
  Journal of Clinical Investigation. 45: 1036. 1966.

  Abstract only. The authors report that medullin isolated from rabbit kidney medullas has biological and chemical properties similar to PGA_2. This is taken as proof that PGA_2 and medullin are identical. (JRH) 0528

0240

LE GUEDES, G. and J. RAULIN
  Un nouveau centre d'interet en nutrition: Les prostaglandines. Derives des acides gras essentiels. [A new center of interest in nutrition: Prostaglandins. Derivatives of essential fatty acids.]
  Annales de la Nutrition et de l'Alimentation. 20: 1-12. 1966.

  The biochemical aspects of prostaglandin research are presented along with sections on their struc-ture, extraction, biosynthesis, distribution in the body and physiological actions. Particular mention is made of their effects on the reproductive system, smooth muscles, central nervous system, pulmo-cardiac system, as well as their role in the mobilization of lipids. Their metabolic processes are also discussed. 28 references are cited. (NES) 0585

0241

MAIN, I.H.M.
  Actions of prostaglandins on the oviduct of the ewe in vivo.
  In: Pickles, V.R. and R.J. Fitzpatrick, eds., "Endogenous Substances Affecting the Myometrium. Proceedings of a Symposium held at the Department of Pharmacy," (Memoirs of the Society for Endocrinology No. 14) University of Bristol, 19-20 July, 1965. p. 105-106. New York, Cambridge University Press, 1966.

  The author briefly reviews a previously reported paper in which intravenous or intra-aortic seminal fluid PG's increased intraluminal oviduct pressure in 3 ewes and lowered it in 2 others. Since it required about 4 times the amount of PG normally found in a single ram ejaculate to cause a response, it was considered unlikely that vaginally absorbed PG's act as hormones in the sheep oviduct smooth muscle. (JRH) 0542

60

0242
McCURDY, R. and J. NAKANO
  Cardiovascular effects of prostaglandin E_1.
  Clinical Research. 14: 428. 1966.

  PGE_1 (0.25-4.0 µg/kg i v) decreased systemic arterial and left and right arterial pressures, while heart rate, pulmonary arterial pressure, cardiac output and myocardial contractile force increased in a dose related manner in anesthetized dogs. PGE_1 (0.1 µg/kg i a) increased blood flow in the coronary, brachial, femoral carotid and renal arteries. No tachyphylaxis could be shown. Propranolol, sufficient to block the effects of norepinephrine, had no effect on the inotropic or vasodilator effects of PGE_1 (4.0 µg/kg) in vagotomized dogs. (JRH) 0578

0243
MERCURI, O., R.O. PELUFFO and R.R. BRENNER
  Depression of microsomal desaturation of linoleic to γ-linolenic acid in the alloxan-diabetic rat.
  Biochimica et Biophysica Acta. 116: 409-411. 1966.

  It is briefly mentioned in the discussion of this paper that a decrease in linoleic acid desaturation seen in diabetes could lead to depressed prostaglandin synthesis. (JRH) 0526

0244
MIYAZAKI, E., T. MICHIBAYASHI, S. SUNANO and Y. MIYAZAKI
  [Pharmacological action of PGE_1 on the isolated ureter of guinea-pig.]
  Japanese Journal of Smooth Muscle Research. 4: 212-219. 1966.

  Article in Japanese. Abstract not available at present. 0515

0245
NUGTEREN, D.H., D.A. van DORP, S. BERGSTROM, M. HAMBERG and B. SAMUELSSON
  Absolute configuration of the prostaglandins.
  Nature. 212: 38-39. 1966.

  The authors briefly review previous work on the structure of prostaglandins. They then present conclusive evidence for the absolute configuration of PG's and suggest a nomenclature system which retains the α and β notation. Structural diagrams are given for PGE_1. (JRH) 0520

0246
NUGTEREN, D.H., R.K. BEERTHUIS and D.A. van DORP
  The enzymic conversion of all-cis 8,11,14-eicosatrienoic acid into prostaglandin E_1.
  Recueil des Travaux Chimiques des Pays-Bas et Belgique. 85: 405-419. 1966.

Several sheep tissue homogenates were examined for their ability to synthesize PGE_1 in vitro. Listed in descending order of activity, they are seminal vesicle, intestine, lung, uterus, thymus, liver, heart, kidney and pancreas. Human myometrium and decidua were also found to contain the enzyme system. The mechanism of PG synthesis was studied in homogenates of sheep vesicular glands. Reduced glutathione caused a highly specific enhancement of the synthesis which other SH-compounds could not duplicate. Low levels of antioxidant (propylgallate or hydroquinone) also enhanced PG production. Omission of either or both of these types of agents reduced the yield of

PGE$_1$ and increased the yield of several byproducts. Cu^{++}, Zn^{++} or Cd^{++} (10^{-5}M) caused a pronounced inhibition of synthesis. With the used ^{16}O$_2$-^{18}O$_2$ mixtures it was shown that both of the oxygens on the cyclopentane ring are derived from one O$_2$-molecule. The oxygen at C-15 is also derived from molecular oxygen. A scheme for the mechanism of the biosynthesis of PGE$_1$ as well as the observed byproducts is described. It involves peroxy-radicals and a cyclic peroxide. (JRH) 0587

0247
PABON, H.J.J., L. van der WOLF and D.A. van DORP
Preparation of primary alcohols related to prostaglandins.
Recueil des Travaux Chimiques des Pays-Bas et Belgique. 85: 1251-1253. 1966.

Primary alcohols related to prostaglandins were prepared by reduction of the methyl ester of PGE$_1$ with lithium tetrahydridoaluminate. The structures of the reaction products were established by mass spectrometry. (Authors modified) 0601

0248
PAOLETTI, R.
Lipid pharmacology.
Journal of the American Oil Chemists' Society. 43: 112A. 1966.

Abstract only. The ability of prostaglandins to block free fatty acid release in low concentrations is mentioned briefly. (JRH) 0595

0249
PICKLES, V.R.
The menstrual stimulant in puberty.
Journal of Physiology. 183: 69P-70P. 1966.

The menstrual stimulant was extracted from the menstrual fluid of a girl during 19 of her first 23 menstrual periods (14-16 years of age). When compared to PGF$_{2\alpha}$ on rabbit jejunum and PGE$_1$ on diestrous guinea pig uterus, the mean total activities per period were 1.5 μg PGF$_{2\alpha}$ and 0.2 μg PGE$_1$. When the remaining acetone soluble lipid was subjected to PG extraction, the same PG's as found in adult menstrual fluid were found. Two other peaks of unidentified compounds were also discovered during chromatography. During anovulator cycles (based on sublingual temperature records) the amount of PGF-like activity was lower. (JRH) 0546

0250
PICKLES, V.R., W.J. HALL, P.C. CLEGG and T.J. SULLIVAN
Some experiments on the mechanism of action of prostaglandins on the guinea-pig and rat myometrium.
In: Pickles, V.R. and R.J. Fitzpatrick, eds., "Endogenous Substances Affecting the Myometrium. Proceedings of a Symposium held at the Department of Pharmacy," (Memoirs of the Society for Endocrinology No. 14) University of Bristol, 19-20 July, 1965. p. 89-103. New York, Cambridge University Press, 1966.

In this paper presented at a symposium, the authors review some of their experiments on the mechanism of action of prostaglandins on the guinea pig and rat myometrium. The prostaglandins produced 5 apparently different effects. The mechanisms of these effects could be: facilitation of the normal changes in cell membrane excitation; facilitation of excitation-contraction coupling; and stabilization of the cell membrane. The movement of Ca^{++} ions may be involved in the first 2 mechanisms. (JRH) 0541

0251
PICKLES, V.R.
  Structure-activity relationships in six prostaglandins.
  Journal of Physiology. 186: 107P. 1966

  Title only. 0519

0252
PIKE, J.E.
  Prostaglandins and essential fatty acids.
  Journal of the American Oil Chemists' Society. 43: 135A. 1966.

  Abstract only. Prostaglandins comprise a hormone-like group of oxygenated lipid acids with activities in a physiological range and which affect smooth muscle, cardiovascular phenomena and lipid mobilization. These humoral agents which occur widely distributed in animal species including man are biosynthesized efficiently from essential fatty acids. Recent findings relating to the biological activities of the prostanoic acids will be discussed with special emphasis on their possible role in fatty acid deficiency. (Author) 0596

0253
RAMWELL, P.W.
  Evoked release of prostaglandin and acetylcholine from the spinal cord.
  Federation Proceedings. 25: 627. 1966.

  PGE_1 and $PGF_{1\alpha}$ were detected in perfusates of the frog spinal cord. Bilateral stimulation of the hind limbs elicited increased release (P < 0.01) of both $PGF_{1\alpha}$ (10 ng/min) and acetylcholine (0.05-0.1 ng/min). This evoked release of prostaglandin was also elicited by addition of DOPA, serotonin or tranylcypromine (5 μg/ml) to the perfusing medium. Chronic treatment with tranylcypromine (200 mg/Kg, 18 hr previously) abolished both the spontaneous and evoked release of prostaglandin. Other substances detected in the perfusates included serotonin, Substance P (fraction Fa), and a series of amino acids. (Author modified) 0566

0254
RAMWELL, P.W., J.E. SHAW, W.W. DOUGLAS and A.M. POISNER
  Efflux of prostaglandin from adrenal glands stimulated with acetylcholine.
  Nature. 210: 273-274. 1966.

  Cat adrenal glands, acutely denervated were perfused in situ with Locke's solution. Addition of acetylcholine to the perfusion fluid resulted in an increase in PG (18.8-250 ng/ml) assayed as $PGF_{1\alpha}$ and catecholamines in the venous effluent from the gland. However, addition of epinephrine or norepinephrine to the perfusion fluid did not cause PG release. When adrenal glands were assayed for PG's, they contained a mean concentration of 15.5 ng/g tissue with the bulk of the PG being in the cortex. (JRH) 0521

0255
RAMWELL, P.W. and J.E. SHAW
  Spontaneous and evoked release of prostaglandins from cerebral cortex of anesthetized cats.
  American Journal of Physiology. 211: 125-134. 1966.

  The surface of the cerebral hemispheres of pentobarbital, chloralose, halothane anesthetized cats was perfused with saline and the perfusate bioassayed for PG content. Direct electrical or radial nerve stimulation increased the release of PG's. Sectioning the corpus callosum abolished the evoked release. Analeptics (i v) increased the release of PG; picrotoxin applied topically potentiated the evoked release. There seems to be some relationship between PG's and nervous activity in the cerebral cortex. (JRH) 0538

0256
RAMWELL, P.W., J.E. SHAW and R. JESSUP
  Spontaneous and evoked release of prostaglandins from frog spinal cord.
  American Journal of Physiology. 211: 998-1004. 1966.

  Perfusates of frog spinal cord contain substances with chemical, chromatographic, and pharmacological properties consistent with their identification as a mixture of PGE_1 and $PGF_{1\alpha}$; increased release of PG's occurred following stimulation of the hindlimbs, or addition of serotonin, dopa, or tranylcypromine (5 μg/ml) to the perfusing fluid. At least four other smooth-muscle-stimulating substances were also detected in perfusates including acetylcholine, serotonin, a polypeptide, and an unknown material. An evoked release of acetylcholine as well as prostaglandin occurred on electrical stimulation. It was concluded that prostaglandin release may be associated with prior release of biogenic amines. (Authors) 0537

0257
SAMUELSSON, B.
  The biosynthesis of prostaglandin-E_1 studied with specifically ^{3}H-labeled precursors.
  Advances in Tracer Methodology. 3: 241-242. 1966.

  In this symposium paper, the author briefly reports on the results of some experiments on the mechanism of PGE_1 biosynthesis studied with tritium labelled precursors (arachidonic acid.). With these techniques, it was shown that the hydrogens at C-8 and C-12 remain in their original position and that the hydrogen at C-11 is retained during the addition of the hydroxyl group. (JRH) 0557

0258
SCHNEIDER, W.P., J.E. PIKE and F.P. KUPIECKI
  Determination of the origin of 9-keto-15-hydroxy-10,13-prostadienoic acid by a double-labeling technique.
  Biochimica et Biophysica Acta. 125: 611-613. 1966.

  By means of a double labeling technique, the authors demonstrate that virtually all of the PGA_1 recovered after incubation of homo-γ-linolenic acid with sheep seminal vesicle homogenates arose by dehydration of PGE_1 during the isolation process. (JRH) 0507

0259
SEN, S., R.R. SMEBY and F.M. BUMPUS
  Isolation of a phospholipid renin inhibitor from dog kidney.
  Abstracts, 152nd Meeting of the American Chemical Society, New York, 11-16 September, 1966.
    p. C 273.

  Abstract only. Lipid extracts of dog kidneys inhibited the angiotensin production of purified dog renin. PGE_1 and PGE_2 had no inhibitory effect on angiotensin production in this test system. (JRH) 0586

0260
SHAW, J.E.
Prostaglandin release from adipose tissue in vitro evoked by nerve stimulation or catecholamines.
Federation Proceedings. 25: 770. 1966.

Abstract only. Rat epididymal fat pad contained 30-100 ng PG/g assayed as PGE_1. When incubated in vitro the fat pads released 2-15 μg/g/30 min of PG. Nerve stimulation or addition of catecholamines (5 μg/ml) caused a 5-20 fold increase in PG secretion. Omission of glucose or addition of 5% albumin had no effect on catecholamine stimulation of PG release, but omission of glucose completely blocked the effects of nerve stimulation. Fasting for 48 hr eliminated spontaneous PG release. Preliminary identification revealed the presence of PGE and PGF compounds (JRH) 0568

0261
STEINBERG, D.
Catecholamine stimulation of fat mobilization and its metabolic consequences.
Pharmacological Reviews. 18: 217-235. 1966.

Prostaglandins are briefly mentioned as being potent inhibitors of lipolysis induced by a variety of agents. (JRH) 0536

0262
STEINBERG, D. and R. PITTMAN
Depression of plasma FFA levels in unanesthetized dogs by single intravenous doses of prostaglandin E_1.
Proceedings of the Society for Experimental Biology and Medicine. 123: 192-196. 1966.

Rapid injection of PGE_1 (30 μg/kg) intravenously into conscious dogs caused a large drop in plasma free fatty acid (FFA) levels, which persisted for up to 40 min, and a transient drop in blood pressure. Similar doses of PGA_1 or injection of nitroglycerine produced a drop in blood pressure similar to that caused by PGE_1 but caused a slight increase in FFA levels. PGE_1 (10 μg/kg) completely blocked the normal lipolytic response to epinephrine and even lowered FFA levels. However, PGE_1 did not alter the increase on glucose levels produced by epinephrine. Thus PGA_1 seems to lack the direct anti-lipolytic properties of PGE_1 but, by stimulating a compensatory sympathetic discharge due to its hemodynamic effect, causes a slight increase in FFA levels. (JRH) 0534

0263
STOCK, K. and E. WESTERMANN
Competitive and non-competitive inhibition of lipolysis by α- and β-adrenergic blocking agents, methoxamine derivatives, and prostaglandin E_1.
Life Sciences. 5: 1667-1678. 1966.

A series of experiments were performed to elucidate the site of action of several lipolytic and antilipolytic drugs on rat epididymal adipose tissue in vitro. PGE_1 was found to be a consistent inhibitor of the lipolytic effects of norepinephrine. (JRH) 0539

0264
STOCK, K. and E. WESTERMANN
Hemmung der lipolyse durch prostaglandin E_1. [The inhibition of lipolysis by prostaglandin E_1.]
Naunyn-Schmiedebergs Archiv fur Pharmakologie und Experimentelle Pathologie. 253: 86-87. 1966.

Abstract only. The effect of PGE_1 on glycerine liberation, a reliable indicator of lipolysis, in epididymal fatty tissue from fasting, alloxan-diabetic, and normally fed (control) rats was studied. Added to the culture media in a concentration of 2.5 g/ml, PGE_1 was found to inhibit the liberation of glycerine in tissues from normal and alloxan-diabetic animals. No effect was noted on the fatty tissues of fasting animals. (MEMH) 0576

0265
STOCK, K. and E. WESTERMANN
Hemmung der lipolyse durch α- und β-sympathicolytica, nicotinsaure und prostaglandin E_1. [Inhibition of lipolysis through α- and β-sympathetolytics, nicotinic acid and prostaglandin E_1.]
Naunyn-Schmiedebergs Archiv fur Pharmakologie und Experimentelle Pathologie. 254: 334-354. 1966.

PGE_1 (2.5 μg/ml) inhibited basal as well as norepinephrine, ACTH, or alloxan diabetes stimulated lipolysis in rat epididymal fat pads in vitro. However, PGE_1 had no effect on the lipolysis stimulated by fasting. (JRH) 0575

0266
STRONG, C.G. and D.F. BOHR
Prostaglandins and vascular smooth muscle.
Physiologist. 9: 298. 1966.

Abstract only. Smooth muscle from the rabbit aorta was contracted in a dose related manner by PGE_1 (10^{-9}-10^{-7} g/ml and higher) in vitro. However, muscle from dog mesenteric arteries (500 μg o.d.) partially contracted with epinephrine, was relaxed by PGE_1 and PGA_1 (10^{-9}-10^{-6} g/ml) and contracted by higher concentrations. This system could serve as an in vitro model for the vasodepressor action of PGE. (JRH) 0532

0267
STRONG, C.G., R. BOUCHER, W. NOWACZYNSKI and J. GENEST
Renal vasodepressor lipid.
Mayo Clinic Proceedings. 41: 433-452. 1966.

A renal vasodepressor lipid (VDL) was extracted from rabbit, pig, human and bovine renal medullas and purified by column and thin layer chromatography. This VDL was inseparable from PGE_1 by chromatography in 5 solvent systems or starch-gel electrophoresis. They both have potent vasodepressor and non-vascular smooth muscle stimulating properties. When acidified to pH 1 or 2 both VDL and PGE_1 were inseparable by chromatographic means from PGA_1, and they were very similar in biological activity to medullin. Neither VDL or PGE_1 had any effect on the production or destruction of angiotensin in vitro. (JRH) 0547

0268

STRUIJK, C.B., R.K. BEERTHUIS, H.J.J. PABON and D.A. van DORP

Specificity in the enzymic conversion of polyunsaturated fatty acids into prostaglandins.
Receuil des Travaux Chimiques des Pays-Bas et Belgique. 85: 1233-1250. 1966.

A series of higher and lower homologues of dihomo-γ-linolenic acid and arachidonic acid (C_{18}-C_{22}) were synthesized. The homologues when allowed to react with a particulate enzyme fraction from sheep vesicular glands were converted to prostaglandin homologues. When compared to the biological activity of PGE_1 and PGE_2, the activity of the homologues was negligible. Since methyl esters of PG precursors were not converted to PG's, it was concluded that a free carboxyl group is essential for the enzymic reaction. (JRH) 0600

0269

SULLIVAN, T.J.

Response of the mammalian uterus to prostaglandins under differing hormonal conditions.
British Journal of Pharmacology and Chemotherapy. 26: 678-685. 1966.

The relative activity of PGE_1 and $PGF_{2\alpha}$ on uterine tissue from rats and guinea pigs under different hormonal conditions was studied in vitro. PGE_1 was more effective on the guinea pig uterus while $PGF_{2\alpha}$ was more effective on the rat uterus. Estrogen had little effect on uterine response to either PG, but progesterone caused a general inhibition of responsiveness to both PG's and oxytocin. PGE_1 caused a persistent enhancement of responsiveness to electrical stimulation in the guinea pig uterus even after it had been washed out. This effect did not occur with $PGF_{2\alpha}$ or in the rat uterus. (JRH) 0522

0270

SWIDERSKA-KULIKOWA, B.

[Studies on hypotensive function of the kidney.]
Polskie Archiwum Medycyny Wewnetrznej. 37: 555-562. 1966.

Article in Polish. Abstract not available at present. 0506

0271

VAUGHAN, M.

Introductory remarks.
Pharmacological Reviews. 18: 215-216. 1966.

It is briefly mentioned that PGE_1 can completely block the effect of epinephrine on lipase, but has very little effect on the stimulation of phosphorylase by epinephrine. (JRH) 0553

0272

VOGT, W., T. SUZUKI and S. BABILLI

Prostaglandins in SRS-C and in a darmstoff preparation from frog intestinal dialysates.
In: Pickles, V.R. and R.J. Fitzpatrick, eds., "Endogenous Substances Affecting the Myometrium. Proceedings of a Symposium held at the Department of Pharmacy," (Memoirs of the Society for Endocrinology No. 14) University of Bristol, 19-20 July, 1965. p. 138-142. New York, Cambridge University Press, 1966.

Guinea pig lungs perfused with bee venom released SRS-C which contained 2 active components causing guinea pig ileum to contract. The first component was composed of oxidized fatty acids and the second was similar to PG on the basis of thin layer chromatography. The PG's in SRS-C do not arise from the cleavage of PG-phospholipid complexes by the phospholipase A in the bee venom, but are apparently synthesized by the cells in response to injury by the bee venom. Frog darmstoff was also found to contain PG's which upon chromatographic analysis were found to be PGE_1, PGE_2 and a PGF, probably PGF_1. The active principles in horse darmstoff were acidic phosphatides and not PG's. The difference in horse and frog darmstoff was most likely due to the differences in extraction procedures and not to any species differences. (JRH) 0543

0273
VONKEMAN, H.
  De biosynthese van prostaglandines. [The biosynthesis of prostaglandins.]
  Chemisch Weekblad. 62: 361-366. 1966.

  The author reviews the literature on prostaglandins. Special consideration is given to the biosynthesis of PG's and its relationship to essential fatty acids. There are 21 references cited. (JRH) 0599

0274
WAITZMAN, M.B. and C.G. KIRBY
  Effects of chromatographically pure compounds extracted from iris on intraocular pressure and pupil size.
  Investigative Ophthalmology. 5: 329. 1966.

  Abstract only. Lipids extracted from rabbit irides were separated by thin layer chromatography. One of the zones of migration was chromatographically and biologically similar to PGE_1, PGE_2 and $PGF_{1\alpha}$. (JRH) 0581

0275
WAITZMAN, M.B. and C.D. KING
  Effects of prostaglandins (PG's) on intraocular pressure and pupil size.
  Investigative Ophthalmology. 5: 329. 1966.

  Abstract only. When 0.1 μg of PGE_1 or PGE_2 was injected into the anterior chamber of the rabbit eye, there was a large elevation of IOP and a miosis developed (threshold 0.01 μg). Intravenous injection of PGE_1 or PGE_2 also caused an increase in IOP in spite of a decrease in blood pressure (i v threshold 0.25 μg). In the cat, injection of 2.5 μg of PGE_1 or PGE_2 did not raise IOP, but did cause a long-lasting miosis even in the presence of atropine. $PGF_{1\alpha}$ (5 μg) did not alter IOP or cause miosis, but 125 μg caused miosis and severe iris inflammation with no change in IOP. (JRH) 0582

0276
ABRAHAMS, O.L. and D.F. HAWKINS
  Lipid-soluble uterine muscle stimulants in human amniotic fluid.
  Journal of Obstetrics and Gynaecology of the British Commonwealth. 74: 235-246. 1967.

  By chromatographic comparison it was demonstrated that the lipid-soluble uterine muscle stimulants isolated from amniotic fluid were not prostaglandins (JRH) 0763

0277
ADAMSON, U., R. ELIASSON, and B. WIKLUND
  Tachyphylaxis in rat uterus to some prostaglandins.
  Acta Physiologica Scandinavica. 70: 451-452. 1967.

  Repeated doses of PGE_1, PGE_2 or $PGF_{1\alpha}$ were administered to rat uterine horns in vitro. PGE_1 produced a tachyphylaxis in 11 of 11 experiments. Tachyphylaxis was observed in 2 of 6 experiments with PGE_2 and 6 of 8 with $PGF_{1\alpha}$. However, no cross-desensitization was observed between PGE_1 and $PGF_{1\alpha}$. After a preparation had become desensitized to a given PG, sensitivity could often be restored by exposing it to another smooth muscle stimulant such as oxytocin or ADH. It would seem that the uterus has different receptors for $PGF_{1\alpha}$ and PGE_1, and perhaps PGE_2. (JRH) 0746

0278
ANGGARD, E. and B. SAMUELSSON
  The metabolism of prostaglandins in lung tissue.
  In: Bergstrom, S. and B. Samuelsson, eds., "Prostaglandins. Proceedings of the Second Nobel Symposium," Stockholm, 1966. p. 97-105. New York, Interscience Publishers, 1967.

  Several experiments are reviewed on the metabolism of prostaglandins by guinea pig and swine lung. Sections on the metabolism of PGE_1, PGE_2, PGE_3 and $PGF_{2\alpha}$; biological properties of PG metabolites; and purification and properties of 15-hydroxyprostaglandin dehydrogenase are presented. 12 references are cited. (JRH) 0724

0279
ANONYMOUS
  Congress Report.
  German Medical Monthly. 12: 552. 1967.

  The experiments of Bohle et al. on the effects of PGE_1 on fat and carbohydrate metabolism are reviewed in a section of this symposium report. (JRH) 0891

0280
ANONYMOUS
  Prostaglandins.
  New England Journal of Medicine. 277. 1091-1092. 1967.

  In this editorial comment, it is suggested that while prostaglandins show great promise as being the antihypertensive substance of the kidney, there may still be other explanations. "The final chapter on the antihypertensive function of the kidney is yet to be written." (JRH) 0756

0281
ANONYMOUS
[Seventy-third meeting of the German society for internal medicine]
Deutsche Medizinische Wochenschrift. 92: 1200-1206. 1967.

Among the reports of research in progress heard by the membership was that of E. Bohle and associates, who are engaged in animal experiments on the effects of PGE_1 on fat and carbohydrate metabolism. In isolated rat epididymal fatty tissue the hormone raises the production of CO_2 from glucose, glucose uptake and the incorporation of glucose into triglycerides and fatty acids. Lipolysis is inhibited. Intraperitoneal injection stimulates lipid and fatty acid synthesis and the incorporation of glucose into the glycogen of the rat diaphragm, as well as the release of fatty acids into the musculature. The effect is comparable to that of insulin. (MEMH) 0874

0282
ARIENTI, S., F. PICCININI and P. POMARELLI
Prostaglandina E_1 (PGE_1) e scambi del calcio a livello della cellula miocardica. [Prostaglandin E_1 (PGE_1) and calcium exchange at the level of the myocardial cell.]
Bollettino della Societa Italiana di Biologia Sperimentale. 43: 521-524. 1967.

The effect of PGE_1 on calcium metabolism in frog myocardium was studied. Frog hearts were perfused with Ringer solution containing 0.2 μC/ml of Ca^{45}. After one hour, normal Ringer solution was substituted and the release of the marked calcium into the normal solution was determined at 2,4,6,10,20,30, and 60 minutes. In some cases, 0.4 μg/ml PGE_1 was added to the normal Ringer solution. About twice as much Ca^{45} had been released into the PGE_1-enriched Ringer solution in the first two minutes as into the control (normal) solution, indicating a mobilizing effect of PGE_1 on myocardium calcium. The release of calcium tapers off after the first few minutes, suggesting that PGE_1 affects primarily free or weakly bound calcium. (MEMH) 0866

0283
AVANZINO, G.L., P.B. BRADLEY and J.H. WOLSTENCROFT
Actions of prostaglandins on brain stem neurons.
In: Bergstrom S., and B. Samuelsson, eds., "Prostaglandins. Proceedings of the Second Nobel Symposium," Stockholm, 1966. p. 261-264. New York, Interscience, 1967.

Previously reported work on the effects of PGE_1, PGE_2 and $PGF_{2\alpha}$ applied iontophoretically to single neurons of the decerebrate cat is summarized and discussed. The responses of the neurons were variable with some neurons being stimulated, others being inhibited and some being unresponsive. Tachyphylaxis could be demonstrated in responsive neurons, but a neuron made insensitive to one PG still responded to a second PG. A comparison of PGE_1 and $PGF_{2\alpha}$ with acetylcholine did not reveal any definite correlation of responses between the two types of compounds. Because of differences in duration of response, and differences in type of response between PG's and norepinephrine, it seems unlikely that PG's produce their effect by releasing norepinephrine. The exact function of PG's in the central nervous system is unknown. (JRH) 0709

0284
AVANZINO, G.L., P.B. BRADLEY and J.H. WOLSTENCROFT
Actions of prostaglandins on brain stem neurones.
In: Kraitchevsky, P., R. Paoletti and D. Steinberg, eds., "Progress in Biochemical Pharmacology," p. 136-138. New York, Karger, 1967.

The effect of iontophoretically applied PGE$_1$ on neurons in the brain stem of unanesthetized, decerebrate cats was investigated. PGE$_1$ excited 20% and inhibited 2% of the neurons while PGF$_{2\alpha}$ excited 26% and inhibited 10%. The onset of PGF$_{2\alpha}$ effects were often slower than those of PGE$_1$. A tachyphylaxis was often observed with both PG's. Preliminary experiments with PGE$_2$ indicate that it has only excitatory actions. (JRH) 0774

0285

BAGLI, J.F. and T. BOGRI

Prostaglandins II-an improved synthesis and structural proof of ($\pm$)-11-deoxyprostaglandin F$_{1\beta}$.
Tetrahedron Letters. 5-10. 1967.

The authors report an improvement in an earlier reported total synthesis of d1-11-deoxy PGF$_{1\beta}$ which also gives further proof that the assigned structure was correct. A step by step synthesis with structures of important intermediates is given. (JRH) 0896

0286

BARTELS, J., W. VOGT and G. WILLE

Prostaglandin release from and formation in perfused frog intestine.
British Journal of Pharmacology and Chemotherapy. 31: 207. 1967.

Title only. 0757

0287

BEAL, P.F. III, J.C. BABCOCK and F.H. LINCOLN

Synthetic approaches in the prostanoic acid series.
In: Bergstrom, S. and B. Samuelsson, eds., "Prostaglandins, Proceedings of the Second Nobel Symposium," Stockholm, 1966. p. 219-230. New York, Interscience Publishers, 1967.

Several approaches to the snythesis of prostaglandins are discussed. One of these procedures led to a product identical to dihydro PGE$_1$ ethyl ester and thus represents a total synthesis of a natural prostaglandin. Step by step structural formulas are given of all intermediates. (JRH) 0704

0288

BENNETT, A., C.A. FRIEDMANN and J.R. VANE

Release of prostaglandin E$_1$ from the rat stomach.
Nature. 216: 873-876. 1967.

Whole and strips of isolated rat stomachs were studied for release of prostaglandins due to vagal and transmural electrical stimulation in vitro. PGE$_1$ was released from the mucosal surface of the stomach in response to stimulation of nerves in the stomach wall. PGE$_1$ was also released in response to stretching of stomach strips. Addition of hexamethonium to the bath fluid almost abolished muscular activity in response to electrical stimulation and reduced the release of PG by 39%. Tetrodoxin produced similar results. Antagonism of muscle responses therefore preceded the reduction in the release of prostaglandin. Prostaglandins may have a physiological role in controlling gastric muscle motility. (JRH) 0809

0289
BERGSTROM, S.
   Isolation, structure and action of the prostaglandins.
   In: Bergstrom, S. and B. Samuelsson, eds., "Prostaglandins. Proceedings of the Second Nobel Sympo-
      sium," Stockholm, 1966. p. 21-30. New York, Interscience Publishers, 1967.

   The literature dealing with the isolation and determination of chemical structure of prostaglandins is
   reviewed. 62 references are cited. (JRH) 0735

0290
BERGSTROM, S.
   The prostaglandins—a new group of hormones.
   In: Leonardi, A., and J. Walsh, eds., "International Symposium on Drugs of Animal Origin," p. 9-28.
      Milan, Ferro Edizione, 1967.

   The general area of prostaglandin research is reviewed. Included are sections on history, structure and
   nomenclature; tissues containing PG's; biosynthesis; mechanism of biosynthesis; physiological action
   on circulatory organs; action on female reproductive organs and metabolic actions. There are 74
   references. (RAP) 0867

0291
BERGSTROM, S.
   Prostaglandins: members of a new hormonal system.
   Science. 157: 382-391. 1967.

   This general review of prostaglandin literature contains sections on: isolation and structure; occurrence and metabolism; biosynthesis; mechanism of biosynthesis; chemical synthesis; physiologic
   action on the circulatory system; action on female reproductive organs; and metabolic actions. There
   are 76 references cited. (JRH) 0860

0292
BERGSTROM, S., L.A. CARLSON and L. ORO
   A note on the cardiovascular and metabolic effects of prostaglandin A_1.
   Life Sciences. 6: 449-455. 1967.

   In dogs anesthetized with pentobarbital, 0.056 or 0.2 μg/kg/min PGE_1 or PGA_1 lowered blood
   pressure and increased heart rate, with PGA_1 usually being the most potent. PGA_1 did not produce
   any significant changes in plasma free fatty acid, glycerol or blood glucose levels during the 20 min
   infusion. In vitro experiments with rat adipose tissue showed PGA_1 to have only 10% of the antilipolytic power of PGE_1. (JRH) 0784

0293
BERTI, F., R. LENTATI and D. GRAFNETTER
   Effetti della somministrazione di prostaglandina E_1 sulla lipasi lipoproteica cardiaca. [The effects of
      prostaglandin E_1 on cardiac lipoprotein lipase.]
   Bollettino della Societa Italiana di Biologia Sperimentale. 43: 515-158. 1967

   Anesthetized male rats and male guinea pigs were perfused with PGE_1 (5.6 g/kg body weight in 10 ml
   physiological solution) for 25 min. (Control animals received only physiological solution.) The animals were then exsanguinated, and the blood was analyzed for glucose and for non-esterified fatty

acids (NEFA). Myocardial tissue was homogenized for determination of cardiac lipolytic activity. Lipoprotein lipase was shown to be significantly activated, in contrast to the block of the hormone-sensitive lipase of fatty tissue induced in the same animals and demonstrated by the drop in plasma NEFA values. The relationship of these phenomena has not yet been elucidated. (MEMH) 0819

0294
BERTI, F., R. LENTATI, M.M. USARDI and R. PAOLETTI
The effect of prostaglandin E_1 on free fatty acid mobilization and transport.
Protoplasma. 63: 143-146. 1967.

PGE_1 (5.6 µg/kg/min iv) reduced free fatty acid (FFA) levels in the blood of rats in which lipolysis had been stimulated by ACTH, norepinephrine or exposure to cold. It also lowered FFA levels in untreated control rats. However, it did not lower FFA levels in rats in which FFA levels were elevated due to fasting. This would indicate that PGE_1 has a specific effect on hormone-activated lipase. (JRH) 0831

0295
BERTI, F., M.K. NAIMZADA, R. LENTATI, M.M. USARDI, P. MANTEGAZZA and R. PAOLETTI
Relations between some in vitro and in vivo effects of prostaglandin E_1.
In: Kraitchevsky, P., R. Paoletti and D. Steinberg, eds., "Progress in Biochemical Pharmacology," p. 110-121. New York, Karger, 1967.

The authors review their own work and the work of others on the mechanism of the pharmacologic effects of PGE_1 on several physiologic processes in vivo and in vitro. There are sections on the effect of PGE_1 on lipolysis, glucose metabolism, the vas deferens of rats and rabbits and isolated heart preparations from several species. 42 references are cited. 0767

0296
BERTI, F., M. PROTO, P. POMARELLI and G. ZAMBIANCHI
Sull'identificazione della sostanza utero-stimulante presente nel liquido amniotico umano. [On the identification of the uterine stimulating substance present in the human amniotic fluid.]
Atti della Accademia Medica Lombarda. 22: 483-484. 1967.

Human amniotic fluid was extracted at neutral pH and the ethereal phase discarded. The water phase was brought to pH 3 and extracted again with ether. It is possible to separate by thin layer chromatography a spot having an Rf similar to PGE_1. The spot has been shown to possess a marked stimulating activity on the guinea-pig ileum analogous AA. to that of PGE_1. (Authors) 0855

0297
BESCHEA, M.
Prostaglandinele. [Prostaglandins.]
Revista Medico-Chirurgicala, Iasi. 73: 545-551. 1967.

Article in Roumanian. Abstract not available at present. 0828

0298
BOGRI, T., J.F. BAGLI and R. DEHENGHI
An improved synthesis of the prostanoic acids.
In: Bergstrom, S. and B. Samuelsson, eds., "Prostaglandins. Proceedings of the Second Nobel Symposium," Stockholm, 1966. p. 231-236. New York, Interscience Publishers, 1967.

The authors report an improved chemical synthesis of racemic 11-deoxy-PGF$_{1\beta}$. Step by step structural formulas of all intermediates are given. Physiochemical evidences of the stereochemical, structural assignments are also given. (JRH) 0705

0299
BOHLE, E., H. RETTBURG, H.H. DITSCHUNEIT, R. DOBERT and H. DITSCHUNEIT
Tierexperimentelle untersuchungen uber die beeinflussing des fettund kohlenhydratstoffwechsels durch prostaglandin E$_1$. [Animal experimentation on the influence of prostaglandin E$_1$ on fat and carbohydrate metabolism.
Verhandlungen der Deutschen Gesellschaft fur Innere Medizin. 73: 797-801. 1967.

The technique of intraperitoneal assay was used on male rats to assess the incorporation of U^{14}-C-glucose into the lipids and fatty acids of epididymal fatty tissue and into the glycogen of the diaphram under the influence of PGE$_1$. It was found that PGE$_1$ stimulates not only lipid and fatty acid synthesis in testicular fatty tissue, but also the incorporation of marked glucose into the glycogen of rat diaphragm. This effect is directly related to dosage (0.01 to 1.00 μg/10g body weight) and is comparable to that of insulin. (Author modified) 0824

0300
BOHLE, E., R. DOBERT and I.M. MERKL
Uber die wirkung von prostaglandin E$_1$ und insulin auf die heparininduzierte triglyceridhydrolyse. [The working of prostaglandin E$_1$ and insulin on heparin induced triglyceride hydrolysis.]
Zeitschrift fur die gesamte experimentelle Medizin. 144: 285-299. 1967.

PGE$_1$ (3 μg/kg) injected into the aorta of starving rabbits reduced the plasma free fatty acid (FFA) levels within 10 min as much as 0.1 μ/kg insulin. However, the insulin produced a slight reduction in blood sugar and a transient reduction in glycerol, while PGE$_1$ had no effect on glycerol and caused a hyperglycemia. Both PGE$_1$ and insulin could inhibit the increase in blood FFA induced by injection of heparin, but not the heparin induced increase in glycerol. In vitro, neither PGE$_1$ nor insulin affected the activity of lipoprotein-lipase of human plasma induced by heparin. The increase in blood level of lipoprotein-lipase induced by heparin in vivo was not altered by insulin or PGE$_1$. Thus, PGE$_1$ does not influence the heparin-induced release or activation of lipoprotein-lipase, but causes an enhanced uptake and re-esterifiction of non-esterified fatty acids. (JRH) 0898

0301
BOHLE, E.
Welche nachweismoglichkeiten gibt es fur prostaglandine? [How can the presence of prostaglandins be demonstrated.]
Deutsche Medizinische Wochenschrift. 92: 409-410. 1967.

The chemical composition and structure of prostaglandins are briefly described. Biological and analytic methods of identifying prostaglandins in tissues are discussed. The latter usually include extraction with organic solvents and bioassay of the resulting extract. Biological tests (of blood pressure changes, in vitro contractility of smooth muscle tissue or uterine tissue) are of limited value, however, as the effects of various prostaglandins sometimes conflict. Special analytical techniques involving spectrophotometry and thin-layer chromatography are presently carried out in only a few specialized laboratories. (MEMH) 0859

0302

BUTCHER, R.W., J.E. PIKE, E.W. SUTHERLAND and C.E. BAIRD
The effect of prostaglandin E_1 on adenosine 3', 5'-monophosphate levels in adipose tissue.
In: Bergstrom, S. and B. Samuelsson, eds., "Prostaglandins. Proceedings of the Second Nobel Symposium," Stockholm, 1966. p. 133-138. New York, Interscience Publishers, 1967.

PGE_1 at low concentrations antagonized the action of epinephrine on cAMP accumulation by the intact rat epididymal fat pad in vitro and also acted as an agonist, i.e. cAMP levels were increased by PGE_1 in the absence of epinephrine. The antagonistic action of PGE_1 was also oberved with isolated fat cells, while the stimulatory effect was absent. $PGF_{1\alpha}$ was less active as an antagonist than PGE_1 and $PGF_{1\beta}$ was without activity on whole fat pads. These data suggest that the anti-lipolytic action of PGE_1 is mediated at least in part by lowering intracellular cAMP. (Authors modified) 0754

0303

BUTCHER, R.W., R.E. SCOTT and E.W. SUTHERLAND
The effects of prostaglandins on cyclic AMP levels in tissues.
Pharmacologist. 9: 172. 1967.

Abstract only. The author discusses the effects of PG's on cAMP. Special attention is given to the observations that PG's inhibit cAMP production in some tissues while they stimulate it in other tissues. (JRH) 0862

0304

BUTCHER, R. W. and E.W. SUTHERLAND
The effects of the catecholamines, adrenergic blocking agents, prostaglandin E_1 and insulin on cyclic AMP levels in the rat epididymal fat pad in vitro. Annals of the New York Academy of Sciences. 139: 849-859. 1967.

PGE_1 (2 μg/ml) prevented the increase in cAMP levels caused by either epinephrine or epinephrine plus caffeine, in rat adipose tissue in vitro. PGE_1 itself caused a slight increase in cAMP in adipose tissue which was attributed to its effect on some cell type other than adipocytes, since no stimulation was found with isolated adipocytes. These results suggested that PGE_1 acts as a non-specific competetive inhibitor of the adenyl cyclase system in adipocytes. (JRH) 0816

0305

BYGDEMAN, M., S. KWON and N. WIQVIST
The effect of prostaglandin E_1 on human pregnant myometrium in vivo.
In: Bergstrom, S. and B. Samuelsson, eds., "Prostaglandins. Proceedings of the Second Nobel Symposium," Stockholm, 1966. p. 93-96. New York, Interscience Publishers, 1967.

The effects of intravenous PGE_1 (0.6-8 μg/min) on uterine activity in 3 midpregnant and 4 term patients in vivo was recorded. In the midpregnant uterus PGE_1 produced mainly increased tonus but some increase in amplitude and frequency of contraction was observed. In 1 of the 3 patients, labor-like contractions were produced. In 1 term patient the increase in amplitude and frequency was most pronounced but in the others the effect on tone was dominant. In all experiments, there was a 15 min latency period and a slow return to preinfusion activity after termination of infusion. At these dose levels, the only observed side effect was a slight increase in heart rate. (JRH) 0725

0306

BYGDEMAN, M. and M. HAMBERG

The effect of eight new prostaglandins on human myometrium.

Acta Physiologica Scandinavica. 69: 320-326. 1967.

Human semen contains 13 different prostaglandins. The effect of 5 of these (PGE_1, PGE_2, PGE_3, $PGF_{1\alpha}$ and $PGF_{2\alpha}$) on the motility of human myometrium in vitro has been described earlier. The new compounds (PGA_1, PGA_2, PGB_1, PGB_2 and their 19-hydroxy derivatives) are derivatives of the PGE-compounds formed by hydroxylation at C-19 and/or dehydration involving the hydroxyl group at C-11. All the compounds inhibited the motility of the myometrium in a similar way as the PGE-compounds. The total effect of the 19-hydroxylated compounds was 4-40 per cent and of the dehydrated compounds 3-30 per cent of that of the PGE-compounds. (Author modified) 0793

0307

BYGDEMAN, M. and B. SAMUELSSON

Prostaglandins in human seminal plasma and their effects on human myometrium.

In: Westin, B. and N. Wiqvist, eds., "Fertility and Sterility, Proceedings of the Fifth World Congress," Stockholm, 16-22 June 1966. p. 681-684 International Congress Series No. 133. New York, Excerpta Medica, 1967.

The authors review the literature on the relative concentrations of different PG's in human seminal fluid. They also discuss the effects of these PG's individually and in combination on the human myometrium in vitro. The general conclusion is that the PG's are inhibitory in the ratio and concentration at which they are found in the seminal fluid. (JRH) 0835

0308

BYGDEMAN, M. and B. SAMULESSON

Prostaglandins in human seminal plasma and their effects on human myometrium.

International Journal of Fertility. 12: 17-20. 1967.

Literature on the prostaglandin content of human seminal plasma is reviewed. Tables are presented that list the relative concentration of the different PG's in semen from men with apparently normal fertility and men with low fertility. Tables are also given which show the relative effectiveness of the different PG's found in seminal plasma on the human myometrium in vitro. (JRH) 0880

0309

BYGDEMAN, M.

Studies of the effects of prostaglandins in seminal plasma on human myometrium in vitro.

In: Bergstrom, S. and B. Samuelsson, eds., "Prostaglandins. Proceedings of the Second Nobel Symposium," Stockholm, 1966. p. 71-77. New York, Interscience Publishers, 1967.

Previously reported experiments on the in vitro effects of prostaglandins found in human seminal plasma on myometrial tissue from pregnant and nonpregnant humans are reviewed. PGE, PGA and PGB compounds caused inhibition of uterine motility in most cases. PGF compounds generally stimulated uterine tone, and the frequency and amplitude of contraction. The myometrium was most sensitive to prostaglandins around the time of ovulation and the most effective PG was PGE_1. Based on relative activity and the ratio of PG's normally found in human seminal fluid, the effects of seminal plasma on the uterus are due mainly to the PGE compounds present. (JRH) 0729

0310
CARLSON, L.A.
Cardiovasular and metabolic effects of prostaglandins.
In: Kraitchevsky, P., R. Paoletti and D. Steinberg, eds., "Progress in Biochemical Pharmacology," p. 94-109. New York, Karger, 1967.

The author reviews experiments on the metabolic and cardiovascular effects of prostaglandins in dogs, rats and man. On the basis of these experiments, the author formulates a series of concepts to explain the observed phenomena. (JRH) 0768

0311
CARLSON, L.A.
Metabolic and cardiovascular effects in vivo of prostaglandins.
In: Bergstrom, S. and B. Samuelsson, eds., "Prostaglandins, Proceedings of the Second Nobel Symposium," Stockholm, 1966. p. 123-132. New York, Intersience Publishers, 1967.

Reviewed are 11 of L.A. Carlson's papers on the metabolic and cardiovascular effects of prostaglandins in vivo, especially those of PGE_1. There are 22 references cited. (JRH) 0721

0312
CARLSON, L.A., L-G. EKELUND and L. ORO
Metabolic and cardio-vascular effects of serotonin.
Life Sciences. 6: 261-271. 1967.

It is briefly mentioned that infusion of serotonin and PGE_1 cause many similar effects. Therefore, they may be mediated by the same mechanism or they may interact with each other. (JRH) 0785

0313
CARR, A.A.
The effect of a prostaglandin on renal function.
Clinical Research. 15: 354. 1967.

Abstract only. PGA_1 (0.99-1.39 µg/kg/min) was infused intravenously into anesthetized dogs. PGA_1 reduced glomerular filtration (14%), increased renal blood flow (62%), decreased solute free water absorption (130%) and at times resulted in free water clearance and reduced sodium excretion. Mean blood pressure was reduced 54%. These results suggest washout of the renal medullary area due to vasodilation. PGA_1 may be important in the renal physiology of the dog. (JRH) 0837

0314
CHANDRASEKHAR, N.
Inhibition of platelet aggregation by prostaglandins.
Blood. 30: 554. 1967.

The effects of PG's on platelet aggregation induced by ADP and collagen were investigated in vivo and in vitro with rat and human blood. All the PG's tested inhibited ADP induced platelet aggregation with PGE_1 being the most potent. PGE_1 (0.1 µg/ml) blocks ADP induced platelet aggregation and lyses already formed platelet aggregates. PGE_1 also inhibits collagen induced aggregation but does not break up already formed aggregates. In vivo PGE_1 (2 mg/kg iv) prevents platelet aggregation in the rat for 30 min. (JRH) 0800

0315
CHIESA, F., R. OBEROSLER, E. SEREN and E. MARTINI
   Differenziazione tra prostaglandine E sostanze neutre liposolubili attive sulla pressione arteriosa.
   [Differentiation between prostaglandin E and an active, neutral liposoluble substance on arterial
   pressure.]
Bollettino della Societa Italiana di Biologia Sperimentale. 43 (supp. 20): abs. 86. 1967.

   Article in Italian. Abstract not available at present. 0886

0316
CLAYTON, J.A. and C.M. SZEGO
   Depletion of rat thyroid serotonin accompanied by increased blood flow as an acute response to
   thyroid-stimulating hormone.
Endocrinology. 80: 689-698. 1967.

The serotonin content of rat thyroids was estimated in a bioassay system based on the isolated heart
of *Venus merceneria.* Several other biologically active compounds that might be found in the thyroid
were also tested in this bioassay system. PGE_1 in concentrations corresponding to approximately 70
$\mu g/g$ of thyroid tissue had no effect on the *Venus* heart. (JRH) 0752

0317
COCEANI, F., C. PACE-ASCIAK, F. VOLTA and L.S. WOLFE
   Effect of nerve stimulation on prostaglandin formation and release from the rat stomach.
American Journal of Physiology. 213: 1056-1064. 1967.

PGE and PGF were found to be present in the rat stomach and to be released spontaneously in vitro.
Cholinergic but not adrenergic nerve stimulation increased significantly the release of PG's and the
increase was dependent on the rate of stimulation. Hyoscine completely abolished the effect of
parasympathetic nerve stimulation. Denervated preparations were still able to form and release PG's.
Anoxia inhibited but ascorbic acid increased the release rate of PG's. PGE_2 and $PGF_{2\alpha}$ were the
major prostaglandins formed and released during nerve stimulation. (Author modified) 0781

0318
COCEANI, F. and L.S. WOLFE
   Pharmacological properties of prostaglandin E_1 and prostaglandins extracted from brain.
   In: Kraitchevsky, P., R. Paoletti and D. Steinberg, eds., "Progress in Biochemical Pharmacology,"
   p. 129-135. New York, Karger, 1967.

Superfused cat brains liberated PG-like activity (bioassayed on rat stomach strips as PGE_1) at a rate
of 0.4-1.6 ng/15 min/cm^2 surface. Cat and ox brain tissue was found to contain 0.01-0.03 μg PG/g
bioassayed as PGE_1. Ox white matter contained 20% of the PG activity. The mechanism of action of
PGE_1 and PG's extracted from the brain on the fat fundus strip was also investigated. The response of
the strip to PG was enhanced by hemicholinium, bretylinium, cocaine, procaine, xylocaine, morphine
and dibenamine. PG's were inhibited by epinephrine, norepinephrine, papverine, cooling to 24°C and
lack of oxygen. Atropine, hyoscine brom-LSD, hexamethonium and papverine had no effect on the
PG response. The enhancing agents could have acted by blocking sympathetic nerves since both
epinephrine and norepinephrine counteracted the PG effect. It appears that PG's act directly on
smooth muscle. (JRH) 0765

0319
COCEANI, F., C. PACE, F. VOLTA and L.S. WOLFE
  Presenza e liberazione di prostaglandine dallo stomaco di ratto. [Presence and release of prostagland-
    ins from rat stomach.]
  Bollettino della Societa Italiana di Biologia Sperimentale. 43: 76-78. 1967.

  Substances identified as prostaglandins were found in rat stomach tissue; chromatography showed
  that prostaglandins of the E type, particularly PGE_2, predominated. At-rest release of prostaglandins
  from the tissue was found to be approximately 0.3 ng/min, and this rate increased in direct proportion to frequency of stimulation. Prostaglandin release was shown to take place in muscle membrane
  which had been deprived of nervous tissue. It is proposed that prostaglandins or their metabolites are
  intimately involved in the phenomenon of gastric motility. (MEMH) 0821

0320
COLLINGWOOD, J.G.
  Oils and fats in the 1970's.
  Chemistry and Industry. 44: 1202-1211. 1967.

  In this editorial review, it is briefly mentioned that essential fatty acids are metabolized into prostaglandins which can stimulate smooth muscle, lower blood pressure and reduce platelet adhesiveness.
  (JRH) 0869

0321
DANIELS, E.G., J.W. HINMAN, B.E. LEACH and E.E. MUIRHEAD
  Identification of prostaglandin E_2 as the principal vasodepressor lipid of rabbit renal medulla.
  Nature. 215. 1298-1299. 1967.

  Acidic lipids were extracted from rabbit renal medulla tissue. They were subjected to silicic acid
  chromatography and were found to be mainly PGE compounds with some PGA. Thin layer chromatography revealed the PGA component to be PGA_2 and the PGE component to be PGE_2. Thus it
  seems that the principal vasodepressor lipid of the rabbit renal medulla is PGE_2. (JRH) 0736

0322
DAVIES, B.N. and P.G. WITHRINGTON
  The effects of prostaglandins E_1 and E_2 on the smooth muscle of the dog spleen and on its responses
    to catecholamines, angiotensin and nerve stimulation.
  British Journal of Pharmacology and Chemotherapy. 32: 136-144. 1967.

  In isolated dog spleen, perfused with blood from other dogs anesthetized with morphine, chloralose
  and urethane, both PGE_1 and PGE_2 (0.5-5.0 μg/min), perfused close intraarterially, caused an immediate fall in splanchic vascular resistance and subsequent delayed increase in spleen volume. PGE_1
  was more potent than PGE_2, but both were less potent than acetylcholine. The responses to PG's
  were not affected by phenoxybenzamine or propranolol infusions. The effects of sympathetic nerve
  stimulation (3/sec) and infusion of epinephrine (E), norepinephrine (NE) or angiotensin were not affected by simultaneous infusion of PGE_1 (2.5 μg/min). Results of 2 experiments with PGE_2 indicated
  that it also did not alter the responses of the spleen to E or NE. (JRH) 0751

0323
DAVIES, B.N., E.W. HORTON and P.G. WITHRINGTON
  The occurrence of prostaglandin E_2 in splenic venous blood of the dog following splenic nerve
    stimulation.
  Journal of Physiology. 188: 38P-39P. 1967.

Isolated dog spleens were perfused with blood from a donor dog. Splenic venous blood collected during and after electrical stimulation (600 impulses at 10/sec) of the splenic nerves, contained PGE_2 (up to 200 ng/ml) while blood prior to stimulation had less than 1 ng/ml. (JRH) 0872

0324
de la RIVA, I. and D.F. BOHR
A new vasoactive substance from dog kidney cortex.
Physiologist. 10: 154. 1967.

It is reported that the vasoactive protein free filtrate obtained from the dog kidney cortex is not a prostaglandin, since it does not contract coronary smooth muscle. (JRH) 0761

0325
DIRKS, J.H. and J.F. SEELEY
Micropuncture studies on the effect of vasodilators on proximal tubule sodium reabsorption in the dog.
Clinical Research. 15: 478. 1967.

Abstract only. The effects of intra-arterial infusion of several vasodilators including PGE_1 (6 μg/min) on proximal reabsorption of sodium was measured by the recollection micropuncture technique. All the vasodilators caused increases in sodium excretion but their effects on vasodilation in the kidney were different. The results indicate that the natriuretic action of various vasodilators may occur at different sites of the nephron or within different groups of nephrons at the same rate. (JRH) 0839

0326
van DORP, D.A.
Aspects of the biosynthesis of prostaglandins.
In: Kraitchevsky, P., R. Paoletti and D. Steinberg, eds., "Progress in Biochemical Pharmacology," p. 71-82. New York, Karger, 1967.

The author reviews several aspects of the biosynthesis of PG's investigated in his laboratory which include: distribution of the PG synthesizing enzyme system in various tissues of animal and human origin; substrate specificity; mechanistic considerations; and possible further links between prostaglandins and essential fatty acids. There are 15 references. (JRH) 0772

0327
van DORP, D.A., G.H. JOUVENAZ and C.B. STRUIJK
The biosynthesis of prostaglandin in pig eye iris.
Biochimica et Biophysica Acta. 137: 396-399. 1967.

The various parts of fresh pig eyes were separated. Lyophilized particulate enzyme fractions were then prepared from the eye parts and allowed to react with radiolabelled all-cis-8,11,14-eicosatrienoic acid. Enzymes extracted from the iris were able to convert significant amounts of the eicosatrienoic acid to porstaglandins (PGE_1 and $PGF_{1\alpha}$). The retina and lens did not produce significant amounts of PG. When the eye parts were analyzed for fatty acid content, the iris was found to contain considerable arachidonic acid, indicating that the iris could synthesize PGE_2 and $PGF_{2\alpha}$. (JRH) 0743

0328

DuCHARME, D.W. and J.R. WEEKS

Cardiovascular pharmacology of prostaglandin $F_{2\alpha}$, a unique pressor agent.

In: Bergstrom, S. and B. Samuelsson, eds., "Prostaglandins. Proceedings of the Second Nobel Symposium," Stockholm, 1966, p. 173-181. New York, Interpublishers, 1967.

The mechanism of the pressor effect of $PGF_{2\alpha}$ was studied in rats and dogs. The pressor effect was not altered by ganglionic blockade or pretreatment with reserpine indicating that the effects are not due to stimulation of the sympathetic nervous system nor to peripheral catecholamine release. $PGF_{2\alpha}$ did produce an increase in vascular resistance when given intraarterially into the dog limb, due mainly to an increase in venous pressure. $PGF_{2\alpha}$ produced an increase in systemic blood pressure and cardiac output in unanesthetized dogs but did not alter total peripheral resistance. In pentobarbital anesthetized dogs the increase in cardiac output was not due to increased myocardial contractility but to increased right arterial pressure. Thus it seems the pressor effects of $PGF_{2\alpha}$ are due to venoconstriction. (JRH) 0716

0329

DuCHARME, D.W. and J.R. WEEKS

Prostaglandin $F_{2\alpha}$, a unique pressor substance.

Federation Proceedings. 26: 681. 1967.

$PGF_{2\alpha}$, unlike most prostaglandins is pressor in rats and dogs. The pressor activity persists in rats following ganglion blockade or reserpine pretreatment. Studies in the perfused limb of the dog demonstrated that the pressor activity was, at least partially, due to increased peripheral resistance. Analysis of segmental vascular pressures revealed that the increased resistance occurred primarily in the venous segments. In unanesthetized dogs, $PGF_{2\alpha}$ increased both cardiac output and blood pressure, while TPR was virtually unchanged. In anesthetized dogs it was demonstrated that the increased cardiac output was not associated with an increased force of myocardial contraction, but rather an increased right atrial pressure. $PGF_{2\alpha}$ also caused a marked increase in pulmonary arterial pressure; however, experiments in which the right ventricle was by-passed by pumping the blood from the right atrium into the pulmonary arteries at constant flow demonstrated that the increased atrial pressure was independent of the increase in pulmonary resistance. It is concluded that the pressor activity of $PGF_{2\alpha}$ is primarily due to an increase in cardiac output resulting from a decrease in venous capacity and perhaps pulmonary capacity. (Authors modified) 0799

0330

ELIASSON, R.

The effect of prostaglandins on the human uterus in vivo.

In: Westin, B. and N. Wiqvist, eds., "Fertility and Sterility, Proceedings of the Fifth World Congress," Stockholm, 16-22 June, 1966. p. 673-674.

International Congress Series No. 133. New York, Excerpta Medica, 1967.

Experiments on the effects of prostaglandins on the human uterus in vivo are reviewed. Most of these studies involve application of extracts of human seminal plasma or seminal plasma to the vagina or uterus of normal and low fertility women and recording its effect on motility. The PG's used were only partially purified and contained a mixture of PG's in approximately the same concentration found in human seminal fluid. (JRH) 0827

0331

ELIASSON, R. and P.L. RISLEY

Potentiated response of isolated seminal vesicles to catecholamines and acetycholine in the presence of prostaglandins.

In: Bergstrom, S. and B. Samuelsson, eds., "Prostaglandins. Proceedings of the Second Nobel Symposium," Stockholm, 1966. p. 85-90. New York, Interscience Publishers, 1967.

Isolated seminal vesicles of guinea-pigs were used to examine the effects of adrenaline, noradrenaline, acetylcholine, and 5 hydroxytryptamine in the presence of PGE_1 or $PGF_{1\alpha}$. The addition of PGE_1 (12.5-2000 ng/ml bath fluid) one minute before the drug caused a marked potentiation of the response. In the presence of PGE_1, the latency period was markedly abbreviated. $PGF_{1\alpha}$ in the same dose-range did not alter the reaction. (Authors) 0727

0332

ELLIS, S.

The effects of sympathomimetic amines and adrenergic blocking agents on metabolism.

In: Root, W.S. and F.G. Hofmann, eds., "Physiological Pharmacology" Vol. 4. p. 179-241. New York, Academic Press, 1967.

Brief mention is made of prostaglandins ability to inhibit the increase in plasma free fatty acid caused by norepinephrine or epinephrine without antagonizing the hyperglycemia produced by these catecholamines. PG's also inhibit the pressor responses to catecholamines through their own depressor activity. (JRH) 0787

0333

EMMONS, P.R., J.R. HAMPTON, M.J.G. HARRISON, A.J. HONOUR and J.R.A. MITCHELL

Effect of prostaglandin E_1 on platelet behaviour in vitro and in vivo.

British Medical Journal. 2: 468-472. 1967.

PGE_1 (2.5 μg/ml) greatly reduced platelet adhesiveness in human whole blood. PGE_1 (0.001-1.0 μg/ml) also inhibited platelet aggregation stimulated by ADP, norepinephrine, serotonin, ATP, thrombin and an extract of connective tissue. PGE_1 (1.0 μg/ml) had no effect on electrophoretic mobility (post contact) but 0.001 μg/ml completely blocked the decrease in electrophoretic mobility induced by ADP. However, addition of Ca^{++} (25mM final concentration) prevented the inhibition of ADP by PGE_1. In vivo PGE_1 reduced or prevented platelet thrombi formation in injured rabbit cerebral arteries when applied topically or given as intravenous injections (5-1000 μg) or infusions (0.2-1.6 μg/kg/min) both before and after injury. In all these experiments PGE_1 was effective at much lower concentrations than other agents that cause similar effects. (JRH) 0810

0334

ERNESTI, M., M.S. RABEN and M.L. MITCHELL

Hyperglycemic properties of prostaglandin and other preparations.

Diabetes. 16: 515. 1967.

Abstract only. Intravenous injection of PGE_1 (0.2 mg) into conscious fasting rabbits caused a striking hyperglycemia, which reached a peak by 30 min then declined. The PG was much more potent as a hyperglycemic agent than growth hormone, posterior pituitary peptide II or diazoxide, and produced its effect in a much shorter period. Also there was no rise in blood free fatty acid levels with the PG as there was with the other agents. When PGE_1 and growth hormone were given together there was no additive effect, in fact there was no elevation of blood sugar levels at all. (JRH) 0813

0335

von EULER, U.S. and R. ELIASSON

Prostaglandins.

In: von Euler, U.S. and R. Eliasson, eds., "Medicinal Chemistry Monographs," Vol. 8, p. 1-147. New York, Academic Press, 1967.

This extensive review of prostaglandin literature contains chapters entitled: historical survey; chemistry; occurrence; biological effects; and relations of prostaglandins to other lipid acids active on smooth muscle. The Bibliography lists 9 reviews and 299 references. (JRH) 0747

0336

von EULER, U.S.

Welcoming address.

In: Bergstrom, S. and B. Samuelsson, eds., "Prostaglandins. Proceedings of the Second Nobel Symposium," Stockholm, 1966. p. 17-20. New York, Interscience Publishers, 1967.

In this introductory address to the Second Nobel Symposium on prostaglandins, the author briefly reviews the history of PG research. (JRH) 0714

0337

FAIN, J.N.

Adrenergic blockade of hormone-induced lipolysis in isolated fat cells.

Annals of the New York Academy of Sciences. 139: 879-890. 1967.

PGE_1 (0.1 μg/ml) inhibited basal lipolysis in rat adipose tissue incubated with glucose. It also inhibited lipolysis stimulated by growth hormone, dexamethason, ACTH, and epinephrine. On the basis of these results it was concluded that PGE_1 is a nonspecific competitive inhibitor of lipolysis and not an α- or β-adrenergic inhibitor. (JRH) 0815

0338

FARRAR, J.T. and A.M. ZFASS

Small intestinal motility.

Gastroenterology. 52: 1019-1037. 1967.

Prostaglandins are briefly mentioned in this extensive review on small intestinal motility as being stimulators of motility. The mechanism of action of PG's is unknown. (JRH) 0834

0339

FASSINA, G. and A.R. CONTESSA

Digitoxin and prostaglandin E_1 as inhibitors of catecholamine-stimulated lipolysis and their interaction with CA^{2+} in the process.

Biochemical Pharmacology. 16: 1447-1453. 1967.

Both PGE_1 (5 X $10^{-7}-10^{-5}$M) and digitoxin were mixed inhibitors of norepinephrine (NE) stimulated lipolysis in rat adipose tissue in vitro. Lowering the Ca^{++} content of the medium enhanced the non-competitive, component of the inhibition of NE. PGE_1 did not affect Na^+-, K^+-, or Mg^{++}-stimulated ATP-ase activity of human erythrocyte membrane. PGE_1 (10^{-5}M) increased the short-circuit current in frog skin when applied to the internal side of the skin. Neither PGE_1 nor digitoxin had any effect on oxidative phosphorylation in rat mitochondria. It was not proven that both PGE_1 and digitoxin operate through the same biochemical mechanism but it seems that calcium movement is involved in the action of both drugs. (JRH) 0826

0340
FASSIO, N., P. P. POMARELLI
Le prostaglandine. [The prostaglandins.]
Bollettino della Societa Italiana di Farmacia Ospedaliera. 13: 194-198. 1967.

The authors briefly review present knowledge of prostaglandins. The chemical structures of the E and F prostaglandins are shown, and their interrelationships and basic chemical properties are outlined. Early isolation and biosynthetic studies are described. The physiological effects of prostaglandins, particularly on human and animal uterine tissue, as vasodilator agents, and on the central nervous system, and their biochemical effects on lipid metabolism are discussed. It is foreseen that future research will be directed especially toward the role of calcium ions in the biological activity of prostaglandins. (MEMH) 0823

0341
FERREIRA, S.H. and J.R. VANE
The detection and estimation of bradykinin in the circulating blood.
British Journal of Pharmacology and Chemotherapy. 29: 367-377. 1967.

A technique for the bioassay of bradykinin in circulating blood is described. Several endogenous substances that might be found in the blood which could affect the bioassay tissue (cat jejunum) were also tested. PGE_2 and $PGF_{2\alpha}$ up to 1 μg/ml had no effect on the bioassay system. (JRH) 0789

0342
FERREIRA, S.H. and J.R. VANE
Prostaglandins: their disappearance from and release into the circulation.
Nature. 216: 868-873. 1967.

Cats anesthetized with ethyl chloride, ether, and chloralose, dogs anesthetized with halothane, chloralose and pentobarbitone, and rabbits anesthetized with pentobarbitone were used to study the origin and disappearance of PG's from the blood. The PG's were assayed by a blood bathed organ technique and in a few cases by chemical extraction and bioassay. PGE_1, PGE_2 and $PGF_{2\alpha}$ were not significantly degraded after 2 min incubations in blood. However, when the PG content of venous blood from various vascular beds was measured after close intraarterial PG infusion, the lungs removed 90-95%, the liver 70-93% and the hind quarters 50-66% of the infused PGE's. Electrical stimulation of splenic nerves adequate to cause contraction of the spleen, caused release of PG-like substances which caused a response in the bioassay tissues similar to that caused by PGE_2 (30 ng/ml). Infusion of epinephrine, norepinephrine or histamine into the spleen also caused contractions and release of PG's (the stronger the contraction, the more PG released). Phenoxybenzamine and phentolamine blocked the contraction and release of PG from the spleen due to nerve stimulation or epinephrine infusion. When stimulated at rates sufficient to maintain contraction there was a steady release of PG equivalent to 15 ng/ml PGE_2. The experiments suggest that PG's are released from spleen smooth muscle and are rapidly degraded by lung and liver tissue, making their role as circulating hormones doubtful. (JRH) 0808

0343
FREDHOLM, B.B., B. OBERG and S. ROSELL
Vascular reactions in canine subcutaneous adipose tissue following prostaglandin E_1 (PGE_1).
Acta Pharmacologica et Toxicologica. 24 (supp. 4): 28. 1967.

A pletysmographic technique which allows analysis of the reactions of various series-coupled vascular sections was adapted to the subcutaneous adipose tissue in the inguinal region of female mongrel dogs. PGE_1 was found to be a potent vasodilator agent and i a injections of 0.1 mμg or more caused a marked rise in total flow. PGE_1 was a more effective vasodilator agent than histamine or acetylcholine. The capillary filtration coefficient rose, indicating an increased capillary permeability and/or rise in the number of patent capillaries. PGE_1 did not have any marked effect on the venous section of the vascular bed. The results indicate that the vascular effect of PGE_1 in subcutaneous adipose tissue is mainly confined to the pre-capillary resistance vessels and to the exchange vessels (capillary section). (Authors modified) 0794

0344
FROESCH, E.R.
The physiology and pharmacology of adipose tissue lipolysis: its inhibition and implications for the treatment of diabetes.
Diabetologia. 3: 475-487. 1967.

The antilipolytic properties of PGE_1 are briefly discussed. The effect of PGE_1, insulin, and 5-methylpyrazole-3-carboxylic acid on glucose and fructose metabolism are compared in adipose tissue. (JRH) 0900

0345
GOTTENBOS, J.J., R.K. BEERTHUIS and D.A. van DORP
Essential fatty acid bio-assay of prostaglandin E_1 in rats and mice.
In: Bergstrom, S. and B. Samuelsson, eds., "Prostaglandins. Proceedings of the Second Nobel Symposium," Stockholm, 1966. p. 57-62. New York, Interscience Publishers, 1967.

The EFA-activity of PGE_1 was determined in EFA-deficient animals. PGE_1 was administered in different single daily doses orally and intravenously to mice and rats and by continuous intravenous infusion during 7 or 14 days to rats. As criteria served the changes in body weight and in water release via skin and respiration. The administration of PGE_1 did not result in changes in these criteria. Consequently, it can be concluded that PGE_1 does not show EFA-activity in rats and mice under the present experimental conditions. (Author modified) 0731

0346
GRAHAM, J.D.P. and H. AL KATIB
Adrenolytic and sympatholytic properties of 2-halogenoalkylamines in the vas deferens of the guinea-pig.
British Journal of Pharmacology and Chemotherapy. 31: 42-55. 1967.

As part of an experiment on the mechanism of action of 2-halogenoalkylamines, guinea pig vas deferens were incubated in Krebs solution containing atropine (10^{-7} g/ml). PGE_1 (2- 5 X 10^{-7} g/ml) caused strong contractions of the vas deferens after a delay which was longer than that noted after addition of norepinephrine (NE). When the preparation was treated with PGE_1 and then washed, subsequent responses to NE were potentiated 5 fold. Compound SY28, a 2-halogenoalkylamine, blocked the contraction caused by both NE and PGE_1 (1-3 X 10^{-7} g/ml) and reduced the response to

PGE$_1$ at 5 X 10^{-7} g/ml. When the bath fluid was changed (still in the presence of SY28 and atropine) a contraction occurred. PGF$_{1\alpha}$ (2 X 10^{-7} g/ml) was inactive and did not potentiate the response to NE. (JRH) 0739

0347
GRANSTROM, E.
On the metabolism of prostaglandin E$_1$ in man. Prostaglandins and related factors 50.
In: Kraitchevsky, P., R. Paoletti and D. Steinberg, eds., "Progress in Biochemical Pharmacology,"
p. 89-93. New York, Karger, 1967.

Tritium labelled PGE$_1$ was infused intravenously into human male subjects. Starting 1 min after completion of PG infusion, samples of blood, urine and feces were collected for 3 days and analyzed for radioactivity. There was a rapid drop in PG content in the blood during the first 60 min to a level which remained relatively constant for the next 48 hr. After 3 days no radioactivity could be detected in the blood. Excretion into the urine and feces was rapid with 60% of the total radio-activity being recovered by these routes (40% urine, 20% feces). Chromatographic analysis of the urine showed none of the radioactivity was in the form of unaltered PGE$_1$ or conjugated PGE$_1$ but was associated with 2 less polar metabolites of PGE$_1$. Acid hydrolysis of both of these metabolites yielded the same less polar product, which may indicate that the two metabolites were different conjugates of the same PGE$_1$ metabolite. (JRH) 0769

0348
GREEN, K., E. HANSSON and B. SAMUELSSON
Synthesis of tritium labeled prostaglandin F$_{2\alpha}$ and studies of its distribution by autoradiography.
In: Kraitchevsky, P., R. Paoletti and D. Steinberg, eds., "Progress in Biochemical Pharmacology,"
p. 85-88. New York, Karger, 1967.

A technique for the synthesis of tritium labelled PGF$_{2\alpha}$ is described. Tritium labelled PGF$_{2\alpha}$ was injected intravenously in 2 mice which were sacrificed and quick frozen 5 and 15 min after PG injection. Autoradiographs were made of median sagital sections of the mice. The PGF$_{2\alpha}$ was found distributed in essentially the same pattern as PGE$_1$ in earlier similar studies. Highest concentrations of radioactivity were in the liver, kidneys and connective tissue with lower but significant amounts in lung tissues. No radioactivity was found in the myocardium, brain, adipose tissue or endocrine glands. (JRH) 0770

0349
GREENBERG, R.A., J.R. CANT and H.V. SPARKS
Prostaglandins and the resistance, capacitance, and capillary filtration coefficient of the hind limb.
Physiologist. 10: 187. 1967.

Abstract only. The effects of PGE$_1$ or PGA$_1$ on consecutive vascular sections of an isolated canine skin-muscle preparation have been studied. Close arterial infusion of 0.01 to 0.1 μg/min of PG resulted in an increased blood flow within 30 sec. Maximum increase, about two fold, resulted from an infusion of 10 μg/min. A rapid increase in volume of the limb associated with the increased flow was interpreted to be increased vascular capacitance. Infusion of 0.01 μg/min of PG caused a slightly increased capillary filtration coefficient (K$_f$) and the maximum effect, a doubling of K$_f$, occurred with a dose of 1 to 10 μg/min. Increased K$_f$ could have resulted from either increased capillary permeability or increased capillary surface area due to decreased precapillary sphincter tone. Because of the characteristics of the response it was concluded that the increased K$_f$ was due to decreased tone. (Authors modified) 0760

0350

GUTMAN, H. and O. ISLER

Vitamin F (Essentielle Fettsauren). [Vitamin F Essential fatty acids].

In: Ullman, F. ed., "Ullman's Encyclopadie der Technischen Chemie, 3rd Edition," Vol. 18, p. 247. Munich, Urban/Schwarzenberg, 1967.

Several references are listed which discuss the role of vitamin F and essential fatty acids in prostaglandin biosynthesis. (JRH) 0885

0351

HAESSLER, H.A. and J.D. CRAWFORD

Insulin-like inhibition of lipolysis and stimulation of lipogenesis by prostaglandin E_1 (PGE_1)

Journal of Clinical Investigation. 46: 1065. 1967.

Abstract only. PGE_1 stimulated the synthesis of neutral fat from acetate-^{14}C by 31% and reduced by 7% the amount of $^{14}CO_2$. Insulin caused a 114% increase in lipid synthesis and a 30% reduction in $^{14}CO_2$. There was no additive effect of insulin and PGE_1 together. PGE_1 did not increase the incorporation of labelled glucose into fats, as insulin did, but the partition of label between glycerol and fatty acids was shifted, the latter being increased 18%. Thus PGE_1 can inhibit fatty acid release both by its known antilipolytic action and by promoting re-esterification and synthesis of new fat. In these actions, PGE_1 is qualitatively similar to insulin. (JRH) 0759

0352

HAMBERG, M. and B. SAMUELSSON

New groups of naturally occurring prostaglandins.

In: Bergstrom, S. and B. Samuelsson, eds., "Prostaglandins. Proceedings of the Second Nobel Symposium," Stockholm, 1966. p. 63-70. New York, Interscience Publishers, 1967.

Previous experiments on 8 new prostaglandin groups found in human seminal plasma are reviewed. The techniques are described and the results obtained discussed. From these experiments it appears that these compounds are normally present in seminal plasma and are not the result of alteration of other PG's during the isolation process. (JRH) 0730

0353

HAMBERG, M. and B. SAMUELSSON

19-hydroxylated prostaglandins.

In: Kraitchevsky, P., R. Paoletti and D. Steinberg, eds., "Progress in Biochemical Pharmacology," p. 83-84. New York, Karger, 1967.

The authors briefly review the results of an already reported isolation of 2 new groups of 19-hydroxylated PG's from human seminal plasma. (JRH) 0771

0354

HAMBERG, M. and B. SAMUELSSON

On the mechanism of the biosynthesis of prostaglandins E_1 and F_1.

Journal of Biological Chemistry. 242: 5336-5343. 1967.

The mechanism of the conversion of 8,11,14-eicosatrienoic acid into PGE_1 and $PGF_{1\alpha}$ has been studied. Incubation of [13D-H^3,3-^{14}C]-and [13L-^{3}H,3-^{14}C] 8,11,14-eicosatrienoic acids showed that the hydrogen removed from C-13 during the conversion into PGE_1 and $PGF_{1\alpha}$ has the L configuration. The conversion of [13L-^{3}H,3^{14}C] 8,11,14-eicosatrienoic acid into PGE$_1$ is accompanied by a hydrogen isotope effect. No conversion of [2-^{14}C] 15L-hydroperoxy-8(*cis*),11(*cis*),13(trans)-eicosatrienoic acid or [2-^{14}C] 15L-hydroxy-8(*cis*),11(*cis*),13(trans)-eicosatrienoic acid into PGE$_1$ or PGF$_{1\alpha}$ could be detected. Incubations of [9-^{3}H,3^{14}C] 8,11,14-eicosatrienoic acid revealed that PGF$_{1\alpha}$ is not formed via PGE$_1$ in the system used. The mechanistic implications of the results obtained are discussed. It is suggested that 11-peroxy-8,12,14-eicosatrienoic acid is the first inter-mediate in the conversion. The peroxy acid is cyclized into an endoperoxide, which is eventually transformed into PGE$_1$ or PGF$_{1\alpha}$ by independent reactions. (Authors modified) 0777

0355
HAMBERG, M. and B. SAMUELSSON
    Oxygenation of unsaturated fatty acids by the vesicular gland of sheep.
    Journal of Biological Chemistry. 242: 5344-5354. 1967.

    The enzyme extract from sheep vesicular glands that is used to produce prostaglandins also produces several monohydroxy acids from the PG precursor acids. The mechanism of synthesis of some of these monohydroxy acids was studied. The results support the view that reactions of the lipoxidase type do occur in animal tissue. Also the mode of formation of one of the monohydroxy acids provides additional evidence for the existence of the proposed endoperoxide intermediate in prosta-glandin biosynthesis. (JRH) 0776

0356
HAMPTON, J.R.
    The study of platelet behaviour and its relevance to thrombosis.
    Journal of Atherosclerosis Research. 7: 729-746. 1967.

    The author reviews the techniques used to investigate platelet behavior. It is briefly mentioned that PGE$_1$ has been shown to inhibit platelet aggregation in vitro and may also be important in vivo. (JRH) 0804

0357
HARDEGGER, E., H.P. SCHENK and E. BROGER
    Synthese der dl-form eines naturlichen prostaglandins. [Synthesis of the dl form of a natural prosta-
        glandin.]
    Helvetica Chimica Acta. 50: 2501-2504. 1967.

    The authors briefly describe their technique for the synthesis of the racemates of natural PGE$_1$-278 (PGB$_1$) and PGE$_1$-237 and describe the physical and chemical properties of the compounds. (MEMH) 0848

0358
HARTLEY, F.
    Hormones and their synthesis.
    Chemistry in Britain. 3: 217-223. 1967.

    A brief section on prostaglandins is included in this editorial review. The chemical nature of PG's is described and some of the biological activities are discussed. (JRH) 0806

0359
HAUGE, A., P.K.M. LUNDE and B.A. WAALER
Effects of catecholamines on pulmonary blood volume.
Acta Physiologica Scandinavica. 70: 323-333. 1967.

The effect of catecholamines on pulmonary hemodynamics was investigated in isolated lungs from rabbits and cats. Epinephrine infused intraarterially caused a decrease in pulmonary vascular resistance and at the same time a decrease in pulmonary blood volume. However, PGE_1 (25 µg) infused by the same, route, caused a similar reduction in pulmonary vascular resistance, but had little or no effect on pulmonary volume. (JRH) 0845

0360
HAUGE, A., P.K.M. LUNDE and B.A. WAALER
Effects of prostaglandin E_1 and adrenaline on the pulmonary vascular resistance (PVR) in isolated rabbit lungs.
Life Sciences. 9: 673-680. 1967.

In blood perfused isolated rabbit lungs, PGE_1 (1-50 µg i a) usually caused a reduction in pulmonary vascular resistance. Compared to epinephrine, PGE_1 was about equipotent as a vasodilator, but the individual responses to both agents varied considerably from preparation to preparation. Neither propranolol nor phentolamine blocked the response of the lung to PGE_1, but they effectively blocked the response to epinephrine. Thus PGE_1 and epinephrine apparently have different sites of action. (JRH) 0783

0361
HERZOG, J., H. JOHNSTON and D. LAULER
Comparatively natriuretic effect of prostaglandin E_1 and E_1-217 in the dog kidney.
Clinical Research. 15: 360. 1967.

Abstract only. PGE_1 or PGA_1 (0.1-2.0 µg/min) infused into the left renal artery of anesthetized dogs did not alter systemic blood pressure or glomerular filtration rate. PGE_1 caused a natriuresis, increased renal plasma flow, and positive free water clearance on the infused side. PGE_1 decreased para amino hippuric acid extraction and urine osmolarity on the infused side. PGA_1 had little or no effect on these parameters when infused at the same rate. (JRH) 0843

0362
HERZOG, J., H. JOHNSTON and D. LAULER
Natriuretic effect of prostaglandin E_2 ("medullin") in the dog kidney.
Clinical Research. 15: 479. 1967.

Abstract only. PGE_2 (0.01 and 2.0 µg/min) was infused into the left renal artery of anesthetized dogs. The PGE_2 produced increases in urinary sodium excretion, volume, renal plasma flow and water excretion that were similar to those caused by PGE_1 in earlier experiments. Systemic blood pressure and glomerular filtration were not altered. Para amino hippuric acid extraction was decreased. These results are consistent with a direct effect of PGE_2 on renal hemodynamics, sodium and water excretion. This is most likely caused by selective arteriolar vasodilation. (JRH) 0838

0363
HIMMS-HAGEN, J.
Sympathetic regulation of metabolism.
Pharmacological Reviews. 19: 367-461. 1967.

Prostaglandins are mentioned several times in this review. They are discussed in their role as general inhibitors of lipolysis and as not being specific inhibitors of the catecholamines. (JRH) 0857

0364
HINMAN, J.W.
The prostaglandins.
Bioscience. 17: 779-785. 1967.

This comparative review covers all aspects of prostaglandin research. Special attention is given to the areas in which research is needed. There are 48 references cited. (JRH) 0803

0365
HOLMES, S.W. and E.W. HORTON
The nature and distribution of prostaglandin in the central nervous system of the dog.
Journal of Physiology. 191: 134P-135P. 1967.

Various portions of the central nervous system were dissected out of exsanguinated pentobarbital anesthetized dogs. The parts were extracted with methanol and the polar acidic lipids separated by chromatographic methods. On the basis of chromatographic properties and effects on bioassay tissues, compounds resembling PGE_1, PGE_2, $PGF_{1\alpha}$, and $PGF_{2\alpha}$ were found fairly evenly distributed in the central nervous tissues with more variation in PG content between individual animals than between different parts of the central nervous system. (JRH) 0864

0366
HORTON, E.W. and I.H.M. MAIN
Central nervous actions and occurrence of the prostaglandins.
In: Leonardi, A., and J. Walsh, eds., "International Symposium on Drugs of Animal Origin," p. 35-37. Milan, Ferro Edizioni, 1967.

The occurrence in the CNS of prostaglandins and the effect of prostaglandins on the CNS are reviewed. Several PG's are discussed (E_1, E_2, E_3 and $F_{2\alpha}$) with respect to dosages, route of administration, species used and results. It is stated that "the possibility that prostaglandins are central transmitters cannot yet be excluded. However, their widespread occurrence and variety of pharmacological actions is suggestive of a modulator role possibly affecting a biochemical pathway which is common to several different tissues." (RAP) 0877

0367
HORTON, E.W. and I.H.M. MAIN
Central nervous actions of the prostaglandins and their identification in the brain and spinal cord.
In: Bergstrom, S. and B. Samuelsson, eds., "Prostaglandins. Proceedings of the Second Nobel Symposium," Stockholm, 1966. p. 253-260. New York, Interscience Publishers, 1967.

The central nervous system actions of PGE_1 and $PGF_{2\alpha}$ were investigated in cats and chickens prepared in various ways in an effort to localize the site of action of these PG's in producing several previously reported effects. PG's were also extracted and identified from several central nervous tissues. The experiments indicated that these PG's act at various sites in the central nervous system. $PGF_{2\alpha}$ and a PGE were found in the cat forebrain, and $PGF_{2\alpha}$ and PGE_2 were found in the chick brain and spinal cord. These results may indicate that PG's act as central nervous system transmitters. However, they could also function by mediating the effect of other transmitters. (JRH) 0708

0368
HORTON, E.W. and I.H.M. MAIN
Further observations on the central nervous actions of prostaglandins $F_{2\alpha}$ and E_1.
British Journal of Pharmacology and Chemotherapy. 30: 568-581. 1967.

$PGF_{2\alpha}$ on intravenous injection in chicks increases gastrocnemius muscle tension. Experiments were made to locate its site of action. The effect was observed in the decapitated chick but was abolished by denervation of the muscle. It could be elicited in the urethanized chick in which reflex contractions were blocked. It is concluded that the site of this action of $F_{2\alpha}$ in the chick is upon the spinal cord. In spinal cats PGE_1 injected intravenously increased gastrocnemius muscle tension also by an action on the spinal cord. The effect was abolished by denervation of the muscle but not by dorsal root section. Close-arterial injection of PGE_1 to the gastrocnemius muscle did not elicit a contraction but did inhibit muscle twitches and acetylcholine-induced contractures. In one experiment PGE_1 applied topically to the spinal cord induced a contraction of the gastrocnemius muscle. Decerebrate rigidity in the cat was potentiated by $PGF_{2\alpha}$ and PGE_1 applied topically to the spinal cord induced rigidity in the cat was potentiated by $PGF_{2\alpha}$ and PGE_1. Crossed extensor reflexes were potentiated by $PGF_{2\alpha}$ in the spinal chick and by PGE_1 in the spinal cat. The patellar reflex in the spinal cat was little affected by PGE_1. In the chloralosed chick $PGF_{2\alpha}$ potentiated but PGE_1 inhibited the crossed extensor reflex. (Authors modified) 0737

0369
HORTON, E.W. and I.H.M. MAIN
Identification of prostaglandins in central nervous tissues of the cat and chicken.
British Journal of Pharmacology and Chemotherapy. 30: 582-602. 1967.

The object of the investigation was to identify the prostaglandins present in the central nervous system of the cat and the chicken. On the basis of their behaviour on solvent partition, silicic acid column chromatography, thin layer chromatography and on biological preparations, $PGF_{2\alpha}$ and PGE_1 have been identified in cat brain (supracollicular) and PGE_2 and $PGF_{2\alpha}$ in chicken brain and spinal cord. The concentration of $PGF_{2\alpha}$ in the three tissues was approximately 10 ng/g whereas the concentrations of PGE_2 in the chicken brain and spinal cord were about 100 ng/g and 400 ng/g respectively. The physiological significance of the prostaglandins in the central nervous system is discussed in the light of the known pharmacological actions. (Authors modified) 0738

0370
ISHIZAWA, M.
Studies on the stimulating action of prostaglandin E_1 on the intestinal smooth muscle.
Sapporo Medical Journal. 31: 31-40. 1967.

In experiments with isolated rabbit and guinea pig intestinal muscle, alteration of concentration or types of ions (Ca, Mg, Na, or K) in the incubation medium caused similar effects on both spontaneous and PGE stimulated activity. Hexamethonium, tubocurarine, tetrodotoxin and botulinum toxin had no effect on the stimulating action of PGE. Thus it seems that PGE does not stimulate intrinsic nerves, but directly affects the muscle membrane and that there is some relationship between spontaneous and PGE stimulated activity. (JRH) 0832

0371
JACKSON, R.T. and R. STOVALL
Constriction of nasal blood vessels by prostaglandins.
Physiologist. 10: 212. 1967.

Abstract only. A rhinometric technique was used to measure the changes in the blood flow in the dog's nasal mucosa. PGE_1, PGE_2, and PGA and $PGF_{1\alpha}$ were compared to epinephrine, norepinephrine and papaverine. The drugs were introduced into the carotid artery in pentobarbital-anesthetized dogs. All the PG's induced a constriction of the blood vessels of the nasal mucosa. The two PGE forms were equipotent to epinephrine but their duration of action was more than seven times as long. The response to PG seemed to have little effect on subsequent responses to epinephrine. Also, the response to PG seemed little affected by prior treatment with vasodilators such as papaverine or aminophylline. The threshold dose of PGE was near 1×10^{-6} mg/kg; within the limits of sensitivity to PGE in other preparations. Commonly, the lower doses of PGE induced a nasal response without a corresponding change in blood pressure. (Authors modified) 0762

0372
JOHNSTON, H.H., J.P. HERZOG and D.P. LAULER
Effect of prostaglandin E_1 on renal hemodynamics, sodium and water excretion.
American Journal of Physiology. 213: 939-946. 1967.

PGE_1 (0.01-2.0 μg/min) was infused into the left renal artery of pentobarbital anesthetized 18- to 25-kg dogs. Urine volume, urinary sodium excretion, free water clearance, and renal plasma flow increased on the infused side as compared to its own control values and simultaneous values from the non-infused kidney. Glomerular filtration rate and mean aortic pressure remained stable. Para amino hippuric acid extraction fell on the infused side. These effects may be mediated by increased renal, possibly noncortical, blood flow or by an anti-ADH action of PGE_1. A direct tubular action cannot be excluded. Despite systemic vasopressin infusion in one group of animals, unilateral increases in free water clearance occurred. This effect could not be dissociated from the hemodynamic and natriuretic effects. (Author modified) 0779

0373
JOHNSTON, H., J. HERZOG and D. LAULER
Reversal of pitressin-induced antidiuresis by prostaglandin E_1.
Clinical Research. 15: 360. 1967.

Abstract only. Pitressin was given in doses adequate to produce and maintain maximum antidiuresis in anesthetized dogs. Infusion of PGE_1 (2.0-6.0 μg/min) into one renal artery caused bilateral increases in sodium excretion, urine volume and positive free water clearance. In dogs not pretreated with pitressin there was no increase on the non-infused side. Thus either pitressin sensitized the non-infused kidney to lower doses of PGE_1 or vasopressin prevented the normal degradation of PGE_1. (JRH) 0844

0374
JUST, G. and C. SIMONOVITCH
A prostaglandin synthesis.
Tetrahedron Letters. 2093-2097. 1967.

The authors described a total synthesis of $d1$-$PGF_{1\alpha}$. They also obtained PGE_1 and its methyl ester, but they have not yet been purified completely. The starting material was cyclopentenol. A step by step synthesis is given along with structures of important intermediates. (JRH) 0899

0375
KADAR, D., P.D. COOPER and F.A. SUNAHARA
Extraction, isolation, and some pharmacological characteristics of rabbit renal medullary extracts.
Canadian Journal of Physiology and Pharmacology. 45: 1071-1080. 1967.

Ethanol soluble long-chain fatty acids were extracted from rabbit kidney medullas. These were separated by chromatographic methods into six vasoactive fractions. The most abundant fraction was identified as PGE on the basis of behavior in several bioassay systems. Another fraction ressembled PGA_2 and a third was similar to PGF. (JRH) 0812

0376

KANDEMIR, E.

A comparative in vivo and in vitro study of the effects of certain biogenic polypeptides, prostaglandin E_1, adrenaline and noradrenaline on the uterine muscle.

Acta Medica Turcica. 4: 39-52. 1967.

PGE_1 (2-16 ng/ml) caused strong contractions of the isolated non-pregnant rat uterus in a dose related manner. It caused even stronger contractions in the pregnant rat uterus. However, in dogs anesthetized with pentobarbital, PGE_1 (1 gamma/kg) did not affect uterine contractility when injected intravenously. (JRH) 0830

0377

KAPLAN, H.R., G.J. GREGA and J.P. BUCKLEY

Evidence of the central cardiovascular action for prostaglandin E_1.

Pharmacologist. 9: 223. 1967.

PGE_1 was injected into the arterial inflow of vascularly isolated neurally intact heads of recipient dogs. A dosage of 5 μg/kg of PGE_1 produced consistent depressor responses in both donor and recipient animals. Raising the dosage level of PGE_1 to 10 μg/kg caused an increased depressor response in the donors; however, the recipients' responsiveness remained unchanged. When dose levels of less than 5 μg/kg were employed, responsiveness varied in accordance with the animals' inherent compensatory abilities. The administration of PGE_1 to recipients with carotid sinus and carotid body areas which were bilaterally denervated induced reversed effects in that 5 μg/kg of PGE_1 produced a marked pressor response in the recipients paralleled by the usual depressor effects in the donor. PGA_1 appeared to give qualitatively similar depressor effects during preliminary studies. (Authors modified) 0879

0378

KARIM, S.M.M., M. SANDLER and E.D. WILLIAMS

Distribution of prostaglandins in human tissues.

British Journal of Pharmacology and Chemotherapy. 31: 340-344. 1967.

Prostaglandins were extracted, purified chromatographically and bioassayed from 23 human tissues and fluids. Tissues were obtained by necropsy from adult patients within 24 hr of death. PGE_1 was found in the thymus and phrenic nerve. PGE_2 and $PGF_{2\alpha}$ were found in thyroid, pancreas, adrenal cortex, adrenal medulla, parotid gland, submandibular salivary gland, cervical sympathetic chain, vagus nerve brachial plexus, bronchi and lung parenchyma. In addition, PGE_2 was found in cardiac muscle, rectus abdominus muscle and psoas muscle. $PGF_{1\alpha}$ was not found in any tissue. No PG's were detected in spleen, liver, kidney subcutaneous fat, milk, urine or venous blood. (JRH) 0741

0379

KARIM, S.M.M.

The identification of prostaglandins in human umbilical cord.

British Journal of Pharmacology. 29: 230-237. 1967.

Human umbilical cords were homogenized and extracted with alcohol. The prostaglandins in the extracts were separated and identified by chromatographic and bioassay techniques. Placental vessels, placental tissue, fetal blood and maternal blood were also extracted and assayed for PG's. Umbilical cords (1,000 g) contained PGE_1 (40 μg), PGE_2 (30 μg), PGF_1 (100 μg) and $PGF_{2\alpha}$ (80 μg) when assayed on guinea pig colon. The placental blood vessels contained the same PG's as the umbilical cord. Placental tissue, maternal and fetal blood contained only very small amounts of PG. Isolated umbilical arteries, in vitro, contracted in response to PGE_2, $PGF_{1\alpha}$, and $PGF_{2\alpha}$, but PGE_1 relaxed the arteries. When a mixture of the 4 PG's in the same proportions as found in the umbilical cord were added, they always produced a contraction. (JRH) 0790

0380
KARIM, S.M.M. and J. DEVLIN
  Prostaglandin content of amniotic fluid during pregnancy and labour.
  Journal of Obstetrics and Gynaecology of the British Commonwealth. 74: 230-234. 1967.

The relative distribution of various prostaglandins in 38 individual specimens of human amniotic fluid obtained during different stages of pregnancy was measured. Of the 16 specimens obtained prior to term from patients 8-35 weeks pregnant and not in labor, only PGE_1 (0.06-3.1 ng/ml) was found in 8 specimens. In one case of spontaneous abortion at 16 weeks of pregnancy, PGE_1 (0.7-4.1 ng/ml) was found in all 11 specimens obtained at or near term (38-42 weeks), but not in labor, and 3 of the specimens also contained PGE_2 (0.11-0.16 ng/ml). All 10 specimens obtained during labor contained $PGF_{2\alpha}$ (5.5-57.0 ng/ml), also 6 contained PGE_1 (0.9-4.0 ng/ml), 9 contained PGE_2 (0.04-2.3 ng/ml) and 5 contained $PGF_{1\alpha}$ (2.3-170.0 ng/ml). The decidua from patients prior to term and not in labor contained only PGE_1 in concentrations 10-30 times higher than in the amniotic fluid. The authors suggest that PG's may play an active role in the initiation of labor and may be produced by the decidua cells. (JRH) 0764

0381
KATAOKA, K., P.W. RAMWELL and S. JESSUP
  Prostaglandins: localization in subcellular particles of rat cerebral cortex.
  Science. 157: 1187-1188. 1967.

Homogenates of rat cerebral cortex contain material corresponding to PGE_1, PGE_2, $PGF_{1\alpha}$ and $PGF_{2\alpha}$ which are concentrated mainly in the light microsomal and mitochondrial fractions. Only the former fraction exhibits significant ability to synthesize PGE_1 and $PGF_{1\alpha}$ from bis-homo-γ-linolenic acid. After subfractionation of the crude mitochondrial fraction, PGE and PGF material is found mainly in the cholinergic and noncholinergic nerve endings. We conclude that the nerve endings are a storage site, whereas the light microsomes are the site of synthesis. (Authors) 0861

0382
KAYLAAP, S.O. and R.K. TURKER
  Release of catecholamines from the adrenal medulla by prostaglandin E_1.
  European Journal of Pharmacology. 2: 175-180. 1967.

Dogs were anesthetized with pentobarbital and their hindquarters were sympathetically denervated and autoperfused. PGE_1 was injected into the aorta at a level proximal to the arterial blood supply of the adrenal glands. As far as possible it was prevented from reaching the perfused vascular bed which served as a non-specific catecholamine detector with adequate sensitivity. PGE_1 (1-4 μg/kg) did not produce any rise but a fall in perfusion pressure when given intravenously, whereas it usually caused a sustained rise in perfusion pressure when given into the aorta. This pressor response was abolished by phenoxybenzamine or dihydroergotamine. The administration of hexamethonium also blocked the pressor response to PGE_1. The pressor response did not occur in spinal dogs with their adrenal glands excluded from the circulation. These results indicate that PGE_1 releases catecholamines from the adrenal medulla in the dog, possibly through a presynaptic action. A reflex indirect action seems highly unlikely. (Authors modified) 0744

0383
KHAIRALLAH, P.A., I.H. PAGE and R.K. TURKER
Some properties of prostaglandin E_1 action on muscle.
Archives Internationales de Pharmacodynamie et de Therapie. 169: 328-341. 1967.

The effects of PGE_1 on a variety of muscle preparations under a variety of physiological conditions was examined in a series of experiments. Muscle strips from rabbit aorta; cat carotid artery, trachea nictitating membrane and gastrocnemius; rat duodenum; and frog rectus abdominus muscle were tested in vitro. Similar preparations from animals pretreated with reserpine were also tested. The drugs used included angiotensin, phentolamine, bromo-lysergic acid, ADH, oxytocin, bradykinin, tyramine, diphenhydramine, epinephrine, norepinephrine, histamine, atropine, propranolol, decamethonium, tubocurarine, acetylcholine and phytostigmine. Some preparations were stimulated electrically. On the basis of results from these experiments it was postulated that some of the actions of PGE_1 are catecholamine mediated, while other indirect ones are due to interaction with receptor sites, leading to hypopolarization of cell membranes, hence enhancement. Larger amounts of PGE_1 cause depolarization, hence response. (JRH) 0796

0384
KISCHER, C.W.
Alterations in the fine structure of developing skin produced by prostaglandins.
Texas Journal of Science. 19: 432. 1967.

Abstract only. PGE_1 and PGB_1 completely block feather organ development in embryo chick skin in vitro while stimulating proliferation and keratinization. Ultrastructure observations indicate (1) the epidermal cells of treated tissues have a marked increase in macrovilli over those of the controls; (2) bundles of prekeratin filaments appear much sooner than normal and are distributed throughout the epidermis; (3) the mitochondrial cristae of treated skins are long, closely apposed and oriented along the longest axis; (4) more collagen appears in the mesenchyme of the treated tissue; (5) throughout the epidermis of treated skins are identified structures which have a striking similarity to the nucleolemata of nucleoli, but are always intra-cytoplasmic. They occur in large number ranging in size similar to that of mitochondria. (Authors modified) 0865

0385
KISCHER, C.W.
Effects of specific prostaglandins on development of chick embryo skin and down feather organ, in vitro.
Developmental Biology. 16: 203-215. 1967.

Addition of PGE_1 or PGB_1 to embryo chick skins (stages 29-34) prevented feather organ development and caused a precocious keratinization and thickening of the epidermis. PGB_1 was the most effective (1 μg/ml minimum and 10 μg/ml optimum effective dose). The optimally effective dose for PGE_1 was 100 μg/ml. Histochemical studies revealed no abnormalities in distribution or concentration of physiological agents which are known to affect feather organ development. (JRH) 0863

0386
KISCHER, C.W.
Unusual structures in epidermal cells of developing skin under treatment by prostaglandins.
Journal of Cell Biology. 35: 69A-70A. 1967.

Abstract only. Organ cultures of developing chick skin were treated with either PGE_1 or PGB_1. Ultrastructural observations of control and treated skin after 1,2, and 3 days of culture indicate the following: (1) the epidermal cells of treated tissues have a marked increase in macrovilli over those of the controls. (2) bundles of prekeratin filaments appear much sooner in treated skin than in normal, and are distributed throughout the epidermis. (3) the mitochondrial cristae of treated skins are long, closely apposed, and oriented along the longest axis. (4) More collagen appears in the mesenchyme of the treated tissues than in the controls. (5) Throughout the stratified and superficial epidermis of treated skins unusual structures occur which are composed of very fine filaments and granules. They have a striking similarity to the nucleolenema of nucleoli, but are always intracytoplasmic. No such structures have been found in control tissues. (Authors modified) 0775

0387
KLAUS, W. and F. PICCININI
Uber die wirkung von prostaglandin E_1 auf den Ca-haushalt isolierter meerschweinschenherzen. [The working of PGE_1 on the internal Ca content of isolated guinea pig hearts.]
Experientia. 23: 556-557. 1967.

The stimulatory effect of PGE_1 on different functions of isolated guinea pig hearts (Langendorff method, Tyrode solution) was coupled with an increase in the rate of ^{45}Ca uptake from the perfusion medium. The total myocardial Ca content and the amount of exchangeable cellular Ca were not affected. This action of PGE_1 on the myocardial Ca metabolism seems to be related to the positive inotropic action of PGE_1 and can most probably be explained by an increase in the membrane permeability to Ca ions (similar to the action of epinephrine). (Authors) 0807

0388
KLOEZE, J.
Influence of prostaglandins on platelet adhesiveness and platelet aggregation.
In: Bergstrom, S. and B. Samuelsson, eds., "Prostaglandins. Proceedings of the Second Nobel Symposium," Stockholm, 1966. p. 241-255. New York, Interscience Publishers, 1967.

The effect of several prostaglandins on platelet adhesiveness and platelet aggregation was measured in platelets from rats, pigs and man. Without ADP neither PGE_1 nor PGE_2 induced platelet adhesiveness or platelet aggregation. However, it appeared that PGE_1 inhibited the ADP-induced platelet adhesiveness in citrated platelet-rich pig plasma, while it seemed to be inactive in human plasma. The related substances PGA_1, dinor-PGE_1 and iso-PGE_1 were found to possess PGE_1-like effects on ADP-induced platelet aggregation. In a roughly quantitative assay and expressed on a dose-level, they showed 5, 1.25 and 1% of PGE_1 activity respectively. (Author modified) 0707

0389
KUPIECKI, F.P.
Effects of prostaglandin E_1 on lipolysis and plasma free fatty acids in the fasted rat.
Journal of Lipid Research. 8: 577-580. 1967.

Contrary to published reports, PGE_1 in vitro and in vivo inhibited fasting lipolysis in rats. Adipose tissue lipolysis was inhibited when the tissue was incubated in the presence of PGE_1 and when the compound was administered intravenously. A biphasic plasma free fatty acid (FFA) response was obtained in fasted rats after intravenous injection of 80 μg of PGE$_1$ per kg body weight; plasma FFA concentrations were lowered at 7 min, elevated at 15 min, and at normal concentrations at 30 min. The FFA depression at 7 min was independent of the animal's nutritional state, but the rebound at 15 min did not occur in fed rats. The plasma FFA rebound in fasted rats at 15 min may be a consequence of rapid inactivation of PGE$_1$, followed by unopposed activity of factors which enhance fasting lipolysis. (Author modified) 0753

0390
LABAY, P. and S. BOYARSKY
Ureteral effect of prostaglandins in the dog.
Clinical Research. 15: 362. 1967.

Abstract only. PGE$_1$ depressed ureteral peristalsis in vivo in doses of 0.1-33 μg/kg i v and in vitro at concentrations of 5-10 μg/ml. Slowed rate, lowered amplitude and contraction waves, and a slight fall in baseline pressure (2-4 mg Hg) reslulted. Effect lasted 2-10 min after 4-10 μg/kg and 26 min after 15 μg/kg i v. Slowing of urine flow occurred. Blood pressure fell. PGF$_{2\alpha}$ had an opposite effect. In five dogs with bladder explants, 12 experiments showed that peristaltic rate increased after 0.5 to 50 μg/kg i v. Antidiuretic urine flow was unaltered. Baseline shifts and amplitude were slight. Frequency rose 4 X or more, effects lasted up to 30 min. (Authors modified) 0842

0391
LAPIDUS, M., N.H. GRANT and H.E. ALBURN
Prostaglandin (PGE$_2$): biosynthesis and purification.
Abstracts of the 154th Meeting of the American Chemical Society, Chicago, 10-15 September, 1967. abs. 68.

Abstract only. An enzymatic system was developed for the production of PGE$_2$ from arachidonic acid by extracts of sheep seminal vesicular glands. This system converts 30-40% of the arachidonic acid into PGE$_2$. The presence of glutathione assures high yields of PGE$_2$. A new procedure for the purification of PGE$_2$ is based on the dialysis of the biosynthesized PGE$_2$ at pH 8 and extraction of the dialyzate at pH 3 with chloroform. This procedure routinely gives PGE$_2$ preparations having a purity of 90-100%. (Authors modified) 0884

0392
LEE, J.B.
Antihypertensive activity of the kidney-the renomedullary prostaglandins.
New England Journal of Medicine. 277: 1073-1079. 1967.

The author reviews his own work and the work of others which has led to the hypothesis that renomedullary prostaglandins may have a physiological role in regulating blood pressure. (JRH) 0755

0393
LEE, J.B.

Chemical and physiological properties of renal prostaglandins: The antihypertensive effects of medullin in essential human hypertension.
In: Bergstrom, S. and B. Samuelsson, ed., "Prostaglandins. Proceedings of the Second Nobel Symposium," Stockholm, 1966. p. 197-210. New York, Interscience Publishers, 1967.

After a brief review of experiments showing medullin to be PGA_2, the effects of medullin on the blood pressure of a patient with fixed hypertension is described. Intraarterial injection of 50 μg of medullin produced a fall in blood pressure after 18 sec which lasted 18 sec and which was due to a decrease in peripheral vascular resistance. There was an increase in heart rate and cardiac output following the injection which may have been reflexive. Prolonged infusion of medullin at 191 or 382 μg/min caused a prolonged, dose-dependent, decrease in blood pressure and an increase in heart rate. There was a marked diuresis accompanying the fall in blood pressure. (JRH) 0702

0394
LEE, J.B., K. CROWSHAW, B.H. TAKMAN, K.A. ATTREP and J.Z. GOUGOUTAS
The identification of prostaglandins E_2, $F_{2\alpha}$, and A_2 from rabbit kidney medulla.
Biochemical Journal. 105: 1251-1260. 1967.

The acidic lipids were extracted from 10 kg of frozen rabbit kidney medullas and separated by silicic acid columns into biologically active fractions. After further purification the 3 acids were identified as PGE_2, $PGF_{2\alpha}$ and PGA_2 by thin-layer chromatographic, spectroscopic and mass-spectral analysis. It was considered likely that at least part of the PGA_2 came from dehydration of endogenous PGE_2 during the long purification process. (JRH) 0791

0395
LEE, J.B., K. CROWSHAW and B.H. TAKMAN
Isolation and identification of prostaglandin-E_2 and prostaglandin-$F_{2\alpha}$ from rabbit kidney medulla.
Journal of Clinical Investigation. 46: 1082-1083. 1967.

Abstract only. Two prostaglandins were isolated from the rabbit kidney medulla. Chromatographic and nuclear magnetic resonance studies revealed them to be identical to PGE_2 and $PGF_{2\alpha}$. No PGE_1, PGE_3, PGF_1 or PGF_3 could be found. While PGE_2 and PGA_2 were vasodilators, $PGF_{2\alpha}$ exhibited pressor activity when injected into vagotomized, pentolinium-treated rats. The renomedullary prostaglandins may have complex intrarenal regulatory functions. (JRH) 0758

0396
LILJESTRAND, G.
Discussion remarks to Anggard and Samuelsson: The metabolism of prostaglandins in lung tissue.
In: Bergstrom, S. and B. Samuelsson, eds., "Prostaglandins. Proceedings of the Second Nobel Symposium," Stockholm, 1966. p. 107-108. New York, Interscience Publishers, 1967.

It is suggested that results obtained so far with prostaglandins and lung tissue would justify a complete investigation to determine their physiological role, if any, in regulating lung hemodynamics. One particular area of investigation would be to see if O_2-tension changes in the lung could lead to changes in PG concentrations in the blood. (JRH) 0723

0397
MANDEL, L.R. and F.A. KUEHL, Jr.
Lipolytic action of 3,3'5-triiodo-L-thronine.
Biochemical and Biophysical Research Communications. 28: 13-18. 1967.

Addition of 3,3'5-triiodo-L-thyronine (T_3) to rat adipocytes causes a lipolysis and potentiates the lipolytic response to epinephrine. The T_3-induced lipolysis is blocked by PGE_1 (0.1 μg/ml) or insulin but not by the β-adrenergic blocking agent Ko 592. The similarity between the effect of these agents on T_3 and theophylline, suggested that they might have a common mechanism. It was subsequently shown that like theophylline, T_3 induces lipolysis by inhibiting phosphodiesterase, thus increasing cAMP levels in the cells. (JRH) 0792

0398

MAXWELL, G.M.

The effect of prostaglandin E_1 upon the general and coronary haemodynamics and metabolism of the intact dog.

British Journal of Pharmacology and Chemotherapy. 31: 162-168. 1967.

In 10 intact dogs anaesthetized with morphine, pentobarbital, allobarbitone, monoethylthiourea and urethane, cardiact output, vascular pressure, coronary flow glucose, free fatty acid (FFA), lactate and pyruvate levels were measured before and after an intravenous infusion of 1.5 µg/kg/min of PGE_1. Comparison with control studies showed a significant increase in respiratory rate, respiratory exchange, cardiac output, and heart rate. Systemic pressure decreased as did systemic and pulmonary vascular resistances. Coronary flow increased as did myocardial oxygen extraction and cardiac oxygen usage. Cardiac efficiency decreased, as did coronary vascular resistance. Blood glucose and FFA decreased, so did arterial pyruvate values. PGE_1 is a potent vasodilator of the circulatory beds studied. (Author modified) 0740

0399

MILLER, R.W., F.R. EARLE and I.A. WOLFF

Search for new industrial oils. XV. Seed oils of borgaginaceae.

Journal of the American Oil Chemists' Society. 44: 138A-139A. 1967.

Abstract only. All-cis 6,9,12, octadeca-trienoic acid which is a precursor for the synthesis of prostaglandins was found in all 27 species of the subfamily Borgainoidae in amounts from 0.2 to 18%. (Authors modified) 0871

0400

MIYAZAKI, E., M. ISHIZAWA, S. SUNANO, B. SYUTO and T. SAKAGAMI

Stimulating action of prostaglandin on the rabbit duodenal muscle.

In: Bergstrom, S. and B. Samuelsson, eds., "Prostaglandins. Proceedings of the Second Nobel Symposium," Stockholm, 1966. p. 278-281. New York, Interscience Publishers, 1967.

PGE extracted from sheep seminal fluid was used to study the mechanism of the stimulating action of PGE on rabbit duodenal muscle. Increase in Mg ion concentration inhibited both spontaneous and PGE stimulated contraction. Substitution of LiCl for NaCl also decreased PGE and spontaneous contractions. The optimal K ion concentration was 10mM and both lower and higher concentrations were inhibitory. With guinea pig colon muscle, PGE caused a 2 to 3-fold increase in spike frequency. In mice, botulinum toxin did not reduce PGE stimulation. Thus PGE does not act by stimulating intrinsic nerves but by stimulating the muscle membrane. There also seems to be some relationship between PGE stimulating activity and spontaneous contraction of the duodenal muscle. (JRH) 0711

0401

MUHLBACHOVA, E., A. SOLYOM and L. PUGLISI

Investigations on the mechanism of the prostaglandin E_1 antagonism to norepinephrine and theophylline-induced lipolysis.

European Journal of Pharmacology. 1: 321-325. 1967.

The effects of PGE_1 on the dose-response curves of glycerol release from adipose tissue by norepinephrine and theophylline have been investigated in vitro. PGE_1 antagonizes the lipid mobilizing action of both drugs; this antagonism is competitive with theophylline but is non-competitive with norepinephrine. The possible interference of PGE_1 with the rate of synthesis and breakdown of 3'5'-adenosine-monophosphate (3'5'-AMP) is discussed. (Authors modified) 0742

0402
MUIRHEAD, E.E., B.E. LEACH, G.B. BROWN, E.G. DANIELS and J.W. HINMAN
Anithypertensive effect of prostaglandin E_2 (PGE_2) in renovascular hypertension.
Journal of Laboratory and Clinical Medicine. 70: 986-987. 1967.

Abstract only. PGE_2 (1-3.3 mg/kg/day) was injected subcutaneously into moderately hypertensive rats (154 mm Hg) for 9 days. Arterial blood dropped to 124 mm Hg after 4 days of treatment and remained depressed during the rest of the treatment period. An average of 3.8 days was required for blood pressure to return to pretreatment levels. In an additional 6 rats with significantly higher hypertension (166 mm Hg) the blood pressure dropped in a similar manner, but returned to pretreatment levels during the treatment period. In 3 rabbits having a mean arterial pressure of 90 to 113 mm Hg there was a 20 mm drop in pressure following intraperitoneal injection of 1 to 2 mg/kg of PGE_2. The pressure returned to pretreatment levels after 1 to 4 days. (JRH) 0814

0403
MUIRHEAD, E.E., B.E. LEACH, B. BROOKS, P.H. SHAW, W.L. BROSIUS, Jr., E.G. DANIELS and J.W. HINMAN
Antihypertensive renomedullary lipid in the hypertensive rabbit.
Revue Francaise d'Etudes Cliniques et Biologiques. 12: 893-898. 1967.

Reference to prostaglandins is confined to the discussion section of this paper. It is briefly mentioned that PGE_2 appears to be the acute vasodepressor lipid of the rabbit kidney. (JRH) 0895

0404
MUIRHEAD, E.E., E.G. DANIELS, J.E. PIKE and J.W. HINMAN
Renomedullary antihypertensive lipids and the prostaglandins.
In: Bergstrom, S. and B. Samuelsson, eds., "Prostaglandins. Proceedings of the Second Nobel Symposium," Stockholm, 1966. p. 183-196. New York, Interscience Publishers, 1967.

The ability of orally administered PGE_2 (15-29 μg/kg/day), PGA_1 (50-100 μg/kg/day), and $PGF_{1\alpha}$ (15-30 μg/kg/day) to prevent renoprival hypertension in dogs was investigated. All 3 PG's prevented hypertension during the first 2 days after kidney removal. PGE_2 and $PGF_{1\alpha}$ were more effective than PGA_1 in reducing hypertension for more than 2 days. Preliminary experiments on the effects of oral PGE_1 (10 μg/kg/day) and PGA_1 (5 and 50 μg/kg/day) on the blood pressure of dogs with renal (Goldblatt or Page) hypertension indicated that both PG's produce a moderate reduction in hypertension during the treatment period. The authors review literature on renomedullary antihypertensive lipids and speculate on relationships of prostaglandins to other active lipids. (JRH) 0715

0405
MULLER, E.
Uber die bestimmung der prostaglandine in menschlichen sperma. [The determination of prostaglandins in human sperm.]
Fresenuis' Zeitschrift fur Analytische Chemie. 229: 156. 1967.

Article in German. Abstract not available at present. 0887

0406
MUSTARD, J.F.
Recent advances in molecular pathology: a review. Platelet aggregation, vascular injury and athero-
sclerosis.
Experimental and Molecular Pathology. 7: 366-377. 1967.

PGE_1 is briefly mentioned in this review as being an inhibitor of ADP induced platelet aggregation.
(JRH) 0817

0407
NAKANO, J. and J.R. McCURDY
Cardiovascular effects of prostaglandin E_1.
Journal of Pharmacology and Experimental Therapeutics. 156: 538-547. 1967.

In pentobarbital anesthetized dogs, PGE_1 (0.25-4.0 $\mu g/kg$ i v) decreased arterial pressure, left atrial
pressure and left ventricular end-diastolic pressure, whereas heart rate, pulmonary arterial pressure,
cardiac output and myocardial contractile force increased in a dose related manner. PGE_1 (0.1 $\mu g/kg$
i a) increased blood flow markedly and decreased peripheral resistances in the brachial, femoral,
carotid and renal arteries without any significant change in mean systemic arterial pressure and
myocardial blood flow and myocardial contractile force. Furthermore, the i a injection of PGE_1
increased both coronary arterial blood flow and myocardial contractile force without any change in
mean systemic arterial pressure. Propranolol (1 mg/kg i v) did not block the positive inotropic and
vasodilator actions of PGE_1 (4 $\mu g/kg$). Thus PGE_1 exerts a positive inotropic action on the heart and
induces multiple hemodynamic effects by direct action on all peripheral vascular beds and indirectly
through reflex sympathetic stimulation. PGE_1 does not appear to block the hemodynamic effects on
norepinephrine in vagotomized dogs. (JRH) 0782

0408
NAKANO, J. and J.R. McCURDY
Effects of prostaglandins E_1 (PGE_1) and A_1 (PGA_1) on the systemic venous return and pulmonary
circulation.
Clinical Research. 15: 409. 1967.

Abstract only. The effect of PGE_1 and PGA_1 on systemic venous return and pulmonary circulation
was studied in pentobarbital anesthetized, open chest dogs, by comparing the hemodynamic effects
of intravenous administration of PG's (0.25-4.0 $\mu g/kg$) to intact dogs with those in which right
cardiac in input was kept constant. On the basis of these experiments, both PGE_1 and PGA_1
increased pulmonary arterial pressure in dogs with intact circulation by increasing markedly the
systemic venous return and right ventricular output. (JRH) 0841

0409
NAKANO, J. and J.R. McCURDY
Hemodynamic actions of prostaglandins.
Oklahoma State Medical Association Journal. 60: 418-419. 1967.

In this brief review, the authors discuss all aspects of prostaglandin research. However, the main
emphasis is on the cardiovascular effects of PGE_1 and $PGF_{2\alpha}$. 6 references are cited. (JRH) 0856

0410
NAIMZADA, M.K.
  Azione della prostaglandina E_1 sulla vesicica urinaria. (Esperienza nella cavia e nel ratto). [Action of prostaglandin E_1 on the urinary vesicle. (Experiments on the guinea-pig and rat)].
  Bollettino della Societa Italiana di Biologia Sperimentale. 43: 518-521. 1967.

  Article in Italian. Abstract not available at present. 0890

0411
NG, K.K.F. and J.R. VANE
  Conversion of angiotensin I to angiotensin II.
  Nature. 216: 762-766. 1967.

  It is briefly mentioned that the lungs are important in the metabolism of peptides, amines and prostaglandins. (JRH) 0892

0412
NISSEN, H.M.
  On lipid droplets in renal interstitial cells.
  Zeitschrift fur Zellforschung und Mickroskopische Anatomie. 83: 76-81. 1967.

  Histochemical studies of rat kidneys revealed the presence of lipid droplets in the interstitial cells which contained simple saturated and unsaturated lipids. It is suggested that if it can be proven that these droplets exert an active functional influence in the renal medulla, there is a probability that they are identical with the isolated lipids (medullin and prostaglandin). (JRH) 0893

0413
NUGTEREN, D.H., R.K. BEERTHUIS and D.A. van DORP
  Biosynthesis of prostaglandins.
  In: Bergstrom, S. and B. Samuelsson, eds., "Prostaglandins. Proceedings of the Second Nobel Symposium," Stockholm, 1966. p. 45-50. New York, Interscience Publishers, 1967.

  The mechanism of the conversion of dihomo-γ-linolenic acid into PGE_1 by enzymes from sheep-vesicular glands was studied. Glutathione caused a large increase in yield. Oxygen is necessary; all 3 oxygen atoms in the PGE_1 are derived from molecular oxygen (based on incubation with $^{18}O_2$). There were several indications that a free radical mechanism was involved. A proposed mechanism of PGE_1 formation is given. (JRH) 0733

0414
NUGTEREN, D.H., H. VONKEMAN and D.A. VAN DORP
  Non-enzymic conversion of all-cis 8,11,14-eicosatrienoic acid into prostaglandin E_1.
  Recueil des Travaux Chimiques de Pays-Bas et Belgique. 86: 1237-1245. 1967.

  It is shown that prostaglandin E_1 is formed during autoxidation of all-cis 8,11,14-eicosatrienoic acid. However, the yield is low, which is due to the occurrence of many side reactions and the complex stereochemistry of the products. This non-enzymic prostaglandin formation may give rise to artifacts in biological studies. (Authors) 0889

0415
ORLOFF, J. and J. GRANTHAM
The effect of prostaglandin (PGE$_1$) on the permeability response of rabbit collecting tubules to vasopressin.
In: Bergstrom, S. and B. Samuelsson, eds., "Prostaglandins. Proceedings of the Second Nobel Symposium," Stockholm, 1966. p. 143-146. New York, Interscience Publishers, 1967.

Individual rabbit kidney collecting tubules were dissected out and arranged so that changes in permeability could be measured during perfusion with various substances. Vasopressin (2.5 μU/ml) caused a large increase in water absorption. PGE$_1$ (10^{-7}M) alone only caused a slight increase in absorption, but when it was added to the vasopressin it greatly inhibited the response to vasopressin. It is speculated that PG's might have a physiological role in the kidney, perhaps as a damper to vasopressin activity. (JRH) 0719

0416
ORLOFF, J. and J. HANDLER
The role of adenosine 3',5'-phosphate in the action of antidiuretic hormone.
American Journal of Medicine. 42: 757-768. 1967.

The author briefly reviews the literature on the effects of prostaglandins on tissue responses to vasopressin. It is suggested that prostaglandins may serve to dampen the effects of ADH in the kidney tubule and prevent large overshoots in permeability which might otherwise occur. (JRH) 0788

0417
OTROVSKY, D., S. SEN, R.R. SMEBY and F.M. BUMPUS
Chemical assay of phospholipid renin preinhibitor in canine and human blood.
Circulation Research. 21: 497-505. 1967.

Previously reported work indicates that neither PGE$_1$ nor PGA$_1$ show any inhibitory effects on renin as does the preinhibitor investigated in this paper. (JRH) 0749

0418
PAOLETTI, R., R.L. LENTATI and Z. KOROLKIEWICZ
Pharmacological investigations on the prostaglandin E$_1$ effect on lipolysis.
In: Bergstrom, S. and B. Samuelsson, eds., "Prostaglandins. Proceedings of the Second Nobel Symposium," Stockholm, 1966. p. 147-159. New York, Interscience Publishers, 1967.

PGE$_1$, at low concentrations, antagonizes the increased lipolysis induced by norepinephrine and theophylline in adipose tissue in vitro (epididymal fat pads, 5.65 × 10^{-7} M PGE$_1$) and in vivo (rabbit, 0.4 μg/kg/min for 10 min). The possible physiological role of PGE$_1$ as a regulator of free fatty acid and glycerol release from adipose tissue is suggested by the greater activity shown by PGE$_1$ in hyperthyroid rats, and by the increased response to lipolytic agents (norepinephrine and theophylline) of adipose tissue obtained from rats deficient in essential fatty acids, which are the biological precursors of prostaglandins in mammalian tissues. (Authors modified) 0718

0419
PATON, D.M. and E.E. DANIEL
Contractile response of uterine horns to prostaglandin E$_1$.
Federation Proceedings. 26: 736. 1967.

The effect of PGE on isolated oestrogenised rat uterine horns was studied at 17-20°C. PGE always produced a contraction; tachyphylaxis was not observed. Atropine and phenoxybenzamine did not prevent responses to PGE. Contractions to PGE were calcium dependent; 20-40 minutes in Ca^{++} free Krebs abolished responses to supramaximal concentrations while responses to acetylcholine could still be elicited. In preparations rendered unresponsive to all agonists by exposure to Ca^{++} free Krebs and Na_2 EDTA, responses were partially restored by exposure to Ca^{++} (2.5 mM for 60 sec); however, the duration of this restoration was less for PGE than for acetylcholine. Responses to PGE occurred in depolarized preparations. Pretreatment with dinitrophenol and iodoacetic acid abolished contractile responses to PGE. Anoxia reduced responses to submaximal concentrations of PGE. These findings suggest that PGE acts on a specific receptor distinct from those for acetylcholine and serotonin; however, its action is not dependent on membrane depolarisation. PGE does not appear to release Ca^{++} from the sequestered Ca^{++} store and its oxygen dependence is greater than that of most other agonists. (Authors) 0798

0420
PATON, D.M. and E.E. DANIEL
    On the contractile response of the isolated rat uterus to prostaglandin E_1.
    Canadian Journal of Physiology and Pharmacology. 45: 795-804. 1967.

    The mechanism of action of PGE_1 on rat uterine horns was analyzed in vitro. Contractile responses to PGE_1 were not due to either release of known neurotransmitters or an interaction with their receptors. Responses could still be elicited in depolarized preparations. PGE_1 contractions were Ca^{++} dependent. Anoxia markedly reduced responses to PGE_1. Neither PGE_1 or acetylcholine were potentiated by elimination of external glucose or by lowering of the external K^+; both were still effective in depolarized horns; and pretreatment with adrenaline or 2,4-dinitrophenol and iodoacetic acid reduced responses to both. It was concluded that PGE_1 acts on a specific receptor (Authors modified) 0811

0421
PENTO, J.T. and R.J. CENEDELLA
    In vitro effects of prostaglandin E_1 and F_1 upon the glucose metabolism of ejaculated and epididymal ram spermatozoa.
    Pharmacologist. 9: 254. 1967.

    Abstract only. The effect of PGE_1 or $PGF_{1\alpha}$ (20 µg/ml) on the metabolism of glucose, and O_2 consumption in ejaculated or epididymal ram sperm was measured in vitro. Both PG's decreased the percentage of glucose oxidized to CO_2 by epididymal sperm. There was no effect on ejaculated sperm on O_2 consumption in either type of sperm. (JRH) 0882

0422
PICCINI, F. and P. POMARELLI
    Prostaglandina E_1 (PGE_1) e trasporto del calcio, (Esperienze eseguite in vitro secondo uno schema di particolare semplicita.) [Prostaglandin E_1 (PGE_1) and calcium transport. (In vitro experimentation in accordance with a particularly simple scheme.)]
    Bollettino della Societa Italiana di Biologia Sperimentale. 43: 1412-1414. 1967.

    The effect of adrenalin and PGE_1 on calcium transport in a membrane model composed of a chloroform-methanol extract of lipid fractions of guinea-pig heart was studied. When a Ringer solution containing Ca^{45} was brought into contact with the chloroform-methanol mixture, only a small quantity of calcium passed from the Ringer solution to the choloroform phase. A considerable quantity of calcium was fixed by the chloroform phase when myocardial mitochondria or microsomes were present. The effects of adrenalin and PGE_1 on this phenomenon were dose-related; adrenalin increased it slightly, PGE_1 more markedly. It is suggested that this dose-dependent effect on calcium transport may be due to the structure of PGE_1 which, as a higher fatty acid, may itself function as a lipid solvent in the presence of calcium. (MEMH) 0853

0423
PICKLES, V.R.
The myometrial actions of six prostaglandins: consideration of a receptor hypothesis.
In: Bergstrom, S. and B. Samuelsson, eds., "Prostaglandins. Proceedings of the Second Nobel Symposium," Stockholm, 1966. p. 79-83. New York, Interscience Publishers, 1967.

9β, 15ξ dihydroxyprost-13-enoic acid, $PGF_{1\alpha}$, $PGF_{1\beta}$, PGE_1, nor-PGE_1 and dihydro-PGE_1 were examined for stimulation and enhancement in the guinea pig uterus; stimulation and inhibition in human myometrial strip; and stimulation of the rabbit jejunum in vitro. Doses ranged between 1,000 and 0.3 ng/ml/ The results indicate the simple hypothesis that tissues contain PGE and PGF_α receptors (analogous to α-and β-adrenergic receptors) is not adequate to explain all the results. Some of the observed effects could be due to physiological effects such as facilitation of Ca^{++} movement or alteration of metabolic processes in the cell as well as pharmacologic effects. (JRH) 0728

0424
PICKLES, V.R.
The prostaglandins.
Biological Reviews. 42: 614-652. 1967.

This extensive review covers all aspects of prostaglandin research with sections on nomenclature, identification, sources, compounds with similar effects, endocrine control, biological actions and possible functions. There are 205 references. (JRH) 0701

0425
PICKLES, V.R.
Prostaglandins in the human endometrium.
In: Westin, B. and N. Wiqvist, eds., "Fertility and Sterility, Proceedings of the Fifth World Congress," Stockholm, 16-22 June 1966. p. 678-680. International Congress Series No. 133. New York, Excerpta Medica, 1967.

The author reviews several experiments which lead to the conclusion that PGF_1 and PGF_2 are the most important components of the menstrual stimulant found in human menstrual fluid. Preliminary experiments suggest that the PG content of the menstrual fluid may be under endocrine control. (JRH) 0833

0426
PICKLES, V.R.
Uterine suction during orgasm.
British Medical Journal. 1: 427. 1967.

It is suggested that human semen could contain a substance which affects the contractility of the uterus. Since prostaglandins are known to occur in the seminal fluid and have definite effects on the uterus, they could cause the "sucking" action reported by others during orgasm. (JRH) 0802

0427
PIKE, J.E., F.P. KUPIECKI and J.R. WEEKS
Biological activity of the prostaglandins and related analogs.
In: Bergstrom, S. and B. Samuelsson, eds., "Prostaglandins. Proceedings of the Second Nobel Symposium," Stockholm, 1966. p. 161-171. New York, Interscience Publishers, 1967.

Routes for preparing some natural prostaglandins and related structures in amounts sufficient for biological characterization are described. The results of the biological assays confirmed much of the earlier information. By suitable modification it was established that certain activities associated with PGE_1 could be eliminated while maintaining others. This could best be seen in the conversion of PGE_1 or PGE_2 to the corresponding 9-oxo-Δ^{10}-prostenoic acids which lost effects on smooth muscle and antilipolytic properties, but in which effects on blood pressure were retained or even increased. (Authors modified) 0717

0428
RAMWELL, P.W. and J.E. SHAW
Prostaglandin release from tissues by drug, nerve and hormone stimulation.
In: Bergstrom, S. and B. Samuelsson, eds., "Prostaglandins. Proceedings of the Second Nobel Symposium," Stockholm, 1966. p. 283-292. New York, Interscience Publishers, 1967.

Experiments on the occurrence of PG's in the brain; release of PG and acetylcholine from the cerebral cortex; differentiation between acetylcholine and PG release; PG release from innervated isolated tissue preparations; the souce of PG efflux; humoral stimulation of PG efflux and the relation of PG's to lipolysis are reviewed. (JRH) 0712

0429
ROBERT, A., J.E. NEZAMIS and J.P. PHILLIPS
Inhibition of gastric secretion by prostaglandins.
American Journal of Digestive Diseases. 12: 1073-1076. 1967.

Dogs with Pavlov or Heidenhain stomach pouches were used to measure the effect of 4 prostaglandins on the secretion of gastric acid and pepsin stimulated by food or histamine. PGE_1 and PGE_2 infused intravenously reduced gastric secretion (acid and pepsin) in all cases. The ED_{50} was about 0.5 μg/kg/min with 1.0 μg/kg/min causing almost total inhibition after 30 to 45 min of infusion. PGA_1 (1 μg/kg/min) did not reduce histamine-induced secretion (ED_{50} 0.08-0.1 μg/kg/min). $PGF_{2\alpha}$ was only tested against histamine-induced secretion and was inactive at 1 μg/kg/min. The highest doses of PG caused vomiting, defecation and urination. PG's might be endogenous regulators of gastric secretion. (JRH) 0795

0430
ROHLE, E.B., H. RETTBERG, H.H. DITSCHUNEIT, R. DOBERT and H. DITSCHUNEIT
Tierexperimentelle untersuchungen uber die beeinflussung des fett-und kohlenhydratstoffwechsels durch prostaglandin E_1. [Animal experiments investigating the influence of fat and carbohydrate metabolism by means of PGE_1.]
Medizinische Welt. 73: 1349. 1967.

Article in German. Abstract not available at present. 0901

0431
RUDDON, R.W. and J.M. JOHNSON
  The effect of prostaglandins on protein and nucleic acid synthesis in a cell free system.
  Life Sciences. 6: 1245-1252. 1967.

  The effects of PGE_1 and $PGF_{1\alpha}$ on protein and nucleic acid synthesis in a cell free system from *Escherichia coli* was studied by measuring the incorporation of radiolabelled phenylalanine into protein and labelled uridine into nucleic acids. At 10^{-4} M, neither PG caused a significant alteration in protein synthesis, but at lower concentrations (10^{-6}-10^{-7} M) there was a reduction in protein synthesis (14-18%). The PG's apparently worked at low concentrations by inhibiting protein synthesizing enzymes directly. These PG's had no effect on DNA or RNA synthesis directed by endogenous template, but did inhibit DNA synthesis directed by DNA primer extracted from calf thymus. PGE_1 and $PGF_{1\alpha}$ have only slight effects on the synthesis of macromolecules in a cell free system and if they have a role in intracellular control mechanisms, it must be at some other level of cellular integration. (JRH) 0786

0432
SAKAGAMI, T.
  [Prostaglandins.]
  Protein, Nucleic Acid, Enzyme. 12: 135-144. 1967.

  This review in Japanese is concerned with the synthesis of prostaglandins and their chemical structure. There are 69 references cited. (JRH) 0854

0433
SAMUELSSON, B.
  Biosynthesis and metabolism of prostaglandins.
  In: Kraitchevsky, P., R. Paoletti and D. Steinberg, eds., "Progress in Biochemical Pharmacology," p.
    59-70. New York, Karger, 1967.

  The author reviews the literature on the mechanism of the biosynthesis of prostaglandins. Based on available evidence a step by step synthesis is described. There is a brief discussion of some experiments on the metabolism of prostaglandins in biological systems. 0773

0434
SAMUELSSON, B.
  Chemistry, biosynthesis and metabolism.
  Pharmacologist. 9: 170. 1967.

  Abstract only. The present status of knowledge of the chemistry and nomenclature of the prostaglandins will be reviewed. The concersion (sic) of unsaturated fatty acid into prostaglandins will be discussed with respect to the mechanism of the conversion and factors influencing the composition of the product. The metabolism of prostaglandins in several systems will be described as well as some data on the biological significance of the transformations. (Author) 0878

0435
SAMULESSON, B., E. GRANSTROM and M. HAMBERG
  On the mechanism of the biosynthesis of prostaglandins.
  Abstracts of the 7th International Congress of Biochemistry, Tokyo, 19-25 August 1967. p. 739.

Abstract only. Stereospecifically tritium labelled precursors (13L-^{3}H-and 13D-^{3}H 8,11,14-eicosatrienoic acid) were prepared and it was shown (through measurement of isotope discrimination) that the initial step of the biosynthetic process involves the removal of the 13L hydrogen. Since neither 15-hydroxy- nor 15-hydroperoxy-8,11,13-eicosatrienoic acid is a precursor of PGE$_1$, the initial transformation seems to consist of a lipoxidase type or reaction with formation of 11-peroxy-8,12,14-eicosatrienoic acid. The latter compound is then converted into the endoperoxide which is converted into PGE$_1$ by removal of a hydrogen from C-9 and into PGF$_{1\alpha}$ through reductive opening of the endoperoxide. Some properties of the enzyme system including the stoichiometry of the cofactor oxidation will also be presented. (Authors modified) 0883

0436

SAMUELSSON, B., E. GRANSTROM and M. HAMBERG

On the mechanism of the biosynthesis of prostaglandins.

In: Bergstrom, S. and B. Samuelsson, eds., "Prostaglandins. Proceedings of the Second Nobel Symposium," Stockholm, 1966. p. 31-44. New York, Interscience Publishers, 1967.

A series of experiments designed to investigate the mechanism of the biosynthesis of PGE$_1$ and PGF$_{1\alpha}$ from 8,11, 14-eicosatrienoic acid are reviewed. A step by step mechanism is outlined with the evidence for each step given. (JRH) 0734

0437

SANBAR, S.S., S.A. MITCHELL, R.F. LOCKEY, E.A. VLCEK and J.A. GREENE, Jr.

Metabolic effects of hypotension induced by hemorrhage and by hypotensive drugs.

Clinical Research. 15: 410. 1967.

Abstract only. The effects of hypotension caused by hemorrhage or hypotensive drugs, including PGE$_1$, on plasma glucose, free fatty acid (FFA) and insulin was investigated in dogs. In all types of hypotension there was an increase in plasma glucose and no alteration in insulin levels. Also in all types of hypotension except that induced by infusion of PGE$_1$ (10 μg/kg/min) into the abdominal aorta, there was an increase in FFA. However, with PGE$_1$ there was a decrease in FFA. (JRH) 0840

0438

SANDBERG, F., A. INGELMAN-SUNDBERG and G. RYDEN

The effect of prostaglandins E$_1$, E$_2$, E$_3$, F$_{1\alpha}$, F$_{1\beta}$, F$_{2\alpha}$, and F$_{2\beta}$ on the human uterus and the fallopian tube.

In: Westin, B. and N. Wiqvist, eds., "Fertility and Sterility, Proceedings of the Fifth World Congress," Stockholm, 16-22 June 1966. p. 675-677. International Congress Series No. 133. New York, Excerpta Medica, 1967.

The fallopian tubes from 26 non-pregnant women were divided into 4 equal segments and the effect of 7 PG's on each segment recorded. PGE$_1$ caused a contraction of the longitudinal muscles in the proximal segments and relaxation in the rest of the tube. No such specific effect was observed in the circular muscles. Of the other PG's tested on the fallopian tubes, PGE$_3$ and PGF$_{2\beta}$ had inhibitory effects. PGF$_{1\alpha}$, PGF$_{1\beta}$ and PGF$_{2\alpha}$ had stimulatory effects with PGF$_{2\alpha}$ being the most potent stimulator. The non-pregnant uterus in vitro was less responsive to all the PG's than the tubes. PGF$_{1\alpha}$ and PGF$_{2\alpha}$ stimulated while PGF$_{1\beta}$ and PGF$_{2\beta}$ weakly inhibited the uterus. All the PGE's caused a pronounced relaxation of the uterus. (JRH) 0829

0439

SANDBERG, F., A., INGELMAN-SUNDBERG, I. JOELSSON and G. RYDEN
  Preliminary investigation on the absorption of prostaglandin E₁ from the human vagina.
  In: Bergstrom, S. and B. Samuelsson, eds., "Prostaglandins. Proceedings of the Second Nobel Sympo-
    sium," Stockholm, 1966. p. 91. New York, Interscience Publishers, 1967.

  Tritium labelled PGE₁ was dissolved in human seminal plasma and placed in the posterior fornix of 5
  human female volunteers. Urine was collected periodically during a 24 hr period through an indwelling catheter in the urinary bladder. The PG's were extracted with n-butanol and the radioactivity
  determined by liquid scintillation counting. On the basis of recovered radioactivity, it was concluded
  that 10-20% of the deposited PGE₁ was absorbed from the vagina and excreted into the urine during
  a 24 hr period. (JRH) 0726

0440

SAXE, B.D.
  Approaches to the synthesis of prostaglandin E₁.
  Dissertation Abstracts. 27: 3870-B. 1967.

  Abstract only. The author proposes the synthesis of PGE₁ starting with a bicyclic precursor (structure given). This intermediate could be derived from bicyclo (7,3,0) dodecan-2,10-dione. (JRH) 0858

0441

SEN, S., R.R. SMEBY and F.M. BUMPUS
  Isolation of a phospholipid renin inhibitor from kidney.
  Biochemistry. 6: 1572-1581. 1967.

  The phospholipid inhibitor of renin activity isolated from dog kidney was not a prostaglandin since
  PGE₁ and PGA₁ were inactive in bioassay system and others have shown that PGA₂ (medullin) is not
  an inhibitor of renin activity. (JRH) 0801

0442

SHAW, J.E. and P.W. RAMWELL
  Prostaglandin release from the adrenal gland.
  In: Bergstrom, S. and B. Samuelsson, eds., "Prostaglandins. Proceedings of the Second Nobel Sympo-
    sium," Stockholm, 1966. p. 293-299. New York, Interscience Publishers, 1967.

  Release of PGF₁α into venous effluents from stimulated adrenal glands was found to be associated
  with efflux of FFA but not with catecholamines. A number of prostaglandins were formed by
  incubating homogenates of rat adrenal glands and formation was increased by both ACTH and ACh.
  These results extend the correlation between release of prostaglandin and FFA from tissues stimulated by either drugs, nerves or hormones. (Authors) 0713

0443

SMEBY, R.R., S. SEN, and F.M. BUMPUS
  A naturally occurring renin inhibitor.
  Circulation. 20/21 (supp. 2): II-129-134. 1967.

  A phospholipid was extracted from dog kidneys which inhibits the production of angiotensin II by
  renin. Production of angiotensin by renin was not inhibited by PGE₁ or PGA₁. (JRH) 0750

0444

SMITH, E.R., J.V. McMORROW, Jr., B.G. COVINO and J.B. LEE
Mechanism of the hypotensive and vasodilator action of prostaglandin E_1.
Clinical Research. 15: 222. 1967.

Abstract only. The vasodilator action of PGE_1 on femoral arterial blood flow was investigated in anesthetized dogs. Comparison of dose response curves obtained following intraarterial injection revealed the following relative molar potencies: PGE_1 > acetylcholine = isoproterenol > histamine. Atropine, propranolol and pyribenzamine did not alter the response to PGE_1. These experiments demonstrate that PGE_1 is a potent peripheral dilator acting directly on vascular smooth muscle without apparent involvement of adrenergic, cholinergic or histaminergic components. (JRH) 0836

0445

SOLYOM, A., E. MUHLBACHOVA and L. PUGLISI
Incorporation of $[^{14}C_6]$ glucose and $[9,10^{-3} H]$ palmitic acid in vitro into lipids of adipose tisse from essential fatty acid-deficient rats.
Biochimica et Biophysica Acta. 137: 427-434. 1967.

Discussion of prostaglandins is confined to the introduction and discussion sections of this paper. Because of their powerful antilipolytic properties, it is suggested that the increased mobilization of free fatty acids observed in rats fed diets free of essential fatty acids could be due to lowered PG content, since essential fatty acids are precursors of PG's. (JRH) 0748

0446

STEINBERG, D. and M. VAUGHN
In vitro and in vivo effects of prostaglandins on free fatty metabolism.
In: Bergstrom, S. and B. Samuelsson, eds., "Prostaglandins. Proceedings of the Second Nobel Symposium," Stockholm, 1966. p. 109-121. New York, Interscience Publishers, 1967.

The role of PGE_1 on mobilization of free fatty acids (FFA) from adipose tissue, lipolysis in adipose tissue, and in vivo FFA levels is reviewed as is the site of action of PGE_1 in the lipase activating system. There are 33 references. (RMS) 0722

0447

STEINBERG, D.
Prostaglandins as adrenergic antagonists.
Annals of the New York Academy of Sciences. 139: 897-909. 1967.

This review is mainly concerned with the inhibitory properties of PG's on adrenergic agents. The overall conclusion is that PG's interfere with the formation of cAMP but not with the action of already formed cAMP. There are sections on: the nature and occurrence of PG's; biological activity; effects on lipolysis in vitro and in vivo; effects on blood glucose levels; effects on blood pressure; mechanism of action; and relationship to essential fatty acid deficiency. (JRH) 0897

0448

STOCK, K., E. BOHLE and E. WESTERMAN
Differential effects of prostaglandin E_1 on lipolysis.
In: Kraitchevsky, P., R. Paoletti and D. Steinberg, eds., "Progress in Biochemical Pharmacology," p. 122-128. New York, Karger, 1967.

To evaluate the effects of PGE_1 on lipolysis it was compared to other antilipolytic agents in rat adipose tissue from fed, fasted and alloxan diabetic rats in vitro. PGE_1 (2.5 μg/ml) and nicotinic acid reduced lipolysis in fed rats, but the β-sympatholytic Ko 592 was inactive. PGE_1 (2.5 μg/ml) also reduced the increased lipolysis in diabetic rats, but was ineffective against lipolysis due to fasting. Only nicotinic acid was effective in fasted rats. PGE_1 like the other antilipolytic agents was effective against norepinephrine induced lipolysis. PGE_1 (0.1 μg/ml) increased the uptake of ^{14}C labelled glucose and also the production of $^{14}CO_2$. (JRH) 0766

0449
STOVALL, R. and R.T. JACKSON
Prostaglandins and nasal blood flow.
Annals of Otology, Rhinology and Laryngology. 76: 1051-1059. 1967.

Rhinometric techniques were used to determine the effects of intraarterially administered PGE_1, PGE_2, PGA_1 or $PGF_{1\alpha}$ on resistance to air flow in the nasal passage of pentobarbital anesthetized dogs. A dose response curve to epinephrine (E) was determined in each dog before administration of PG. The animals showed considerable individual variability in magnitude and duration of response to both E and PG's. All the PG's caused an increase in nasal patency (vasoconstriction) with the PGE's being about equipotent to E on a weight basis in the magnitude of response but the responses to PG's were of a much longer duration (7 fold). PGA_1 and $PGF_{1\alpha}$ were about 100 times less potent than PGE's or E. Administration of aminophylline or papaverine, short acting vasodilators, ended the response to the long acting PG but pretreatment with these drugs had no effect on subsequent application of PG. Pretreatment with PG also had no effect on subsequent injection of E. The PG's were active when given intravenously but their effects on systemic blood pressure were much greater. (JRH) 805

0450
STRONG, C.G. and D.F. BOHR
Effects of prostaglandins E_1, E_2, A_1 and $F_{1\alpha}$ on isolated vascular smooth muscle.
American Journal of Physiology. 213: 725-733. 1967.

This study explores the effects of PGE_1, PGE_2, PGA_1 and $PGF_{1\alpha}$ on isolated smooth muscle from various levels of the vascular tree of rabbits, rats and dogs. Isolated aortic and coronary smooth muscle contracts in response to these PG's, but smooth muscle of small arteries (200-1,000 μ o d) from other sites shows a diphasic dose-response relationship with these compounds. Helical strips from small renal, muscular, and mesenteric arteries, partially contracted by catecholamines, plasma contracting factor, or KCI are relaxed by prostaglandins in low concentrations and contracted further by PG's in high concentrations. These effects are not abolished by α- and β-adrenergic blockade, atropine, LSD-25, or antihistamine. PG's mimic the effect of decreased Ca^{++} concentrations on the rate of rhythmic contraction of isolated rat portal vein and dog ureter strips. (Authors modified) 0780

0451
STRONG, C.G. and D.F. BOHR
Effects of prostaglandins E_1, E_2, A_1 and $F_{1\alpha}$ on isolated vascular smooth muscle.
Circulation. 35/36 (supp. 2): II-244-245. 1967.

Abstract only. PGE_1, PGE_2, PGA_1 and $PGF_{1\alpha}$ contracted aortic and coronary smooth muscle in vitro. However, small arteries (200-1,000 μ o.d.) from renal and mesenteric beds; partially contracted by catecholamines, plasma contracting factor or KCI; showed a biphasic response, relaxing with low

PG concentrations and contracting with higher concentrations. These effects were not blocked by α- or β-adrenergic blocking agents, d-LSD or antihistamine. PG's mimicked the effect of low Ca^{++} concentrations on rhythmic contractions of isolated rat portal vein and dog ureter strips. (JRH) 0818

0452
STRUIJK, C.B., R.K. BEERTHUIS and D.A. van DORP
   Specificity in the enzymatic conversion of poly-unsaturated fatty acids into prostaglandins.
   In: Bergstrom, S. and B. Samuelsson, eds., "Prostaglandins. Proceedings of the Second Nobel Symposium," Stockholm, 1966. p. 51-56. New York, Interscience Publishers, 1967.

The specificity of prostaglandin synthetase from sheep vesicular glands was studied by allowing it to react with dihomo-γ-linolenic and arachidonic acid as well as higher and lower homologs and positional isomers of these PG precursors. The rate of conversion of the various substrates into PG's was also measured. The natural precursors showed the highest rate of conversion to PG's, but several of the other fatty acids were converted to PG's with the percentage being determined by the chain length and position of double bonds. When the PG's produced were compared to authentic PGE_1 by bioassay on the guinea pig ileum, only the ones that corresponded to PGE_1 and PGE_2 showed biological activity similar to PGE_1. Thus the response of the tissue to PG seems to be much more specific than the enzyme system which produces the PG's. (JRH) 0732

0453
SWIDERSKA-KULIKOWA, B.
   Prostaglandyny. [Prostaglandins.]
   Polskie Archiwum Medycyny Wewnetrznej. 39: 837-840. 1967.

This brief review in Polish covers all aspects of prostaglandin research. Special consideration is given to the pharmacology of the prostaglandins. There are 21 references cited. (JRH) 0852

0454
SWIDERSKA-KULIKOWA, B.
   Badania nad mechanizmem dzialania substancji hipotensyjnej z rdzenia nerek. [Studies of the mechanism of action of the renal medullary hypotensive principle.]
   Polskie Archiwum Medycyny Wewnetznej. 39: 465-474. 1967.

Article in Polish. Abstract not available at present. 0851

0455
VAUGHAN, M.
   An effect of prostaglandin E_1 on glucose metabolism in rat adipose tissue.
   In: Bergstrom, S. and B. Samuelsson, eds., "Prostaglandins. Proceedings of the Second Nobel Symposium," Stockholm, 1966. p. 139-142. New York, Interscience Publishers, 1967.

Both PGE_1 (0.4 or 1.0 $\mu g/ml$) and insulin stimulated glucose uptake in rat epididymal fat pads in vitro. In the presence of epinephrine (E) or TSH, both of which also stimulated glucose uptake, the effects of PGE_1 and E could still be seen. PGA_1 (1.0 $\mu g/ml$) did not alter glucose uptake or free fatty acid (FFA) release. At 5 $\mu g/ml$, in the presence of E, PGA_1 significantly inhibited FFA release but did not alter glucose uptake. Both PGE_1 and insulin produced an increase in incorporation of glucose carbon (^{14}carbon labelled glucose) into glycerides, mainly by increasing its incorporation into the

FFA fraction and having relatively no effect on the glycerol fraction. This effect could be seen in the presence of E. PGA_1 (1.0 $\mu g/ml$) under the same conditions was inactive. Both PGE_1 and insulin caused an increase in glucose carbon incorporation into glycogen. In all of these experiments, PGE_1 was much less effective than insulin in producing these effects. The similarity of action of both agents suggests that their mechanisms may be related. (JRH) 0720

0456

VERGROESEN, A.J., J. de BOER and J.J. GOTTENBOS

Effects of prostaglandins on perfused isolated rat hearts.

In: Bergstrom, S. and B. Samuelsson, eds., "Prostaglandins. Proceedings of the Second Nobel Symposium," Stockholm, 1966. p. 211-218. New York, Interscience Publishers, 1967.

Isolated rat hearts were perfused with different concentrations of a range of prostaglandins. The criteria used were the contractile force of the myocard, the frequency of the heart beat and the amount of fluid perfused through the coronary vascular system. PGE_1, PGE_2, PGA_1, $PGF_{1\alpha}$ and $PGF_{1\beta}$ appeared to be active in concentration of 0.01 $\mu g/ml$ and higher. The three PG's acted mainly as vasodilators, PGE_1 being the most active, and had little or no effect on the frequency of the heart beat. The effect of PGE_1 on the contractile force was variable; mostly a moderate increase was observed but sometimes a definite decrease occurred. On the other hand the perfusion with $PGF_{1\alpha}$ had practically no vasodilating effect but resulted in a sharp increase in mechanical activity of the myocard without causing a change in the frequency of the heart beat. $PGF_{1\beta}$ was much less active than the same concentration of $PGF_{1\alpha}$. Several other isomers and homologues of PGE_1 and PGE_2 were tested and found to be inactive in this experimental set-up. (Authors modified) 0703

0457

VOGT, W.

Release of prostaglandins by venoms and by endogenous mechanisms.

In: Leonardi, A. and J. Walsh, eds., "International Symposium on Drugs of Animal Origin," p. 29-33. Milan, Ferro Edizione, 1967.

The author reviews his own work and the work of others with SRS-C, including discussion of the available evidence that SRS-C contains prostaglandins. Also mentioned is evidence about the origin of PG's from tissue phosphatides, the mechanism of formation and the release of PG's by endogenous processes. It is concluded that prostaglandins are released when venoms act on lung tissue and that the liberation of prostaglandins is a two step process involving a phospholipase and endogenous enzymes which appears to be the same whether the release of PG's is in response to SRS-C or endogenous processes. (RAP) 0894

0458

VOGT, W. and B. DISTELKOTTER

Release of prostaglandin from frog intestine.

In: Bergstrom, S. and B. Samuelsson, eds., "Prostaglandins. Proceedings of the Second Nobel Symposium," Stockholm, 1966. p. 237-240. New York, Interscience Publishers, 1967.

Whole isolated frog gastro-intestinal tracts were incubated in frog Ringer solution or distilled water for up to 180 min. The intestines were extracted for PG's and the incubation medium was extracted in the same way. PG was released from the intestine into the Ringer solution for about 1 hr, reaching a maximal level equivalent to 17 μg PGE_1/g tissue which is about 3 times the amount that could be extracted from the tissue before incubation. Thus most of the released PG was formed during incubation. Incubation in distilled water led to a prolonged release of PG (180 min) and a greater total release of PG (27 μg PG/g tissue). When the frogs were given a paralyzing dose of cocaine (10 mg) 10 to 15 min before decapitation and removal of the intestine, the release of PG was much lower (5 μg PG/g tissue) and the PG content of the intestinal tissue was much greater, 12 μg/g compared to 6 μg/g in tissue from untreated controls. It is suggested that PG formation is triggered by activation of phospholipase A. (JRH) 0706

0459
WAITZMAN, M.B., W.R. BAILEY, Jr. and C.G. KIRBY
Chromatographic analysis of biologically active lipids from rabbit irides.
Experimental Eye Research. 6: 130-137. 1967.

Extraction of lipids from rabbit irides and subsequent gas-liquid and thin-layer chromatographic analysis of these extracts revealed the presence of a very low concentration of a biologically active substance(s). The active substance(s) represents one or more derivatives of prostanoic acid containing one or more hydroxyl groups in the molecule. Because these materials had little effect on ocular pressure, but did reduce pupil size in rabbits, it is likely that the active material is in the prostaglandin F family of compounds. (Authors) 0888

0460
WAITZMAN, M.B. and C.D. KING
Mechanisms of prostaglandin influences on ocular pressure and pupil size.
Federation Proceedings. 26: 655. 1967.

Abstract only. PGs, when injected into the anterior chamber (100ng dose range) of cat, rabbit and monkey eyes, cause a long acting miosis but elevate ocular pressure only in the rabbit eye. A physiological antagonism appears to exist between the PGs and adrenergic agonists, with the latter generally predominating but both likely acting at different receptor sites. Studies with adrenergic antagonists in combination with adrenergic agonists and PGs suggest different mechanisms in PG's ocular pressure and iris muscle action. Neither propranolol nor phenoxybenzamine interfered with the ocular pressure elevation in rabbit, but propranolol prevents the PG-miosis in rabbits. PG, then, may have some β-adrenergic receptor sites. (Authors modified) 0797

0461
WAITZMAN, M.B. and C.D. KING
Prostaglandin influences on intraocular pressure and pupil size.
American Journal of Physiology. 212: 329-334. 1967.

PGE_1, 0.10 μg or PGE_2 administered into the anterior chamber of the urethane anesthetized rabbit eye causes a large elevation in intraocular pressure (IOP) and also miosis. The threshold dose for both effects is about 0.01 μg. $PGF_{1\alpha}$ in doses of 5 μg, administered in a similar manner, causes none of the above effects although 125 μg $PGF_1{}^{\alpha}$ so administered causes miosis but no significant IOP elevation. In cats, 2.5 μg PGE's given the same way do not affect IOP but do elicit a long-lasting miosis, even in the atropinized eye. Intravenously, PGE's raise the IOP while simultaneously lowering blood pressure; the iv threshold dose is about 0.25 μg. The miotic effect caused by PG is interpreted as meaning some direct effect on iris smooth muscle, and the IOP change in the rabbits is interpreted as meaning an elevation of aqueous humor production with intraocular vasodilation possibly playing some contributory role. PG does not appear to reduce the rate of outflow of aqueous humor and, in fact, may cause a slight increase in this rate. (Authors modified) 0778

0462
WEEKS, J.R.
Pharmacology of the prostaglandins with emphasis upon the cardiovascular system.
Pharmacologist. 9: 171. 1967.

Abstract only. Both vagal and transmural stimulation of the isolated rat stomach increased the release of PG (assayed as PGE_1) from the serosal surface. Hyoscine completely blocked the vagally-induced increase in PG secretion. When both sympathetic and parasympathetic fibers were stimulated simultaneously, the increase normally seen with parasympathetic stimulation was blocked. Anoxia was the only condition which reduced basal secretion of PG from the stomach. It was proven that PGE_2 and $PGF_{2\alpha}$ are synthetized in the stomach. While nerve stimulation was not necessary for basal secretion of PG, it seems likely that both adrenergic and cholinergic stimulation may modulate formation and release. (JRH) 0876

0463
WEEKS, J.R. and R.A. WALK
Reduction of variability between replicates in assay of hormone stimulated lipolysis: a study of prostaglandin E_1 (PGE_1) inhibition.
Pharmacologist. 9: 254. 1967.

Abstract only. A technique for measuring lipolysis in rat adipose tissue based on diced fat pads rather than snips of tissue is described. The new technique as compared to the old method showed much less variability, and a higher average lipolytic rate. With epinephrine stimulated lipolysis, PGE_1 (3.2-100 ng/ml) produced a dose related inhibition of lipolysis (up to 41%). (JRH) 0881

0464
WIQVIST, N., M. BYGDEMAN and S. KWON
Inverkan av prostaglandin E_1 pa den gravida uterus. [Action of prostaglandin E_1 on the gravid uterus.]
Nordisk Medicin. 77: 677. 1967.

Article in Swedish. Abstract not available at present. 0820

0465
WOLFE, L.S., F. COCEANI and C. PACE-ASCIAK
Brain prostaglandins and studies of the action of prostaglandins on the isolated rat stomach.
In: Bergstrom, S. and B. Samuelsson, eds., "Prostaglandins. Proceedings of the Second Nobel Symposium," Stockholm, 1966. p. 265-275. New York, Interscience Publishers, 1967.

PG-like compounds were released from the superfused cat cerebellum in vivo. Cerebral cortical tissue contained chiefly $PGF_{2\alpha}$. The tissue content of PG's decreased in severe hypoxia. Incubation of cerebral cortex homogenates or microsomes in the presence of oxygen increased 5- to 7-fold the amount of $PGF_{2\alpha}$ present but had little effect on the amount of $PGF_{1\alpha}$. Experiments with labelled arachidonic acid indicated that direct conversion to brain PG's was very small, and the formation of PG's in the brain arose almost entirely from endogenous tissue precursors. PGE_1 acted directly on the smooth muscle of the rat stomach. In circumstances when the sympathetic fibers, terminals or receptors were inhibited, PG action was potentiated. Catecholamines inhibited PG contractions. PG's did not contract the muscle in the absence of oxygen. The action of PG's is closely linked to enzyme reactions which require oxygen. The dose response curves for PG action on the stomach muscle preparation in high potassium and high and low calcium environments indicated that the ultimate effect of PG's is in some way associated with the release of calcium to initiate contraction. (Authors modified) 0710

0466
WOLFE, L.S., F. COCEANI, C. PACE-ASCIAK
The relationship between nerve stimulation and the formation and release of prostaglandins.
Pharmacologist. 9: 171. 1967.

Abstract only. The relative activity of several PG's is reviewed. Special attention is given to their effects on the circulatory system. (JRH) 0870

0467

ABE, K., T. MOURI, T. SEKI, M. SUZUKI, T. TAKANO and K. YOSHINAGA
An improved bioassay method for kinins.
Experientia. 24: 455-457. 1968.

A technique for the bioassay of kinins was developed using the rabbit hind-quarter. The effects of other vasoactive agents including PGE_1 were studied in this preparation. PGE_1 (200 ng) produced a long lasting vasodilation. Only 1 or 2 ng of bradykinin were needed to produce a similar vasodilation in this test system. (JRH) 1185

0468

ABE, K.
[Prostaglandins.]
Japanese Journal of Clinical Medicine. 26: 1901-1907. 1968.

Article in Japanese. Article not available at present. 1262

0469

ADOLPHSON, R.L., T.J. KENNEDY and R.G. TOWNLEY
Relative effects of theophylline and isoproterenol on respiratory smooth muscle in man and other species.
Clinical Research. 16: 471. 1968.

Abstract only. The responses of human tracheal muscle strips in vitro to isoproterenol and theophylline were compared with the responses of similar preparations from rats, dogs and guinea pigs. Isoproterenol showed considerable variability in its ability to relax contracted muscle strips from the various species, while theophylline was equally effective in all species. $PGF_{2\alpha}$ (10^{-5} M) partially inhibited the relaxing effects of isoproterenol and theophylline in human tissue. Potentiation of methacholine contraction with $PGF_{2\alpha}$ (10^{-5}-10^{-7} M) occurred in the human and rat, but not in the guinea pig. The marked species variation with the effect of isoproterenol in contrast to the universal effect of theophylline implies different modes of action. (JRH) 1203

0470

ALFIN-SLATER, R.B. and L. AFTERGOOD
Essential fatty acids reinvestigated.
Pharmacological Reviews. 48: 758-784. 1968.

A section on prostaglandins reviews about 30 documents on: sites of prostaglandin synthesis, conversions of EFA to prostaglandins, and the physiological roles attributed to these compounds. (RMS) 1064

0471

AMBACHE, N. and H.C. BRUMMER
A simple chemical procedure for distinguishing E from F prostaglandins with application to tissue extracts.
British Journal of Pharmacology and Chemotherapy. 33: 162-170. 1968.

This technique for distinguishing PGE from PGF depends on the keto group present in the PGE group and absent in PGF compounds. Water soluble hydrazine derivatives are reacted with the keto group using Girard's reagent T (trimethylammonium-acetohydrazide chloride) which makes the resultant hydrazones more water and less ether soluble than unreacted PGE compounds. Non-ketonic PGF does not react and therefore does not change in solubility. The T reagent is not soluble in the ether and therefore does not interfere with the bioassay test for the presence of prostaglandin. (JRH) 1012

0472

AMBACHE, N. and M.A. FREEMAN
 Atropine-resistant longitudinal muscle spasms due to excitation of non-cholinergic neurones in
  Auerbach's plexus.
 Journal of Physiology. 199: 705-727. 1968.

Using isolated guinea pig ileum muscle with an intact Auerbach's plexus it was shown that two types of neurones are present. The first type produces a rapid twitch response which can be blocked by cholinergic blocking agents such as atropine and a second delayed tetanic response which could not be blocked by atropine. It was concluded that the tetanic response is not mediated by prostaglandin because a prostaglandin blocking agent (patulin) was able to block the effects of applied PGE_2, and $PGF_{2\alpha}$ did not reduce tetanic response. In fact, PGE_2 potentiated the tetanic responses even in the presence of the inhibitor. (JRH) 1037

0473

ANGGARD, E., F. MATSCHINSKY and B. SAMUELSSON
 Enzymatic assay of the prostaglandins.
 British Journal of Pharmacology and Chemotherapy. 34:190P. 1968.

Abstract only. In the present communication the general priniciples for the enzymatic analysis of prostaglandins are discussed. Using direct fluorimetric measurement of the generated NADH, 10^{-10}-10^{-8} moles of prostaglandins have been assayed. By coupling the prostaglandin dehydrogenase catalysed reaction to an amplifying system (Lowry, Passoneau, Schultz & Rock, 1961), the sensitivity of the assay is further increased to a lower limit of about 10^{-12} moles. (Authors modified) 1045

0474

ANONYMOUS
 Essential fatty acids: structural requirements for biological activity.
 Nutrition Reviews. 26: 187-189. 1968.

This review mentions that the role of both linoleate and linolenate as precursors of prostaglandins has been clearly established but that PG's have no curative effect on the skin symptoms of essential fatty acid deficiency. (RAP) 1233

0475

ANONYMOUS
 Hemodynamic balance and the kidney.
 Journal of the American Medical Society. 206: 336. 1968.

In this editorial review the likely involvement of prostaglandins and renal hypertension are discussed. One reference to this effect is cited. (JRH) 1065

0476
ANONYMOUS
Pharmacology of the carcinoid syndrome.
Lancet. 1: 404-405. 1968.

It is mentioned that prostaglandins may be involved in the flushing and gastrointestinal symptoms observed in patients with carcinoid tumors. (JRH) 1067

0477
ANONYMOUS
Prostaglandins.
British Medical Journal. 4: 657-658. 1968.

Prostaglandin research is briefly reviewed with special attention given to the use of prostaglandins for the induction of labor. Also discussed is the physiological role of prostaglandins in natural labor. (JRH) 1004

0478
ANONYMOUS
Les Prostaglandines. [Prostaglandins.]
Concours Medical. 90: 829-834. 1968.

After a brief introduction including the history and structure of prostaglandins, various areas of research are reviewed. The role of prostaglandins in labor, their antihypertensive effect on the kidneys and their relationship to cAMP are mentioned. Two references are cited. (NES) 1312

0479
ANONYMOUS
Prostaglandins.
Journal of the American Medical Society. 203: 592-593. 1968.

This is a brief editorial review of the field of prostaglandin research; 4 references are cited. (JRH) 1063

0480
ANONYMOUS
Prostaglandins.
Lancet 1: 30-32. 1968.

This leading article briefly reviews prostaglandin research with special consideration of prostaglandins and the central nervous system. (JRH) 1071

0481
ANONYMOUS
Upjohn chemists and prostaglandins.
Chemical and Engineering News. 46: 7. 1968.

In describing the total synthesis of prostaglandins by E. J. Corey in a previous issue of *Chemical and Engineering News*, no mention was made of earlier syntheses of prostaglandins by other chemists. This is used as an example of the care that reporters must exercise in the sensitive area of giving proper credit in scientific news coverage. (JRH) 1332

0482
ANONYMOUS
 Utilization of glucose and palmitic acid in adipose tissue of rats deficient in essential fatty acids.
 Nutrition Reviews. 26: 49-51. 1968.

 Since essential fatty acids are prostaglandin precursors, it is suggested that the effects of an essential fatty acid free diet could be due to the reduced synthesis of prostaglandins. (JRH) 1196

0483
ARFORS, K-E, H. C. HINT, D. P. DHALL and N. A. MATHESON
 Counteraction of platelet activity at sites of laser-induced endothelial trauma.
 British Medical Journal. 4: 430-431. 1968.

 The effect of PGE_1 and several other anticoagulant drugs on platelet activity following endothelial trauma was studied in rabbits. A small arteriole (20-29 μ in diameter) was traumatized with a biolaser. The number of microemboli formed were counted. PGE_1 (25 and 125 μg/kg) given intravenously had no detectable effect on platelet function. (JRH) 1171

0484
BARBORIAK, J. J., R. C. MEADE, J. OWENBY and R. A. STIGLITZ
 Blockers of free fatty acid release and postprandial lipemia.
 Archives Internationales de Pharmacodynamie et de Therapie. 176: 249-254. 1968.

 Prostaglandins are mentioned as being compounds which can inhibit the catecholamine stimulated release of fatty acids from adipose tissue. (JRH) 1180

0485
BARRY, E. and W. J. HALL
 Prostaglandin E_1 and sodium ion movement across frog skin.
 Irish Journal of Medical Sciences. 1: 327. 1968.

 Abstract only. Prostaglandin E_1 when added to the solution bathing either the inside or the outside of a sheet of frog skin, increases the short-circuit current. In eleven experiments prostaglandin E_1 in a dose of 0.2 g. per ml. of inside solution produced a range of increases in short-circuit current varying from 18 to 108% of the resting values with a mean of 50. In most experiments the response reached its peak within thirty minutes of applying the drug and was sustained. An increase in short-circuit current does not necessarily imply an increased sodium ion influx. For example C1-secretion by skin glands could be responsible. However, in five experiments prostaglandin E_1 added to the inside solution caused large and sustained increases in short-circuit current; the calculated increase in ion transport from the short-circuit current agreed, within the limits of experimental error, with the calculated increase in gross sodium ion movement inwards from ^{24}Na studies. It is concluded that the increase in short-circuit current, using prostaglandin E_1, is due solely to an increase in sodium ion movement inwards across the skin. (Authors modified) 1302

0486
BARTELS, J., W. VOGT, and G. WILLE
 Prostaglandin release from and formation in perfused frog intestine.
 Nauyn-Schmeidebergs Archiv fur Pharmakologie and Experimentelle Pathologie. 259: 153-154. 1968.

 At a perfusion rate of 0.5 ml/min the average content of prostaglandin in the perfusate of isolated frog intestine ranged from 1 to 30 μg/min, depending on the season. Injection or infusion of acetycholine (10 μg/ml) increased the amount 1.9 fold. Dimethyl-phenyl-piperazinium iodide raised the concentration in the same way. From the results described above, no decision can be made whether the increase in prostaglandin after pharmacological stimulation is due simply to liberation from stores or to new formation. However, a quick formation is possible. Injection of arachidonic acid into perfused frog intestine preparation leads to a tenfold or even greater increase in prostaglandin content of effluent. Linoleic acid and lysolecithin have no comparable effect. After injection of phospholipase A, the prostaglandin release increases 20 to 40 times. This supports the hypothesis that prostaglandins are formed in a two-step raction and that the rate of formation of prostaglandins in frog intestine depends mainly on the level of substrate, i.e. on the activity of endogenous phospholipase. (JRH) 1224

0487

BASS, P. and D. R. BENNETT

Local chemical regulation of motor action of the bowel-substance P and lipid-soluble acids.

In: Code, C. F., ed., "Handbook of Physiology," p. 2193-2212. Washington, American Physiological Society, 1968.

This review article contains a section on prostaglandins. The subheadings are: identification and chemistry; extraction and bioassay; occurrence, biosynthesis, metabolism; biological actions; and significance. (JRH) 1090

0488

BECK, L., A. A. POLLARD, J. N. HARBO and T. M. SILVER

Does prostaglandin mediate sustained dilatation?

In: Ramwell, P. W. and J. E. Shaw, eds., "Prostaglandin Symposium of the Worcester Foundation for Experimental Biology," p. 295-307. New York, Interscience, 1968.

Electrical stimulation of the left lumbar sympathetic chain of dogs resulted in a vasodilation in the left hind leg which could be divided into two components. The first or transient component appeared at low frequencies (0.1-0.3 cps) of stimulation and resulted in a rapid vasodilation. The second or sustained component appeared at frequencies of 0.6 cps and above and persisted throughout the period of stimulation. The sustained dilation was different from the transient dilation since it was not blocked by atropine, antihistamine, antiserotonin, or β-adrenergic blocking agents. PGE_1 (0.0108-0.224 μg/min) or PGE_2 (0.0108-0.615 μg/min) infused intra-arterially produced a dilation similar to that of the sustained type. Furthermore, constant infusion of PGE_1 at basal levels greatly reduced the sustained dilations induced by sympathetic stimulation or infusion of PGE_1 itself, while the transient dilation induced by nerve stimulation showed only a proportional reduction. These results suggest that PGE_1 or PGE_2 may be mediators of the sustained dilation component of the response to sympathetic nerve stimulation. (JRH) 1145

0489

BEDWANI, J. R. and E. W. HORTON

The effects of prostaglandins E_1 and E_2 on ovarian steroidogenesis.

Life Sciences. 7: 389-393. 1968.

Chopped rabbit ovaries were divided into three flasks for incubation. A mixture of human chorionic gonadotropin and pregnant mare's serum was added to 2 of the flasks. PGE_1 (1 or 10 μg/ml) or PGE_2 (1or 2 μg/ml) was then added to 1 of the flasks with gonadotropins. After incubation the tissue was homogenized and the steroids extracted and analyzed by thin-layer chromatography. There were significant increases in production of progesterone and 20α-hydroxypregn-4-en-3-one in response to the gonadotropins. Neither PGE_1 or PGE_2 caused any inhibition of steroid production and in 5 out of 6 experiments PGE_2 caused a substantial increase in 20α-hydroxypregn-4-en-3-one. It is concluded that PGE_1 and PGE_2 do not inhibit ovarian cAMP. (JRH) 1078

0490
BEHRMAN, S. J. and W. BURCHFIELD
The intrauterine contraceptive device and myometrial activity.
American Journal of Obstetrics and Gynecology. 100: 194-202. 1968.

Prostaglandins are mentioned only in the discussion section of this paper. It is noted that they may be responsible for the increased uterine contractility seen during the menstrual period. (JRH) 1119

0491
BERMAN, H.J. and G.R. SIGGINS
Neurogenic factors in the microvascular system.
Federation Proceedings. 27: 1384-1390. 1968.

The retrolingual membrane of MS-222 anesthetized frogs was exposed by dissection so that the small arterioles could be observed microscopically. The effects of several topically applied vasoactive substances including PGE_1 on the retrolingual arterioles were measured. Of all the agents studied, PGE_1 was by far the most potent. As little as 10^{-11} g/ml produced threshold dilation. The latent period for PGE_1 (10^{-11} to 10^{-10} g/ml) was 6 to 15 sec, and the dilation lasted 24 to 47 sec. PGE_1 at 10^{-8} g/ml caused a dilation that usually persisted for an hour or more. (JRH) 1251

0492
BELL, C.
Dual vasoconstrictor and vasodilator innervation of the uterine arterial supply in the guinea pig.
Circulation Research. 23: 279-289. 1968.

PGE_1 is mentioned in the discussion section as being a possible non-cholinergic vasodilatory substance. (JRH) 1038

0493
BENNETT, A., K.G. ELEY and G.B. SCHOLES
Effect of prostaglandins E_1 and E_2 on intestinal motility in the guinea-pig and rat.
British Journal of Pharmacology. 34: 639-647. 1968.

PGE_1 and PGE_2 affected intestinal activity both in vitro and in vivo. Serosal application of prostaglandin (0.1-1.0 μg/ml) to guinea-pig isolated ileum stimulated the longitudinal muscle but reduced peristaltic contractions of the circular muscle and the propulsion of fluid through the gut. Intraluminal application (0.5-10 μg/ml) had little effect. Injection of prostaglandin (1-2 μg/kg) into the bloodstream of urethane anaesthetized rats stimulated the longitudinal muscle of the ileum and increased the intraluminal pressure. A similar response sometimes occurred in the guinea-pig, but in general the effect was variable. Release of prostaglandin in the gut wall, but probably not into the blood or into the lumen of the gut, may play a part in controlling intestinal motility. (Author's modified) 1013

122

0494

BENNETT, A., K.G. ELEY and G.B. SCHOLES

Effects of prostaglandins E_1 and E_2 on human, guinea-pig and rat isolated small intestine.

British Journal of Pharmacology. 34: 630-638. 1968.

PGE$_1$ and PGE$_2$ (0.01 to 20 μg/ml) contracted the longitudinal muscle of human guinea-pig and rat isolated ileum. The site of action varied with the species. In the rat and in some strips of human tissue prostaglandin appeared to have only a direct action on or in the muscle cells. In the other strips of human tissue and in guinea-pig ileum the prostaglandins seemed to stimulate both the intrinsic cholinergic nerves and the muscle cells. In contrast to the longitudinal muscle, the circular muscle of human, guinea-pig and rat isolated ileum was usually inhibited by prostaglandin, apparently by an action directly on the muscle cells. Prostaglandins may play a part in the control of intestinal motility. (Authors modified) 1015

0495

BENNETT, A., J.G. MURRAY and J.H. WYLLIE

Occurrence of prostaglandin E_2 in the human stomach, and a study of its effects on human isolated gastric muscle.

British Journal of Pharmacology and Chemotherapy. 32: 339-349. 1968.

The authors previously isolated a smooth muscle stimulating substance in extracts of human stomach mucosa; which is shown in this paper to be PGE$_2$ by bioassay and a combination of chromatographic techniques. The concentration present in mucosa was of the order of 1 μg/g of wet tissue. Concentrations were similar in mucosa of the body and the pyloric part of the stomach. Lesser amounts were found in the submucosa and only traces were found in the stomach muscle. When isolated strips of circular and longitudinal muscle were exposed to PGE$_1$ and PGE$_2$ in vitro, both PG's inhibited the spontaneous and acetylcholine stimulated contraction in circular muscle but stimulated the contractions of longitudinal muscle. Doses as low as 4-10 ng/ml were effective in producing inhibition in circular muscle and only 2 ng/ml caused contractions in longitudinal muscle. Longitudinal muscle from the antrum gave a more varied response than that from the body of the stomach. Experiments with blocking agents indicated that the PG exerts its effect directly on the muscle cells. (JRH) 1022

0496

BENNETT, A., J.G. MURRAY and J.H. WYLLIE

A prostaglandin in the human stomach.

Gastroenterology. 55: 140-141. 1968.

Abstract only. By combined chromatographic and bioassay techniques the human gastric mucosa was shown to contain PGE$_2$ (0.01-2.4 μg/g wet tissue). The PGE$_2$ content in the mucosa from the gastric body and antrum were similar, but the mucosa contained 4 times as much PGE$_2$ as the sub-mucosa and the muscle layers contained only traces. When circular and longitudinal muscle strips from the antrum and body of the stomach were exposed to PGE$_1$ and PGE$_2$ the spontaneous circular muscle activity was depressed as was its response to acetycholine and KC1. In contrast these prostaglandins excited all longitudinal muscle strips from the body although those from the antrum showed a more variable response. This excitation was not blocked by acetylcholine and serotonin antagonists, nor by hexamethonium and cocaine. These results indicate that prostaglandins act directly on stomach muscle. (JRH) 1098

0497
BENNETT, A.
Relationship between in-vitro studies of gastrointestinal muscle and motility of the alimentary tract in vivo.
American Journal of Digestive Diseases. 13: 410-414. 1968.

Mention is made of the possible roles of prostaglandins in intestinal motility. (JRH) 1123

0498
BERGSTROM, S. and B. SAMUELSSON
The prostaglandins.
Endeavour. 27: 109-113. 1968.

The history of prostaglandin research is briefly reviewed. Sections are included on their chemical, biochemical and physiological properties; 6 references are cited. (NES) 1177

0499
BERGSTROM, S., L.A. CARLSON and J.R. WEEKS
The prostaglandins: a family of biologically active lipids.
Pharmacological Reviews. 20: 1-48. 1968.

This review article contains sections on all aspects of prostaglandin research. This includes chemistry; actions on smooth muscles in vitro; smooth muscle organs; reproductive system; nervous system; cardiovascular system; and relationships to lipid and carbohydrate metabolism. 344 references are listed. (JRH) 1001

0500
BERNIER, J.J., J.C. RAMBAUD, D. CATTAN and A. PROST
Medullary carcinoma of the thyroid associated with diarrhoea: Report of five cases.
Gut 9: 726. 1968.

Abstract only. It is possible that the diarrhea found in many cases of medullary carcinoma of the thyroid could be due to secretion of prostaglandins, however, this will require further confirmation. (JRH) 1219

0501
BERTI, F., C. GALLI, P. POMARELLI and M.M. USARDI
Attivita' della prostaglandina E_1 (PGE$_1$) sull'automatismo della vena porta "in vitro." [Effect of prostaglandin E_1 (PGE$_1$) on the mechanism of the portal vein "in vitro."]
Rassegna Medica Sarda. 71: 445-450. 1968.

PGE$_1$ effect on portal vein isolated "in vitro" from cat, rabbit, guinea pig and rat has been studied. Having stressed the species specificity of activity of this compound, the authors briefly discuss its possible mechanism of action. (Authors) 1257

0502
BIRON, P.
Vasoactive hormone metabolism by the pulmonary circulation.
Clinical Research. 16: 112. 1968.

Abstract only. The effect of the pulmonary circulation on several vasoactive substances was investigated. The effects of injecting vasoactive hormones into the jugular vein was compared

124

with the effect of injecting them into the ascending aorta of rats, rabbits, cats and dogs. The pulmonary extraction percentages for prostaglandin and bradykinin exceeded 80%. Since other vasoactive substances were either not removed or were activated by the pulmonary circulation, it was concluded that one of the functions of the pulmonary circulation is selective removal of some vasoactive substances. (JRH) 1211

0503
BIZZI, A., A.M. CODEGONI, A. LIETTI and S. GARATTINI
Different responses of white and brown adipose tissue to drugs affecting lipolysis.
Biochemical Pharmacology. 17: 2407-2412. 1968.

Rats were treated with several drugs known to affect lipolysis. At various intervals after treatment plasma was collected and the epididymal (white) and scapular (brown) fat removed for subsequent chemical measurement of free fatty acid (FFA) content. It was found that white adipose tissue was much more sensitive to both antilipolytic and lipolytic agents in vivo than was brown adipose tissue. The authors state that PGE_1 infusion causes a decrease in FFA in white but not in brown tissue. Their data show that infusion of PGE_1 (0.14 mg/kg) significantly reduces the increase in FFA content in white adipose caused by 150 mg/kg of theophylline but does not significantly alter the response in brown fat. In vitro it was found that 1 μg/ml PGE_1 significantly inhibited the stimulation of FFA release caused by both norepinephrine and theophylline in white fat but inhibited only the stimulation due to theophylline in brown fat. (JRH) 1271

0504
BLAKELEY, A.G.H., G.L. BROWN, D.P. DEARNALRY and R.I. WOODS
The use of prostaglandin E_1 in perfusion of the spleen with blood.
Journal of Physiology. 198: 31P-32P. 1968.

Experiments were performed to determine if PGE_1 could prevent the formation of platelet thrombi which cause a rise in arterial perfusion pressure during perfusion of cat spleens with heparinized blood due to occlusion of small arterioles. Both the spleen and blood donors were treated with 70 μg of PGE_1. Another 70 μg was added to the mixed blood in the perfusion circuits. This procedure prevented occlusion of the vessels completely. Blood flow and transmitter overflow with nerve stimulation was greater with PGE_1 than without. The standard errors of transmitter overflows were also less with PGE_1 which the authors attributed to improved uniform blood flow. No consideration was given to possible direct effects of the PGE_1 on the blood vessels or nerves. 1008

0505
BOHLE, E. and B. MAY
Metabolic effects of prostaglandin E_1 upon lipid and carbohydrate metabolism.
In: Ramwell, P.W. and J.E. Shaw, eds., "Prostaglandin Symposium of the Worcester Foundation for Experimental Biology," p. 115-129. New York, Interscience, 1968.

The effects of PGE_1 in vivo and in vitro on rat tissues, suggested that this compound possesses a partial insulin-like action upon intermediary lipid and carbohydrate metabolism. PGE_1 decreased the cleavage of triglycerides in adipose tissue, not only by a glucose-independent inhibition of the activation of lipolysis, but also by an increase in reesterification, supplying α-glycerophosphate by an augmentation of glucose utilization. PGE_1 also possesses insulin-like activity as regards the lipogenic action in adipose tissue and the enhancement of glycogen synthesis in muscle. The increase in blood gulcose after application of PGE_1 is due, at least in part, to a stimulation of glycogenolysis in the liver. (JRH) 1159

0506
BOWEN, D.M. and N.S. RADIN
Hydroxy fatty acid metabolism in brain.
In: Paoletti, R. and D. Kraitchevsky, eds., "Advances in Lipid Research," p. 255-272. New York, Academic Press, 1968.

A brief section on prostaglandins is included in this review. Prostaglandins have been found in the brain and are concentrated in nerve endings. Also, they have biosynthesized by subcellular fractions of the cerebral cortex. When injected into the cerebral ventricles of cats, PG's produce sedation. It seems likely that PG's have a physiological role in the brain. 61 references are cited. (JRH) 1114

0507
BOYARSKY, S., and P. LABAY
The effect of prostaglandins upon ureteral peristalsis.
In: Ramwell, P.W. and J.E. Shaw, eds., "Prostaglandin Symposium of the Worcester Foundation for Experimental Biology," p. 175-181. New York, Interscience, 1968.

The effects of PGE_1 and $PGF_{2\alpha}$ on ureteral peristalsis were studied in dogs both in vitro and in vivo. PGE_1 (0.1-33 μg/min iv) decelerated and stopped peristalsis without significantly changing urine flow in vivo. In vitro PGE_1 (1-30 μg/ml) also inhibited peristalsis. $PGF_{2\alpha}$ (0.5 μg/kg iv) accelerated peristalsis without changing urine flow in vivo. PG's seem to exert a direct effect on ureteral musculature. (JRH) 1155

0508
BRADLEY, P.B.
Synaptic transmission in the central nervous system and its relevance for drug action.
International Review of Neurobiology. 11: 1-56. 1968.

This extensive review contains a brief section citing 5 references on the possible role of prostaglandin as a synaptic transmitter. (JRH) 1197

0509
BRENNER, R.R., R.O. PELUFFO, O. MERCURI and M.A. RESTELLI
Effect of arachidonic acid in the alloxan-diabetic rat.
American Journal of Physiology. 215: 63-69. 1968.

The PGE content of the seminal vesicles of normal and alloxan-diabetic rats fed normal and arachidonic acid-free diets was measured. The diabetic rats showed atrophy of the seminal vesicles which was prevented by a supplement of ethyl arachidonate. However, the vesicles of diabetic rats contained higher concentrations of PGE (1.23 mg/g dried tissue) than diabetic rats receiving ethyl arachidonate (0.58 mg/g) or normal rats (0.70 mg/g). The total amount of PGE per rat was constant: 0.034mg for diabetic rats receiving ethyl arachidonate and 0.032 mg for the normal rat. This apparently indicates that the higher concentration found in diabetic rats was due to a decrease in weight of the gland. (JRH) 1088

0510
BROWN, J.D. and D.B. STONE
The mechanisms of action of antilipolytic agents: a comparison of the effects of PGE_1, insulin, tolbutamide, and phenformin or lipolysis induced by dibutyryl cyclic AMP.
Diabetes. 17: 304-305. 1968.

Abstract only. In isolated rat fat cells in vitro, insulin (10 μU/ml) or PGE_1 (10 mg/ml) decreased lipolysis induced by low concentrations of dibutyryl cyclic AMP (0.1 mM) plus theophylline (10^{-5} M) (P<0.0005). Insulin (1,000 μU/ml) did not inhibit the lipolytic effects of high concentrations of dibutyryl cyclic AMP (1 mM) plus theophylline (4×10^{-3} M) (P<0.25). High concentrations of PGE$_1$ (10 μg/ml) caused only slight but significant inhibition of the lipolytic effects of high concentrations of dibutyryl cyclic AMP plus theophylline. At low concentrations of dibutyryl cyclic AMP and theophylline, the formation of intracellular cyclic AMP is probably a rate-limiting step in lipolysis. In contrast, the rate-limiting step with high concentrations of dibutyryl cyclic AMP plus theophylline is probably not intracellular cyclic AMP formation but rather lipase activation or action. The results with insulin and PGE$_1$ are therefore consistent with the hypothesis that these substances inhibit adenyl cyclase activity, decrease cyclic AMP formation, and thus decrease lipolysis. The results are also consistent with the hypothesis that PGE$_1$ may enhance the activity of phosphodiesterase. (Authors modified) 1184

0511
BRUNDIN, J.
The effect of prostaglandin E$_1$ on the response of the rabbit oviduct to hypogastric nerve
stimulation.
Acta Physiologica Scandinavica. 73: 54-57. 1968.

The effect of PGE$_1$ on the response of the oviduct to hypogastric (sympathetic) nerve stimulation was measured in pentobarbital anesthetized rabbits that had been pretreated with atropine. Changes in resistance to perfusion flow from a catheter which was inserted into the lumen of the oviduct was measured following electrical stimulation of the nerve. PGE$_1$ and norepinephrine were infused through the ear vein. PGE$_1$ (1 μg/kg) inhibited the constriction caused by nerve stimulation and norepinephrine and also caused a reduction in systemic blood pressure. (JRH) 1035

0512
BUTCHER, R.W., C.E. BAIRD and E.W. SUTHERLAND
Effects of lipolytic and antilipolytic substances on adenosine 3',5'-monophosphate levels in
isolated fat cells.
Journal of Biological Chemistry. 243: 1705-1712. 1968.

Prostaglandins are mentioned as having antilipolytic properties in isolated fat cells. (JRH) 1086

0513
BUTCHER, R.W. and C.E. BAIRD
Effects of prostaglandins on adenosine 3',5'-monophosphate levels in fat and other tissues.
Journal of Biological Chemistry. 243: 1713-1717. 1968.

The action of prostaglandins on cyclic AMP levels in rat epididymal fat pads, isolated fat cells, and other tissues was studied. Fat pads incubated with prostaglandin E$_1$(PGE$_1$) contained higher levels of cyclic AMP than controls but lipolysis was not stimulated. However, PGE$_1$ did not cause increased cyclic AMP levels in isolated fat cells, suggesting that the stimulatory action occurred in another cell type. PGE$_1$ at very low concentrations antagonized the action of epinephrine on cyclic AMP levels in both intact fat pads and isolated fat cells. PGE$_1$ was considerably more effective as an inhibitor of epinephrine than PGF$_{1\alpha}$; PGF$_{1\beta}$ was without effect. PGE$_1$ lowered cyclic AMP levels in isolated fat cells very rapidly, and was effective against adrenocorticotropic hormone, glucagon, and thyroid stimulating hormone as well as the catecholamines. Incubation of intact cell preparations of lung, spleen, diaphragm, kidney, and other tissues with 2.8 μM PGE$_1$ resulted in increased levels of cyclic AMP. (Authors) 1085

0514
BUTCHER, R.W., C.E. BAIRD and E.W. SUTHERLAND
Effects of prostaglandins on cyclic AMP levels in fat and other tissues.
In: Ramwell, P.W. and J.E. Shaw, eds., "Prostaglandin Symposium of the Worcester Foundation for Experimental Biology," p. 93-101. New York, Interscience, 1968.

In the presence of caffeine, PGE_1 (2.8 μM) antagonized the stimulatory effects of epinephrine, ACTH, glucagon and TSH on cAMP accumulation in rat epididymal fat pad adipocytes in vitro. A 50% inhibition of the stimulatory effects of 5.5 μM epinephrine on cAMP accumulation was produced by 0.004 μM PGE_1 and 0.00004 μM PGE_1 produced significant inhibition. The amount of cAMP in adipocytes incubated with epinephrine and caffeine was reduced by 50% within 2 min of the addition of PGE_1. $PGF_{1\alpha}$ was less potent than PGE_1 and $PGF_{1\beta}$ was inactive. In the diaphragm, lung, spleen, kidney, testis and certain other tissues, PGE_1 stimulated cAMP synthesis in vitro. It is suggested that the antilipolytic properties of PGE_1 are due to its inhibition of cAMP synthesis in adipocytes. (JRH) 1161

0515
BUTCHER, R.W.
Role of cyclic AMP in hormone actions.
New England Journal of Medicine. 279: 1378-1384. 1968.

The role of prostaglandins as antilipolytic substances are discussed in this review article; 6 references are cited. (JRH) 1041

0516
BUTCHER, R.W., G.A. ROBISON, J.G. HARDMAN and E.W. SUTHERLAND
The role of cyclic AMP in hormone actions.
In: Weber, G., ed., "Advances in Enzyme Regulation." Vol. 6: p. 357-389. New York, Pergamon Press. 1968.

This review article has one section devoted to the effects of prostaglandins on cAMP. Special attention is given to effects in adipose tissue. (JRH) 1227

0517
BYGDEMAN, M., S.U. KWON, T. MUKHERJEE and N. WIQVIST
Effect of intravenous infusion of prostaglandin E_1 and E_2 on motility of the pregnant human uterus.
American Journal of Obstetrics and Gynecology. 102: 317-326. 1968.

PGE_1 or PGE_2 was infused intravenously into 8 midpregnant (14-22 weeks) and 7 term (34 to 40 weeks) patients. PGE_1 (0.6 to 1.2 μg/min) caused an increase in uterine tone but had no effect on contractility. At higher doses (2.0 to 9.0 μg/min), PGE_1 caused an increase in the amplitude of contraction as well as in increase in uterine tone. PGE_2 (4.0 and 8.0 μg/min) caused an increase in both tone and contractility in 3 patients. 2 of the 8 patients were used in both the PGE_1 and PGE_2 experiments. The 2 prostaglandins seemed to be equally potent with the threshold dose being about 2 μg/min. The only side effect was vomiting in 1 patient receiving 8 μg/min PGE_2. 4 of the 7 term patients received PGE_1 (2.5 to 8.0 μg/min) and 3 received PGE_2 (1.0 to 8.0 μg/min). The threshold dose was about 4 to 8 μg/min. The sensitivity of the uterus in terms of elevation of tone seems to be higher at midpregnancy than at term. No side effects were observed in the term patients but in 2 cases the infusion had to be terminated because of nonphysiological uterine hypertonus and fetal bradycardia. The authors conclude that the PGE's are not suitable for induction of labor, but that they might be of value in cases of uterine atonia after delivery or miscarriage. (JRH) 1116

128

0518
CAREN, R. and L. CORBO
Origin of plasma arachidonic acid.
Clinical Research. 16: 158. 1968.

Abstract only. Total hepatectomy was performed on 5 dogs. The plasma lipids were determined hourly by column chromatography followed by gas liquid chromatography. There was a progressive fall of the arachidonic acid percent of the unesterified fatty acids. This was the only significant change in fatty acid content of the plasma. Since arachidonic acid was not replenished by adipose tissue lipolysis as were the other fatty acids, it was concluded that the liver is the only source of plasma arachidonic acid. (JRH) 1210

0519
CAREN, R. and L. CORBO
The origin of plasma arachidonic acid in dogs.
Metabolism 17: 1043-1050. 1968.

Removal of the liver from fasted dogs led to a gradual decrease in plasma arachidonic acid. This was the only fatty acid found to decline. The source of plasma arachidonic acid, which is a precursor of prostaglandins, is the liver in fasted dogs. (JRH) 1113

0520
CARLSON, L.A. and D. HALLBERG
Basal lipolysis and effects of norepinephrine and prostaglandin E_1 on lipolysis in human subcutaneous and omental adipose tissue.
Journal of Laboratory and Clinical Medicine. 71: 368-377. 1968.

Samples of both subcutaneous and omental tissue were obtained from 20 patients undergoing elective abdominal surgery. Basal glycerol release from the 2 adipose tissues was measured in vitro. PGE_1 (0.01-1.0 $\mu g/ml$) and/or norepinephrine were added to the incubation medium. The basal release of glycerol was higher in subcutaneous adipose than in omental tissue and there was a positive correlation between the rate of lipolysis in the 2 adipose tissues. In most cases (9 out of 11), PGE_1 caused a significant reduction in basal lipolysis in subcutaneous adipose tissue but the effect on omental fat was much less (3/13). NE stimulated lipolysis in omental tissue to a greater extent than in subcutaneous tissue. PGE_1 inhibited the NE-induced stimulation in both types of adipose tissue. (JRH) 1189

0521
CARLSON, L.A., L-G. EKELUND and L. ORO
Clinical and metabolic effects of different doses of prostaglandin E_1 in man.
Acta Medica Scandinavica. 183: 423-430. 1968.

PGE_1 was infused intravenously into 8 male volunteers in doses of 0.032 to 0.58 $\mu g/kg/min$ while clinical effects were recorded. Flushing, headaches and abdominal cramps were experienced by all volunteers. In most cases the headaches and/or abdominal cramps became so severe that infusion had to be terminated before reaching the highest doses. Flushing was generally the first effect seen and in some cases it was followed by pallor and feeling of coldness as the dose was increased. Other effects were tiredness, nausea, vomiting, visual symptoms, dyspnea, back pain, leg cramps and restlessness. All clinical effects disappeared within 15 to 30 min of termination of infusion. The free fatty acid (FFA) turnover rate was determined by infusion tritium labelled palmitic acid. PGE_1 infusion caused a rise in blood FFA levels in all cases at some time during the infusion of PGE_1. In some cases there was also a fall in FFA levels. There was little change in FFA turnover rates. There was no correlation between FFA levels and clinical effects nor between FFA levels and PGE_1 dose levels. (JRH) 1217

0522
CARLSON, L.A., E. IRION and L. ORO
Effect of infusion of prostaglandin E_1 on the aggregation of blood platelets in man.
Life Sciences. 7: 85-90. 1968.

PGE_1 infused intravenously to healthy persons at rates of 0.05 to 0.1 μg/kg/min for 30 minutes had no discernible effect on platelet aggregation. (Authors) 1080

0523
CARR, A.A.
The effect of prostaglandin A_1 on urinary concentration.
In: Ramwell, P.W. and J.E. Shaw, eds., "Prostaglandin Symposium of the Worcester Foundation for Experimental Biology," p. 163-173. New York, Interscience, 1968.

Hydropenic, pentobarbital anesthetized dogs were used to study the effects of PGA_1 (0.95-2.4 μg/min) infused into the renal artery or intravenously on kidney function when variables of sodium steroid and ADH were kept constant by infusion of supramaximal desoxycorticosterone and vasopressin. Osmotic diuresis was produced by manitol infusion. PGA_1 decreased tubular solute free water absorption during osmotic diruesis by decreasing the renal medullary interstitium tonicity. The exact mechanism of this decrease cannot be determined from this data. (JRH) 1156

0524
CARR, A.A. and A.L. HUMPHRIES
A mechanism for inhibition of a solute-free water reabsorption (T^CH_2O) by a prostaglandin.
Clinical Research. 59: 16a. 1968.

Abstract only. Peripheral vein infusion of PGA_1 0.95-1.5 μg/kg per min in hydropenic dogs during mannitol diuresis inhibited T^CH_2O in association with a significant decrease in femoral artery blood pressure and a significant rise in renal plasma flow (RPF). Direct renal artery infusion of PGA_1, 0.7-16 μg/min, which did not lower femoral artery blood pressure, resulted in an inhibition of T^CH_2O. At the highest renal artery infusion rate of PGA_1 16 μg/min, there was a dramatic increase in medullary-papillary hypertonicity and thus inhibition of T^CH_2O. (JRH) 1052

0525
CARR, A.A. and A.L. HUMPHRIES
A mechanism for inhibition of solute-free water reabsorption by a prostaglandin.
Journal of Clinical Investigation. 47: 16a. 1968.

Abstract only. Peripheral vein infusion of PGA_1 0.95-1.5 μg/kg per min in hydropenic dogs during mannitol diuresis inhibted T^CH_2O in association with a significant decrease in femoral artery blood pressure and a significant rise in renal plasma flow (RPF). There was an insignificant drop in glomerular filtration rate (GFR) along with a small drop in sodium and potassium excretion. Direct renal artery infusion of PGA_1, 0.7-16 μg/min, which did not lower femoral artery blood pressure, resulted in an inhibition of T^CH_2O but insignificant changes in GFR and sodium excretion. There was a rise in RPF in each experiment, but over all, the changes were insignificant. At 16 μg/min, there was a dramatic increase in RPF and a rise in sodium excretion. Medullary-papillary tonicity was found to be decreased in the PGA-infused kidney as compared with the contralateral kidney. Thus PGA_1 caused a decrease in medullary-papillary hypertonicity and thus inhibition of T^CH_2O by an unknown mechanism. Perhaps changes in medullary-papillary blood flow resulted in either dilution or washout of the area. A decrease in delivery of sodium to the ascending loop of Henle or inhibition of sodium transport at that site could give the same results. (RMS) 1246

130

0526

CAVALLITO, D.J.

Some relationships between chemical structure and pharmacological activities.
Annual Review of Pharmacology. 8: 39-66. 1968.

In this review, prostaglandins are mentioned as being a new group of compounds with interesting pharmacological properties. (JRH) 1102

0527

CHANCE, B.

Biological membranes: Regulatory functions.
Science. 160: 1261-1266. 1968.

This review of a symposium briefly mentioned that a paper was presented on the interactions of prostaglandins and cAMP. (RAP) 1267

0528

COCEANI, F., C. PACE-ASCIAK and L.S. WOLFE

Studies of the effect of nerve stimulation on prostaglandin formation and release in the rat stomach.

In: Ramwell, P.W. and J.E. SHAW, eds., "Prostaglandin Symposium of the Worcester Foundation for Experimental Biology," p. 39-45. New York, Interscience, 1968.

The PG content of rat stomach tissue and the rate of PG release into a perfusion medium following adrenergic and cholinergic nerve stimulation was measured in vitro. The stomach released PGE_1, PGE_2, $PGF_{1\alpha}$ and $PGF_{2\alpha}$ at basal levels even when denervated by cold storage. Vagal (cholinergic) nerve stimulation (up to 10 pulses/sec) caused a large increase in release of PG's, especially PGE_2 and $PGF_{2\alpha}$ which was dependent on the rate of stimulation. Addition of hyoscine to the medium prior to vagal stimulation blocked the increase in PG release. Stimulation of sympathetic (adrenergic) nerves did not alter basal secretion of PG but did reduce the increase caused by vagal stimulation. Transmural stimulation of the stomach, even at low frequencies, also led to large increases in PG release. The PG's released resulted from biosynthesis in the stomach rather than release of stored PG's. Carbamylcholine caused a variable increase in PG release in fresh stomachs but after denervation by cold storage the increase became constant. Serotonin also caused a variable increase in secretion which may have been due to the unknown differences in sensitivity of different parts of the stomach of serotonin. It seems that PG's are spontaneously released from muscle and that release may be influenced by activation of specific receptors within the cell membrane. (JRH) 1166

0529

COLE, B. and J. NAKANO

Effects of prostaglandin $F_{2\alpha}$ ($PGF_{2\alpha}$) on the systemic venous return and splanchnic circulation.
Pharmacologist. 10: 175. 1968.

Abstract only. The effects of $PGF_{2\alpha}$ on systemic venous return and splanchnic circulation was studied in anesthetized, open-chest dogs. Injection of 2-10 µg/kg of $PGF_{2\alpha}$ into a femoral vein increased systemic arterial pressure (SAP), pulmonary arterial pressure (PAP) cardiac output, myocardial contractile force (MCF) and portal venous pressure (PVP) essentially in proportion to dose. In contrast, the injection of the same doses of $PGF_{2\alpha}$ into the portal vein decreased SAP,PAP, and MCF but increased heart rate and PVP. The magnitude and duration of the increment of PVP were greater and more prolonged with injection into the portal vein, than with injection into a femoral vein. In dogs, in which cardiac input was kept constant, $PGF_{2\alpha}$ decreased systemic venous return and increased PAP. $PGF_{2\alpha}$ increases both SAP and PAP by its direct vasoconstrictor action. $PGF_{2\alpha}$ increases PVP and causes pooling of blood in the splanchnic vascular beds, thereby decreasing systemic venous return significantly. (Authors modified) 1321

0530
COLLIER, H.O.J. and W.J.F. SWEATMAN
Antagonism by fenamates of prostaglandin $F_{2\alpha}$ and slow reacting substance on human bronchial muscle.
Nature. 219. 864-865. 1968.

The effect of flufenamate and meclofenamate on the constriction of human bronchial muscle caused by $PGF_{2\alpha}$, histamine, slow reacting substance A (SRS-A) and acetycholine was measured in vitro. Both fenamates caused a relaxation of the muscle and were highly effective against $PGF_{2\alpha}$, moderately effective against SRS-A, slightly effective against histamine and ineffective against acetylcholine. The fenamates caused a shift toward higher doses in the dose response curve of $PGF_{2\alpha}$ but had no effect on the dose response curve of acetylcholine. Doses of meclofenamate as low as 100 ng/ml caused an inhibition of the response to $PGF_{2\alpha}$ without affecting the response to SRS-A. Meclofenamate was more effective than flufenamate. A few tests were performed with phenylbutazone and aspirin, but both of these drugs were less effective than the fenamates against $PGF_{2\alpha}$ and SRS-A, with aspirin being the least effective. The fenamates at similar doses were ineffective in suppressing the relaxation of smooth muscle caused by PGE_1 and PGE_2. It is suggested that the mechanism of action of these anti-inflammatory drugs might be through their inhibition of $PGF_{2\alpha}$. (JRH) 1248

0531
COLLIER, H.O.J., L.C. DINNEEN, A.C. PERKINS and P.J. PIPER
Curtailment by aspirin and meclofenamate of hypotension induced by bradykinin in the guinea pig.
Nauyn-Schmeidebergs Archiv fur Pharmakologie und Experimentelle Pathologie. 259: 159-160. 1968.

Guinea pigs were anesthetized with pentobarbital. Intraarterial injection of bradykinin (0.5-4 μg) or PGE_1 (0.5-5 μg) induced hypotension lasting 1-17 min. By this route, these doses produced no evidence of bronchoconstriction. Aspirin (4 or 8 mg/kg) administered intravenously as the sodium salt, curtailed the hypotensive response to bradykinin (P<0.05), but not that to PGE_1. Aspirin did not reduce the maximal fall in blood pressure. (JRH) 1223

0532
COLLINS, P., C.J. JUNG and R. PAPPO
Prostaglandin studies. The total synthesis of dl-prostaglandin B_1.
Israel Journal of Chemistry. 6: 839-841. 1968.

Starting with 9-ketodecanoic, a total synthesis of $dl-PGB_1$(x) is described. Chemical and physical properties of some compounds in the synthesis are given and diagrams for all steps are shown. The authors point out that their method is also adaptable to the synthesis of 11-hydroxylated derivatives closely related to PGE_1. (RAP) 1244

0533
COREY, E.J., I. VLATTAS, N.H. ANDERSEN and K. HARDING
A new total synthesis of prostaglandins of the E_1 and F_1 series including 11-epiprostaglandins.
Journal of the American Chemical Society. 90: 3247. 1968.

The authors describe a total synthesis of PGE_1 and PGF_1 prostaglandins which can modified to yield the C_{11} and C_{15} epimers as well as the natural prostaglandins. Structural formulas of key intermediates as well as the steps in the synthesis are given. The author reports that racemic 11, 15-epimers of PGE_1 are about twice as active in stimulating the rat uterus as racemic PGE_1, but much less potent as vasodepressors. (JRH) 1243

0534

COREY, E.J., N.H. ANDERSEN, R.M. CARLSON, J. PAUST, E. VEDEJS, I. VLATTAS and R. E. K. WINTER
Total synthesis of prostaglandins. Synthesis of the pure dl-E_1, -$F_{1\alpha}$ -$F_{1\beta}$, -A_1 and -B_1 hormones.
Journal of the American Chemical Society. 90: 3245-3247. 1968.

The total synthesis of pure racemic PGE_1, $PGF_{1\alpha}$, $PGF_{1\beta}$, PGA_1, and PGB_1 is described. Step by step procedure for synthesis is given along with the structural formulas for all key intermediates. (JRH) 1173

0535

COVINO, B.G., J.B. LEE and J. McMORROW
Circulatory effects of renal prostaglandins.
Circulation 37/38 (supp.6): VI-60. 1968.

Abstract only. The comparative effects of intravenously administered renal prostaglandins (PGA_1, PGA_2, PGE_1 and PGE_2) on systemic arterial pressure and carotid, femoral, renal and mesenteric artery blood flow were determined in pentobarbitalized dogs. Arterial pressure was measured by a pressure transducer attached to an indwelling femoral artery cannula. Blood flow in the various beds were measured by an electromagnetic flowmeter. One $\mu g/kg$ of PGA_1, PGA_2 and PGE_1 caused a decrease in mean arterial pressure of 19.6±1.9 to 21.6±2.7 mm Hg. PGE_2 caused a 9.4±1.6 mm Hg fall in pressure which was significantly less than that of the other renal prostaglandins ($P<0.01$). Concomitant with the hypotension, mesenteric blood flow increased 71.3±14.3 to 108.5±8.9 ml/min following administration of PGA_1, PGA_2 and PGE_1. Carotid flow also increased (49.2±9.8 to 54.0±14.1). However, the increase in carotid flow was significantly less than the increase in mesenteric flow ($P<0.05$). No significant change in either renal or femoral flow occurred. The data indicate that of the four renal prostaglandins studied, PGE_2 is the least hypotensive and that vasodilation of the mesenteric vascular bed appears to be primarily responsible for the hypotensive action of the renal prostaglandins. (Authors) 1049

0536

CRAWFORD, J.D. and H.A. HAESSLER
Insulin-like activities of prostaglandin E_1.
In: Ramwell, P.W. and J.E. Shaw, eds., "Prostaglandin Symposium of the Worcester Foundation for Experimental Biology," p. 103-113. New York, Interscience, 1968.

The effects of insulin and PGE_1 on free fatty acid (FFA) release and synthesis was studied in the presence and absence of glucose in epinephrine stimulated adipose tissue from fed and unfed rats. PGE_1 (0.1 $\mu g/ml$) caused a reduction in glycerol release and an increase in reesterification of FFA. The effect was greater in tissue from fed rats in the presence of glucose. Insulin caused a similar effect in this test system. Studies of lipid synthesis from glucose and labelled acetate indicates that PGE_1 produces a small increase in lipid synthesis and insulin produces a larger increase. Both PGE_1 and insulin caused a sharp increase in the percentage of glucose transformed to fatty acid with a reduction in glucose appearing as glyceride glycerol. The effects of PGE_1 plus insulin on total lipolysis were at least additive. PGA_1 lacked antilipolytic activity and had no effect on lipid synthesis from acetate. PGE_1 and insulin exhibit certain common effects but it seems that they have different mechanisms. (JRH) 1160

0537

vanCREVELD S. and C.N. PASCHA

Abnormality in the aggregation of blood platelets in various morbid conditions and the influence of prostaglandins upon this abnormality.

Thrombosis et Diathesis Haemorrhagica. 20: 180-189. 1968.

The effects of PGE_1 and PGE_2 on ADP-induced aggregation of platelets from normal human blood and from patients with various hemorrhagic conditions were measured in vitro by optical methods. Addition of ADP to platelet rich plasma of patients with von Willebrand's disease, hemophilia A, hemophilia B or uremia resulted in a rapid platelet aggregation followed within a few minutes by a very strong and rapid disaggregation. PGE_1 (0.5γ/ml) strongly inhibited the ADP-induced aggregation of both the normal and abnormal platelets. PGE_2 (0.5γ/ml) stimulated the aggregation of both types of platelets and blocked or reduced the disaggregation of the abnormal platelets. In two cases of uremia, the platelets were much more sensitive to both PG's than normal platelets, with doses as low as 0.05 and 0.01γ/ml being effective in altering ADP-induced effects. Platelets from patients with Glanzmann's thrombasthnia were not aggregated by ADP even at higher doses and PGE_1 and PGE_2 had no effect on these platelets. Overproduction or insufficient catabolism of PGE_1 or a relative deficiency of PGE_2 could be involved in the pathology of the above diseases. (JRH) 1221

0538

van CREVELD, S. and C.N. PASCHA

Influence of the prostaglandins E_1 and E_2 on aggregation of blood platelets.

Nature. 218: 361-362. 1968.

The authors investigated the influence of PGE_1 and PGE_2 on cases of abnormal platelet aggregation in humans in vitro. In all cases PGE_1 decreased the aggregation by ADP whereas PGE_2 increased the aggregation. PGE_2 was also added to the PRP of a patient with Glanzmann's thrombasthenia in an attempt to induce aggregation by ADP. No aggregation occurred, even when the concentration of PGE_2 was increased from 0.5 to 5 μg/ml of plasma. (RAP) 1250

0539

van CREVELD, S., C.N. PASCHA, L. KERKMEESTER and C.J. MCEWAN-VAN DER HORST

Ziekte van Von Willebrand en enige andere toestanden, gecombineerd met een „nieuwe" afwijking in de functie der bloedplaatjes. Invoed van prostaglandinen. [Von Willebrand's disease and some other conditions combined with a "new" anomaly of the function of the thrombocytes. The influence of prostaglandins.]

Nederlands Tijdschrift voor Geneeskunde. 112: 646-651. 1968.

In 2 young patients belonging to the same family both suffering from Von Willebrand's disease a "new" anomaly of the function of the blood platelets was observed, viz. a normal, rapid initial aggregation of the platelets after addition of ADP, which after a few minutes was followed by rapid disaggregation. The inhibiting action of PGE_1 on the aggregation by ADP could be confirmed. PGE_2 was found to increase the aggregation by ADP in the cases investigated by us. It had, however, no effect on the absence of aggregation by ADP in Glanzmann's thrombasthenia. (From English summary) 1261

0540
CRONBERG, S.
   Investigations in haemorrhagic disorders with prolonged bleeding time but normal number of
      platelets. With special reference to platelet adhesiveness.
   Acta Medica Scandinavica. Supp. 486: 2-54. 1968.

   The authors cite 3 references indicating that PGE_1 inhibits platelet reactions in vitro.
   (RMS) 1200

0541
CROWSHAW, K., J.B. LEE and K.A. ATTREP
   Origin of the renomedullary prostaglandins.
   Abstracts of the 156th Meeting of the American Chemical Society, Atlantic City, 8-13
      September, BIOL. abs.21. 1968.

   Abstract only. The biosynthesis of prostaglandins from arachidonic acid by rabbit kidney
   medulla slices and homogenates was attempted. Chromatographic analysis of the acidic lipid
   extracts of the homogenized incubation mixture indicated that only small quantities of PGA_2
   and PGE_2 were formed. No PG's were detected in control incubations without arachidonic acid.
   Incubation of homogenates of the medulla with arachidonic acid in the presence of
   gluthathione and hydroquinone did not produce any PG. These results suggest that the
   renomedullary prostaglandins may not be synthesized from arachidonic acid within the medulla
   itself. (JRH) 1331

0542
DANIEL, E.E.
   Pharmacology of the gastrointestinal tract.
   In: Code, C.F., ed., "Handbook of Physiology," p. 2267-2324. Washington, American
      Physiological Society, 1968.

   This extensive review contains a brief section on prostaglandins which outlines their discovery
   and their pharmacological effects on gastrointestinal muscle. A table is presented which
   summarizes the relative biological activities of the PGE's and PGF's on various preparations
   from a variety of animal sources. (JRH) 1091

0543
DANIELS, E.G., W.C. KRUEGER, F.P. KUPIECKI, J.E. PIKE and W.P. SCHNEIDER
   Isolation and characterization of a new prostaglandin isomer.
   Journal of the American Chemical Society. 90: 5894-5895. 1968.

   The isolation and structure determination of 8-iso-PGE_1 following the bioconversion of
   8,11,14-eicosatrienoic acid to PGE_1 in a biological system is reported. The mother liquors
   obtained from the crystalization of PGE_1 were separated chromatographically and analyzed by
   infrared sepctrophotometry and nuclear magnetic resonance. The presence of 8-iso-PGE_1 was
   confirmed. (JRH) 1182

0544
DANIELS, E.G. and J.E. PIKE
   Isolation of prostaglandins.
   In: Ramwell, P.W. and J.E. Shaw, eds., "Prostaglandin Symposium of the Worcester Foundation
      for Experimental Biology," p.379-387. New York, Interscience, 1968.

A technique for large scale biosynthesis and purification of prostaglandins is described. The starting materials are sheep seminal vesicle homogenates (75 kg), an appropriate fatty acid (37 g), reduced glutathione, hydroquinone and EDTA. After a 60 min reaction period the PG's are extracted with acetone then Skellysolve B and finally methylene chloride. This produces 8-12 g of PG and represents a 15-30% yield. The PG's are purified by column chromatography, first on silica gel followed by silver impregnated Amberlyst 15. This procedure produces prostaglandins of high purity in quantity amounts. (JRH) 1137

0545
DAUGHARTY, T.M., L.J. BELLEAU, J.A. MARTINO and L.E. EARLEY
Interrelationship of physical factors affecting sodium reabsorption in the dog.
American Journal of Physiology. 215: 1442-1447. 1968.

Experiments to measure the effect of renal vasodilation on kidney function were performed in pentobarbital anesthetized dogs. PGE_1 (1 μg/min) was used to produce vasodilation in one kidney by infusion into the renal artery. PGE_1 caused an increase in urine volume within seconds; sodium excretion increased 335% and total renal blood flow increased 41%. This was similar to the change caused by acetylcholine. (JRH) 1229

0546
DAUGHERTY, Jr., R.M., J.M. SCHWINGHAMER, S. SWINDALL and F.J. HADDY
The effects of local and systemic infusions of prostaglandin E_1 on the skin and vasculature of the dog forelimb.
Journal of Laboratory and Clinical Medicine. 72: 869. 1968.

Abstract only. Venous blood outflows from the skin and muscle of isolated dog forelimbs were used to compare the effects of infusion of PGE_1 (0.5-10 μg/min) into the brachial artery of the isolated limb with intravenous infusion of PGE_1 (0.97-19.4 μg/min) into the systemic circulation. The low intravenous (systemic) infusion rate (0.97 μg/min) of PGE_1 caused an increase in both skin and muscle outflows but had little effect on aortic blood pressure. Increased infusion rates (1.94-19.4 μg/min) caused a dose related decrease in blood pressure but the outflows remained elevated until blood pressure fell below 20% of the control value. Infusion of PGE_1 into the brachial artery produced a large increase in both outflows. The effect on muscle outflow was greater than the effect on skin outflow, even when the blood flow into the limb was held constant with a perfusion pump. Thus PGE_1 given systemically or directly into the limb lowers vascular resistance in both these vascular beds. PGE_1 did not affect the response of the limb vasculature to epinephrine, levarterenol, or angiotensin. In the naturally perfused limb a slight, sudden increase in limb weight was recorded each time the PGE_1 infusion rate was increased, suggesting that the PGE_1 infusion had little effect on net water efflux, which could be due to proportional dilation of both arteries and veins. (JRH) 1191

0547
DAVIES, B.N., E.W. HORTON and P.G. WITHRINGTON
The occurrence of prostaglandin E_2 in splenic venous blood of the dog following splenic nerve stimulation.
British Journal of Pharmacology and Chemotherapy. 32: 127-135. 1968.

In several experiments, the spleen from one dog was perfused with blood from a second dog. The splenic nerve was stimulated electrically at 10 pulses per sec. Samples of effluent blood were extracted chemically and chromatographically and then bioassayed. Electrical stimulation caused a release of PGE_2. Phenoxybenzamine blocked PGE_2 release but increased norepinephrine release. There was no increase in free fatty acids. (JRH) 1011

0548
DAWSON, W., P.W. RAMWELL and J. SHAW
Metabolism of prostaglandins by the rat isolated liver.
British Journal of Pharmacology and Chemotherapy. 34: 668P-669P. 1968.

Radiolabelled PGE_1 and isolated blood perfused rat liver preparations were used to study the metabolsim of PGE_1 in the liver. Rapid uptake of $1-^{14}C$ PGE_1 was shown since only 3-9% of the PGE_1 appeared in the effluent blood. Very little radioactivity was found in the bile. When the blood was recirculated through the liver, 15-20% was detected as $^{14}CO_2$ after 10-15 min, suggesting β oxidation at the Cl-8 side chain. The remaining radioactivity in the liver was found with lecithin and C16-22 fatty acids. When $5-6^3H$ PGE_1 was used, 19-30% of the radioactivity appeared in the bile. Oxidation followed by biliary secretion from the liver might be a significant route of prostaglandin metabolism. (JRH) 1053

0549
Del TACCA, M., S. LECCHINI, G.M. FRIGO, A. CREMA and G. BENZI
Antagonism of atropine towards endogenous and exogenous acetycholine before and after sympathetic system blockade in the isolated distal guinea-pig colon.
European Journal of Pharmacology. 4: 188-197. 1968.

It is briefly mentioned in the discussion section of this paper that the resistance of some tissues to atropine stimulation might be due to the release of other substances such as prostaglandins. (JRH) 1029

0550
De WIED, D., A. WITTER, D.H.G. VERSTEEG and A.H. MULDER
Release of ACTH by substances of central nervous system origin.
Endocrinology. 85: 561-569. 1968.

The effect of several substances on pituitary ACTH release was studied. In rats pretreated with pentobarbital and chloropromazine. PGE_1 and PGE_2 produced an increase in ACTH secretion; $PGF_{1\alpha}$ and $PGF_{2\alpha}$ were not effective. (JRH) 1040

0551
DISTELKOTTER, B. and W. VOGT
Spontane neubildung von prostaglandin im isolierten froschdarm. [Spontaneous formation of prostaglandin in the isolated frog intestine.]
Naunyn-Schmiedebergs Archiv fur Pharmakoloige und Experimentelle Pathologie. 260: 324-329. 1968.

The release of prostaglandin from isolated gastro-intestinal tracts of frogs (*Rana esculenta*) was estimated, and was compared with the prostaglandin content of the tissue. During one hour the intestinal tract liberates about 4 times as much prostaglandin into the surrounding bath fluid as is contained in the tissue, i.e. there is a formation de novo during the incubation. The formation of prostaglandin is increased by immersing the intestine into distilled water instead of Ringer's solution, it is inhibited by cocaine. This indicates that the biosynthesis of prostaglandin is variable and can be enhanced by stimuli from the intramural nervous system. (Authors) 1202

0552
DITSCHUNEIT, H.
[Hormonal regulation of lipogenesis. Hyperinsulinism in prediabetes and obesity]
Acta Diabetologica Latina. 5: 364-378. 1968.

PGE_1 and nor-PGE_1 are mentioned as lipogenic factors. Their in vitro effects on glucose oxidation, glycerol release, glucose uptake, and lipid synthesis are indicated in 5 figures, but the data are not discussed in the text. (RMS) 1305

0553
DOUGLAS, W.W.
Stimulus-secretion coupling: the concept and clues from chromaffin.
British Journal of Pharmacology. 34: 451-474. 1968.

In this review article, it is mentioned that prostaglandins were released from adrenal glands exposed to acetylcholine. (JRH) 1025

0554
DUCHARME, D.W., J.R. WEEKS and R.G. MONTOGOMERY
Studies on the mechanism of the hypertensive effect of prostaglandin $F_{2\alpha}$.
Journal of Pharmacology and Experimental Therapeutics. 160: 1-10. 1968.

The present investigation was undertaken to determine the mechanism of the pressor activity of $PGF_{2\alpha}$. The pressor activity in rats persisted after ganglion blockade or reserpine pretreatment. Administration of $PGF_{2\alpha}$ into the perfused limb of the dog demonstrated that the pressor activity resulted, at least in part, from an increase in peripheral resistance, and that the increase in resistance occurred primarily in the venous segment. In unanesthetized, trained dogs $PGF_{2\alpha}$ increased both cardiac output and systemic blood pressure, leaving calculated total peripheral resistance virtually unchanged. The increase in cardiac output was not caused by a direct effect of the agent on myocardial contractility, as was demonstrated in pentobarbital anesthetized dogs. Further evidence that the pressor activity of $PGF_{2\alpha}$ is primarily the result of venoconstriction was obtained from experiments in which a perfusion pump was used to by-pass the right heart and a constant pressure stabilizing reservoir was placed in the venous return. (Authors modified) 1074

0555
DUDA, P., E.W. HORTON and A. McPHERSON
The effects of prostaglandins E_1, $F_{1\alpha}$, and $F_{2\alpha}$ on monosynaptic reflexes.
Journal of Physiology. 196: 151-162. 1968.

Ventral root potentials evoked by stimulation of the ipsilateral dorsal root were recorded in cats anesthetized with halothane, nitrous oxide and chloralose. Those with a latent period corresponding to a monosynaptic pathway were used primarily in this investigation. PGE_1 (3.5-17.8 μg/kg) injected into the aorta reduced the amplitude of monosynaptic responses in 6 out of 10 cats for about 3 hours. In 1 cat, PGE_1 caused potentiation of the reflex and in 3 cats there was no effect. $PGF_{1\alpha}$ (2.4-3.5 μg/kg) inhibited the monosynaptic response in 4 cats but in 1 of these (19.6 μg/kg) greatly potentiated the reflex. $PGF_{2\alpha}$ (1.4-17.8 μg/kg) injected into the aorta was followed by significant but variable changes in monosynaptic response. In 1 experiment an intravenous injection of 30.3 μg/kg was followed by a long-lasting potentiation of the reflex response. It is concluded that prostaglandins, 2 of which have previously been identified in cat brain extracts, have pronounced and long-lasting effects on monosynaptic spinal reflexes. (Authors modified) 1024

138

0556
DUKE, H.N. and J.R. VANE
  An adverse effect of polyvinylchloride tubing used in extracorporeal circulation.
  Lancet. 2: 21-23. 1968.

  $PGF_{2\alpha}$ response of isolated cat lungs was not affected by circulation of the blood through
  polyvinylchloride tubing as were some other biological responses. (JRH) 1062

0557
DUNHAM, E. T. ROLEWICZ and B. ZIMMERMAN
  Prostaglandin as the possible mediator of cutaneous sympathetic vasodilation.
  Federation Proceedings. 27: 536. 1968.

  Abstract only. The effects of drugs on the vasodilation caused by sympathetic nerve stimulation
  were studied in isolated, perfused dog hind paws or gracilis muscles. The data indicated that
  this vasodilator system innervates the cutaneous bed to a relatively greater degree than the
  muscle bed. Results of analyses of hind paw skin and gracilis muscle for prostaglandin-like
  activity indicated also that skin possessed much greater activity. (JRH) 1131

0558
DYER, D.C. and E.J. WALASZEK
  Occurrence and pharmacologic properties of a lipid substance obtained from rabbit kidney.
  Journal of Pharmacology and Experimental Therapeutics. 160: 360-366. 1968.

  The authors report the isolation and testing of a lipid-like substance from rabbit kidneys which
  is not a prostaglandin but which has smooth muscle stimulating properties. (JRH) 1075

0559
EGGENA, P., R. WALTER and I.L. SCHWARTZ
  Relationship between hydro-osmotic flow and the inhibited response of the toad bladder to
    vasopressin.
  Life Sciences. 7: 59-63. 1968.

  PGE_1 is mentioned in the introduction as being a potent inhibitor of the effects of vasopressin
  on the toad bladder. (NES) 1081

0560
ELIASSON, R.
  Biochemical analyses of human semen in the study of the physiology and pathophysiology of
    the male accessory genital glands.
  Fertility and Sterility. 19: 344-350. 1968.

  The author cites 2 references on the prostaglandin content of semen as measured by the split
  ejaculate technique, 2 references on the effect of prostaglandins on the human uterus, and 3
  references on the correlation of seminal prostaglandin content with human fertility.
  (RMS) 1061

0561

ELIASSON, R., R.N. MURDOCH and I.G. WHITE

The metabolism of human spermatozoa in the presence of prostaglandin E_1.

Acta Physiologica Scandinavica. 73: 379-382. 1968.

The oxygen uptake and metabolism of human spermatozoa has been measured using concentrated suspensions of cells incubated with radioactive fructose in small Warburg flasks. The oxygen uptake was about 14 μl per 10^8 sperm per hr and represented true respiratory activity since radioactive carbon dioxide was produced as well as lactic acid. The presence of 100 μg of prostaglandin E_1 per ml had no effect on oxygen uptake, the amount of fructose metabolized or the amount of lactic acid accumulating. (Authors) 1036

0562

ELKELES, R.S., J.R. HAMPTON, A.J. HONOUR, J.R.A. MITCHELL and J.S. PRITCHARD

Effect of a pyrimido-pyrimidine compound on platelet behaviour in vitro and in vivo.

Lancet. 2: 751-754. 1968.

It is briefly mentioned that PGE has been shown by others to inhibit platelet aggregation. (JRH) 1058

0563

EMBREY, M.P. and D.L. MORRISON

The effect of prostaglandins on human pregnant myometrium in vitro.

Journal of Obstetrics and Gynaecology of the British Commonwealth. 75: 829-832. 1968.

A study of the effects of some individual prostaglandins on the spontaneous motility of human pregnant myometrium in vitro is described. Stimulation of myometrial contractility was observed using both the E prostaglandins and those of the F series; undoubted spasmogenic responses were being recorded consistently with PGE_2 (80-100 ng/ml) and $PGF_{2\alpha}$ (50-100 ng/ml). These findings contrast with the reported behaviour of non-pregnant myometrium, whose contractility is usually stimulated by PGF and inhibited by PGE. A selective action of the prostaglandins on the pregnant uterus was evidenced by the relative inactivity of the lower segment compared with the upper segment. (Authors modified) 1073

0564

von EULER, U.S.

Prostaglandins.

Clinical Pharmacology and Therapeutics. 9: 228-239. 1968.

This review of the distribution and biological effect of prostaglandins contains sections on occurrence, biological effects, antagonisms, synergisms, and tachyphylaxis to prostaglandins. There are 53 references. (JRH) 1179

0565

FAIN, J.N.

Antilipolytic effect of prostaglandin E_1 on free fat cells.

In: Ramwell, P.W. and J.E. Shaw, eds., "Prostaglandin Symposium of the Worcester Foundation for Experimental Biology," p.67-77. New York, Interscience, 1968.

The presence or absence of Ca^{++} or K^+ had little effect on PGE_1 (10 ng/ml) inhibition of basal lipolysis or lipolysis stimulated by a combination of theophylline, growth hormone and dexamethasone in rat white fat adipocytes in vitro. PGE_1 (100 ng/ml) or insulin greatly reduced glycerol release stimulated by a combination of dexamethasone, theophylline and growth hormone, but addition of norepinephrine (NE) caused a further increase in lipolysis which was not lowered by PGE_1. PGE_1 also had no effect on lipolysis in white adipocytes stimulated by dibutyryl cAMP. With brown fat adipocytes PGE_1 (100 ng/ml) reduced the free fatty acid (FFA) and glycerol release stimulated by theophylline, but only inhibited FFA release stimulated by NE (glycerol release unchanged). In contrast PGE_1 increased FFA but not glycerol release in the presence of dibutyryl cAMP. Only 1 ng/ml PGE_1 was required to produce maximal inhibition of theophylline stimulated lipolysis in brown adipocytes. The antilipolytic effects of PGE_1 are not due to stimulation of phosphodiesterase, but may be due to inhibition of adipocyte adenyl cyclase. (JRH) 1163

0566
FAIN, J.N.
  Stimulation by insulin and prostaglandin E_1 of glucose metabolism and inhibition of lipolytic action of theophylline on fat cells in the absence of K^+.
  Endocrinology. 83: 548-554. 1968.

Insulin and PGE_1 (10 ng/ml) inhibited the lipolytic action of theophylline on white fat cells and stimulated the metabolism of glucose by cells isolated and incubated in K^+-free buffer. Fat cells incubated in K^+-free buffer metabolized more glucose than in regular buffer and under certain conditions a marked increase in glucose metabolism due to insulin was best seen in the presence of theophylline. The addition of tetraphenylborate, a K^+ complexing agent, did not block the action of insulin in K^+-free buffer. These results indicate that the mechanisms by which insulin and PGE_1 affect the fat cell metabolism are not dependent upon extracellular K^+. (Author modified) 1033

0567
FLACK, J.D., R. JESSUP and P.W. RAMWELL
  Effect of prostaglandins on the pituitary-adrenal axis in the intact and hypophysectomised rat.
  In: Proceedings of the International Union of Physiological Sciences, 24th, Washington, D.C. 1968. Vol 7. Abstracts of volunteer papers and films. p. 137. Washington, D.C. Federation of American Societies for Experimental Biology, 1968.

Abstract only. Adrenal glands from hypophysectomised, and intact female rats, (120-240 g) which were killed under conditions known to prevent endogenous release of ACTH, contained prostaglandin-like material which after chromatography was identified with PGE_1 and $PGF_{2\alpha}$ (105 and 368 ng respectively/g adrenal gland). Ether stress or treatment with ACTH (sc 5 IU/kg) was found to significantly reduce ($P<0.01$) the prostaglandin content of adrenal gland examined 30 min later. The time course of release of prostaglandin-like material, free fatty acids (FFA) and corticosterone (B) was also studied by superfusing decapsulated adrenal glands with Krebs bicarbonate buffer for 4-5 hr. During the first 60 min the release of $PGF_{1\alpha}$ and $PGF_{2\alpha}$ from adrenal glands taken from intact and hypophysectomised animals was high (272 ng $PGF_{2\alpha}$/g/hr) but during the first 60 min of perfusion with ACTH, prostaglandin efflux was significantly lower (51.5 ng $PGF_{2\alpha}$/g/hr) although both FFA and B were significantly increased. These results indicate either inhibition of prostaglandin synthesis or increased prostaglandin metabolism by ACTH. In vivo and in vitro methods using PGE_1, $PGF_{2\alpha}$ and their precursors have been employed to investigate the effect of prostaglandins on corticosteroidogenesis. (Authors) 1325

0568
FREDHOLM, B.B. and S. ROSELL
Metabolic effects of prostaglandin E_1 in canine adipose tissue in situ.
Acta Physiologica Scandinavica. 74: 6A. 1968.

Subcutaneous dog adipose tissue was perfused with defibrinated blood in situ. PGE_1 $(0.4 \times 10^{-9}$ to $0.6 \times 10^{-6} M)$ caused an increased uptake of glucose. The effect on lipolysis was complex. There was an initial increase in glycerol and free fatty acid release with PGE_1 concentrations between 0.5×10^{-8} and $0.6 \times 10^{-7} M$. When infusion was stopped, there was a second transient increase in lipolysis. Larger doses of PGE_1 caused a decreased basal lipolysis. PGE_1 was found to inhibit sympathetic nerve stimulation of lipolysis with the threshold dose being about $10^{-9} M$ and $0.6 \times 10^{-7} M$ producing a 50% reduction. Complete inhibition was obtained with higher doses. It is concluded that PGE_1 has a dual effect on lipolysis depending on dose. (JRH) 1019

0569
FRIED, J., S. HEIM, P. SUNDER-PLASSMANN, S.J. ETHEREDGE, T.S. SANTHANAKRISHNAN, and J. HIMIZU
Synthesis of 15-desoxy-7-oxaprostaglandin $F_{1\alpha}$ and related substances.
In: Ramwell, P.W. and J.E. Shaw, eds., "Prostaglandin Symposium of the Worcester Foundation for Experimental Biology," p. 351-363. New York, Interscience, 1968.

The synthesis of 7-oxa $PGF_{1\alpha}$ and several related compounds is described. The starting material is *cis*-cyclopentene-3,5-diol. Structural formulas and step-by-step procedures are given. (JRH) 1140

0570
FRIED, J., S. HEIM, S.J. ETHEREDGE, P. SUNDER-PLASSMANN, T.S. SANTHANAKRISHNAN, J. HIMIZU and C.H. LIN
Synthesis of $(\pm)$-7-oxaprostaglandin $F_{1\alpha}$.
Chemical Communications. 634-635. 1968.

The synthesis of $(\pm)$-7-oxaprostaglandin $F_{1\alpha}$ from all-cis-1,2-epoxycyclopentane-3.5-diol is described. Chemical and physical properties of some intermediates and the final product are given in addition to structural formulas for each step in the synthesis. (RAP) 1307

0571
GAJDOS, A.
Less prostaglandines et leurs effets biologiques. [Prostaglandins and their biological effects.]
La Presse Medicale. 76: 513-516. 1968.

The biochemical and biological actions of the prostaglandins are reviewed. Nomenclature, structural diagrams, biosynthesis and metabolism are discussed briefly. Also mentioned are their effects on uterine and smooth muscle, the cardiovascular system, cell metabolism in relation to cAMP, and the central nervous system. (NES) 1242

0572
GANTT, C.L., L.R. KIZLAITIS, D.R. THOMAS and J.G. GRESLIN
Fluorescence of prostaglandin E_1.
Analytical Chemistry. 40: 2190-2191. 1968.

142

PGE$_1$ was dissolved in 70% sulfuric acid-water (v/v) and incubated at 65°C for 30 min. Fluorescence was determined on a spectrophotoflurometer. The samples were activated at 366 mμ and read at 402 mμ. The acid-water reagent deteriorated in a few days making it necessary to use freshly prepared reagent. The higher the concentration of PGE$_1$, the greater the fluorescence. The limits of detection for accurate measurement are in the range of 0.2 to 0.5 μg of PGE$_1$ C°. PGE$_2$, PGE$_3$, PGA$_1$ and PGB$_1$ fluoresce at absorption and emmission spectra very similar to that of PGE$_1$, and have about the same intensity of fluorescence. The emission spectrum of PGF$_{1\alpha}$ appears to be about 30 mμ longer than those of the other PG's. This technique could be used to estimate prostaglandins, but should be used only with relatively pure PG's to avoid contaminating compounds that also fluoresce. (JRH) 1194

0573
GILMORE, N., J.R. VANE and J.H. WYLLIE
Prostaglandin output from the spleen.
British Journal of Pharmacology and Chemotherapy. 32: 425P-426P. 1968.

Isolated dog spleens were perfused with Krebs soultion and the splenic nerve was stimulated (5-50 shocks/min). The effluent perfusion fluid was bioassayed for prostaglandin and catecholamine contents. The prostaglandins were also extracted chemically to remove other pharmacologic substances and then bioassayed. Slow rates of stimulation led to release of blood but no prostaglandin. Higher rates of stimulation produced release of prostaglandin (1-500 ng/ml assayed as PGE$_2$). When the sinusoids had been emptied of blood by repeated stimulation, high levels of prostaglandins could still be produced by rapid stimulation. It seems that upon nerve stimulation, the prostaglandins are either released from splenic muscle or some other tissue in the spleen, but not from stored cells. (JRH) 1056

0574
GILMORE, N., J.R. VANE and J.H. WYLLIE
Prostaglandins released by the spleen.
Nature. 218: 1135-1140. 1968.

Isolated dog spleens were used to determine the source and kinds of prostaglandins released from the spleen after contraction by nerve stimulation or infusion of epinephrine. The prostaglandins released were PGE$_2$ and PGF$_{2\alpha}$. The prostaglandins were not released from nerve cells, but from smooth muscle or other adrenergic innervated cells. Since infusion of prostaglandins into the spleen did not cause a contraction, it seems unlikely that they are mediators of smooth muscle contractions. (JRH) 1178

0575
GLAVIANO, V. and T. MASTERS
Inhibitory action of prostaglandin E$_1$ on myocardial lipolysis.
Circulation 37/38 (supp. 6): VI-83. 1968.

Abstract only. The action of prostaglandin E$_1$ (PGE$_1$) on cardiac lipid metabolism was studied in 15 anesthetized dogs. Myocardial uptake of plasma free fatty acids (FFA) and glucose were determined immediately after the intracoronary administration of 0.5 μg/kg/min of PGE$_1$ for 10 minutes. Following the infusion, cardiac muscle was excised and analyzed for triglycerides. Continuous recordings were made of mean arterial and coronary perfusion pressures, force of ventricular contraction, coronary sinus flow and ECG. PGE$_1$ caused mean arterial and coronary perfusion pressure to decline. Coronary flow decreased from 56 to 39 ml/100g/min, and ventricular force of contraction decreased 12%. PGE$_1$ depressed myocardial uptake on FFA from 18 to 5 μEq/100g/min while cardiac muscle triglycerides rose from 8 to 40 mEq/kg. The blocking action of PGE_1 on lipolysis of triglycerides was accompanied by a rise in myocardial uptake of glucose from -0.5 to 35 mM/100g/min. Since triglyceride formation requires the synthesis of β-glycerophosphate from glucose, the elevation in cardiac triglycerides may have resulted from the increase in glucose uptake. Thus the inhibitory action of PGE_1 on basal lipolysis in the heart appears similar to that reported for adipose tissue. (Authors) 1044

0576
GRANSTROM, E., W.E.M. LANDS, and B. SAMUELSSON
Biosynthesis of 9α, 15-dihydroxy-11-ketoprost-13-enoic acid.
Journal of Biological Chemistry. 243: 4104-4108. 1968.

Incubation of 8,11,14-eicosatrienoic acid with homogenates of sheep vesicular gland yielded a new product, which has been identified as 9α, 15-dihydroxy-11-ketoprost-13-enoic acid. The identification was based on mass spectrometry of the methyl ester and the methoxime derivative, and on infrared spectroscopy and ultraviolet spectroscopy, before and after treatment with alkali. The mechanism of the conversion of 8.11.14-eicosatrienoic acid into prostaglandins is discussed. The identification of 9α, 15-dihydroxy-11-ketoprost-13-enoic acid and the finding that ^{3}H at C-9 is retained during its formation futher support the involvement of an endoperoxide in the biosynthesis of prostaglandins. (Authors modified) 1082

0577
GRANT, J.K., J.W. DOBBIE, A.M. MacKAY, T. SYMINGTON and C. RILEY
The production and secretion of adrenocortical steroids in man.
In: Dorfman, R.I., K. Yamasaki and M. Dorfman, eds., "Biogenesis and Action of Steroid Hormones," p. 74-92. Los Altos, Geron-X. 1968.

Administration of ACTH daily for 3 or 4 days prior to surgical removal of the adrenal glands from human patients resulted in a decrease in arachidonic acid content of the glands. Since this acid is a known precursor of PGE_2, the possible role of prostaglandins in adrenal function is discussed. (JRH) 1306

0578
GRANTHAM, J.J. and J. ORLOFF
Effect of prostaglandin E_1 on the permeability response of the isolated collecting tubule to vasopressin, adenosine 3',5'-monophosphate, and theophylline.
Journal of Clinical Investigation. 47: 1154-1161. 1968.

The effect of PGE_1 on the water permeability response to vasopressin, theophylline and cAMP of isolated, perfused collecting tubules of the rabbit kidney was investigated in vitro. In the collecting tubule, PGE_1 elicited a minimal increase in net water absorption along an osmotic gradient. However, when administered in association with a concentration of vasopressin (2.5 μU/ml^{-1}) selected to induce a submaximal increment in water absorption, the effect of the latter was reduced by approximately 50%. Theophylline (5 $\times$ 10^{-3} M) also increased net water absorption. This effect was potentiated by the simultaneous addition of PGE_1. In contrast, PGE_1 did not influence the increase in net water absorption induced by cAMP (10^{-2} M). Since cAMP is responsible for the permeability effect of vasopressin in renal tissue, the present results are consistent with the view that PGE_1 interferes with the action of the octapeptide by competing with it at a site which influences the generation of cAMP. (Authors modified) 1192

0579

GREEN, K. and B. SAMUELSSON

On the excretion of dinor-prostaglandin $F_{1\alpha}$ in the rat.

In: Ramwell, P.W. and J.E. Shaw, eds., "Prostaglandin Symposium of the Worcester Foundation for Experimental Biology," p.389-394. New York, Interscience, 1968.

The metabolism of $PGF_{1\alpha}$ was followed quantitatively by measuring the production of dinor-$PGF_{1\alpha}$ in the urine of rats with isotope dilution techniques. Duterium labelled $PGF_{1\alpha}$ was injected into rats and the urine collected for 24 hr. The dinor-$PGF_{1\alpha}$ isolated indicated a production of 11.7 mμg of dinor-$PGF_{1\alpha}$/24 hr/200 g/rat. (JRH) 1136

0580

GULBENKIAN, A., L. SCHOBERT, C. NIXON, and I.I.A. TABACHNICK

Metabolic effects of pertussis sensitization in mice and rats.

Endocrinology. 83: 885-892. 1968.

The release of free fatty acids (FFA) from the epididymal fat pads of normal rats and rats sensitized with *Bordetella pertussis* vaccine was compared in vitro. Normal fat pads showed little FFA release while pertussis sensitized (PS) fat pads released FFA at a rate similar to normal fat pads stimulated with 0.25 μg/ml norepinephrine (NE). PGE_1 (1.0 μg/ml) caused a 50% inhibition of NE stimulated FFA release in normal fat pads but had on effect on FFA release from PS adipose tissue. Insulin (0.02 U/ml) caused a 79% reduction in FFA release from the PS fat pads but had no effect on NE stimulated lipolysis in normal fat. Dichlorisoproterenol inhibited NE-induced FFA release by 95% in normal fat pads and also inhibited PS pad release but only by 36%. These data imply that the mechanism of lipolysis in the fat pad from PS rats differs from catecholamine induced lipolysis in the fat pad of the normal rat. (JRH) 1201

0581

HALL, L. J. and R.T. JACKSON

Effects of alpha and beta adrenergic agonists on nasal blood flow.

Annals of Otology, Rhinology and Larynology. 77: 1120-1130. 1968.

The effects of various adrenergic agonists and antagonists on the nasal vasculature of pentobarbital anesthetized dogs were measured by recording nasal airway resistance. In this system, vasoconstriction of nasal vasculature results in decreased nasal resistance, while vasodilation results in an increase in nasal resistance. $PGF_{1\alpha}$ was used as a nonadrenergic vasoconstrictor for comparison with known adrenergic agents. The vasoconstriction caused by intraarterial $PGF_{1\alpha}$ was not altered by phenoxybenzamine and α-adrenergic antagonist, or propranolol, a β-adrenergic antagonist. (JRH) 1193

0582

HAMBERG, M.

On the absolute configuration of 19-hydroxy-prostaglandin B_1.

European Journal of Biochemistry. 6: 147-150. 1968.

19-Hydroxy-prostaglandin B_1, isolated from human seminal plasma, was treated with tritium-labeled acetic anhydride and degraded by oxidation with permanganate in acetone. Tritium-labeled 3-acetoxybutanoic acid was isolated from the oxidation product and diluted with unlabeled *dl*-3-acetoxybutanoic acid. The morphine salt of the acid was resolved. These experiments showed that the tritium-labeled 3-acetoxybutanoic acid had the *d*-configuration and consequently the parent 19-hydroxy-prostaglandin B_1 has the *d*-configuration at C-19. The configuration at c-15 is discussed. (Authors) 1188

0583
HAMBERG, M.
   Metabolism of prostaglandins in rat liver mitochondria.
   European Journal of Biochemistry. 6: 135-146. 1968.

   A series of prostaglandins has been incubated with the β-oxidizing enzyme system of rat liver. All of the compounds tested were oxidized. Prostaglandin E_1, prostaglandin B_1, prostaglandin $F_{1\alpha}$ and prostaglandin $F_{1\beta}$ were converted into the corresponding C_{18} homologues. Prostaglandin A_1, 11α, 15-dihydroxy-9-keptoprostamoic acid and 11α-hydroxy-9,15-diketoprostanoic acid yielded mixtures of the C_{16} and C_{18} homologues. Nor-prostaglandin $F_{1\alpha}$ and homo-prostaglandin $F_{1\alpha}$ were both converted into trinor-prostaglandin $F_{1\alpha}$. (Authors) 1190

0584
HAMBERG, M.
   Studied on the biosynthesis and metabolism of prostaglandins.
   Dissertation, Department of Chemistry, Karolinska, and Department of Medical Chemistry, Royal Veterinary College, Stockholm. 35p. 1968.

   The author reviews 7 papers published between 1966 and 1968 by himself and collaborators on prostaglandin biosynthesis, configuration and metabolism. (RMS) 1314

0585
HANDLER, J.S., R. BENSINGER and J. ORLOFF
   Effect of adrenergic agents on toad bladder response to ADH, 3',5'-AMP, and theophylline.
   American Journal of Physiology. 215: 1024-1031. 1968.

   It is mentioned in the introduction of this paper that prostaglandins have been reported to inhibit cAMP in the toad bladder. (JRH) 1094

0586
HANDLER, J.S. and J. ORLOFF
   The role of cyclic AMP in the renal responses to hormones.
   Pharmacologist. 10: 146. 1968.

   Abstract only. Prostaglandins are mentioned as being able to inhibit renal responses to vasopressin by inhibiting its effect on adenylcyclase. (JRH) 1320

0587
HANDSCHIN, U., H.P. SIGG and Ch. TAMM
   Zur biosynthese von brefeldin A. [On the biosynthesis of brefeldin A.]
   Helvetica Chimica Acta. 51: 1943-1965. 1968.

   One of the proposed mechanisms in the biosynthesis of brefeldin A suggests the involvement of O_2 in closing the carbon ring; such an involvement has been shown to occur in the formation of prostaglandins. The mechanism postulated for the closing of the ring in the case of the prostaglandins is not, however, applicable to that of brefeldin A, as it would lead to a different distribution of the oxygen functions. (MEMH) 1092

0588
HARRY, J.D.
The action of of prostaglandin E_1 on the guinea-pig isolated intestine.
British Journal of Pharmacology and Chemotherapy. 33: 213-214. 1968.

The effects of PGE_1 (1.5 to 40 ng/ml) on isolated guinea pig intestine were measured with transmural electrical stimulations and in the presence of several compounds which are known to alter smooth muscle activity. Atropine (0.01 to 0.1 μg/ml), procaine (5 μg/ml) and hexamethonium (50 μg/ml) reduced the response to PGE_1 while diisopropylflurophosphate (5 μg/ml) potentiated the response. PGE_1 was also potentiated by electrical stimulation since subthreshold doses of PGE_1 caused an increase in tone due to electrical stimulation. It is concluded that PGE_1 has two sites of action, one involving intramural nerves and the other directly on the smooth muscle cells. It is also suggested that the effect may be due to PGE_1 potentiating postganglionic transmitters, i.e. acetycholine. (JRH) 1054

0589
HAWKINS, D.F.
Revelance of prostaglandins to problems of human subfertility.
In: Ramwell, P.W. and J.E. Shaw, eds., "Prostaglandin Symposium of the Worcester Foundation for Experimental Biology," p. 1-10. New York, Interscience, 1968.

The prostaglandin content of semen samples from 49 subfertile men was bioassayed on rabbit intestine. Comparison of PG content with other factors such as sperm count or motility did not reveal any signficant relationships. However, when the PG content was compared with subsequent fertility, the results suggested a positive relationship between PG content and fertility. Bioassay (guinea pig ileum) of semen from oligospermic (20 million/ml) men from this group showed a much more positive relationship between fertility and PG content, even when compared to the other subfertile men. Analysis of individual prostaglandins (PGE_1, PGE_2, PGE_3, $PGF_{1\alpha}$ and $PGF_{2\alpha}$) revealed a proportionate reduction in all PG's rather than a deficiency in one particular PG in oligospermic men. (JRH) 1169

0590
HAWKINS, R.A., R. JESSUP and P.W. RAMWELL
Effect of ovarian hormones on response of the isolated rat uterus to prostaglandins.
In: Ramwell, P.W. and J.E. Shaw, eds., "Prostaglandin Symposium of the Worcester Foundation for Experimental Biology," p. 11-19. New York, Interscience, 1968.

The sensitivity of the rat uterus under specifically defined hormonal conditions to PGA_1, PGE_1, $PGF_{1\alpha}$ and $PGF_{2\alpha}$ was measured in vitro and compared with sensitivity to oxytocin, serotonin and bradykinin. The rats were either bilaterally ovariectomized or ovariectomized and adrenalectomized 14 days before removal of the uterus. Some of the animals were treated with progesterone (2 mg/day s.c.) or estradiol (6 μg/day s.c.) for four days prior to removal of the uterus. Pretreatment of the ovariectomized animals with estradiol greatly reduced the sensitivity (by 16-30 fold) of the uteri to all the PG's tested. Progesterone pretreatment had little effect on PG sensitivity. Adrenalectomy had no effect except that the sensitivity to $PGF_{1\alpha}$ was decreased. Neither estradiol nor progesterone pretreatment altered the response to serotonin or bradykinin. Estrogen pretreatment greatly enhanced (45 fold) the response to oxytocin. However, 2 of the 6 controls and 3 of 9 progesterone treated rats were insensitive to oxytocin and were omitted from the calculations. When 20 μg/ml progesterone or estradiol was added to the organ bath with uteri from ovariectomized rats, the response to PGE_1 and oxytocin was inhibited. (JRH) 1168

0591

HEDQVIST, P.

Reduced effector response to nerve stimulation in the cat spleen after administration of prostaglandin E_1.

Acta Physiologica Scandinavica. 74: 7A. 1968.

Experiments were carried out in the isolated, perfused cat spleen to study the effects of prostaglandins on the contraction of vascular smooth muscle in response to sympathetic nerve impulses. Electrical stimulation of the splenic nerves at 10/sec for 20 sec produced a large rise of the perfusion pressure in the isolated spleen. Intrarrterial infusion of PGE_1 in doses ranging from 0.013 to 1.3 µg/ml of perfusion fluid markedly reduced this response. The inhibitory effect progressively increased with the PGE_1 dose. Experiments were also carried out to test the effect of PGE_1 on the response to intraarterial injection of noradrenaline (NA). PGE_1 was found to inhibit the pressor response caused by injection of NA as readily as that resulting from nerve stimulation. It appears conceivable that local release of PGE_1 may play a modulatory role in the vascular sympathetic neuro-effector system. (Author modified) 1018

0592

HENDERSON, P. Th., E.J. ARIENS and A.M. SIMONIS

Differentiation of various types of cholinergic and other spasmogenic actions on the isolated guinea-pig ileum.

European Journal of Pharmacology. 4: 62-70. 1968.

A variety of smooth muscle spasmogens were tested on isolated guinea pig ileum in the presence of various inhibitors. PGE_1, histamine, barium and serotonin were classed as having mainly non-cholinergic action on the basis of these experiments. PGE_1 (1.4×10^{-6}M) caused contraction of the ileum which was not affected by a variety of inhibitors and conditions including lachesine, hexamethonium, hemicholinium, morphine, procaine, cooled aging, and neobenodine. The addition of extra Mg^{++} or the absence of Ca^{++} caused a dramatic inhibition of the action of PGE_1 on this tissue. (JRH) 1030

0593

HERMAN, T.S. and I.S. EDELMAN

The effect of prostaglandin E_1 (PGE_1) on the action of vasopressin and aldosterone in the toad bladder.

In: Proceedings of the International Union of Physiological Sciences, 24th, Washington, D.C., 1968 Vol. 7. Abstracts of volunteer papers and films. p. 188. Washington, D.C. Federation of American Societies for Experimental Biology, 1968.

Abstract only. At a concentration of 2.5×10^{-7}M PGE_1, an inhibitor of the enzyme adenyl cyclase, had no detectable effect on the osmotic flow of water across the isolated urinary bladder of the toad, although it did completely inhibit the usual osmotic flow of water induced by submaximal concentrations of vasopressin (1 mUnit/ml). We extended these studies to test the ability of PGE_1 (2.5×10^{-7}M) to inhibit the increase in short-circuit current (scc) produced by vasopressin (1 mUnit/ml). PGE_1 alone increased scc 13 ± 7% in 1 hour. However, the rise in scc was the same in hemibladders tested with PGE_1 and vasopressin as in hemibladders treated with vasopressin alone. These results indicate that PGE_1 can inhibit the effect of vasopressin on osmotic water flow without changing its effect on Na transport. The data support the hypothesis of Petersen and Edelman that in the toad bladder vasopressin stimulates synthesis of cyclic 3',5'-AMP at 2 separate sites. It has been suggested that cyclic 3',5'-AMP is involved in the mechanism of action of aldosterone. Paired hemibladders were treated with either a submaximal concentration (7×10^{-9}M) of aldosterone alone or with aldosterone and PGE$_1$ (2,5 X 10^{-7}M). There was no significant difference in the increment of scc on PGE$_1$-treated and control hemibladders. (Authors modified) 1326

0594

HERZOG, J.P., H. JOHNSTON and D. LAULER
Effects of prostaglandins E$_1$, E$_2$ and A$_1$ on renal hemodynamics, sodium and water excretion in the dog.
Clinical Research. 16: 386. 1968.

Abstract only. When infused into the left renal artery of the dog in doses of 0.1 to 2.0 μg/min, PGE$_1$, PGE$_2$ and PGA$_1$ all caused a significant increase in renal plasma flow, urine sodium excretion, osmolar excretion, urine volume, and free water clearance from the infused side. Mean systemic blood pressure and glomerular filtration rate remained stable. The changes are not blocked by atropine, dibenzyline, propanalol, or exogenous pitressin. No statistically significant differences on free water clearance between PGE$_1$, PGE$_2$ and PGA$_1$ were demonstrated. Since PGA$_1$ does not appear to inhibit the adenyl-cyclase system in the isolated fat pad, the increases in free water clearance seen with these compounds may not be due to a direct anti-ADH effect in the canine kidney, but rather may reflect increased solute clearance, vasodilation, a direct tubular effect, or a combination of these factors. (Authors modified) 1206

0595

HERZOG, J.P., H.H. JOHNSTON and D.P. LAULER
Effects of prostaglandins E$_1$, E$_2$, and A$_1$ on renal hemodynamics, sodium and water secretion in the dog.
In: Ramwell, P.W. and J.E. Shaw, eds., "Prostaglandin Symposium of the Worcester Foundation for Experimental Biology," p. 147-161. New York, Interscience, 1968.

Sodium pentobarbital anesthetized dogs were used to investigate the effects of infusion of PGE$_1$, PGE$_2$ and PGA$_1$ into a renal artery on kidney function and hemodynamics. All three PG's (1.0-2.0 μg/min) caused significant increase in renal plasma flow, urine sodium excretion, osmolar excretion, urine volume and free water clearance. At these dosage levels systemic blood pressure and glomerular filtration rate were unchanged. These changes were not blocked by atropine or exogenous ADH. Since PGE's are found in high concentration in the kidney they may be important in intrarenal control of blood flow, sodium and water excretion. (JRH) 1158

0596

HERZOG, J.P., H.H. JOHNSTON, D.P. LAULER
Renal roles of prostaglandins.
New England Journal of Medicine. 278: 219-220. 1968.

In this letter to the editor the authors briefly review their own work on the effects of prostaglandins on kidney function and hemodynamics which are in agreement with those of Lee whose article appeared in an earlier issue of the journal. (JRH) 1048

0597

HICKLER, R.B.
The identification and measurement of prostaglandin in human plasma.
In: Ramwell, P.W. and J.E. Shaw, eds., "Prostaglandin Symposium of the Worcester Foundation for Experimental Biology," p. 279-293. New York, Interscience, 1968.

A technique for extraction of prostaglandins from human plasma is described. The initial extraction was made with a 1:5 mixture of ethanol and methylal. The separation of PG's was performed on column of silica gel. Quantification was performed with a bioassay system. It was found that incubation of 25 μg of PGE$_1$ in 1 ml of plasma at room temperature for 90 min did not result in any loss of biological activity. Prostaglandins seem to be relatively stable in plasma. (JRH) 1146

0598
HILLIER, K. and S.M.M. KARIM
Effects of prostaglandins E$_1$, E$_2$, F$_{1\alpha}$, F$_{2\alpha}$ on isolated human umbilical and placental blood vessels.
Journal of Obstetrics and Gynaecology of the British Commonwealth. 75: 667-673. 1968.

The effects of PGE$_1$, PGE$_2$, PGF$_{1\alpha}$ and PGF$_{2\alpha}$ were investigated on human umbilical and placental vessels obtained at 13-24 weeks gestation and 34-42 weeks gestation. Using term preparations, PGE$_1$ produced a relaxation of 24 of 32, but doses up to 5 μg/ml had no effect on the remaining 8. PGF$_{1\alpha}$ produced contractions in 18 of 24 term preparations. No effect was seen in 6 at up to 8 μg/ml, but they did respond to 5-HT. PGE$_2$ and PGF$_{2\alpha}$ were essentially equiactive. Material from early (13-24 week) pregnancies yielded 35 artery and 10 vein preparations. Only 2 responded to 5-HT, neither responded with up to 1 μg/ml of PGE$_2$, PGF$_{2\alpha}$ or PGE$_1$. Preparations from 2 spontaneous abortions at 19 and 20 weeks did not respond to 5-HT or to prostaglandins. (RMS) 1107

0599
HINMAN, J.W. and R.M. MORRELL
Non-steroidal hormones and their antagonists.
In: Cain, C.K., ed., "Annual Reports of Medicinal Chemistry," p. 184-199. New York, Academic Press. 1968.

Contained in this review is a section on miscellaneous biologically active substances in which it is mentioned that prostaglandin E$_2$ and A$_2$ have been isolated from rabbit kidney. (JRH) 1225

0600
HOLDEN, K.G., B. HWANG, K.R. WILLIAMS, J. WEINSTOCK, M. HARMAN and J.A. WEISBACH
Synthetic studies on prostaglandins.
Tetrahedron Letters. 1569-1574. 1968.

The practical utility of a total synthesis of PGF$_{1\alpha}$ and PGE$_1$ previously described by Just and Simonovitch was tested. No detectable PGE$_1$ was formed and only a low yield of PGF$_{1\alpha}$ isomers were obtained with this technique. Structural formulas of the products formed at each step of the synthesis are given. A modification of the synthesis which leads directly to PGB$_1$ is also described. It was concluded that this synthesis is not a practical method for the synthesis of PGF$_{1\alpha}$ or PGE$_1$, but with modification it can yield PGB$_1$. (JRH) 1329

0601
HOLLENBERG, M., R.S. WALKER and D.P. McCORMICK
Cardiovascular responses to intracoronary infusion of prostaglandin E$_1$, F$_{1\alpha}$, and F$_{2\alpha}$.
Archives Internationales de Pharmacodynamie et de Therapie. 174: 66-73. 1968.

150

The direct cardiac actions of PGE_1, $PGF_{1\alpha}$ and $PGF_{2\alpha}$ were studied using intracoronary infusions in unanesthetized and pentobarbital anesthetized dogs. PGE_1, when infused directly into the left anterior descending coronary artery (2 to 8 μg/min) greatly increases myocardial blood flow (29 to 216%) while it increases contractile force only slightly (8 to 24%). $PGF_{1\alpha}$ and $PGF_{2\alpha}$ (4 to 12 μg/min) have only slight or no effects on myocardial blood flow and no effects on myocardial contractility. None of the compounds have any electrocardiographic effects nor do they block the positive inotropic action of norepinephrine. Thus, in the canine heart, the prostaglandins have mainly coronary vasodilating properties which, moreover, appear to be of doubtful physiologic importance. (Authors modified) 1126

0602
HOLMES, S.W. and E.W. HORTON
The distribution of tritium-labelled prostaglandin E_1 injected in amounts sufficient to produce central nervous effects in cats and chicks.
British Journal of Pharmacology. 34: 32-37. 1968.

After injection of tritiated PGE_1 into the lateral cerebral ventricle, the carotid and vertebral arteries and the aorta in cats, only a small proportion of the radioactivity was recovered from the central nervous system. Similar results were obtained when the prostaglandin was given intravenously and intra-arterially to chicks. The results of this study confirm work in other species and suggest that PGE_1 exerts it central pharmacological actions in small concentrations. (Authors modified) 1014

0603
HOLMES, S.W. and E.W. HORTON
The identification of four prostaglandins in dog brain and their regional distribution in the central nervous system.
Journal of Physiology. 195: 731-741. 1968.

Dogs were anesthetized with pentobarbital, exsanguinated and the brains removed. The brains were dissected and similar parts from 19 brains pooled. Prostaglandins were extracted and purified chromatographically and bioassayed. The prostaglandins found were PGE_1, PGE_2, $PGF_{1\alpha}$ and $PGF_{2\alpha}$. Each of these prostaglandins was distributed throughout all regions of the central nervous system investigated (cortex, hippocampus, caudate nucleus, hypothalamus, cerebellum, medulla and pons, cortical white matter and spinal cord). (JRH) 1010

0604
HOLMES, S.W., E.W. HORTON and M.J. STEWART
Observations on the extraction of prostaglandins from the blood.
Life Sciences. 7: p. 349-354. 1968.

The efficiency of standard prostaglandin extraction techniques for recovering PGE_1 and PGE_2 was examined using dog blood. The commonly used ethanol extraction only recovered 5 to 10% of the PGE_1 or PGE_2 added at a concentration of 1 μg/ml. When the ethanol extraction step was omitted and the blood extracted directly with ethyl acetate, 55% of the PGE_1 added at a concentration of 0.1 to 1.0 μg/ml was recovered. When the PGE_1 concentration was lowered to 20 to 50 ng/ml, the ethyl acetate recovery became more variable with recovery being reduced to 25% in some cases. Incubations of up to 3 hr did not significantly alter the concentration of PGE_1 in the blood. When blood, incubated for 30 min with 5 μg/ml PGE_1, was subjected to electrophoresis and the various protein zones analyzed for PGE_1 content, all of the PGE_1 activity was found in the zones corresponding to the α_1 and β_2 globulin zones. The authors suggest that when extracting PG's from blood, the initial ethanol extraction step should be omitted. The authors feel that the poor recovery of PG when ethanol is used may be due to its precipitation of the protein to which the PG is bound, namely the α globulins. (JRH) 1238

0605

HOLMES. S.W. and E.W.HORTON
Prostaglandins and the central nervous system.
In: Ramwell, P.W. and J.E. Shaw, eds., "Prostaglandin Symposium of the Worcester Foundation for Experimental Biology," p. 21-37. New York, Interscience, 1968.

All parts of the dog brain and spinal cord were found to contain PGE and PGF compounds. Assay of perfusates from the cerebral ventricular system showed PGE_1, PGE_2, and $PGF_{1\alpha}$ and $PGF_{2\alpha}$ to be present in measurable quantities in the perfusate. PGE_1 (1 mg/kg s c or i v) caused a ptosis and passing of liquid feces in mice. $PGF_{2\alpha}$ (1 mg/kg) did not produce ptosis. Both PGE_1 and PGE_2 increased hexobarbitone sleeping time in mice. PGE_1 (1 mg/kg) antagonized convulsions induced by leptazol, electro-shock and strychnine. $PGF_{2\alpha}$ potentiated the convulsions caused by leptazol. PGE_1-induced ptosis could be prevented by pretreatment of the mice with imipramine. Convulsions due to picrotoxin were unaffected by PGE_1. In the species tested PGE_1 has a sedative-tranquilizer action. $PGF_{2\alpha}$ seems to facilitate motor pathways. (JRH) 1167

0606

HOPKIN, J.M., E.M. HORTON and V.P. WHITTAKER
Prostaglandin content of particulate and supernatant fractions of rabbit brain homogenates.
Nature. 217: 71-72. 1968.

The prostaglandin content of subcellular fractions of homogenized whole rabbit brains was determined by bioassay and chromatographic methods. Ninety percent of the PG activity in homogenates was due to PGE_2 and $PGF_{2\alpha}$. Each gram of tissue was estimated to contain 500 ng of $PGF_{2\alpha}$ and 75 ng of PGE_2. After centrifugation 80 to 90% of the PG activity was in the supernatant. Between 5 and 10% of the activity was found in the particulate fraction comprised of mitochondria and nerve endings. No bound prostaglandin could be eluted from the particulate fraction. (JRH) 1017

0607

HOROWITZ, J.D. and M.L. MASHFORD
Vasoactive substances in plasma.
Australian Journal of Experimental Biology and Medical Science. 46: P-37. 1968.

Abstract only. The vasoconstriction observed when small amounts of rabbit or human plasma are rapidly injected into isolated perfused vessels is not due to prostaglandins. (JRH) 1121

0608

HOWOWITZ, J.D. and M.L. MASHFORD
Vasoactivity of human plasma and plasma protein fractions.
Experientia. 24: 1126-1127. 1968.

The vasoconstrictor responses seen in the rabbit ear following injection of plasma or Cohn fractions of plasma could be distinquished from the responses to prostaglandins and other vasoactive substances by the use of antagonists, or by considering the nature of the rabbit ear vein preparation, which is relatively insensitive to most vasoactive substances. (JRH) 1186

0609
HORTON, E.W., I.H.M. MAIN, C.J. THOMPSON and P.M. WRIGHT
Effect of orally administered prostaglandin E_1 on gastric secretion and gastrointestinal motility in man.
Gut. 9: 655-658. 1968.

Healthy male subjects, having fasted over night, were given various doses of PGE_1 orally. Two-fifths of the total dose of PGE_1 (or solvent control) diluted in 25 ml water was then drunk by the subject. 15 minutes later, residual fluid in the stomach was aspirated and one-fifth of the total PG dose dissolved in 5 ml was administered through a stomach tube. Pentagastrin, 6 μg/kg, was then injected subcutaneously. The stomach was emptied as completely as possible at 15, 30, 45, 60 and 75 minutes after the injection. Two further doses, each amounting to one-fifth of the total dose, of PGE_1 dissolved in 5 ml water were administered via the tube after the 15 and 30 minute samples were withdrawn. PGE_1 given in doses of 10-40 μg/kg did not inhibit normal or pentagastrin stimulated gastric secretion. The only effect on gastric secretion content was the appearance of bile in the stomach after administration of PGE_1 at higher dose levels. Prostaglandin also increased intestinal motility resulting in loose faeces. (JRH) 1220

0610
HORTON, E.W.
Problems in the identification of submicrogram amounts of prostaglandins.
Biochemical Journal. 107: 12P. 1968.

The difficulty in identifying prostaglandin extracted from biological tissues when the amount extracted is less that 1 μg is discussed. No chemical techniques are available for absolute determination of such small quantities. The best method is to use as many parallel assays (chromatographic, biological, etc.) as possible. The greater the agreement among these assays, the greater the probability that a correct identification has been made. (JRH) 1118

0611
HORTON, E.W. and I.H.M. MAIN
Prostaglandins.
Lancet. 1: 478. 1968.

Sir,—Your leading article (Jan 6, p.30) credits us with observations which were in fact made by others. The release of prostaglandins from the cerebral cortex, cerebellum, and spinal cord was reported by Ramwell and Shaw and Coceani and Wolfe. Furthermore in our original paper we did not exclude the possibility that prostaglandins may also act upon γmotor neurones. (Entire article cited) 1066

0612
HORTON, E.W.
The prostaglandins.
In: Stacey, R.S. and J.M. Robson, eds., "Recent Advances in Pharmacology," p. 185-212. London, Churchill. 1968.

This paper reviews several aspects of prostaglandins but is primarily concerned with the pharmacological effects on the various systems. There are 123 references. (RAP) 1265

0613

HUMES, J.L., L.R. MANDEL and F.A. KUEHL, Jr.
Studies on the mechanism of the antilipolytic action of prostaglandin E_1 in vitro.
In: Ramwell, P.W. and J.E. Shaw, eds., "Prostaglandin Symposium of the Worcester Foundation for Experimental Biology," p. 79-91. New York, Interscience, 1968.

The mechanism of the antilipolytic action of PGE_1 was investigated in rat epididmyal fat pads and isolated adipocytes. With intact fat pads PGE_1 (0.32 μg/ml) blocked the stimulation of cAMP by epinephrine or theophylline with norepinephrine (NE). PGE_1 (8.0 μg/ml) itself increased cAMP concentration in fat pads. With free adipocytes, PGE_1 (0.4-0.8 μg/ml) blocked the increase in cAMP due to NE with theophylline. However, unlike the intact fat pad experiments, PGE_1 (up to 5.0 μg/ml) did not stimulate cAMP in free adipocytes. PGE_1 had no significant effect on lipolysis due to applied cAMP in free adipocytes. PGE_1 acted synergistically with Ko 592 in blocking lipolysis due to NE and theophylline. The antilipolytic properties of PGE_1 are apparently due to its inhibition of adenyl cyclase synthesis. (JRH) 1162

0614

HYDOVITZ, J.D.
Prostaglandins and diarrhea.
New England Journal of Medicine. 278: 915. 1968.

This letter to the editor points out that the findings of Williams et al on the association of prostaglandin synthesis with medullary carcinoma of the thyroid have not been discussed in recent published material on thyrocalcitonin-like activity in medullary thyroid cancer. (RMS) 1047

0615

HYMAN, A.L.
Active responses of pulmonary veins.
Clinical Research. 16: 71. 1968.

Abstract only. Catheter techniques were designed to pump-perfuse a hemodynamically separated lobe (PPL) of intact dog lungs at constant rates, permitting direct measurement of active responses. After a 20F balloon catheter was passed from a jugular vein into a left lobar pulmonary artery (LPA), the balloon was inflated and the lobe was perfused with arterial blood at constant rates. With transseptal techniques, simultaneous pressures were obtained in the main PA and LPA, the PPL-small pulmonary veins (LPV) and both atria. Blood volume and flow in normally perfused lobes (NPLs) and blood volume of PPL were measured by dye techniques. Active responses were identified in the PPL, and modulated responses in NPLs. $PGF_{2\alpha}$ actively constricted LPVs and LPAs, while PGE_1 dilated these vessels. Pressure changes in vessels of NPLs were related to changes in output. The experiments demonstrated the active responses of intact pulmonary veins to vasoactive substances, and indicate the ability of these vessels to modulate cardiopulmonary hemodynamics. (JRH) 1214

0616

HYMAN, A.L.
Active responses of pulmonary veins to pharmacological stimuli in intact dogs.
Circulation 37/38 (supp. 6): VI-105. 1968.

Abstract only. The active pharmacological responses of small pulmonary veins (SPV) were studied in intact dogs by pump-perfusing shunted systemic arterial or venous blood through special pulmonary arterial (PA) balloon-perfusion catheters, at constant flow rates, into hemodynamically-separated lung segments. $PGF_{2\alpha}$ actively constricted PA and SPV, whereas both PGE_1 and isoproterenol actively dilated these vessels. The pulmonary veins are capable of active responses to vasoactive substances and can modulate cardio-pulmonary hemodynamics. (RAP) 1300

0617

HYNIE, S., D. MISEKOVA, K. ELISOVA
   Inhibitory effect of the beta sympatolytic drug propanolol on lipolysis induced by noradrenaline and ACTH in normal and hyperthyroid rats.
   Physiologia Bohemoslovaca. 17: 191-198. 1968.

In the discussion section, a hypothesis concerning the composition of adenyl cyclase is briefly mentioned. According to the authors, the regulatory effects of prostaglandins on lipolysis and various hormones can be explained by this hypothesis. (RAP) 1291

0618

INGELMAN-SUNDBERG, A., F. SANDBERG and G. RYDEN
   Experimentos en la absorcion de prostaglandinas en la vagina de mujeres no embarazadas.
   [Experiments on the absorption of prostaglandins in the vagina of non-pregnant women.]
   Ginecologia y Obstetricia de Mexico. 23: 491-495. 1968.

Tritium marked PGE_1 was placed for 6 hours in the vaginal fundus in 7 patients with normal menstruation. 10% to 25% of the administered radioactivity was recovered in the urine between 24 and 32 hours post administration. The application of a cervical cap had no influence on the absorption which is grossly estimated at 20 to 50%. The physiological significance of this absorption is as yet considered uncertain. (Translation of authors' summary) 1292

0619

INGELMAN-SUNDBERG, A.
   The prostaglandins.
   In: Behrman, S.J. and R.W. Kistner, eds., "Progress in Infertility," p. 639-648., Boston, Little, Brown, 1968.

This review contains sections on chemistry, occurrence, metabolism, biosynthesis, biological action and pathophysiological significance of prostaglandins. There are 41 references. (JRH) 1122

0620

ISHII, T., R.D. OLIVER and J. NAKANO
   Effect of dihydroergotamine (DHE) and nicotinic acid (NIA) on dibutryl cyclic 3',5'-AMP(DB-AMP)-induced lipolysis in isolated rat fat cells.
   Pharmacologist. 10: 212. 1968.

Abstract only. The antilipolytic properties of PGE_1 were compared to those of DHE and NIA in vitro. PGE_1 $(3 \times 10^{-6}$ to $10^{-8}M)$ added to rat adipocytes competetively blocked the lipolytic effects on norepinephrine and ACTH. However, PGE_1 did not block DB-AMP induced lipolysis. DHE and NIA did block DB-AMP-induced lipolysis. (JRH) 1309

0621
IWATA, H. and S. FUJIMOTO
[Physiological and pharmacological significance of prostaglandins.]
Saishin Igaku. 23: 1719-1729. 1968.

Article in Japanese. Abstract not available at present. 1303

0622
JACKSON, R.T. and R. STOVALL
Vasoconstriction of nasal blood vessels induced by prostaglandins.
In: Ramwell, P.W. and J.E. Shaw., ed., "Prostaglandin Symposium of the Worcester Foundation
for Experimental Biology," p. 329-333. New York, Interscience. 1968.

The effects of injection of (5×10^{-6}-10^{-2} mg/kg) PGE_1, PGE_2 $PGF_{1\alpha}$ or PGA_1 into the
carotid artery on nasal patency was measured in pentobarbital anesthetized dogs. A dose
response curve for epinephrine was determined in each dog prior to injection of PG. All 4 PG's
produced a vasoconstriction with the PGE's being about 100 times more effective than $PGF_{1\alpha}$
or PGA_1. Considerable variation in response was found from dog to dog, both in sensitivity and
duration of response. However, dogs that proved sensitive to low doses of epinephrine were also
sensitive to PG's. The amount of vasoconstriction produced by epinephrine and PG was about
the same but the duration of response was much longer with PG's (about 7 times longer). The
large degree of variability found in responses made comparison of drugs and tachyphylaxis hard
to assess. However, PGE_1 or PGE_2 did not seem to affect the response to epinephrine or be
affected by a prior response to papaverine. (JRH) 1142

0623
JESKE, W.
[Preliminary pharmacological evaluation of prostaglandins.]
Farmacja Polska. 24: 667-672. 1968.

Article in Polish. Abstract not available at present. 1260

0624
JESKE, W.
[Prostaglandins]
Farmacja Polska. 24: 921-925. 1968.

Article in Polish. Abstract not available at present. 1258

0625
KALEY, G. and R. WEINER
Microcirculatory studies with prostaglandin E_1.
In: Ramwell, P.W. and J.E. Shaw, eds., "Prostaglandin Symposium of the Worcester Foundation
for Experimental Biology," p. 321-327. New York, Interscience, 1968.

Topical administration of PGE_1 (0.1-1.0 μg in 0.1 ml) to the rat mesocecum in vivo caused a
diffuse dilation of metarterioles, precapillary sphincters and muscular venules within 3-5 sec
which dissappeared within 1 min. Topical application of a single dose of PGE_1 (0.1-1.0 μg)
produced a partial or complete blockage of the constriction of mesocecal microvessels caused
by subsequent application of epinephrine, norepinephrine or angiotensin which persisted for

20-30 min. The vasoconstriction due to serotonin and the vasodilation due to histamine and bradykinin were unaffected by PGE_1. The effects of PGE_1 on vascular permeability were measured in rat and rabbit skin by dye diffusion techniques and in the rat cremasteric muscle by a carbon labelling technique. PGE_1 caused increases in vascular permeability in these test systems which were similar to the increases caused by histamine and serotonin. In the rat cremasteric muscle, PGE_1 was more effective on a weight basis than the other agents. It is reasonable to suggest that PGE_1 may modulate vascular permeability changes and may be involved in acute inflammatory reaction. (JRH) 1144

0626
KARIM, S.M.M.
Appearance of prostaglandin $F_{2\alpha}$ in human blood during labour.
British Medical Journal. 4: 618. 1968.

Blood samples from over 70 pregnant women were examined for the presence of 4 prostaglandins. Samples obtained from women not in labour at different gestation periods and at term contained no detectable amounts of prostaglandins. Prostaglandin $F_{2\alpha}$ was present in samples of blood obtained during normal spontaneous labour. The appearance of this substance in the blood preceded the uterine contraction. Whether prostaglandins play a part in the process of normal labour is still conjectural. (Author) 1002

0627
KARIM, S.M.M., K. HILLIER and J. DEVLIN
Distribution of prostaglandins E_1, E_2, $F_{1\alpha}$ and $F_{2\alpha}$ in some animal tissues.
Journal of Pharmacy and Pharmacology. 20: 749-753. 1968.

A survey of the distribution of prostaglandins E_1, E_2, $F_{1\alpha}$ and $F_{2\alpha}$ in 14 tissues from the dog, cat, rat, rabbit, guinea-pig and chicken has been made. 1 or more of these prostaglandins are present in varying amounts in most tissues with PGE_2, $PGF_{2\alpha}$ occurring most commonly. (Authors) 1198

0628
KARIM, S.M., R.R. TRUSSELL, R.C. PATEL and K. HILLIER
Response of pregnant human uterus to prostaglandin-$F_{2\alpha}$-induction of labour.
British Medical Journal. 4: 621-623. 1968.

$PGF_{2\alpha}$ was used to induce labor in 10 women in the 34th and 44th week of pregnancy. In 2 cases intrauterine death had occurred prior to induction of labor. The $PGF_{2\alpha}$ was infused intravenously at 0.025-0.05 $\mu g/kg/min$ until delivery occurred. In all cases labor-like contractions developed within approximately 20 min with complete relaxation between contractions. The average induction delivery interval was 6 hr 46 min. Of the 8 patients with live babies, induction of labor was successful in 7. The 1 unsuccessful case was terminated by caesarian section. All the babies were normal and showed no signs of stress before and after delivery. No objective side effects were observed. (JRH) 1003

0629
KARIM, S.M.M. and K. HILLIER
A sensitive method for the assay of prostaglandins E_1, E_2, $F_{1\alpha}$ and $F_{2\alpha}$.
European Journal of Pharmacology. 4: 205-210. 1968.

The authors describe a bioassay technique using fruit bat ileum. The average threshold doses for the following prostaglandins were; PGE_1 (0.4 mg/ml), PGE_2 (0.35 ng/ml), $PGF_{1\alpha}$ (0.75 ng/ml) and $PGF_{2\alpha}$ (0.25 ng/ml). Only jird colon of the other commonly used bioassay tissues approached fruit bat ileum in sensitivity to all of these prostaglandins. (JRH) 1228

0630
KAYAALP, S.O. and R.J. McISSAC
Absence of effects of prostaglandins E_1 and E_2 on ganglionic transmission.
European Journal of Pharmacology. 4: 283-288. 1968.

The effects of PGE_1 and PGE_2 on ganglionic transmission were assessed using postganglionic action potentials elicited by preganglionic repetitive stimulation of the cervical sympathetic trunk in the chloralose anesthetized cat. PGE_1 (10-100 μg) given intravenously or intra-arterially did not affect the ganglionic transmission to any detectable extent. PGE_2 (50-100 μg) was also found to have no effect on the ganglionic transmission. A transient facilitiation of transmission by these drugs injected during stimulation with higher frequency was actually a non-specific action due to the vehicle, a weak alcoholic solution. The results obtained do not support any physiological role of these naturally occurring substances in the sympathetic ganglion. (Authors modified) 1006

0631
KAYLAAP, S.O. and R.K. TURKER
Effect of hemicholinium (HC-3) on the catecholamine releasing action of prostaglandin E_1.
European Journal of Pharmacology. 3: 139-142. 1968.

The hindquarters of pentobarbital-anesthetized dogs were sympathetically denervated and autoperfused. PGE_1 (4μg/kg), acetylcholine and diphenyl-piperazinium, injected into the aorta, caused a rise in the perfusion pressure due to the release of catecholamines from the renal medulla. The pressor activity of only PGE_1 was abolished by intravenous hemicholinium. No significant potentiation of the cholinergic drug action was found with PGE_1, indicating that PGE_1 releases catecholamines through a presynaptic action rather than by sensitizing the chromaffin cells of the adrenal medulla to acetylcholine released at the nerve endings. (RMS) 1005

0632
KIRSNER, J.B.
Controlling gastric secretion.
Postgraduate Medicine. 44: 76-79. 1968.

In this editorial about controlling gastric secretion in ulcer patients prostaglandins are put forth as a possibility since they may reduce gastric secretion occurring in response to histamine. (RAP) 1245

0633
KISCHER, C.W.
Fine structure of the down feather during its early development.
Journal of Morphology. 125: 185-203. 1968.

Mentioned in the discussion section is that prostaglandins have been shown to completely block feather development, but stimulate skin development in embryonic chicken skin. (JRH) 1104

0634

KOROLKIEWICZ, Z.

[Investigations on the mechanism of action of some fever producing drugs.]
Acta Biologica et Medica. 13: 5-48. 1968.

Investigations were performed on 50 rabbits, treated with pyrogen and DNP as main representants of fever producing agents. There was also introduced a group of rabbits treated with theophylline and PGE_1 acting on the lipid metabolism. Results of experiments indicate clearly that there is a correlation between the hyperthermising action of drugs and lipomobilization. PGE_1 plays an important role in the energy production homeostasis and thus influences the thermoregulation too. (From English summary) 1256

0635

KUPIECKI, F.P., N.C. SEKHAR and J.R. WEEKS

Effects of infusion of some prostaglandins in essential fatty acid-deficient and normal rats.
Journal of Lipid Research. 9: 602-605. 1968.

Infusion of 1 mg/kg per day of prostaglandins E_1 (PGE_1) for 2 and 7 wk failed to correct the dermal signs of essential fatty acid (EFA) deficiency in rats despite the known conversion of EFA to certain prostaglandins. PGE_1 caused no significant changes in serum cholesterol, triglycerides, or phospholipids or in liver neutral lipids in EFA-deficient or normal rats. In normal rats epinephrine-induced lipolysis was greater in fat pads from infused than from untreated rats. The effect on epinephrine-induced lipolysis was greater after the 7 wk infusion than after the 2 wk infusion. The 7 wk infusion also lowered plasma free fatty acid (FFA) concentrations. Infusion of PGE_2 and $PGF_{2\alpha}$ in combination for 4 wk had no significant effect on either dermal signs of EFA deficiency, lipolysis, or plasma FFA concentrations. (Authors) 1039

0636

LANDS, W.E.M. and B. SAMUELSSON

Phospholipid precursors of prostaglandins.
Biochimica et Biophysica Acta. 164: 426-429. 1968.

To determine if prostaglandins could by synthesized by direct cyclization of phospholipids or after hydrolysis, 1-palmitoyl-2-([2'-^{14}C] eicosatrienoyl) glycero-3-phosphorycholine was incubated with enzyme preparations from sheep vesicular glands. Since no appreciable amount of labeled prostaglandin esterified to phospholipid was found, it is assumed that the observed production of labeled prostaglandin followed hydrolysis of the labeled phospholipid precursor. (JRH) 1027

0637

LAPIDUS, M., N.H. GRANT and H.E. ALBURN

Enzymatic preparation and purification of prostaglandin E_2.
Journal of Lipid Research. 9: 371-373. 1968.

An enzymatic system has been developed for the production of prostaglandin E_2 (PGE_2) from arachidonic acid by extracts of sheep seminal vesicular glands. The presence of glutathione insures high yields. A new procedure for the purification of PGE_2 was also developed, based on the dialysis of the biosynthesized product at pH 8 and extraction of the dialysate at pH 3 with chloroform. This procedure routinely gives yields of PGE_2 of 25-37% (from arachidonic acid) with a purity of 90-100%. Additional analytical proof of the identity of PGE_2 was provided by physicochemical characteristics of the crystalline thiosemicarbazide derivative, which can be readily prepared under mild conditions. (Authors) 1034

0638

LAPIDUS, M., N.H. GRANT and H.E. ALBURN

Prostaglandin E_2: biosynthesis, purification, and derivatives.

In: Ramwell, P.W. and J.E. Shaw, eds., "Prostaglandin Symposium of the Worcester Foundation for Experimental Biology," p. 365-370. New York, Interscience, 1968.

A technique for the biosynthesis of PGE_2 from arachidonic acid by homogenates of sheep vesicular glands is described. Homogenized frozen vesicular glands (2 kg) were allowed to react with arachidonic acid (2 g) for 60 min at 37°. The reaction mixture was frozen and lyophilized. The lyophilized powder was reconstituted with water and the PGE_2 extraction by a series of solvents (chloroform, phosphate buffer and methylene chloride). The crude yield was 30-40% and after purification 26-37% with a purity of 70-100%. (JRH) 1139

0639

La RAIA, P.J. and W.J. REDDY

Adenosine 3',5'-monophosphate in aorta: levels and regulation.

Circulation. 37/38 (supp. 6): VI 122. 1968.

Abstract only. The cAMP levels of rat aortas incubated in the presence of various possible mediators of cAMP synthesis and C^{14} adenine was measured. PGE_1 (1 µg/ml) was found to increase cAMP levels at 5 min. PGA_1 and $PGF_{2\alpha}$ at the same dose levels had no consistent effect. (JRH) 1043

0640

LEE, J.B.

Cardiovascular implications of the renal prostaglandins.

In: Ramwell, P.W. and J.E. Shaw, eds., "Prostaglandin Symposium of the Worcester Foundation for Experimental Biology," p. 131-145. New York, Interscience, 1968.

PGA_2 (38.2-382 µg/min) was infused into a patient with essential hypertension. There was a decrease in peripheral resistance and systemic blood pressure and an increase in cardiac output and pulse rate. Within 30 sec of termination of infusion these parameters returned to near preinfusion levels followed in 90 sec by a gradual secondary relative hypotension which lasted 3 hr. None of the other usual side effects of PG infusion were seen. The relative activities of PGE_1, PGE_2, PGA_1 and PGA_2 as vasodepressors were compared in dogs. The PGA's were equipotent and were 1.5 and 3 times more potent than PGE_1 and PGE_2 respectively. The relative effects of PGE_1, PGE_2, PGA_1, and $PGF_{2\alpha}$ on blood flow in the femoral, renal, mesenteric and carotid arteries were compared. Each prostaglandin had individual effects on each of these vascular beds. Infusion of PGE_1, PGE_2, PGA_1 or PGA_2 (0.64 µg/min/dog) into the renal artery caused an initial decrease in blood flow followed by a gradual increase. There was also a gradual increase in urine production accompanied by an increase in free water and osmolar clearance due to marked natriuresis. Several possible mechanisms by which PG's could control systemic blood pressure in vivo are discussed. (JRH) 1157

0641

LEE, J.B.

Prostaglandins: Evidence of their role in hypertension.

Clinical Research. 16: 196. 1968.

Title only. 1253

0642
LEE, J.B., and K. CROWSHAW
Prostaglandins.
Lancet. 1: 247. 1968.

The incorrect use of prostaglandin nomenclature as published in a previous article is discussed. Disagreement with the statement that the primary effect of PGE_2 and PGA_2 is an increase in heart rate is voiced. Since the rate increase can be prevented by sympathetic blocking drugs, vasodilation is the first effect, while the heart rate is a reflex. (NES) 1069

0643
LEWIS, G.P. and J. MATTHEWS
The mobilization of free fatty acids from rabbit adipose tissue in situ.
British Journal of Pharmacology. 34: 564-578. 1968.

The epigastric adipose tissue of urethane anesthetized rabbits has been prepared so that the effects of close arterial injections and infusions on blood flow and release of free fatty acids (FFA) can be studied. The effects of pharmacologically active agents and hormone preparations have been investigated. Injections of PGE_1 gave sustained increases in blood flow, and inhibited FFA release when stimulated by growth hormone. (Authors modified) 1230

0644
LINZELL, J.L. and B.P. SETCHELL
The output of spermatozoa and fluid by, and the metabolism of, the isolated perfused testis of the ram.
Journal of Physiology. 195: 25P-26P. 1968.

Isolated goat and ram testes were perfused with autologous or homologous blood and blood flow measured. Several factors were able to alter blood flow rates. However, prostaglandin (1-5 μg) had no effect. (JRH) 1009

0645
MAIER, F. and M. STAEHELIN
Adrenal hyperaemia caused by corticotrophin.
Acta Endocrinologica. 58: 613-618. 1968.

One reference (Bergstrom, 1967) is cited in support of the hypothesis that hyperemia seen following injection of ACTH could be due to the formation of prostaglandins. The lipolytic properties of ACTH could cause the release of prostaglandin precursors and the subsequent biosynthesis of prostaglandins. (JRH) 1124

0646
MAIER, R., and M. STAEHELIN
Adrenal responses to corticotrophin in the presence of an inhibitor of protein synthesis.
Acta Endocrinologica. 58: 619-629. 1968.

The effect of ACTH on adrenal blood flow could be due to its stimulation of the formation of prostaglandin via the liberation of unsaturated fatty acids. Cleavage of the cholesterol esters of the adrenal, which are particularly rich in tetraenoic acids, results in release of considerable amounts of unsaturated fatty acids. Since these can easily be converted to prostaglandin, the latter might constitute the link between the action of ACTH in cholesterol-ester cleavage and its action on the adrenal blood flow. (JRH) 1125

0647
MARX, R.
Die thrombozytenfunktion und ihre bedeutung fur den hamostasemechanismus. [Thombocyte function and its significance for hemostasis mechanism.]
Therapiewoche. 18: 2103-2108. 1968.

PGE is mentioned in this review as an inhibitor of platelet aggregation. (MEMH) 1268

0648
MASORO, E.J. ed.
Fat mobilization.
In: Physiological Chemistry, Vol. 1: "Lipids in Mammals." p. 202-210, Philadelphia, Saunders. 1968.

In this review, the author mentions that prostaglandins can modulate fat mobilization. Whether or not they are important in modulating fat mobilization under physiological conditions has yet to be established. (RAP) 1255

0649
MAY, B., H.G. OMONSKY and E.BOHLE
Uber die blutzuckersteigernde wirkung von prostaglandin E_1. [Mechanism of PGE_1-induced hyperglycemia.]
Zeitschrift fur die gesamte experimentelle Medizin. 148: 99-107. 1968.

Intraperitoneal injection of PGE_1 (0.1-1.0 mg/kg) in fasting rats produced a long-lasting and dose-dependent increase in blood sugar but showed no influence upon the level of serum free fatty acids. The catecholamine content of the adrenal glands was significantly decreased 60 to 120 min after the intraperitoneal injection of PGE_1 (0.5-1.0 mg/kg). PGE_1 in vitro did not liberate any histamine from isolated peritoneal mast cells of the rat. The glycogen content of rat liver slides was significantly decreased by incubation of the tissue in vitro with PGE_1 (75-100 μg/ml) for 30 min at 37°C. The results favor the assumption that the increase in blood sugar induced by PGE_1 is the result of first a direct glycogenolytic effect of PGE_1 within the liver and second an indirect action by liberation of catecholamines from the adrenal glands. (Authors) 1281

0650
MAY, H.E. and P.B. McCAY
Reduced triphosphopyridine nucleotide oxidase-catalyzed alterations of membrane phospholipids.
Journal of Biological Chemistry. 243: 2296-2305. 1968.

Brief mention is made of the similarity between the microsomal membrane, TPNH-dependent enzyme system investigated in this paper and the one which synthetizes prostaglandins. Both systems require a donor of reducing equivalents, consume O_2, utilize polyunsaturated fatty acids and produce stoichiometrically small quantities of malondialdehyde. (JRH) 1105

0651

MELLANDER, S. and B. JOHANSSON

Control of resistance, exchange and capacitance functions in the peripheral circulation.
Pharmacological Reviews. 20: 117-196. 1968.

This extensive review of peripheral hemodynamics cites 3 references on the role of prostaglandins as potent vasodilators of unknown significance. (RMS) 1109

0652

MILLER, R.W., F.R. EARLE, I.A. WOLFF and A.S. BARCLAY

Search for new seed oils. XV. Oils of boraginaceae.
Lipids. 3: 43-45. 1968.

In a search for a preferred source of γ-linolenic (all-cis-5,9,12-octadecatrienoic) acid, seed oils of 33 species of boraginaceae were examined. This triene has become of interest as a starting material for the synthesis of all-cis-8,11,14-eicosatrienoic acid, which is then converted to prostaglandins. The desired triene was found primarily in the subfamily boraginoideae in amounts ranging from 0.2 to 18%. (Authors modified) 1111

0653

MILLER, W.L. and J.J. KRAKE

Metabolism of C^{14}-prostaglandin E_1 by rats: exhaled $C^{14}O_2$ and urinary excretion of C^{14}-activity.
Federation Proceedings. 27: 241. 1968.

Abstract only. Male rats were each prepared with an indwelling right jugular cannula 5 days before use, C^{14}-PGE$_1$ was infused in a dosage which would be equivalent to about 1/2 the dose of PGE$_1$ required to produce minimal vasodepressor response. Rats were housed in glass metabolic chambers which allowed measurement of expired $C^{14}O_2$ and urine collection during infusion. Expired $C^{14}O_2$ accounted for 49±9% of the C^{14}-dose given non-fasted rats. In other experiments, exhaled $C^{14}O_2$ accounted for 68±8% of the C^{14}-dose given overnight (15 hr) fasted rats. About 7-8% of the C^{14}-activity was recovered in the 72 hr urine of a fed rat, 81% of this excretion being completed in 24 hr. Major urinary excretion products found in urine would appear to be fragments of C^{14}-PGE$_1$ no longer containing the original terminal carboxyl group. (Authors modified) 1128

0654

MITCHELL, J.R.A.

Platelets and thrombosis.
Scientific Basis of Medicine: Annual Reviews (London). 266-288. 1968.

In a section on prostaglandins and platelet behavior the author reviews the effects of PGE$_1$ on platelet aggregation and platelet adhesiveness in several species citing 5 references. (RAP) 1247

0655

MIYAZAKI, Y.

Isolation of prostaglandin E like substances from the mucous membrane layer of large intestine of pig.
Sapporo Medical Journal. 34: 141-154. 1968.

The occurrence of a prostaglandin like substance (PG-S) which shows smooth muscle stimulating and vasodepressor activities in the large intestine of a pig was investigated. The large intestine of a pig was separated into mucous membrane and muscle layers and each layer was subjected to the procedure of crude prostaglandin extraction as set forth by Samuelsson. The crude materials obtained from the pig large intestine showed smooth muscle stimulating and vasodepressor activities, while those from the muscle layer showed a mere 1/3 strength of activity of the substances from the mucous membrane layer. On a thin layer chromatogram, it was found that the materials obtained from the mucous membrane layer were already fractionated in some spots, one of which was accorded with Rf of authentic PGE_1 in two different solvent systems and was observed to have a large part of the activity in the crude material. This substance (PG-S_2) was further characterized by various experiments. The stimulating effect of PG-S_2 was affected by changes in the extracellular potassium concentration. The stimulating effect of PG-S_2 was unaffected by botulinum toxin treatment of the rabbit duodenum, in contrast with the effect of nicotine. Methysergide, atropine and antihistamine had no effect on the stimulating activity of PG-S_2. The results agreed with those found with PGE_1. From the results described above it may be said that the substance isolated from the mucous membrane layer of the pig large intestine is definitely PGE_1. (Authors modified) 1282

0656
MIYAZAKI, Y.
[Occurence of prostaglandin E_1 in the mucous membrane layer of swine large intestine.]
Sapporo Medicial Journal. 34: 321-334. 1968.

PG like substances from the mucous membrane of swine large intestine were separated by silicic acid column chromatography and further purified by thin layer chromatography. Biological activities were assayed on various smooth muscle preparations. From these observations on chromatography and bioassay it was concluded that the major part of the biologically active lipids reported previously in swine large intestine was PGE_1. (RAP) 1280

0657
MONTI, A.
Prostaglandina. [Prostaglandin.]
Minerva Medica. 59: 645. 1968.

The sources and activity of prostaglandins are very briefly described, with particular attention to their effects on smooth muscle, the female reproductive system, and the blood. (MEMH) 1289

0658
MORIN, R.B., D.O. SPRY, K.L. HAUSER and R.A. MUELLER
Approaches to the chemical synthesis of the prostaglandins.
Tetrahedron Letters. 6023-6028. 1968.

Attempts to synthesize compounds related to PGE_1 are reported. The idea was to attach one or both side chains to an appropriately substituted benzene ring and subsequently convert the aromatic nucleus to the cyclopentane system. So far this approach has not produced good yields. The structural formulas of the products that have been obtained are given. (JRH) 1330

0659

MOSKOWITZ, J. and J.N. FAIN
  Hormonal regulation of lipolysis and glycogenolysis.
  Clinical Research. 16: 553. 1968.

Abstract only. The effects of various lipokinetic agents on isolated human adipose cells were measured. Large differences were found in basal and theophylline induced lipolysis by fat cells from various surgical patients. However, in all cases, low doses of PGE_1 (1 ng/ml) antagonized theophylline and norepinephrine induced release of glycerol and free fatty acids. A stimulation of human fat cell lipolysis was obtained after treatment of the cells with dibutyryl cAMP. PGE_1 did not significantly inhibit the lipokinetic response of dibutyryl cAMP alone or with added theophylline. Glycogen phosphorylase levels were measured after treatment with PGE_1, theophylline, norepinephrine and dibutyryl cAMP. The changes in glycogen phosphorylase paralleled those of lipolysis. This suggests that in man, as in experimental animals, lipolysis and glycogenolysis are regulated through processes involving cAMP. PGE_1 appears to exert its antilipolytic effect in human adipose tissue by interfering with the accumulation of cAMP at the level of adenyl cyclase. (Author modified) 1204

0660

MUIRHEAD, E.E., B. LEACH, B. BROOKS, G.B. BROWN, E.G. DANIELS and J.W. HINMAN
  Antihypertensive action of prostaglandin E_2.
  In: Ramwell, P.W. and J.E. Shaw, eds., "Prostaglandin Symposium of the Worcester Foundation
    for Experimental Biology," p.183-199. New York, Interscience, 1968.

Experimentally hypertensive rats were injected subcutaneously twice daily with either 100, 200, or 400 μg of PGE_2 (.75-3.3 mg/kg/day). The lowest dose was ineffective but with the higher doses the animals fell into two groups. In the first group, rats with moderate hypertension (153 mm Hg), arterial pressure dropped significantly during the treatment period and returned to near pretreatment levels after injections were stopped. In the second group, rats with more severe hypertension (168 mm Hg), there was a drop in blood pressure at the start of treatment followed by an increase to above pretreatment levels even though the rats were still receiving daily injections of PGE_2. These rats also were found to have fibrinoid necrosis of the arterioles and focal necrosis on the myocardium which were absent in the first group. 5 hypertensive rabbits were injected intraperitoneally once or twice with 3 mg of PGE_2; 3 of the 5 showed a decrease and 1 failed to respond. The blood pressure remained depressed for 5 days in the responsive animals. It seems that the effectiveness of PGE_2 in lowering experimental hypertension depends on the grade of hypertension. (JRH) 1154

0661

MUIRHEAD, E.E., B.E. LEACH, E.D. DANIELS and J.W. HINMAN
  Lapine renomedullary lipid in murine hypertension.
  Archives of Pathology. 85: 72-79. 1968.

The effects of neutral renal lipids extracted from the rabbit renal medulla on the blood pressure of experimentally hypertensive rats and dogs were measured. In one part of the experiment, a mixture of neutral lipids and acidic lipids (mainly PGE_2) was given to 4 rats. There was no difference in the responses of these animals and those of rats which received only the neutral lipids. In both cases, there was a prominent lowering of the aortic pressure after several days of treatment followed by a return to pretreatment levels about 3 days after termination of treatment. The possible physiological roles of prostaglandins and the neutral renal lipids are discussed. (JRH) 1172

0662
NAKANO, J. and B. COLE
Cardiovascular effects of prostaglandin $F_{2\alpha}$.
Clinical Research. 16: 242. 1968.

Abstract only. The cardiovascular responses to: the administration of $PGF_{2\alpha}$ (10 µg/ml) into the right atrium (RA), the administration into the left atrium (LA) or into the portal vein (PV) were compared in pentobarbital-anesthetized, open-chested dogs. It was found that RA and LA administration of $PGF_{2\alpha}$ increased SAP, PAP, PV pressure (PVP) and myocardial contractile force (MCF), and decreased RA pressure (RAP). The magnitude of the increases of PAP and MCF was respectively, smaller and greater after the LA administration of $PGF_{2\alpha}$ than after RA administration. The PV administration of $PGF_{2\alpha}$ decreased SAP, PAP and RAP, and increased PVP more markedly. The magnitude of the increase in PVP was greater after the PV administration than after either the LA or RA administration. In addition, the LA administration of $PGF_{2\alpha}$ increased (HR) significantly, whereas neither the RA nor PV administration caused any significant changes in HR. The present study indicates that $PGF_{2\alpha}$ causes a direct vasoconstriction in systemic, pulmonary arterial and portal venous vascular beds, and also increases the splachnic vascular capacitance. Furthermore, this study suggests the $PGF_{2\alpha}$ is mostly metabolized in the liver, and slightly but not significantly in the lungs. (Authors modified) 1209

0663
NAKANO, J.
Effect of prostaglandins E_1, A_1, and $F_{2\alpha}$ on cardiovascular dynamics in dogs.
In: Ramwell, P.W. and J.E. Shaw, eds., "Prostaglandin Symposium of the Worcester Foundation for Experimental Biology," p. 201-213. New York, Interscience, 1968.

Open-chest pentobarbital anesthetized dogs were used to study the effects of PGE_1, PGA_1 and $PGF_{2\alpha}$ on several hemodynamic parameters. PGE_1 or PGA_1 (0.25-4.0 µg/kg i v) increased heart rate, pulmonary arterial pressure, myocardial contractile force and cardiac output while decreasing systemic arterial pressure, total and pulmonary vascular resistance in a dose related manner. $PGF_{2\alpha}$ increased systemic and pulmonary arterial pressure, myocardial contractile force, cardiac output, total and pulmonary peripherial resistances. When given intraarterially (0.1 µg/kg) PGE_1 and PGA_1 behaved as vasodilators and $PGF_{2\alpha}$ acted as a vasoconstrictor in the various vascular beds involved. Their effects were not blocked by α- or β-adrenergic blocking agents. PGE_1 (4 µg/kg) did not block the hypertensive and positive inotropic effects of 0.5 µg/kg norepinephrine. PG's seem to produce their hemodynamic effects by direct action on the blood vessels. (JRH) 1153

0664
NAKANO, J., T. ISHII and A. GIN
Effect of prostaglandins E_1 (PGE_1), A_1 (PGA_1) and $F_{2\alpha}$ ($PGF_{2\alpha}$) on norepinephrine induced lipolysis.
Clinical Research. 16: 30. 1968.

Abstract only. The relative effect of PGE_1, PGA_1 and $PGF_{2\alpha}$ on the lipolysis induced by norepinephrine (NE) was studied in adipose cells isolated from fasted rat epididymal fat pads. The isolated fat cells were incubated for 1 hour after adding NE with or without graded concentration (0.001-1.0 µg/ml) of the three PG's. Free fatty acids (FFA) were determined by Duncombe's colorimetric method. NE (0.1 µg/ml) increased significantly the release of FFA from the fat cells. However, PGE_1 (more than 0.001 µg/ml), PGA_1 (more than 0.1 µg/ml) or

$PGF_{2\alpha}$ (more than 0.01 μg/ml) reduced the FFA release by NE (0.1 μg/ml) significantly. The anti-lipolytic effect of the PG's was essentially in proportion to the concentration. PGE_1 is the most potent inhibitor of NE-induced lipolysis and PGA_1 is the least effective. (JRH) 1215

0665
NAKANO, J., M. PERRY and D. DENTON
  Effect of prostaglandins E_1 (PGE_1), A_1 (PGA_1) and $F_{2\alpha}$ ($PGF_{2\alpha}$) on the peripheral
  circulation.
  Clinical Research. 16: 110. 1968.

Abstract only. The precise effect of PGE_1, PGA_1 and $PGF_{2\alpha}$ on the peripheral circulation was studied in pentobarbital anesthetized dogs. Intraarterial administration of PGE_1 or PGA_1 (0.025-0.1 μg/kg) increased blood flow and decreased the peripheral resistances in the brachial, femoral, carotid, renal and superior mesenteric arteries without any significant change in systemic arterial pressure and myocardial contractile force. The vasodilator action of PGE_1 was greater in magnitude and more prolonged in duration than that of PGA_1. Intraarterial administration of $PGF_{2\alpha}$ (0.025-0.1 μg/kg) decreased blood flow and increased peripheral resistance. The vasodilator action of PGE_1 and PGA_1 was not blocked by the prior I.A. administration of atropine (50 μg/kg), propranolol (25 μg/kg), diphenyhydramine (250 μg/kg) or methysergide (1 μg/kg). The vasoconstrictor action of $PGF_{2\alpha}$ was not blocked by phenoxybenzamine (0.5 mg/kg) or methysergide (1 μg/kg). The hypotensive action of PGE_1 and PGA_1 and the hypertensive action of $PGF_{2\alpha}$ seem to be due to their direct actions on the peripheral vasculatures. (Authors modified) 1212

0666
NAKANO, J., J.R. McCURDY and M. MAZZOLINI
  Effect of prostaglandin $F_{2\alpha}$ ($PGF_{2\alpha}$) on the cardiovascular dynamics.
  Clinical Research. 16: 30. 1968.

Abstract only. The cardiovascular effects of $PGF_{2\alpha}$ on several hemodynamic parameters was measured in open-chest dogs anesthetized with sodium pentobarbital. Intravenous administration of 1-8 μg/kg of $PGF_{2\alpha}$ increased systemic and pulmonary arterial pressures, cardiac output and myocardial contractile force. In addition, $PGF_{2\alpha}$ increased total and pulmonary peripheral resistances significantly. Left atrial pressure and heart rate decreased slightly. In a preparation in which pulmonary arterial blood flow was kept constant, intraarterial injection of 8 μg/kg $PGF_{2\alpha}$ into the pulmonary artery increased pulmonary arterial perfusion pressure considerably, indicating its direct vasoconstrictor action on the pulmonary arterial trees. The multiple hemodynamic changes produced by $PGF_{2\alpha}$ are most likely due to its direct vasoconstrictor action on the peripheral vasculatures and to its direct action on the myocardium. (JRH) 1216

0667
NAKANO, J., B. COLE and T. ISHII
  Effects of disodium EDTA on the cardiovascular responses to prostaglandin E_1.
  Experientia. 24: 808-809. 1968.

The influence of the calcium chelating agent ADTA on the cardiovascular effects of PGE_1 was measured in open-chest, pentobarbital anesthetized dogs. EDTA was infused continuously at 2 or 5 mg/kg/min. A single dose of PGE_1 (1 μg/kg) was given intravenously prior to and after 15 min of EDTA infusion. PGE_1 decreased mean systemic arterial pressure and increased heart rate and myocardial contractile force in all dogs. EDTA decreased heart rate, mean systemic arterial pressure and myocardial contractile force essentially in proportion to the dose. EDTA also decreased all the effects of PGE_1 except the effects on blood pressure. Thus Ca chelation causes a decrease in the positive chronotropic and inotropic actions of PGE_1. The cardio-dynamic actions of PGE_1 could be influenced by the intracellular concentration or availability of Ca in dogs. (JRH) 1187

0668
NAKANO, J.
Effects of prostaglandins E_1, A_1 and $F_{2\alpha}$ on the coronary and peripheral circulations.
Proceedings of the Society for Experimental Biology and Medicine. 127: 1160-1163. 1968.

The effects of PGE_1, PGA_1, and $PGF_{2\alpha}$ on the coronary and peripheral circulations were studied in pentobarbital anesthetized dogs. The i a injection of either PGE_1 or PGA_1 (0.1 µg/kg) increases blood flows and decreases the peripheral resistances in the coronary, brachial, femoral, carotid, and renal arteries. On the other hand, $PGF_{2\alpha}$ (0.1 µg/kg) decreases blood flows and increases peripheral resistance in the brachial, femoral, and renal arteries. The present study indicates that PGE_1 and PGA_1 dilate and $PGF_{2\alpha}$ constricts directly the regional blood vessels. (Author modified) 1057

0669
NAKANO, J. and J.R. McCURDY
Hemodynamic effects of prostaglandin $F_{2\alpha}$.
Circulation 37/38(supp. 6): VI 146. 1968.

Abstract only. The present study was undertaken to elucidate the mechanism of the hemodynamic actions of $PGF_{2\alpha}$ in anesthetized dogs. It was found that the i v injection of 1-10 µg/kg of $PGF_{2\alpha}$ increased systemic and pulmonary arterial pressures, portal vein pressure, cardiac output and myocardial contractile force essentially in proportion to the dose. In contrast, the injection of the same doses of $PGF_{2\alpha}$ into the portal vein increased portal vein pressure, but decreased pulmonary and systemic pressures. The i a injection of 0.1-0.5 µg/kg of $PGF_{2\alpha}$ into a femoral, brachial or superior mesenteric artery decreased blood flow in each artery without producing any significant change in systemic arterial pressure. In dogs in which cardiac input was kept constant by means of a Sigmamotor pump, the injection of $PGF_{2\alpha}$ decreased systemic venous return. This study indicates that $PGF_{2\alpha}$ decreases systemic venous return by pooling of blood in the splanchnic vascular beds. The hypertensive effect of $PGF_{2\alpha}$ is most likely due mainly to its direct vasoconstrictor action. (Authors modified) 1042

0670
NAKANO, J. and J.R. McCURDY
Hemodynamic effects of prostaglandins E_1, A_1, and $F_{2\alpha}$ in dogs.
Proceedings of the Society for Experimental Biology and Medicine. 128: 39-42. 1968.

Pentobarbital anesthetized dogs were used to determine the hemodynamic effects of intravenously administered PGE_1, PGA_1, and $PGF_{2\alpha}$. All three PG's were found to increase heart rate, myocardial contractile force, cardiac output and pulmonary arterial pressure while decreasing left atrial pressure when given in graded doses of 0.25 to 8.0 µg/kg. PGE_1 and PGA_1 are vasodilators and were found to reduce systemic blood pressure while $PGF_{2\alpha}$ is a vasoconstrictor and increased systemic pressure. The authors conclude that the hemodynamic changes caused by these PG's are due to their direct effect on the myocardium as well as their effects on the peripheral vessels. (JRH) 1240

0671
NAIMZADA, M.K.
Effects of prostaglandins (PGE$_1$, PGE$_2$, PGA$_1$ and PGF$_{1\alpha}$) on the hypogastric nerves, vas deferens and seminal vesicle preparations of the guinea pig.
Quatriemes Rencontres de Chimie Therapeutique. 3: 405. 1968.

Abstract only. The effects of some naturally occurring prostaglandins (PGE$_2$, PGA$_1$ and PGF$_{1\alpha}$) have been studied on the hypogastric nerves, vas deferens and seminal vesicle preparations of the guinea pig in vitro and compared with that of PGE$_1$. It has been shown that, like PGE$_1$, PGE$_2$, PGA$_1$ and PGF$_{1\alpha}$ are capable of markedly increasing the effects of the sympathetic stimulation and of the sympathetic amines (adrenaline and noradrenaline on the vas deferens and seminal vesicle of the guinea pig. However, PGE$_2$ has 1/10, and PGA$_1$ and PGF$_1$ 1/100 the potency of PGE$_1$. It is suggested that they may have a specific action in the guinea pig, enhancing the sensitivity of the semen-excreting apparatus to nervous stimuli. (NS) 1284

0672
NEKRASOVA, A.A., E.N. NIKOLAEVA and V.V. KHUKHAREV
[Prostaglandin and depressor function of the kidneys.]
Kardiologiia. 8: 16-22. 1968.

The authors have elaborated a method of quantitative determination of prostaglandin in the kidneys. The partial separation of prostaglandins from the medullary layer of the kidneys was done by way of their consecutive extraction with 50 per cent solution of alcohol, ether in acid medium, phosphate buffer and again with ether. Quantitative determination of prostaglandins in experimental tests was carried out by the biological method. The authors studied the content of prostaglandins in the renal medullary layer of 32 healthy rabbits and in 23 rabbits with hypertension caused by ischemia of one kidney. It was found in rabbits sacrificed 30 days after the operation the prostaglandin content in intact kidneys did not differ from that of the control group; in ischemic kidneys it was reduced. In the group of rabbits sacrificed 150 days after the operation the prostaglandin in intact and ischemica kidneys increased in comparison with the control group. This rise was particularly marked in the kidneys of rabbits sacrificed 270 days after the operation, when the arterial pressure in them reversed to normal values. Increase of the prostaglandin content in the kidneys of this group of rabbits was associated both with augmented synthesis of prostaglandins in the kidneys and with marked hypertrophy of the intact kidney. Apparently, the increased formation of prostaglandins in the kidneys is one of the factors conditioning the benign nature and reverse development of hypertension caused by ischemia on one kidney. (Authors) 1304

0673
NISSEN, H.M., and H. ANDERSEN
On the localization of a prostaglandin-dehydrogenase activity in the kidney.
Histochemie. 14: 189-200. 1968.

Using the kidneys from white rats a method for the localization of a prostaglandin-dehydrogenase activity is presented. The activity demonstrated showed NAD dependence and displayed a high substrate specificity. The most pronounced activity was observed in the thick ascending limb of the loop of Henle and in the distale tubule. Lesser activity was found in the collecting tubules of the inner medulla, in the interstitial cells of the medulla, in the epithelial cells of the pelvis, in the tunica media of the cortical arteries and arterioles and in the visceral epithelium of the renal corpuscles. The pitfalls of the cytochemistry of the co-enzyme linked dehydrogenases are discussed and correlated to the present study. Similarly the observations noticed are discussed in relation to the metabolism and biological activity of the prostaglandins. (Authors) 1100

0674
O'BRIEN, J.R.
Prostaglandins and platelets.
Lancet. 1: 149. 1968.

This letter to the editor points out that in a previously published article the possible role of prostaglandin in platelet physiology was omitted. (NES) 1070

0675
OCHOA, E.
Farmocologia de algunas substancias vasoactivas. [Pharmacology of several vasoactive substances.]
Medicina. 28: 129-143. 1968.

Prostaglandins are discussed in a review of several vasoactive substances. Particular emphasis is given to the effects of PGE_1, PGE_2 and $PGF_{2\alpha}$ on smooth muscle, the cardiovascular system, metabolism, the kidney and the nervous system. There are 33 references listed referring to prostaglandins. (NES) 1286

0676
OHLOFF, G. and W. GIERSCH
Sauerkatalysierte isomerisierung von α, β-cyclopropyl-oxiranen. [Acid-catalysed isomerization of α, β-cyclopropyl-oxiranes.]
Helvetica Chimica Acta. 51: 1328-1342. 1968.

Using the epoxycarane derivatives, the behavior of α, β-epoxycyclopropyl compounds in the presence of acids has been examined by IR, NMR and mass spectroscopy and by gas chromatography, and the main products elucidated. Proposals for the reaction mechanisms are made. A reaction analogous to several of those studied in this series has been found to occur as one stage in the formation of prostaglandins. (Authors modified) 1093

0677
OSOL, A.
Prostaglandins: a new hormonal system?
American Journal of Pharmacology. 140: 22-23. 1968.

History, structure, sources and some proposed uses of the prostaglandins are briefly reviewed; 1 reference is cited. (JRH) 1273

0678
PACE-ASCIAK, C., K. MORAWSKA, F. COCEANI and L.S. WOLFE
The biosynthesis of prostaglandins E_2 and $F_{2\alpha}$ in homogenates of the rat stomach.
In: Ramwell, P.W. and J.E. Shaw, eds., "Prostaglandin Symposium of the Worcester Foundation for Experimental Biology," p. 371-378. New York, Interscience, 1968.

Incubation of rat stomach strips in vitro produced a 20-fold increase in PGE_2 content and a 60-fold increase in $PGF_{2\alpha}$ content. Similar experiments with ox cerebral cortex showed a 6-fold increase in $PGF_{2\alpha}$ with only small amounts of PGE_2 being found. When the rat stomach tissue was homogenized, there was a very rapid synthesis of prostaglandins (mostly PGE_2) even in ice cold media. There was also a rapid appearance of polyunsaturated fatty acids due to selective hydrolysis of phospholipids. (JRH) 1138

170

0679
PACE-ASCIAK, C. and L.S. WOLFE
  Inhibition of prostaglandin synthesis by oleic, linoleic and linolenic acids.
  Biochimica and Biophysica Acta. 152: 784-787. 1968.

  Prostaglandin synthesizing enzymes were extracted from sheep vesicular glands or from rat stomach. The enzyme preparations were incubated with labeled arachidonic acid in the presence of several fatty acids. It was found that oleic (18:1, ω 9), linoleic (18:2, ω 6), and linolenic acid (18:3, ω 3) inhibited PGE_2 biosynthesis by more than 80% in an irreversible and apparently competitive manner. Linolenic acid (1.8mM) was the most effective inhibitor, but its methyl ester did not inhibit. (JRH) 1026

0680
PAOLETTI, R., L. PUGLISI and M.M. USARDI
  The lipolytic action of catecholamines and ACTH and the interaction of prostaglandin E_1.
  In: Black, N., L. Martini and R. Paoletti, eds., "Advances in Experimental Medicine and
    Biology," p. 425-439. New York,Plenum Press, 1968.

  Norepinephrine (NE), theophylline (THEO) or ACTH were used to stimulate lipolysis in rat adipose tissue or adipose cell preparations in vitro. PGE_1 (0.1-1.0 $\times$ 10^{-7}M) was more antilipolytic against THEO than against NE. Dose response experiments with NE and THEO indicated that PGE_1 (10^{-7} and 10^{-6}M) inhibited NE noncompetitively and inhibited THEO competitively. In other experiments, rats were maintained on essential fatty acid (EFA) deficient diets to deplete their supply of prostaglandin precursors. There was a signicantly higher level of free fatty acids (FFA) and glycerol in the plasma of EFA- deficient rats than in normal controls. Adipose tissue from EFA-deficient rats was much more responsive to the lipolytic effects of NE. Isolated fat cells from EFA-deficient rats showed an increased sensitivity to NE and THEO. In these preparations, low doses of PGE_1 were still highly antilipolytic, perhaps even more so than in normal tissue. In vivo the plasma of EFA-deficient rats injected with ACTH showed significantly higher levels of FFA and glucose after 2 hr than control EFA-deficient. (JRH) 1115

0681
PAOLETTI, R.
  Quelques remarques sur le role des prostaglandines dans l'homeostase. [Some remarks on the
    role of prostaglandins in homeostasis.]
  Revue de l'Atherosclerose. 10: 57-67. 1968.

  The prostaglandins are considered as physiological regulators of lipolysis, capable of antagonizing the excessive liberation of fatty acids resulting from stimulation of the sympathetic nervous system or from exogenous catecholamines, and stimulating at the same time carbohydrate mobilization. This fact represents a rather typical example of the role played by physiological substances like prostaglandins which are modulators of the metabolic effects of mediators. However, one may put forward the hypothesis that a limited number of mediators: acetylcholine, catecholamines may exert on the various tissues, and in various experimental and clinical conditions, very numerous and varied effects, according to their interaction with known physiological modulators like prostaglandins, or other yet unknown ones, which are prefabricated or synthetized in the tissues, while the mediators are liberated. (From the English summary) 1252

0682
PARKINSON, T.M., J.C. SCHNEIDER, Jr., J.J. KRAKE and W.L. MILLER
Intestinal absorption and metabolism of prostaglandin E_1-1-^{14}C by thoracic duct and bile duct cannulated rats.
Life Sciences. 7: 883-889. 1968.

Approximately 58% of a dose of PGE_1-1-^{14}C administered orally to thoracic duct cannulated rats was excreted as $^{14}CO_2$ in expired air in 24 hours; an additional 0.5-5.0% was found in thoracic duct lymph. About 50% of the radioactivity in expired air was accounted for in two peaks of maximum specific activity occurring 30-40 minutes and 5-7 hours after dosing. The second peak did not occur in the $^{14}CO_2$ profile of a bile duct cannulated rat and may represent metabolism of radioactive products in the enterohepatic circulation derived from PGE_1-1-^{14}C. (Authors) 1076

0683
PATON, D.M.
The contractile response of the isolated rat uterus to noradrenaline and 5-hydroxytrytamine.
European Journal of Pharmacology. 3: 310-315. 1968.

Uterine horns from rats pretreated daily for 6-12 days with diethylstibestrol were used to study the contractile response of the uterus to norepinephrine (NE) and serotonin (5-HT) in vitro. PGE_1 (2.5 µg/ml) was added to the organ bath. When a maximal contraction occurred, usually within 2 min, the PG was washed out and either NE or 5-HT was added. The pretreatment with PGE_1 greatly potentiated and prolonged the response of the uterine horn to subsequent exposure to NE or 5-HT. (JRH) 1007

0684
PATON, D.M.
Pharmacology of isolated rabbit detrusor muscle.
Clinical Research. 16: 393. 1968.

Abstract only. The mode of action of several drugs including PGE_1 was investigated in muscle strips from the rabbit urinary bladder in vitro. Nicotine and serotonin caused contractions which were potentiated by eserine and partially antagonized by atropine. Eserine and atropine had no effect on the contraction caused by PGE_1, angiotensin, histamine, oxytocin and norepinephrine. The contractions produced by PGE_1, angiotensin and oxytocin were not antagonized by either mepyramine or phentoamine. It was concluded that these substances have separate receptors in the rabbit detrusor muscle. (JRH) 1205

0685
PAUERSTEIN, C.J., J.D. WOODRUFF and A.S. ZACHARY
Factors influencing physiologic activities in the fallopian tube; the anatomy, physiology, and pharmacology of tubal transport.
Obstetrical and Gynecological Survey. 23: 215-243. 1968.

In a section on prostaglandins, the authors review 5 papers on the effects of the various PG's on the human oviduct. (RAP) 1254

0686

PAWAR, S.S. and H.C. TIDWELL

Effect of prostaglandin and dietary fats on lipolysis and esterification in rat adipose tissue in vitro.

Biochimica et Biophysica Acta. 164: 167-171. 1968.

A possible role of PGE_1 in lipid metabolism was investigated. It was found to inhibit the release of free fatty acids and glycerol from adipose tissue of fat-fed rats, the amount depending upon the nature of the fat ingested. The release of less glycerol and free fatty acids from tissue of rats fed a polyunsaturated fat(corn oil), as compared to those fed lard, was in effect similar to that in the presence of prostaglandin. Preincubation of tissue with prostaglandin produced a similar response. There was also a greater incorporation of $[I\text{-}^{14}C]$ palmitic acid into the lipids and triglycerides of this tissue in the corn oil-fed rats and in prostaglandin-treated tissues. An inhibition of lipolysis and increased rate of removal of fatty acids from the medium by adipose tissue in the presence of prostaglandin or by this tissue of corn oil-fed rats, both associated with an increased esterification of fatty acids in the tissue, suggests the possibility that the effect of the latter may be the promotion of the formation by these rats of greater amounts of prostaglandin from the unsaturated fatty acids ingested. (Authors modified) 1032

0687

PAWAR, S.S. and H.C. TIDWELL

In vitro effect of prostaglandin (PGE_1) on the release of glycerol and the metabolism of palmitic acid in rat adipose tissue.

Lipids. 3: 383-384. 1968.

The effect of PGE_1 on glycerol release from rat epididymal fat pads, some of which had been preincubated with PGE_1, was measured in vitro. PGE_1 was antilipolytic in all cases. The effect of PGE_1 on the metabolism of $I\text{-}^{14}C$-palmitic acid was also studied. PGE_1 increased esterification of palmitic acid, especially in the triglycerides. PGE_1 may be important in triglyceride breakdown and reesterification. (JRH) 1112

0688

PETERSON, M.J., C. PATTERSON and J. ASHMORE

Effects of antilipolytic agents on dibutyryl cyclic AMP induced lipolysis in adipose tissue.

Life Sciences. 7: 551-560. 1968.

Lipolysis was measured by determining the rate of glycerol and free fatty acid release from epididymal fat pads of Wistar rats in vitro. It was found that both isoproterenol (1×10^{-6}M) and dibutyryl cAMP (5×10^{-4} to 6×10^{-3}M) caused a similar significant increase in lipolysis over control values. Insulin (2×10^{-7}M), nicotinic acid (1×10^{-4}M), PGE_1 (2.3×10^{-7}M), pronethalol (1×10^{-3}M) and phentolamine (1×10^{-3}M), all known antilipolytic agents, had no significant effects on basal lipolysis but reduced by 75% the stimulation of isoproterenol. However, none of the agents at the above doses significantly altered the stimulation of lipolysis caused by dibutyryl cAMP. It is concluded that these antilipolytic agents exert their effects at some site before the increase in cAMP and have no effects on events subsequent to cAMP formation in the lipolytic process. (JRH) 1077

0689

PETERSON, M.J., C. PATTERSON and J. ASHMORE

Effects of antilipolytic agents on dibutyryl cyclic AMP-stimulated lipolysis.

Federation Proceedings. 27: 241. 1968.

Abstract only. Several antilipolytic agents including PGE_1 were tested for their effect on dibutyryl cAMP stimulated lipolysis in epididymal fat pads in vitro. PGE_1, at concentrations which antagonized an isoproterenol-stimulated lipolysis, had no affect on either basal or dibutyryl stimulated lipolysis. PGE_1 appears to affect, either directly or indirectly, the production of cAMP. (JRH) 1127

0690
PHARRISS, B.B., L.J. WYNGARDEN and G.D. GUTKNECHT
Biological interactions between prostaglandins and lutetropins in the rat.
In: Rosemberg, E., ed., "Gonadotropins," p. 121-129. Los Altos, Geron-X, 1968.

Ovaries from 6 day pseudopregnant rats were incubated in vitro with $PGF_{2\alpha}$ (0.1-10.1 $\mu g/ml$) and/or LH to study the effects of these compounds on luteal progesterone production. $PGF_{2\alpha}$ and LH exhibited similar luteotropic properties in that both produced similar increases in progesterone content. When both compounds were administered in maximal amounts no additive effects were seen. When LH and $PGF_{2\alpha}$ were compared in vivo by the ovarian ascorbic acid depletion assay technique, the curves obtained were different. $PGF_{2\alpha}$ (4 mg/day) had no effect on pituitary LH content in ovariectomized rats but it did significantly increase the FSH content. $PGF_{2\alpha}$ definitely has luteotropic effects in vitro. The meaning of the stimulation of pituitary FSH is not known. (JRH) 1333

0691
PHILLIS, J.W. and A.K. TEBECIS
Prostaglandins and toad spinal cord responses.
Nature. 217: 1076-1077. 1968.

Spinal cords of the toad *Bufo marinus* were isolated, hemisected and perfused with oxygenated amphibian Ringer solution. The eighth dorsal root was stimulated every 10 sec and a.c. or d.c. recordings were made from the ninth dorsal and ventral roots. The three prostaglandins, PGE_1, $PGF_{1\alpha}$ and $PGF_{2\alpha}$, all produced distinct changes in reflexes and polarization level when perfused over the toad spinal cord for periods of 0.5-2 min. Typically, the prostaglandins caused a slowly developing, long-lasting depolarization of both dorsal and ventral roots. PGE_1 (5×10^{-5} g/ml.) caused a hyperpolarization during its 50s application, followed by a depolarization which lasted for several minutes. The dorsal root polarization level had not reached control value 20 min after the application. During the hyperpolarizing phase of the response, polysynaptic reflex discharges in the ventral root were markedly reduced. PGE_1 had the most pronounced effects. $PGF_{1\alpha}$ was usually more potent than $PGF_{2\alpha}$. Desensitization occurred with repeated applications. The extremely slow rate of recovery (their effects sometimes last for 0.5 hr or more) may mean that they are not transmitters but that they affect the release of transmitter as suggested by Ramwell et al. (JRH) 1016

0692
PICKLES, V. R.
Prostaglandins.
Lancet. 1: 309. 1968.

In reference to a Lancet article of Jan. 6, 1968, it is suggested that great care be taken before concluding that a given tissue contains a particular prostaglandin, since large amounts of tissue and considerable technical skill are required to make an analysis. (JRH) 1068

0693

PRICE, W.E., Z. SHEHADEH, D.A. NEUMANN and E.D. JACOBSON
Vasoactive agents in the mesenteric circulation.
Federation Proceedings. 27. 282. 1968.

Abstract only. To observe local vascular responses and minimize systemic circulatory effects, 4 naturally occurring vasoactive agents were infused directly in superior mesenteric arteries of 10 dogs. The drugs used were bradykinin, acetycholine, histamine and $PGF_{2\alpha}$. Dosages administered ranged from 0.01 to 2.0 $\mu g\ min^{-1}\ kg^{-1}$. Bradykinin, acetycholine and histamine increased mesenteric blood flow and decreased vascular resistance. $PGF_{2\alpha}$ caused a 37% maximum in mesenteric perfusion. Responses to these 4 drugs increased in proportion to the increase in dosage. (JRH) 1130

0694

RAMWELL, P.W.
Biological substances.
Science. 161: 458. 1968.

The author gives a concise review of prostaglandins in general in discussing a book on prostaglandins by von Euler and Eliasson. (RAP) 1264

0695

RAMWELL, P.W. and J.E. SHAW
Interactions between prostaglandin E_1 and 3'5' AMP on ion transport and metabolism of gastric mucosa.
In: Proceedings of the International Union of Physiological Sciences, 24th, Washington, D.C. 1968. Vol. 7. Abstracts of volunteer papers and films. p. 359. Washington, D.C., Federation of American Societies for Experimental Biology, 1968.

Abstract only. The H^+ and PG content of mucosal perfusates of anesthetized rats were determined by titration and bioassay of ether extracts of the acidified perfusate (pH 3) respectively; the spontaneous H^+ secretion and PG release (0.1-1.0 $m\mu g/PGE_1$ equivalents/min) were augmented following i v injection of histamine (500 $\mu g/kg$) or pentapeptide (ICI 50, 123, 4 $\mu g/kg$), vagal stimulation, or perfusion of cAMP and theophylline (10^{-4} M). Application of PGE_1 to the mucosal surface (0.5-1.0 $\mu g/min$) decreased spontaneous H^+ secretion and reversibly reduced (by 30-90%) the H^+ response to the aforementioned secretagogues. (Authors modified) 1328

0696

RAMWELL, P.W. and J.E. SHAW
Prostaglandin inhibition of gastric secretion.
Journal of Physiology. 195: 34P-36P. 1968.

Rat gastric mucosa was found to contain PGE_1, PGE_2, $PGF_{1\alpha}$ and $PGF_{2\alpha}$. When the stomach was perfused and the H^+ and prostaglandin of perfusion fluid measured, the spontaneous release of prostaglandin was equivalent to 0.1 to 1.0 $m\mu g/min\ PGE_1$. Stimulation of H^+ secretion by histamine (i v), pentagastrin i v or vagus nerve stimulation caused a 3 to 5 fold increase in release of prostaglandin activity. The increase was completely accounted for by PGE_1, PGE_2, $PGF_{1\alpha}$ and $PGF_{2\alpha}$. Perfusion of the mucosal surface with PGE_1 (0.5-1.0 $\mu g/min$) reduced H^+ secretion stimulated by histamine, pentagastrin or vagal nerve stimulation by 30-80%. Addition of cAMP caused a temporary increase in H^+ secretion which was increased by theophylline and inhibited by concomitant perfusion of PGE_1. (JRH) 1055

0697
RAMWELL, P.W., J.E. SHAW, G.B. CLARKE, M.F. GROSTIC, D.G. KAISER and J.E. PIKE
Prostaglandins.
In: Holman, R.T., ed., "Progress in the Chemistry of Fats and Other Lipids," Vol. 9. p. 231-273. Oxford, Pergamon. 1968.

The biochemical aspects of prostaglandin research are extensively reviewed. There are 205 references cited. (JRH) 1174

0698
RETTBERG, H.
Untersuchungen uber die wirkung von prostaglandin E_1 (PGE$_1$) und insulin auf das epididymale fettgewebe und diaphragma bei ratten nach intraperitonealer injektion von U-C-^{14}-glukose. [Studies on the effect of PGE$_1$ and insulin on the epididymal fatty tissue and diaphragm in rats following intraperitoneal injection of C^{14}-glucose.
Dissertation, Medizinischen Universitatsklinik, Johann Wolfgang Goethe Universitat, Frankfurt am Main. 51p. 1968.

The incorporation of intraperitoneally applied C^{14}-glucose into the lipids and fatty acids of the epididymal fatty tissues and into the glycogen of the hemidiaphragm under the influence of PGE$_1$ (1.0-0.01 μg/rat) was studied and compared with the influence of 500 μE/rat of insulin. PGE$_1$ was found to favor the incorporation of the marked glucose into fatty-tissue lipids, raising the rate 60-90% over the basal rate (50% more than insulin). When PGE$_1$ and insulin were given together, the effects were additive. PGE$_1$ also raised the rate of fatty-acid synthesis from glucose fragments more quickly and strongly than insulin. A direct influence of PGE$_1$ on glucose utilization is considered possible. (MEMH) 1313

0699
RIGNON-MACRI, P. and I. MACRI
Effectto della prostaglandina E_1 sulla respirazione de fettine di corteccia cerebrale di ratto. [The effect of prostaglandin E_1 on respiration in slices of rat cerebral cortex.]
Bollettino della Societa Italiana de Biologia Sperimentale. 44: 1461-1463. 1968.

Effects of PGE$_1$ on the respiration of slices of the cerebral cortex of rats were studied in order to provide material for an explanation of the diverse effects of this substance on the central nervous system. Rats were injected with PGE$_1$ or control solution. PGE$_1$, in quantities of 100 μg had a small, but significant inhibitory effect on the respiration of slices of rat cerebral cortex. In quantities of 50 μg, a very slight, but not statistically significant, inhibitory effect was observed. It is not yet possible to formulate hypotheses concerning the physiological importance of these substances on the central nervous system. (MEMH) 1287

0700
RISLEY, P.L. and F.W. HUI
Hamster seminal vesicle contractile responses to catecholamines and acetycholine in the presence of prostaglandin E_1.
Anatomical Record. 160: 492. 1968.

Abstract only. Using hamster seminal vesicles in vitro, more rapid and potentiated responses to adrenaline and noradrenaline occurred when as low as 10 nanograms of PGE$_1$ were present per ml of bath solution. With dosages of PGE$_1$ between 10 and 100 nanograms, dose response curves were obtained when constant doses of adrenaline or noradrenaline were applied, but higher doses of PGE$_1$ were relatively less effective. Responses to acetycholine in the presence of this prostaglandin were not potentiated as effectively as those to adrenaline or noradrenaline, as they were in guinea-pig seminal vesicles. Hamster seminal vesicles are sensitive organs for the assay of these effects of prostaglandins. (Authors modified) 1133

0701
ROBERT, A.
Antisecretory property of prostaglandins.
In: Ramwell, P.W. and J.E. Shaw, eds., "Prostaglandins Symposium of the Worcester Foundation for Experimental Biology," p. 47-54. New York, Interscience, 1968.

Gastric secretion was stimulated by either food, pentagastrin, 2-deoxyglucose or histamine in dogs with stomach pouches or gastric fistulas. PGE$_1$, PGE$_2$, PGA$_1$ or PGF$_{2\alpha}$ was then infused intravenously and their effects on gastric secretion measured. Rats with pyloric ligatures were used to test the ability of PGE$_1$ (1 and 2 μg/kg/min) infused subcutaneously for 21 hr after ligation to prevent ulcer formation. PGE$_1$ inhibited gastric secretion stimulated by all four agents (1 μg/kg/min causing almost total inhibition). PGE$_2$ caused a similar inhibition of food and histamine stimulation (not tested against pentagastrin or 2-deoxyglucose). PGA$_1$ caused a strong inhibition of food stimulated secretion. With PGA$_1$, PGE$_1$ and PGE$_2$ about 2 hr were required for secretion to return to preinfusion levels after cessation of PG infusion. PGF$_{2\alpha}$ (1 μg/kg/min) had no effect on histamine stimulated secretion. PGE$_1$ greatly reduced ulcer formation and mortality in pylorus ligated rats in a dose dependent manner. (JRH) 1165

0702
ROBERT, A., J.E. MEZAMIS and J.P. PHILLIPS
Effect of prostaglandin E$_1$ on gastric secretion and ulcer formation in the rat.
Gastroenterology. 55: 481-487. 1968.

Gastric secretion was collected and analyzed in rats after pylorus ligation. PGE$_1$, administered subcutaneously either by injection (100 to 800 μg/kg given in two doses, one at the time of ligation and the second 2 hr later) or by constant infusion (0.5 to 2 μg/kg/min in a total volume of 0.54 ml/hr) inhibited gastric secretion (volume, acid concentration and output, pepsin output, fucose output). The degree of inhibition was dose-dependent. PGE$_1$ also strongly inhibited two types of ulcers in the rat: those produced by pylorus ligation (Shay ulcers), and those produced by administration of prednisolone (steroid-induced ulcers). (Authors modified) 1099

0703
ROBERT, A., J. P. PHILLIPS and J.E. NEZAMIS
Inhibition by prostaglandin E$_1$ of gastric secretion in the dog.
Gastroenterology. 54: 685-686. 1968.

Title only. 1084

0704
ROBERT, A., J. P. PHILLIPS and J.E. NEZAMIS
Inhibition by prostaglandin E$_1$ of gastric secretion in the dog.
Gastroenterology. 54: 1263. 1968.

Abstract only. We now report the effect of PGE_1 on secretion from gastric pouch (Heidenhain: H; Pavlov: P) and gastric fistula (F) dogs stimulated by various agents. PGE_1, administered either subcutaneously, I.V., or by constant I.V. infusion, inhibited volume, acid concentration and output, and pepsin output of gastric juice stimulated with histamine (H, P), food (H, P), pentagastrin (H, P, F), 2-deoxyglucose (F), carbachol (F) and reserpine (F). The inhibition was dose dependent, was maximal 30 minutes after either injection or the start of infusion, and persisted throughout infusion. Values returned to pretreatment levels about 2 hours after injection, or after infusion had stopped. The ED_{50} was the same (about 0.5 $\mu g/kg/minute$) regardless of the stimulus used. Since prostaglandins, when given at high doses sometimes induce vomiting, dogs stimulated with histamine were given, in other experiments, antiemetics (atropine, 0.08 mg/kg S.C. or I.V., chlorpromazine, 2 mg/kg, S.C.,) and PGE_1 was administered one hour later. The antisecretory effect of the latter was as marked although there was no sign of nausea. This result indicates that the antisecretory effect of PGE_1 is not due to inducement of nausea. (Authors modified) 1097

0705
ROBISON, G.A., R.W. BUTCHER and E.W. SUTHERLAND
Cyclic AMP.
Annual Review of Biochemistry. 37: 149-174. 1968.

Prostaglandins are only briefly mentioned in this extensive review on cAMP. A 1967 review by Butcher and Sutherland on cAMP and a review on PG's (1967) by Bergstrom are cited. (JRH) 1103

0706
ROBISON, G.A., R.W. BUTCHER and E.W. SUTHERLAND
The role of cyclic AMP in adipose tissue and smooth muscle.
Pharmacologist. 10: 145-146. 1968.

Abstract only. Prostaglandins are mentioned as being antilipolytic agents which prevent the accumulation of cAMP in response to lipolytic hormones. (JRH) 1323

0707
ROSENTHALE, M.E., A. DERVINIS, A.J. BEGANY, M. LAPIDUS and M.I. GLUCKMAN
Bronchodilator activity of the prostaglandin PGE_2.
Pharmacologist. 10: 175. 1968.

Absract only. Using the anesthetized guinea pig lung overflow technique, intravenous administration of PGE_2 at doses of 4-8mcg/kg was able to completely prevent bronchocon-striction induced by histamine, serotonin, acetylcholine, or bradykinin. Bilateral vagotomy, adrenalectomy, double pithing or pretreatment with reserpine or pronethalol did not affect the bronchodilating properties of PGE_2. In unanesthetized guinea pigs bronchoconstriction due to aerosols of histamine in normal, and horse serum in passively sensitized animals, was prevented by PGE_2 aerosol. In the guinea pig, PGE_2 is as active a bronchodilator as isoproterenol, but has a shorter duration. In the anesthetized dog and rhesus monkey, PGE_2 aerosol at doses of 5 to 100 mcgs was able to significantly block the effects of histamine spray on pulmonary resistance and compliance. PGE_2 caused considerably greater bronchodilation in the monkey than the dog, in contrast to isoproterenol which was equiactive in both. Hypotension of short duration was evident in both species after PGE_2. Hexamethonium, atropine or propranolol did not influence the bronchodilating properties of PGE_2 in the monkey. PGE_2 appears to directly relax bronchial smooth muscle when administered either intravenously or as an aerosol. (Authors) 1322

0708

SAID, S.I., O. MUREN and B.J. KIRBY

Some respiratory effects of prostaglandins E_2 and $F_{2\alpha}$.

Clinical Research. 16: 90. 1968.

Abstract only. The effects of intravenous PGE_2 and $PGF_{2\alpha}$ (2-22 μg/kg) on respiratory function in anesthetized dogs were measured. Both agents raised pulmonary arterial and pulmonary venous pressure, without changing left atrial pressure; cardiac output increased moderately, then decreased. Dynamic lung compliance fell sharply and alveolar ventilation decreased with $PGF_{2\alpha}$, whereas PGE_2 caused hyperventilation. The PGE_2 also lowered systemic arterial pressure markedly, while with $PGF_{2\alpha}$ systemic pressure either did not change or increased slightly. The magnitude and duration of the effects were roughly related to the doses. (JRH) 1213

0709

SAID, S.I., B.J. KIRBY and O. MUREN

Some respiratory effects of prostaglandins E_2 and F_2 [α] .

Clinical Research. 16: 374. 1968.

Abstract only. The effects of intravenous PGE_2 and $PGF_{2\alpha}$ (0.7-22 μg/kg) on respiratory function was investigated in anesthetized dogs. Both agents raised pulmonary arterial and pulmonary venous pressure, without changing left atrial pressure; cardiac output increased moderately then decreased. Dynamic lung compliance fell sharply and alveolar ventilation decreased with $PGF_{2\alpha}$ whereas PGE_2 usually caused hyperventilation. PGE_2 also lowered systemic arterial pressure markedly, while with $PGF_{2\alpha}$ systemic pressure either did not change or increased slightly. The magnitude and duration of the effects were roughly related to the doses. It is apparent that PG's strongly influence various aspects of respiratory function. If released from the lung or other organs, these compounds could mediate certain abnormal cardio-pulmonary responses. (Authors modified) 1207

0710

SAID, S.I.

Some respiratory effects of prostaglandins E_2 and $F_{2\alpha}$.

In: Ramwell, P.W. and J.E. Shaw, eds., "Prostaglandin Symposium of the Worcester Foundation for Experimental Biology," p. 267-277. New York, Interscience, 1968.

Spontaneously breathing or open-chest, pentobarbital anesthetized dogs were used to study the effects of PGE_2 or $PGF_{2\alpha}$ (0.66-22 μg/kg) on pulmonary function. Both PG's lead to increased pulmonary arterial pressure without change in left atrial pressure. Both cause a slight or moderate increase in cardiac output, followed by a decline. Both can produce tachypnea, but dynamic compliance of the lungs falls with $PGF_{2\alpha}$. PGE_2 probably causes hyperventilation. Systemic pressure falls with PGE_2 in a dose dependent manner. With $PGF_{2\alpha}$, blood pressure either does not change or increases slightly. (JRH) 1147

0711

SAMUELSSON, B.

The prostaglandins.

In: Jarnefelt, J., ed., "Regulatory Functions of Biological Membranes," p. 216-221. Amsterdam, Elsevier, 1968.

The biosynthesis of prostaglandins and their possible regulatory function in some biological processes are discussed. 46 references are cited. (Author) 1195

0712
SANDBERG, F., A. INGELMAN-SUNDBERG, G. RYDEN and I. JOELSSON
The absorption of tritium labelled prostaglandin E_1 from the vagina of non-pregnant women.
Acta Obstetrica Gynecologica Scandinavica. 47: 22-26. 1968.

Tritium labelled PGE_1 was dissolved in 2 drops of ethanol and diluted with 2 ml of normal seminal plasma. This solution was deposited in the posterior fornix of normally menstruating volunteers. Urine was collected after 1, 2, 4, 6, 12, 24 and 32 hours. In a later series, a cervical cup was fitted over the cervix before the administration of PGE_1. The vagina was washed after 6 hours. Labelled PGE_1 and its metabolites were extracted and the tritium activity was determined. Within 24-32 hours, 10-25% of the radioactivity administered was excreted irrespective of the phase of the menstrual cycle, or whether the external os had been covered or not. As the absorption was interrupted after 6 hours and there was still an excretion of radioactivity 24 hours later, the elimination of the prostaglandin metabolites through the kidney must be rather slow. The absorption was estimated to be 20-50%. The physiological significance of this absorption is still uncertain. (JRH) 1120

0713
SANDLER, M., S.M.M. KARIM and E.D. WILLIAMS
Prostaglandins in amine-peptide-secreting tumours.
Lancet. 2: 1053-1055. 1968.

Prostaglandins E_1, E_2, F_1, and $F_{2\alpha}$ were assayed in tumor tissue taken from 11 neural-crest tumours, 3 endocrine tumours deriving from the foregut, and two from the midgut. High levels of prostaglandin $F_{2\alpha}$ were found in a wide variety of these amine-peptide-secreting tumours—8 out of the 16 studied. Elevated levels of E_2 were present in 3 neural-crest tumours. Of the 4 patients with the highest tumour levels of $F_{2\alpha}$, 3 showed diarrhoea. A hydroxy fatty acid, which might be a hitherto unidentified prostaglandin, was found in the 2 cases of mid-gut carcinoid. (Authors) 1059

0714
SANNER, J.H., L.F. ROZEK and P.S. CAMMARATA
Comparative smooth muscle and cardiovascular actions of prostaglandins E_2, $F_{2\alpha}$, and $F_{2\beta}$.
In: Ramwell, P.W. and J.E. Shaw, eds., "Prostaglandin Symposium of the Worcester Foundation for Experimental Biology," p. 215-224. New York, Interscience, 1968.

The effects of PGE_2, $PGF_{2\alpha}$ and $PGF_{2\beta}$ were compared in several bioassay systems both in vivo and in vitro. In vitro, $PGF_{2\alpha}$ was several times more potent than PGE_2 in stimulating the rabbit duodenum and the estrogen pretreated rat uterus. It was much less effective that PGE_2 in stimulating the guinea pig ileum. In vivo, PGE_2 (i v) lowered the blood pressure of urethane anesthetized rats, while $PGF_{2\alpha}$ raised it. $PGF_{2\beta}$ was about equipotent to PGE_2 in stimulating the guinea pig ileum, but was less potent in the other in vitro assays. Both $PGF_{2\alpha}$ and $PGF_{2\beta}$ acted as vasodepressors in vivo but $PGF_{2\alpha}$ was much more potent. (JRH) 1152

0715
SASAMORI, S. and Y. MIYAZAKI
Occurrence of prostaglandin E like substances in dog large intestine.
Sapporo Medical Journal. 34: 121-127. 1968.

Prostaglandin like substances (PG-S) which show a smooth muscle stimulating activity and a blood pressure lowering effect in rabbits, were extracted from the large intestine of the dog. An attempt to identify these substances with prostaglandins was made by use of thin layer chromatography and bioassay. Small and large intestines were separated into mucous membrane layer and muscle layer, and each was subjected to the procedure for prostaglandin extraction reported by Samuelsson. The crude materials, thus obtained, showed smooth muscle stimulating activity, although a considerably large difference was noted among them. It was found that the material obtained from the mucous membrane layer of the large intestine showed a large activity, while that from others showed only a weak activity. From the results obtained by thin layer chromatography, it was confirmed that almost all of the activity of the materials was derived from a substance which corresponded to Rf of authentic PGE_2. The thin layer chromatographic spot containing this substance gave no ninhydrin reaction and no lipid phosphorus colour reaction was seen. The material also lowered the blood pressure of the rabbit. The occurrence of PG-S in the muscle layer of the small and large intestines, and in the mucous membrane layer of the small intestine could not be clearly detected by thin layer chromatography under this condition. (Authors modified) 1283

0716
SCHNATZ, P. T., J.B. LEE, K. CROWSHAW and V.K. VANCE
Human renomedullary vasodepressor substances.
Clinical Research. 16: 248. 1968.

Abstract only. A crude lipid extract of human renal medullas (autopsy material) produces profound and prolonged hypotension in pentolium pretreated rats. Purification of salicic acid columns or by extraction with diethyl ether reveals the major activity in the nonpolar lipid fraction and only minor activity attributable to the prostaglandins. (Authors modified) 1208

0717
SCHNIEDER, W.P., U. AXEN, F.H. LINCOLN, J.E. PIKE and J.L. THOMPSON
The total synthesis of prostaglandins.
Journal of the American Chemical Society. 90: 5895-5896. 1968.

A total synthesis of crystalline dl-PGE_1 and dl-8-iso-PGE_1 involving solvolysis of bismesylates was performed. Additionally, the synthesis of dl-$PGF_{1\alpha}$ and dl-$PGF_{1\beta}$ methyl esters described previously by Just and Simonvitch was confirmed. All steps in the synthesis are described and structures of all important intermediates are given. (JRH),1183

0718
SCHOLES, G.B., K.G. ELEY and A. BENNETT
Effect of prostaglandins on intestinal motility.
Gut. 9: 726. 1968.

Abstract only. The effects of prostaglandins E_1 and E_2 on strips of human and guinea-pig isolated ileum, and secondly their action on guinea-pig intestinal activity both in vitro and in vivo have been studied. The prostaglandins caused contractions of the isolated longitudinal muscle in both species. In contrast, the circular muscle strips were generally inhibited. Peristalsis in isolated guinea-pig ileum was affected similarly, but the major effects were inhibition of circular muscle contraction and a diminution of propulsion of fluid through the gut lumen. In anaesthetized guinea-pigs, serosal application of prostaglandin always contracted the longitudinal muscle whereas injection into the blood stream caused either contraction or relaxation. The effect of blood-borne prostaglandin on the intraluminal pressure of the intestine was also variable: sometimes the pressure increased and sometimes it remained unchanged. This finding, together with the facts that E-type prostaglandins cause hypotension and are rapidly cleared by the liver and lungs, makes it unlikely that circulating E-type prostaglandins are normally involved in gastrointestinal motility. Our evidence indicates, however, that prostaglandin released locally within the gut wall may have such a role. (Authors modified) 1218

0719
SEKHAR, N.C., J.R. WEEKS and F.P. KUPIECKI
Antithrombotic activity of a new prostaglandin 8-iso-PGE_1.
Circulation 37/38(supp. 6): VI 23. 1968.

Abstract only. Platelet aggregation and thrombolysis were studied by the revolving plastic loop method of Chandler. Both 8-iso-PGE_1 and PGE_1 inhibit aggregation of platelet-rich rat and human plasma and also lyse ADP-induced human platelet-thrombi in vitro at 0.05 μg/ml and higher. In rats, platelet aggregation is inhibited in blood samples drawn 10 min after 3 mg/kg i v of either prostaglandin. In contrast, the potency of 8-iso-PGE_1 as a spasmogen on rabbit duodenum in vitro and cat intestine in vivo and as a vasodepressor agent in the rat, dog and cat ranges between 1 and 10% of PGE_1. As an inhibitor of epinephrine-induced lipolysis in rat epididymal fat, its potency is 24% of PGE_1. These results suggest that 8-iso-PGE_1 is a potent antithrombotic agent like PGE_1 but may have less side effect on the cardiovascular system or smooth muscle organs. (Authors modified) 1050

0720
SEN, S., R.R. SMEBY and F.M. BUMPUS
Antihypertensive effect of an isolated phospholipid.
American Journal of Physiology. 214: 337-341. 1968.

It is briefly mentioned in the discussion section of this paper that prostaglandins have been isolated from the kidney and other tissues. However, prostaglandins are vasodepressors and not antihypertensive as are the phospholipids investigated in this paper. (JRH) 1096

0721
SHEARD, P.
The effect of prostaglandin E_1 on isolated bronchial muscle from man.
Journal of Pharmacy and Pharmacology. 20: 232-233. 1968.

The effect of PGE_1 on human bronchial muscle in vitro was investigated. PGE_1 (0.25-8.0 μg/ml) caused relaxation of the inherent tone of bronchial strips. Isoprenaline (5 and 10 ng/ml) produced similar relaxations. PGE_1, 1 and 2 μg/ml, caused 36% and 100% inhibition respectively of the stimulation of bronchial muscle caused by 4 μg/ml histamine. Isoprenaline, 2 and 5 ng/ml, caused 32% and 100% inhibition respectively. Propranolol (0.1 μg/ml), which completely inhibits the effect of isoprenaline, was ineffective against PGE_1 as was phenoxybenzamine (10 μg/ml). Because of relatively small amounts of PG that have been found in bronchial tissue compared to the relatively large amounts of PG required to produce an effect in vitro, it is difficult to envision a physiological role of PG's in the normal human lung. (JRH) 1110

0722
SHAW, J.E. and P.W. RAMWELL
Inhibition of gastric secretion in rats by prostaglandin E_1.
In: Ramwell, P.W. and J.E. Shaw, eds., "Prostaglandin Symposium of the Worcester Foundation for Experimental Biology. p. 55-65. New York, Interscience, 1968.

The perfusate from the stomachs of urethane anesthetized rats was found to contain PGE_1, PGE_2, $PGF_{1\alpha}$, and $PGF_{2\alpha}$. PGE_1 (0.1-1.0µg/min) added to the perfusion medium inhibited gastric H^+ secretion. Induced by either pentapeptide, histamine, vagal stimulation, perfusion with dibutyryl cAMP or theophylline. When the stomach was stimulated with pentagastrin or by vagal stimulation there was an increase in the PG content of the perfusate. The inhibitory effects of PGE_1 may inhibit the ion transport mechanism by directly reducing intracellular cAMP. (JRH) 1164

0723
SHAW, J.E. and P.W. RAMWELL
Release of a substance P polypeptide from the cerebral cortex.
American Journal of Physiology. 215: 262-267. 1968.

It is briefly mentioned in the introduction of this paper that a smooth muscle stimulating polypeptide as well as a mixture of prostaglandins was released from the somatosensory cortex of anesthetized cats. This paper describes a characterization of this polypeptide. (JRH) 1089

0724
SHAW, J.E. and P.W. RAMWELL
Release of prostaglandin from rat epididymal fat pad on nervous and hormonal stimulation.
Journal of Biological Chemistry. 243: 1498-1503. 1968.

Acidic lipids with smooth muscle-stimulating properties were extracted from rat epididymal fat pads and were shown to co-chromatograph with PGE_1, PGE_2 and $PGF_{1\alpha}$. A basal efflux of PGE and PGF compounds was observed following incubation of adipose tissue in vitro. Increased efflux of prostaglandin was detected under conditions reported to enhance free fatty acid release, i.e. addition of epinephrine, norepinephrine histamine, acetylcholine, ACTH and serotonin, prior fasting of the animal, or stimulation of the epididymal nerve. These results, together with a decreased release of prostaglandin-like material in the presence of insulin, suggested that the efflux may be associated with lipolysis resulting from activation of a hormone-sensitive lipase. (Authors modified) 1087

0725
SINCLAIR, A.J. and F.D. COLLINS
Fatty livers in rats deficient in essential fatty acids.
Biochimica et Biophysica Acta. 152: 498-510. 1968.

It is briefly mentioned that the effects of an essential fatty acid-free diet on concentrations of free fatty acid in the plasma may be due to decreased level of prostaglandins since the essential fatty acids are precursors of prostaglandins. (JRH) 1028

0726
SJOSTRAND, N.O. and G. SWEDIN
Potentiation by smooth muscle stimulants of the hypogastric nerve - vas deferens preparation from normal and castrated guinea-pigs.
Acta Physiologica Scandinavica. 74: 472-479. 1968.

The ability of a variety of smooth muscle stimulants to potentiate hypogastric nerve stimulation of the guinea pig vas deferens in vitro was investigated. PGE_1 (0.1-30 µg), PGF_1 (0.1-10 µ) and a PG crude preparation (0.1-50 Units) produced inconsistent results which showed no correlation with dosage or prostaglandin used. (JRH) 1021

0727
SMITH, C.G.
Cyclic AMP: role of the medicinal chemist.
In: Weber, G., ed., "Advances in Enzyme Regulation," Vol. 6.: p. 353-355. New York, Pergamon Press. 1968.

Prostaglandins are simply mentioned as having a pharmacological connection with cAMP. (RAP) 1310

0728
SMITH, E.R., J.V. McMORROW, Jr., B.G. COVINO and J.B. LEE
Studies on the vasodilator action of prostaglandin E_1.
In: Ramwell, P.W. and J.E. Shaw, eds., "Prostaglandin Symposium of the Worcester Foundation for Experimental Biology," p. 259-266. New York, Interscience, 1968.

Changes in femoral artery blood flow in pentobarbital anesthetized dogs produced by intraarterial injection of PGE_1, serotonin, acetylcholine, isoproterenol and histamine were compared. The effects of inhibitors of the latter three agents on the PGE_1 response was measured. On the basis of molar concentrations, the vasodilator activity of these agents was PGE_1 > acetylcholine = isoproterenol > histamine > serotonin. The three blocking agents tested, atropine, propranolol or tripelennamine, at doses which inhibited the effects of acetylcholine, isoproterenol or histamine respectively, were completely ineffective against PGE_1. PGE_1 apparently acts directly on the blood vessels and does not involve adrenergic, cholinergic or histaminergic vasodilator systems. (JRH) 1148

0729
SOLOMON, L.M., L. JUHLIN and M.B. KIRSCHENBAUM
Prostaglandin on cutaneous vasculature.
Journal of Investigative Dermatology. 51: 280-282. 1968.

The effect of PGE_1 was studied on the cutaneous vasculature of 16 guinea pigs, 6 rabbits, 8 healthy subjects, and 20 patients with various dermatoses. PGE_1 was found to be a potent vasodilator after injection in guinea pig and human skin. It was not absorbed through intact epidermis. PGE_1 (1.0-5.0 μg) in human skin results in a prolonged erythema, up to 10 hours. The erythema is blocked by norepinephrine and epinephrine but is unaffected by anesthetics. In patients with atopic dermatitis, PGE_1 appears to have a vasodilatory effect in contrast to other pharmacological agents which have little or paradoxical effect. $PGF_{2\alpha}$ (76 μg) had no effect on rabbit and guinea pig skin. (Authors modified) 1175

0730
SONNENSCHEIN, R.R. and F.N. WHITE
Systemic circulation: local control.
Annual Review of Physiology. 30: 147-170. 1968.

This review article on local control of systemic circulation cites one reference which presented evidence for dilator nerves in the sympathetic chain of dogs which by pharmacological analysis are neither adrenergic, cholinergic, nor histaminergic and whose action is similar to that of PGE. (JRH) 1134

0731
STEINBERG, A.B., G.B. CLARKE and P.W. RAMWELL
Effects of maternal essential fatty-acid deficiency on neonatal rat brain.
Developmental Psychobiology. 1: 225-229. 1968.

Since essential fatty acids are the precursors of prostaglandins, the authors discuss the possibility that the effects of maternal essential fatty-acid deficiency on brain lipids and development might be due to a lack of prostaglandin. (JRH) 1232

0732
STOCK, K., A. AULICH and E. WESTERMANN
Studies on the mechanism of antilipolytic action of prostaglandin E_1.
Life Sciences. 7: 113-124. 1968.

Rat epididymal tissue was incubated with either norepinephrine, ACTH or theophylline alone and in the presence of PGE_1 (1×10^{-9} to 8×10^{-8}M). PGE_1 did not abolish spontaneous (basal) lipolysis. PGE_1 was a competitive inhibitor for all 3 lipolytic agents with theophylline being more sensitive to PGE_1 inhibition than the hormones. It is concluded that PGE_1 inhibits the formation of cAMP rather than inhibiting the action of formed cAMP. (JRH) 1079

0733
STRONG, C.G. and D.F. BOHR
Effects of several prostaglandins on isolated vascular smooth muscle.
In: Ramwell, P.W. and J.E. Shaw, eds., "Prostaglandin Symposium of the Worcester Foundation for Experimental Biology," p.225-245. New York, Interscience, 1968.

The effects of PGE_1, PGE_2, PGA_1 and $PGF_{1\alpha}$ on isolated smooth muscle from various levels of the vascular trees on rabbits, rats and dogs were studied in vitro. All these PG's produced contraction of rabbit aorta strips. In dog coronary artery strips, all four PG's caused contraction with, $PGF_{1\alpha}$ being much less potent than the others. In dog resistance vessel (kidney, skeletal muscle and mesenteric arteries) strips there was a biphasic response, with low doses (10^{-9}-10^{-7} g/ml) causing relaxation and high doses (10^{-6}-10^{-5} g/ml) causing contraction. The order of potency in producing contraction in mesenteric arteries was $PGA > PGE_1 > PGE_2 > PGF_{1\alpha}$. Atropine, d-lysergic acid, antihistamine, α- and β-adrenergic blocking agents had no effect on PG-induced relaxation or contraction. PGE_1 and PGA_1 altered contractility in rat portal vein and dog ureter in manner similar to alterations caused by low Ca^{++} concentrations. PG's apparently act directly on blood vessels. They may alter spontaneous smooth muscle activity by altering the effective Ca^{++} concentration. (JRH) 1151

0734
STURDE, H.-C.
Experimentelle untersuchungen zur frage der prostaglandine und ihrer beziehungen zur mannlich fertilitat. [Experimental studies on prostaglandins and their relation to male fertility.]
Arzneimittel-Forschung. 18: 895-900, 1158-1163, 1298-1310. 1968.

A group of 100 test persons comprised of men who had lived in involuntarily childless matrimony for years and in whose wives no pathologic findings could be stated to be responsible for the sterility. From these patients a total of 155 ejaculates were examined as to their effect on smooth muscles and this compared with other seminal properties, e.g. volume of semen, number of spermatozoa, motility, percentage of physiologic forms, and fructose value. The pharmacodynamic activity of the seminal liquid was found to be based on the content of prostaglandins. The level of prostaglandins in the seminal plasm was directly proportional to the density of spermatozoa and therefore probably the most important factor of fertility. Following the administration of testosterone (and chorionic gonadotropin), an increase of prostaglandins, fourfold on the average, was noted independent from the intial value. In accordance with these findings the number of spermatozoa/ml could on the average be doubled in the cases of oligo- and hypospermias. Consequently, the prostaglandin level in the seminal plasm represents a sensitive indicator of the androgenic stimulation of the organism. 37 men with normospermia were found with a relatively low prostaglandin content compared with 24 normal men who definitely had children of their own. The average prostaglandin content of the seminal plasma of this group of "fertile men" was higher by a multiple than that of the probands from sterile marriages (1.15). 9 out of these 25 obviously fertile men had an oligospermia or hypospermia but without exception showed a prostaglandin level above the average. It seems advisable to determine the prostaglandin activity in future fertility examinations of men. (From English Summary) 1276

0735
STURDE, H.-C.
Zur bedeutung der prostaglandine im humanen seminalplasm. [The significance of the prostaglandins in human seminal plasma.
Abstracts of the 13th International Congress on Dermatology, Munich, 1967, p. 670-671. Berlin, Springer, 1968.

The comparative prostaglandin content of seminal plasma samples collected from 80 men was determined by application to the intestinal tissue of guinea pigs. The subjects were 50 men with normal sperm counts, but who were childless (average duration of childless marriages, 4.1 years) (Group I); and 30 men with children (Group II). The average prostaglandin value (histamine quotient) of the Group I samples was 1.23, while that of Group II was 3.65, or nearly 3 times as great. In the author's opinion, these findings indicate a close relationship between male fertility and seminal prostaglandin content, although its physiological significance is yet unknown. (MEMH) 1334

0736
SUNAHARA, F.A. and D. KADAR
Effects of ouabain on the interaction of autonomic drugs and prostaglandins on isolated vascular tissue.
In: Ramwell, P.W. and J.E. Shaw, eds., "Prostaglandin Symposium of the Worcester Foundation for Experimental Biology," p.247-257. New York, Interscience, 1968.

Longitudinal or helical strips from dogs and rats killed by stunning or by sodium pentobarbital overdose were tested for response to PGE_1 or $PGF_{1\alpha}$. Results indicate a direct action of vascular smooth muscle and the importance of K^+ concentration to PG effects. PGE_1 at 100 ng/ml inhibited spontaneous concentrations of dog superior mesenteric vein (SMV) while $PGF_{1\alpha}$ enhanced spontaneous activity. These effects were not influenced by adrenergic blocking agents. In the rat, PGE_1 slightly enhanced spontaneous contractions of the SMV while $PGF_{1\alpha}$ showed little or no effect. Phenylephrine (PE)-induced contractions in dog SMV were inhibited by PGE_1 and enhanced by $PGF_{1\alpha}$; in rat SMV, the PE effect was slightly enhanced by PGE_1 and $PGF_{1\alpha}$ was inactive. $PGF_{1\alpha}$ enhanced or initiated spontaneous activity and increased tone and PGE_1 reduced or blocked spontaneous activity of PG-stimulated dog SMV. Effects were greatest in 4,8, and 1mM K^+ media in that order and least in 16mM. $PGF_{1\alpha}$ always enhanced norepinephrine (NE) effects on dog SMV while PGE_1 was inhibitory. Similar qualitative results were obtained using PE and superior mesenteric artery tissue strips. After more than 2 hours exposure to 1.5×10^{-7}M ouabain, dog SMV did not respond to $PGF_{1\alpha}$, but relaxed in response to PGE_1 at 4 or 8 mM K^+. At less than 2 hours exposure, PGE_1 inhibited and response to $PGF_{1\alpha}$ was variable. PGE_1 and $PGF_{1\alpha}$ effects were abolished in 5.8mM K^+ by 1×10^{-5} ouabain after 2 hours. (RMS) 1149

0737
SUTHERLAND, E.W., G.A. ROBISON and R.W. BUTCHER
Some aspects of the biological role of adenosine 3',5'monophosphate (cyclic AMP).
Circulation. 37: 279-293. 1968.

Prostaglandins are mentioned several times in this extensive review on cAMP. Special attention is given to the antilipolytic properties of the prostaglandins. It is suggested that prostaglandins may function by altering cAMP levels. (JRH) 1135

0738
SWEATMAN, W.J.F. and H.O.J. COLLIER
Effects of prostaglandins on human bronchial muscle.
Nature. 217: 69. 1968.

The effects of $PGF_{2\alpha}$, PGE_1, and PGE_2 and also mixtures of PGE_1 or PGE_2 with $PGF_{2\alpha}$ on strips of human bronchial smooth muscle in vitro were measured. $PGF_{2\alpha}$ contracted the muscle and PGE_1 relaxed it. PGE usually realized it but was not as effective as PGE_1. Neither atropine nor mepyramine blocked the effect of $PGF_{2\alpha}$. PGE_1 and PGE_2 reduced the contraction caused by $PGF_{2\alpha}$ after being made insensitive to slow reacting substance (SRS) after repeated exposure. Nor could a cross tachyphylaxis be demonstrated within the muscle insensitive to $PGF_{2\alpha}$. Thus it was shown that SRS and prostaglandin are different substances. It was also concluded that $PGF_{2\alpha}$ does not produce its effect by cholinergic nerve stimulation. (JRH) 1020

0739
SWIDERSKA-KULIKOWA, B.
Hypotensive substance from the renal medulla: isolation and some of its chemical properties.
Bulletin de L'Academie Polonaise des Sciences. 16: 603-608. 1968.

A substance depressing the arterial blood pressure and stimulating the smooth intestinal muscles is present in the medulla of pig kidneys. It is probably a higher unsaturated hydroxy fatty acid of the prostaglandin group. This substance is present in the kidneys probably conjugated with cholesterol. During storage at room temperature with free air access, the substance is partially inactivated; this is chemically expressed by the disappearance in the chromatogram of the spot with $Rf = 0.55$-0.56, the disappearance of double bonds and the decrease in the maximum corresponding to the hydroxyl group, and biologically by a less marked vasodepressor effect. It is at present difficult to answer the question whether the vasodepressor substance from the renal medulla is PGE_2, PGA_2 or a mixture of the two prostaglandins. (Author modified) 1279

0740
SWIDERSKA-KULIKOWA, B.
Studies on action of hypotensive substance from renal medulla on the heart and smooth muscle.
Bulletin de l'Academie Polonaise des Sciences. 16: 787-790. 1968.

The author concludes that the lipid substance which was isolated from hog renal medulla tissue is probably a prostaglandin since it produces similar effects in bioassay. (JRH) 1278

0741
TALALAY, P.
Biological compounds.
Science. 161: 1124. 1968.

The author reviews a book on the Proceedings of the Second Nobel Symposium, the first international symposium on prostaglandins. (RAP) 1263

0742
THIBAULT, Ph.
Les Prostaglandines. [Prostaglandins.]
La Presse Medicale. 76: 501-502. 1968.

The history and structure of prostaglandins are briefly reviewed. Also included are short discussions of some physiological effects attributed to the prostaglandins. There is special emphasis on their antihypertensive effect on the kidney. Eight references are cited. (NES) 1241

0743
THOMPSON, J.H. and M. ANGULO
The effect of prostaglandins on gastrointestinal serotonin in the rat.
European Journal of Pharmacology. 4: 224-227. 1968.

Total serotonin levels were measured spectrophotofluorometrically in the stomach fundus (SF), pyloric antrum (PA) and mid-jejunum (MJ) of adult male Sprague-Dawley rats (200-380 g) following injection of PGE_1 and E_2. PGE_1 100 μg/kg IV did not significantly effect JM serotonin levels at various time periods from 0.5-60.0 minutes. Furthermore, PGE_1 in doses of 10-900 μg/kg SC did not significantly effect total MJ serotonin levels 30 minutes after injection. PGE_1 and PGE_2 in doses of 200 μg/kg SC were without effect on serotonin levels in SF, PA and MJ. Similar results were obtained following a 60 minute subcutaneous infusion of PGE_1 1 μg/kg/min. It is concluded that the gastric secretory depression and smooth muscle stimulation produced by PG are not mediated via the release of serotonin. (Authors modified) 1023

0744
THURAU, K., H. VALTIN and J. SCHNERMANN
Kidney.
Annual Review of Physiology. 30: 441-524. 1968.

This extensive review of renal physiology contains one brief section on PGE_1 as a medullary vasodepressor substance. (JRH) 1108

0745
TIDWELL, H.C. and S.S. PAWAR
Effects of prostaglandin (PGE_1) on lipolysis and on esterification of C^{14}-palmitic acid during incubation with rat adipose tissue.
Federation Proceedings. 27: 819. 1968.

Abstract only. About 300 mg of epididymal fat pads of normally fed male albino rats were incubated for 3 hours in 2 ml of Krebs-Ringer bicarbonate buffer (pH, 7.4) containing bovine serum albumin and C^{14}-palmitic acid in the presence and absence of PGE_1. Some tissue samples were preincubated for 15 minutes in bicarbonate buffer with and without PGE_1. After thoroughly washing, tissues were similarly incubated for 2 hours in the absence of PGE_1. Duplicate samples of tissues were taken for zero time values. At the end of incubation determination of glycerol in the media indicated that PGE_1 inhibits the release of glycerol. Examination of tissue lipid activities showed that there was more incorporation of palmitic acid into lipids due to PGE_1. Most of the lipid activity was found in triglyceride fraction. Preincubation with PGE_1 also showed similar effects both on lipolysis and esterification. It appears that PGE_1 has an effect on lipolysis as well as esterification in rat adipose tissue. (Authors) 1132

0746

TURKER, R.K., S. KAYMAKCALAN and I.H. AYHAN
Effect of prostaglandin E_1 (PGE_1) on the vascular responsiveness to norepinephrine and angiotensin in the anesthetized rat.
Arzneimittel-Forschung. 18: 1310-1312. 1968.

The interaction of PGE_1 with the pressor responses to angiotensin and norepinephrine were studied on pentobarbital anesthetized and atropine pretreated rats. PGE_1 (1mg/kg) produced a marked blood pressure depression when given i.v. and no tachphylaxis was observed after repeated applications. After a single i.v. injection of PGE_1, the pressor response to angiotensin decreased significantly but NE response did not change. Our present knowledge cannot explain the mechanism of the interaction of PGE_1 with angiotensin. (Authors modified) 1277

0747

TUTTLE, R.S. and M.M. SKELLY
Interactions of prostaglandin E_1 and ouabain on contractility of isolated rabbit atria and intracellular cation concentration.
In: Ramwell, P.W. and J.E. Shaw, eds, "Prostaglandin Symposium of the Worcester Foundation for Experimental Biology," p.309-320. New York, Interscience, 1968.

The inotropic effects of PGE_1 and/or ouabain as well as their effects on potassium and sodium ion concentrations were studied in isolated rabbit atria. Ouabain (10^{-8}M) did not alter the K^+ or Na^+ concentration but at 10^{-6}M was significantly reduced. PGE_1 (3×10^{-11} or 3×10^{-10}M) alone reduced K^+ and Na^+. Addition of ouabain (10^{-8}M) prevented the depletion of K^+ by PGE_1. PGE_1 (3×10^{-11}M) or ouabain (10^{-8}M) prevented the time-dependent inotropic deterioration normally seen in incubated rabbit atria. Ouabain (10^{-7}M) produced positive inotropic stimulation while 10^{-6}M caused toxic inotropic decreases. PGE_1 increased the response of the heart muscle to ouabain causing threshold doses (10^{-8}M) to produce therapeutic doses (10^{-7}M) to produce toxic effects. PGE_1 did not alter the toxic effects of high doses of ouabain (10^{-6}M). (JRH) 1143

0748

UEHLEKE, H.
Pharmakologische grundlagen der prophylaxe und therapie von fettstoffwechselstorungen.
[Pharmacological bases of the prophylaxis and treatment of lipid metabolism disturbances.]
International Journal of Clinical Pharmacology, Therapy and Toxicology. Supp. 1: 21-28. 1968.

A number of different substances are discussed as regulators of lipid metabolism. Prostaglandins, particularly PGE_1, are mentioned as possessing anti-lipolytic properties. It is known that cyclic-AMP levels are reduced, but the mechanism is not well understood. (MEMH) 1315

0749
VANDER, A.J.
Direct effects of prostaglandin on renal function and renin release in anesthetized dog.
American Journal of Physiology. 214: 218-221. 1968.

PGE_1 or PGE_2, 10-5,000 ng/min, was infused directly into the right renal artery of pentobarbital anesthetized antidiuretic dogs. There occurred no changes in mean or pulsatic arterial pressure, heart rate, or plasma sodium. Left renal function did not change. In the right kidney, there occurred significant increases in urine flow, sodium excretion, and total renal plasma flow. Para aminohippuric acid extraction ratio was significantly reduced; glomerular filtration rate did not change. Threshold dose for these responses was approximately 20 ng/min. Increased sodium excretion always occurred, but was highly variable both in magnitude and dose-response pattern. It was not dependent upon nor did it correlate with the changes in renal plasma flow. Prostaglandin produced no detectable effect on renin release. It is concluded that prostaglandin may play a normal physiological role within the kidney. (Author modified) 1269

0750
VANE, J.R.
The alteration or removal of vaso-active substances by the pulmonary circulation.
In: Tedeschi, D.H. and R.E. Tedeschi, eds., "Importance of Fundamental Principles in Drug Evaluation," p. 217-235. New York, Raven Press, 1968.

A brief section citing 17 references on the role of the lungs in removing prostaglandins from the blood is contained in this review. (JRH) 1101

0751
VENTURA, W.P., M. FREUND and F. KNAPP
Motility of the vagina and uterine body and horns in the guinea pig.
Fertility and Sterility. 19: 462-474. 1968.

It is mentioned in the introduction that prostaglandins have been isolated from ram and human seminal fluid. (JRH) 1060

0752
VERGROESEN, A.J. and J. De BOER
Effects of prostaglandins E_1 and $F_{1\alpha}$ on isolated frog and rat hearts in relation to the potassium-calcium ratio of the perfusion fluid.
European Journal of Pharmacology. 3: 171-176. 1968.

Inotropic and chronotropic effects were observed in isolated frog and rat hearts exposed to 2 μg/ml PGE_1 under conditions of potassium arrest. A correlation was demonstrated between K^+/Ca^{++} ratio and the PGE_1 effect. PGE_1 seemed to have slightly more effect than epinephrine on normally functioning frog and rat hearts. $PGF_{1\alpha}$ at 2 μg/ml had no effect on frog heart, but a mildly stimulating effect at 5-10 μg/ml. Both PGE_1 and potassium caused an increase in coronary flow. It was concluded that PGE_1 can cause chronotropic and inotropic effects and further assumed that PGE_1 antagonizes K^+ effects rather than enhancing Ca^{++} effects. (RMS) 1234

0753
VONKEMAN, H. and D.A. van DORP
The action of prostaglandin synthetase on 2-arachidonyl-lecithin.
Biochimica et Biophysica Acta. 164: 430-432. 1968.

Substrate requirements of prostaglandin synthetase were examined using a particulate fraction of sheep vesicular glands as a source of prostaglandin synthetase. When 1-stearoyl-2-[^{3}H] arachidonyl-lecithin was used as a substrate for the enzyme system, no PG was produced until lecithinase A was added to the system. This result indicates that a free carboxyl group is necessary for the enzymatic reaction. When fatty acids extracted from the particulate fraction of sheep vesicular glands were used as a substrate, similar results were obtained in that no PG was produced unless the lipids were preincubated with lecithinase A. These results are in agreement with those of other researchers who have also demonstrated the requirement of free fatty acids as a substrate for PG biosynthesis. (JRH) 1249

0754
WAITZMAN, M.B.
Influences of prostaglandin and adrenergic drugs on ocular pressure and pupil size.
Investigative Ophthalmology. 7: 1212. 1968.

Abstract only. Prostaglandin E (PGE) causes a long-lasting increased intraocular pressure in the rabbit and miosis in several animal species. Physiological antagonism takes place when catecholamines in molar concentrations higher than that of PGE are present. Adrenergic receptor antagonists that block these antagonistic effects of catecholamines on PGE do not interfere with the PGE effects on the eye. (Author) 1222

0755
WAITZMAN, M.B.
Influences of prostaglandin and adrenergic drugs on ocular pressure and pupil size.
In: Ramwell, P.W. and J.E. Shaw, eds., "Prostaglandin Symposium of the Worcester Foundation for Experimental Biology," p. 335-349. New York, Interscience, 1968.

The effects of intraarterial, intraocular, intravenous and topical epinephrine (E), norepinephrine (NE), phenylephrine, phenoxybenzamine (PBA), propranolol and isoproterenol on the pupil responses in urethane anesthetized cat and rabbit, and increase in intraocular pressure (IOP) in rabbit caused by injection of PGE$_1$ (3×10^{-4}-7×10^{-3} μM) into the anterior chamber of the eyes was measured. The relative potencies of PGE$_1$, PGE$_2$, PGA$_1$ and PGF$_{1\alpha}$ in raising IOP were also determined in rabbits. Phenylephrine (topical and i o) and E (topical) were able to block the effects of i o PGE$_1$ with E being most potent. NE caused a mydriasis not blocked by even high doses of PGE$_1$ (1.4×10^{-2} μM). PBA i v or i o had little effect on the response to PGE$_1$ but prevented the inhibition by PGE$_1$ by NE. NE i o (3×10^{-3} μM) blocked the miosis caused by 7×10^{-3} μM PGE$_1$, but didn't block the increase in IOP caused by PGE$_1$. Isopropranolol and propranolol had little or no effects on the response of the eye to PGE$_1$ except that propranolol blocked the miosis in the rabbit. PGE$_1$ and PGE$_2$ were about equipotent in raising IOP, but PGA$_1$ and PGF$_{1\alpha}$ were much less effective both in amount of increase and duration of response. The site of action of PGE$_1$ seems to be independent of adrenergic and cholinergic agents. Results from other experiments indicate that PGE effects are also non-muscarinic. The effects of PGE on the iris seem to be independent of their effects on IOP. The possible mechanisms of PG action in the eye are discussed. (JRH) 1141

0756
WAITZMAN, M.B.
Physiological antagonism between prostaglandin and catecholamines relative to ocular pressure and pupil size in normal and sympathectomized animals.
In: Proceedings of the International Union of Physiological Sciences, 24th, Washington, D.C. 1968. Vol 7. Abstracts of volunteer papers and films. p. 457. Washington, D.C. Federation of American Societies for Experimental Biology, 1968.

Abstract only. PGE_1 causes elevation of ocular pressure in rabbits. A physiological antagonism between PGE_1 and various adrenergic agonists, relative to ocular effects, has been demonstrated in the rabbit. If the animal is pretreated with the specific blocker (i.e. phenoxybenzamine or propranolol) of the adrenergic agonist, then the ocular effects of PGE_1 are unaltered. The physiological antagonism between PGE_1 and catecholamines has now been confirmed in the cat and monkey. Destruction of cervical sympathetic innervation affects aqueous humor dynamics and pupil size in operated animals causing then to be super sensitive to adrenergic agonists. The PGE_1-catecholamine interaction was studied in sympathectomized animals. With respect to ocular pressure and pupil size, certain interspecies variations were demonstrated on the influence of normal sympathetic innervation on this interaction. (JRH) 1327

0757
WALK, R.A., J.R. SCHULTZ and J.R. WEEKS
Evaluation and control of variability in hormone-stimulated lipolysis in rat adipose tissue.
Journal of Pharmacy and Pharmacology. 20: 400-402. 1968.

A technique for reducing the variability in experiments with rat adipose tissue is described. The technique involves chopping the tissue into fine cubes, rather than using large pieces. To demonstrate this technique, three replicate experiments with pooled rat epidymymal tissue were performed. Lipolysis was stimulated with epinephrine, and PGE_1 (0.1-320 ng/ml) was added as an inhibitor. Tissue samples for the replicate curves were weighed consecutively, without subdividing the pooled tissue. The first two curves were virtually identical, but the third showed less lipolytic activity. Control vessels from subsequent assays for antilipolytic activity demonstrated that subdivision and resuspension of the tissue prevented this loss of activity. (JRH) 1237

0758
WAREMBOURG, H., G. BISERTE, J. JAILLARD, G. SEZILLE and M. BERTRAND
Variations qualitatives et quantitatives des acides gras non esterifies du plasma humain sous l'effet de perfusions intraveineuses d'hormone adrencorticotrope. [Qualitative et quantitative variations of human plasma non-esterified fatty acids as a result of intravenous perfusions of adrenocorticotropic hormone.]
Pathologie-Biologie. 16: 609-617. 1968.

In this study of the lipid mobilizing effect of ACTH in human plasma, PGE_1 is mentioned as an inhibitor of monoglyceride lipase and as an antagonist to the freeing of fatty acids by noradrenaline. PGE_1 acts in very small quantities and has definite vaso-depressive properties. (NES) 1275

0759
WEEKS, J.R., J.R. SCHULTZ and W.E. BROWN
Evaluation of smooth muscle bioassays for prostaglandins E_1 and $F_{1\alpha}$.
Journal of Applied Physiology. 25: 783-785. 1968.

Isolated smooth muscles, gerbil colon, rat stomach fundus, rabbit duodenum, and guinea pig ileum, were evaluated as test organs for bioassay of prostaglandins E_1 and $F_{1\alpha}$; 2 identical concentration-effect curves were determined for each prostaglandin on each tissue. Gerbil colon was the most suitable for bioassay. The mean responses and slopes of the two curves were almost identical (no tachyphylaxis), and the index of precision (λ, standard deviation/slope) was consistently low (0.12 or less). The suitability of the other tissues were, in descending order, rat stomach fundus, guinea pig ileum, and rabbit duodenum. (Authors) 1181

0760

WEEKS, J.R., N.C. SEKHAR and F.P. KUPIECKI
Pharmacological profile of a new prostaglandin, 8-iso-prostaglandin E_1.
Pharmacologist. 10: 212. 1968.

Abstract only. The effect of 8-iso-PGE_1 was compared to natural PGE_1 in several animal test systems. The 8-iso-PGE_1 was much less potent in stimulating rabbit duodenum in vitro and cat intestine in vivo, and was also much less effective as an inhibitor of epinephrine-induced lipolysis in rat epididymal fat and as a vasodepressor in rats, dogs and cats. The 2 PG's were about equipotent in lowering free fatty acid levels in fasted rats and as inhibitors of platelet aggregation in vivo. PGE_1 was slightly more effective inhibiting platelet aggregation in vitro. The relative effectiveness of the 2 compounds in lowering blood pressure showed considerable variation in the species tested. (JRH) 1319

0761

WEINER, R. and G. KALEY
Effects of prostaglandin E_1 on the microcirculation.
Federation Proceedings. 27: 282. 1968.

Abstract only. In vivo microscopy was utilized to evauate the vasomotor actions of prostaglandin E_1 (PGE_1), as well as its effects on vascular smooth muscle reactivity in the rat mesoceal microcirculation. Topically administered PGE_1 produced a transient dilation of all muscular microvessels and temporarily increased blood flow in the capillary network. Microvascular smooth muscle reactivity to the constrictor agents angiotensin, epinephrine, norepinephrine and vasopressin was inhibited by PGE_1. Local pretreatment with PGE_1 did not modify the constrictor activity 5HT or the dilator activity of histamine(H) or bradykinin(B). In addition, the influence of PGE_1 on vascular permeability was studied in skin and cremasteric muscle of rats. In skin, PGE_1 was as active as B and more active that H or 5HT in increasing vascular permeability. PGE_1 was more effective than 5HT or H in increasing permeability of post-capillary and muscular venules in cremasteric muscle. It is suggested that prostaglandins may play a role in microcirculatory regulation. (Authors) 1129

0762

WENDT, R.L.
Pulmonary vascular responses to angiotensin and to prostaglandins.
Thesis, University of Cincinnati. 57p. 1968.

Intra-pulmonary doses of angiotensin and PGE_1, PGE_2, $PGF_{1\alpha}$, and $PGF_{2\alpha}$ were studied in the perfused lower lobe of the dog lung. In these preparations, flow was controlled and pulmonary venous pressure and left atrial pressure were measured. Of the four PG's studied, PGE_2 (0.1-10.0 μg/kg), $PGF_{1\alpha}$ (0.1-3.0 μg/kg) and $PGF_{2\alpha}$ (0.01-3.0 μg/kg) increased the total pulmonary vascular resistance whereas, PGE_1, up to doses of 10.0 μg/kg, had no effect on the pulmonary circulation. Of the three active agents, $PGF_{2\alpha}$ was found to be the most potent vasoconstrictor. These compounds constricted both the pulmonary arterial and venous vasculature. The finding that the responses to $PGF_{2\alpha}$ remained unchanged after diphenhydramine in the same animal in which the effects of histamine were completely antagonized appears to rule out the possibility of histamine release after $PGF_{2\alpha}$. (JRH) 1311

0763

WENZEL, E., B. MAY, W. JAGER, K. BREDDIN and E. BOHLE

Neue ergebnisse uber den einfluss von prostaglandin E_1 auf die plattchenfunktion und-aggregation. [New findings on the influence of PGE_1 on platelet function and aggregation.]

Verhandlungen der Deutschen Gesellschaft fur Innere Medizin. 74: 961-964. 1968.

In vivo and in vitro experiments were performed on rabbits; PGE_1 in concentrations of 1 ng/ml was found to inhibit platelet aggregation and function. This effect is reversible; platelets washed twice in PGE_1 free plasma or glucose solutions return to normal. The effect of PGE_1 seems to be related to the presence of plasma protein, and to operate on the platelet membrane. In plasma from 36 human patients with pathological PAT-values, similar results were obtained. (MEMH) 1274

0764

WILLEBRANDS, A.F., and S.J.A. TASSERON

Effect of hormones on substrate preference in isolated rat heart.

American Journal of Physiology. 215: 1089-1095. 1968.

The effect of hormones on $^{14}CO_2$ production from ^{14}C-labeled substrates and on substrate preference was studied in isolated rat hearts beating at a fixed frequency. In hearts fed 11.1 mM glucose, insulin (10mU/ml) increased glucose oxidation approximately 50%, epinephrine (1 μg/ml) 300-400%, and PGE_1 and $PGF_{1\alpha}$ (0.5 μg/ml) both approximately 60%. In hearts fed 1 mM palmitic acid, insulin had no effect on the oxidation of this acid, but PGE_1 or $PGF_{1\alpha}$ produced a significant increase. Both prostaglandins also stimulated oxygen consumption considerably. In hearts fed glucose plus palmitic acid, the preference for fatty acid was not influenced by insulin or by the prostaglandins. (Authors modified) 1095

0765

WILLIAMS, E.D., S.M.M. KARIM, and M. SANDLER

Prostaglandin secretion by medullary carcinoma of the thyroid.

Lancet. 1: 22-23. 1968.

Appreciable amounts of prostaglandins were present in tumour tissue from 4 out of 7 cases of medullary carcinoma of the thyroid. Raised blood-levels were detected in 2; concentrations were considerably higher in blood draining the tumour than in peripheral venous blood. Both had diarrhea which could be due to prostaglandin production by the tumour. (Authors) 1072

0766

WILLOUGHBY, D.A.

Effects of prostaglandins $PGF_{2\alpha}$ and PGE_1 on vascular permeability.

Journal of Pathology and Bacteriology. 96: 381-387. 1968.

$PGF_{2\alpha}$, a venoconstrictor, has been found to antagonize increased venular permeability to plasma protein induced by histamine, serotonin, bradykinin and lymph-node permeability factor in rats. Pre-treatment with low doses (10-59 μg/kg) of this PG will suppress increased vascular permeability after thermal injury and the early phase of increased vascular permeability after chemical injury. By contrast PGE$_1$, which lacks the vaso-activity of PGF$_{2\alpha}$, is without effect on either mediators or inflammatory response. (Author modified) 1106

0767
WIQVIST, N., M. BYGDEMAN, S.U. KWON, T. MUKHERJEE and U. ROTH-BRANDEL
Effect of prostaglandin E$_1$ on the midpregnant human uterus.
American Journal of Obstetrics and Gynecology. 102: 327-332. 1968.

The effect of PGE$_1$ on uterine motility was measured in 22 midpregnant (14 to 21 weeks) patients by recording its effect on intraamniotic pressure. The PGE$_1$ was given in single doses intravenously (16 patients), intramuscularly (3 patients), intraamniotically (3 patients) and intravaginally (6 patients). Given intravenously, PGE$_1$ produced an increase in tone and contractions, which in some cases resembled the contractions of labor. The dosage range was 5-150 μg, with the threshold dose being about 20 μg. The uterine response was rapid and dose dependent. In 5 cases a subsequent dose of PGE$_2$ (5 to 50 μg) was given after the uterine tone had returned to preinjection levels. From the limited number of cases it appears that both these PG's produce the same effects both qualitatively and quantitatively. When given intramuscularly, 100 μg of PGE$_1$ produced a more prolonged response of lower intensity than the same amount given intravenously. When given intraamniotically (75 μg) or intravaginally (200-1,000 μg), PGE$_1$ produced no observable effects. The only side effects mentioned were moderate nausea and a slight increase in pulse rate (15-20 beats/min) with dose of 75 μg or more give intravenously. The authors discuss the possible pharmacologic uses of PGE$_1$ and PGE$_2$. (JRH) 1236

0768
YAMAMURA, Y. and T. KOKUBA
[Prostaglandins.]
Naika. 22: 891-899. 1968.

Article in Japnaese. Abstract not available at present. 1270

0769
YEUNG, D. and I.T. OLIVER
Induction of phosphopyruvate carboxylase in neonatal rat liver by adenosine 3',5'-cyclic monophosphate.
Biochemistry. 7: 3231-3239. 1968.

The activity of cytoplasmic phosphopyruvate carboxylase normally develops in liver following birth or premature delivery of the fetus. The enzyme activity can be precociously induced in utero by intraperitoneal injection of cAMP into fetuses. Repression of phosphopyruvate carboxylase synthesis in premature rat liver is achieved by injection of PGE$_1$ (0.25 μg/fetus intraperitoneal), insulin or ergotamine tartrate. (JRH) 1170

0770
AACH, R., and J. KISSANE
Medullary carcinoma of the thyroid with hypocalcemia and diarrhea.
American Journal of Medicine. 46: 961-971. 1969.

The case history of a man who was seen intermittently between 1954 and 1967 for treatment of metatastic thyroid medullary carcinoma is given. Discussion of the case by several physicians follows. The study of Williams et al. (1968) in which increased prostaglandin secretion was noted in 4 of 7 patients with tumors similar to the case under discussion is mentioned, and the role of prostaglandins in the diarrhea of medullary carcinoma of thyroid is discussed. (RMS) 1498

0771
ABIKO, Y.
[Effect of PGE_1 on the heart, blood pressure and the coronary vessel in the dog.]
Folia Pharmacologica Japonica. 65: 86-87. 1969.

Article in Japanese. Abstract not available at present. 1720

0772
ADOLPHSON R.L., T.J. KENNEDY, R. REEB, and R.G. TOWNLEY
Effect of beta-adrenergic blockade and prostaglandin $F_{2\alpha}$ on isoproterenol and theophylline bronchodilation.
Journal of Allergy. 43: 176-177. 1969.

Abstract only. The pharmacologic mechanisms of drugs used in bronchial asthma were investigated by observing the actions of various smooth muscle mediators and β-adrenergic blockers in a muscle bath. The muscle relaxing effect of isoproterenol showed marked species variation. It was incomplete in the rat and dog but more complete in the guinea pig and the human being. The relaxing effect of theophylline was universal. It has been suggested that prostaglandins may function as either β-blockers or as antagonists of theophylline. In human tissue, prostaglandin $F_{2\alpha}$ partially inhibited the relaxing effect of isoproterenol and theophylline. (GW) 1449

0773
AEBI, H.
Biochemische aspekte der atherosklerose: chloesterinstoffwechsel und fettsauremuster. [Biochemical aspects of atherosclerosis: cholesterol metabolism and fatty acid design].
Bibliotheca Nutritio et Dieta. 12: 23-36. 1969.

The author mentions that the prostaglandins are released in the nervous system by the stimulation of an active substance and cause, for example, contraction of smooth muscle as well as a decrease of blood pressure resulting from peripheral vasodilation. In fat tissue they cause in vivo and in vitro inhibition of lipolysis. (GW) 1642

0774
ALBRO, P.W. and L. FISHBEIN
Determination of prostaglandins by gas-liquid chromatography.
Journal of Chromatography. 44: 443-451. 1969.

Two procedures are described for the determination of prostaglandins B, E and F by gas-liquid chromatography. In the first, keto prostaglandins are quantitatively dehydrated, esterified, and acetylated to give products stable to gas chromatography, while F prostaglandins are rapidly analyzed as trimethylsilyl ether-trimethylsilyl esters. In the second procedure, a mixutre of prostaglandins is separated into B, E and F class fractions by column chromatography. The B class of prostaglandins, both that present originally and that formed from E prostaglandins by base treatment, is resolved according to degrees of unsaturation by gas chromatography of their acetylated methyl esters. The F prostaglandins are similarly resolved by gas chromatography of their trimethylsilyl ether-methyl esters. (Authors) 1570

0775
ANDERSEN, N.H.
Dehydration of prostaglandins: study by spectroscopic method.
Journal of Lipid Research. 10: 320-325. 1969.

In the general method for quantification of 9-oxo-prostaglandins, the molecule is dehydrated and then induced to rearrange in the presence of alkali to give the conjugated dienone PGB_1. This sequence has been studied under both acidic and basic conditions by ultraviolet (UV) absorption and optical rotatory dispersion (ORD) spectra of PGE_1, PGA_1, PGB_1 and their naturally derived 15-epimers with the aim of developing accurate and sensitive spectroscopic assays. Conditions for quantitative conversion of PGE_1 to PGA_1 are described. (ART) 1417

0776
ANDERSEN, N.H.
Preparative thin-layer and column chromatography of prostaglandins.
Journal of Lipid Research. 10: 316-319. 1969.

The author outlines analytical and preparative chromatographic methods for monounsaturated prostaglandins developed specifically for the separation of various hydroxy epimers of PGE_1 and F_1 that are effective for separation of known natural prostaglandins. Apparent $R_f \times 100$ values for 15 prostaglandins and epi-prostaglandins are given. (RMS) 1418

0777
ANGGARD, A.
The effect of prostaglandins on nasal airway resistance in man.
Annals of Otology, Rhinology and Laryngology. 78: 657-662. 1969.

The effects of prostaglandins (PGE_1, PGE_2 and $PGF_{1\alpha}$) were investigated regarding nasal airway resistance in a group of 7 normal persons, age 19 to 31, 4 male and 3 female. About 0.24 ml. of the test substance was introduced into each nostril with a nebulizer. Prostaglandins were pure compounds dissolved in physiological saline in a concentration of $50\mu g/ml$. A simultaneous recording was made of the airflow through the nose and the pressure gradient between the oropharynx and the external nares without introducing any instrument through the nostrils. The expiratory air flow at a pressure difference of 10 mm of water was used as a measure of the nasal passage. Naphazoline was used as a control. Findings were that in 4 subjects, 2 male and 2 female, PGE_1 and PGE_2 induced an increase in nasal patency, maximal after 10 minutes and lasting 30 to 60 minutes. Three subjects did not react to any prostaglandin but did to naphazoline. No $PGF_{1\alpha}$ effect was noted. (PR) 1539

0778
ANGAARD, E.
Pharmacology of the prostaglandins.
Abstracts, 4th International Congress on Pharmacology. Basle, 14-18 July, p. 11. 1969.

Abstract only. In this symposium introduction, the author gives a brief history of the discovery of prostaglandins and their pharmacological effects. He states that this symposium focuses on biosynthesis and metabolism, pharmacological and physiological control of their biosynthesis and release, and some interactions of the prostaglandins with various endocrine systems. (RAP) 1582

0779
ANGAARD, E. and B. SAMUELSSON
A prostaglandin dehydrogenase from pig lung.
In: Colowick, P., ed., "Methods in Enzymology," Vol. 14. p. 215-219.
New York, Academic Press 1969.

Two assay methods for isolating a prostaglandin dehydrogenase from pig lung are described in detail. Method A is based upon the development of a strong but transient chromophore with λ max at 490-500μM upon treatment of the product of the oxidation of PGE with alkali. It is faster but less accurate than method B and is therefore useful for the rapid analysis of a large number of fractions. In method B, the activity of the enzyme is determined from the initial reaction rate of PGE_1 and NAD^+ followed in a spectrometer at 340 μM. For each chromatographic step in the purification process, per cent yield and amount of specific activity are given. A final section deals with the physical and chemical properties of the enzyme. (RAP) 1766

0780
ANGGARD, E., F.M. MATSCHINSKY, and B. SAMUELSSON
Prostaglandins: enzymatic analysis.
Science. 163: 476-480. 1969.

By means of a specific nicotinamide-adenine dinucleotide-dependent prostaglandin dehydrogenase from swine lung, and enzymatic method has been developed for the assay of prostaglandins. The method permits analysis with a lower limit of 10^{-12} mole of prostaglandin. (Authors) 1626

0781
ANONYMOUS
Inhibition of prostaglandin synthesis by unsaturated fatty acids.
Nutrition Reviews. 27: 318-320. 1969.

Oleic, linoleic, and, especially, linolenic acid inhibit prostaglandin synthesis from arachidonate by extracts of acetone powders of rat stomach or sheep vesicular glands. (Author) 1736

0782
ANONYMOUS
Prostaglandins.
Canadian Medical Association Journal. 100: 37-38. 1969.

This short article summarizes the research on prostaglandins and suggests some possible areas for further research. The author reviews 17 articles. (JRH) 1540

0783
ANONYMOUS
Prostaglandins proposed for ulcer therapy.
Journal of the American Medical Association. 207: 481. 1969.

This news-type article briefly quotes several Upjohn scientists on experimental results of the use of prostaglandins in control of ulcers. Mention is also made of a new synthesis of PGE_1 and of the luteolytic effect of $PGF_{2\alpha}$. (RMS) 1439

0784
ANONYMOUS
Samen gegen samen: prostaglandin $F_{2\alpha}$ und decapazitationsfaktor.
[Seed vs. seed: prostaglandin $F_{2\alpha}$ and decapacitation factor.]
Euromed 9: 1285. 1969.

The population explosion is no longer a mere threat. For the future, prostaglandin $F_{2\alpha}$ offers promise for a delayed-action contraceptive. $PGF_{2\alpha}$ can be incorporated into a silicon polymer which could be applied intravaginally or subcutaneously. A precise dose of prostaglandin would then be released daily for up to 5 years. Some observers report that $PGF_{2\alpha}$ directs corpus luteum regression and repulsion of the endometrium in monkeys, rabbits and rats. Others assume that the prostaglandin diminishes the local acid supply of the corpus luteum leading to luteolysis. (GW) 1729

0785
ARFORS, K.E., D.P. DHALL, J. ENGESET, H. HINT, N.A. MATHESON, and O. TANGEN
In vivo quantitation of platelet activity using biolaser-induced endothelial injury.
Bibliotheca Anatomica. 10: 502-506. 1969.

A method for in vivo quantification of platelet activity using biolaser endothelial injury is described. Its advantages over the other methods available are discussed. In a preliminary study dextran 70 (macrodex; Mw 70,000) was found to be highly effective in counteracting platelet activity at sites of endothelial injuries. Dipyridamole had a transient effect. Heparin and prostaglandin E_1 were found to have no effect. (Authors) 1595

0786
ASPINALL, R.L. and P.S. CAMMARATA
Effect of prostaglandin E_2 on adjuvant arthritis.
Nature. 224: 1320-1321. 1969.

The author conducted tests to show that PGE_2 has definite prophylactic and therapeutic effect on adjuvant arthritis which indicates high potency and specificity for the processes involved. Rats were inoculated intradermally with heat killed *Myobacterium butyricum*. One group was subcutaneously injected with PGE_2 for 15 days while the control group received only a buffer. The circumference of the tibiotarsal joint was used as a measure of the severity of the disease. PGE_2 in twice daily doses of 200-500 μg reduced the development of the disease. The therapeutic effect of PGE_2 was also tested. A 500 μg dose twice daily was effective in reducing the swelling of established arthritis. PGE_2 failed to reduce the edematous swelling caused by carrageenin and cotton dental pellet implanted subcutaneously in the rat paw. It is suggested that PGE_2 delays hypersensitivity and/or the subsequent processes which preface joint inflammation since it does not act on the inflammatory process. (GW) 1413

0787
AURELL, M.
Renal response in man to plasma volume expansion and angiotension.
Scandinavian Journal of Clinical and Laboratory Investigation. 24 (supp. 112): 1-59. 1969.

In the discussion section, the author briefly mentions that PGE_1 has been shown to increase renal blood flow and to promote sodium excretion in small doses in dogs. (RAP) 1740

0788
AXEN, U., F.H. LINCOLN, and J.L. THOMPSON
A total synthesis of prostaglandin E_1 and related substances via endo-bicyclohexane intermediates.
Chemical Communications. 303-304. 1969.

Conversion of the less polar glycol isolated from the exo-configuration into the bismethane sulphonate followed by solvolysis in 2:1 acetone-water at room temperture gave 17-18% yield of (±) PGE_1 methyl ester and 19% of (±)-15 epi PGE_1 methyl ester. The more polar glycol gave 19% yield of (±)-PGE_1 methyl ester and 16-17% yield of the 15-epi-isomer. (GW) 1638

0789
BAGLI, J.F., and T. BOGRI
Biologically active analogs of prostaglandins.
Abstracts, 158th Meeting of the American Chemical Society. New York, 8-12 September, MEDI abs. 4. 1969.

Abstract only. The present communication describes a simple and flexible method of synthesis of 9,15-oxygenated prostanoic acid derivatives. Irradiation of a mixture of 2-(6-carbmoethoxyhexyl)-cy-clopent-2-ene-1-one and 1-chlorooct-1-ene-3-one by high pressure Hanovia mercury arc lamp over a period of 35-40 hours, generates a photoadduct in 35% yield. This adduct on treatment with zinc and acetic acid yielded 9,15-dioxoprostanoic acid methyl ester in approximately 50% yield. This compound has been transformed into various reduced products under different conditions (Authors modified) 1786

0790
BAGLI, J.F. and T. BOGRI
Prostaglandins III-± 11-deoxy-13,14-dihydroprostaglandin $F_{1\alpha}$ and $F_{1\beta}$-a novel synthesis of prostanoic acids.
Tetrahedron Letters. 1639-1644. 1969.

The synthesis of ± 11-deoxy-13,14-dihydroprostaglandin $F_{1\alpha}$ and $F_{1\beta}$ from cyclopentenone and chlorovinyl ketone by the photoaddition between the α-β-unsaturated ketones. Some chemical and physical properties were given at each step as well as diagrams. The methyl esters and acids of the isomeric diols obtained lowered arterial blood pressure in normotensive cats at 1 to 0.5 mg/kg levels intravenously. The esters also showed vasodepressor activity in hypertensive rats, when administered intraperitoneally. (RAP) 1784

0791

BAKHLE, Y.S., A.M. REYNARD, and J.R. VANE
Metabolism of the angiotensins in isolated perfused tissues.
Nature. 222: 956-959. 1969.

Only brief mention is made of prostaglandins in this study of the metabolism of angiotensin I and angiotensin II in isolated vascular beds perfused with Krebs solution. Throughout these experiments the angiotensins have been assayed by their contractor effect on the rat colon. The distinction between prostaglandins and angiotensin may best be seen in chick rectal caecum. No prostaglandin was released after infusion of angiotensin I or angiotensin II through the lung. An infusion of angiotensin I through the isolated lung contracted the rat colon but not the chick rectal caeca which responded to both prostaglandin E_2 and $F_{2\alpha}$ at a concentration of 1 ng/ml. Contraction of the rat colon was therefore due to angiotensin II. (GW) 1601

0792

BALDAUF, J., and K. GEBHARDT
Ein substanz P-standardpraparat. [A standard preparation of substance P.]
Naunyn-Schmiededbergs Archiv fur Pharmakologie und Experimentelle Pathologie. 265: 278-286. 1969.

Fraction Fa of crude substance P (SP), made from brain and gut, causes, in contrast to the SP peptide Fb, a relaxation of the isolated rat duodenum. Fraction Fa is therefore not a suitable Sp standard preparation. By means of sephadex gel filtration, a relaxing component can be separated from the contracting Fa peptide. The relaxing Fa component is acidic, but is neither a peptide nor prostaglandin $F_{2\alpha}$. Purified Fa material from brain is recommended as SP standard preparation. (Author) 1614

0793

BARAC, G.
Action de la prostaglandine E_1 sur la direses et de debit sanguin renal chez le chien. [Action of prostaglandin E_1 on diruesis and renal blood flow in the dog.]
Comptes Rendus des Seances de la Societe de Biologie et de ses Filiales. 163: 1233-1238. 1969.

In a dog anesthetized with chloralose, prostaglandin E_1 injected intravenously or intraarterially at a dose of 8-9 γ/kg caused a marked hypotension and anuria, sometimes followed by oliguira, even when the arterial pressure was reestablished. With subhypotensive doses, E_1, infused intravenously does not affect diuresis, though when infused in the aorta of the kidney in the neck, it increases diuresis. E_1 increases the blood flow of the perfused kidney, at a constant pressure, with normal blood or the blood of a burned dog. An elevated rate of "erythrotonin" can inhibit the renal vasodilatory action of E_1. (Authors) 1750

0794

BARRY, E. and W.J. HALL
Stimulation of sodium movement across frog skin by prostaglandin E_1.
Journal of Physiology. 200: 83P. 1969.

The authors studied the effects of prostaglandin E_1 on salt movement across frog skin. Ventral skin was used to study short circuit current before and after the addition of prostaglandin E_1. PGE_1 increased the short circuit current by amounts ranging from 14 to 76 μA with a mean of

34 μA. By using [4]Na as a tracer, it was shown that sodium movement across PGE$_1$-stimulated skin is sufficient to account for the increase in short circuit current. (GW) 1427

0795
BEDWANI, J.R.
Prostaglandins and ovarian steroidogenesis.
Journal of Physiology. 201: 98P. 1969.

Title only. 1438

0796
BEITCH, B.R. and K.E. EAKINS
The actions of various prostaglandins on intraocular pressure.
Federation Proceedings. 28: 678. 1969.

Abstract only. The effects of prostaglandins E$_1$, E$_2$, F$_{1\alpha}$ and F$_{2\alpha}$ on the intraocular pressure (IOP) were studied in rabbits anaesthetized ith urethane. Intracameral injection of solutions of prostaglandins in a constant dose volume (10 μl) were made into one eye while the contralateral control eye received an equal dose of the vehicle. PGE$_1$ (5 ng-1 μg), PGE$_2$ (1 ng-1 μg) and PGF$_{2\alpha}$ (50 ng-1 μg) produced a well maintained elevation of IOP. PGF$_{1\alpha}$ at low doses produced a small rise in IOP but larger doses were without effect. A second injection of the same doses had a markedly reduced effect on IOP. With PGE$_1$ and PGE$_2$ a good correlation was found between the rise of IOP and the protein content of the aqueous humor. Intra-arterial infusion of polyphloretin phosphate (0.5 μg/min) antagonizes the rise in IOP produced by PGE$_2$. Findings support the conclusion that the rise in IOP produced by these prostaglandins is mediated in part by an increase in the permeability of the blood-aqueous barrier. (PR) 1546

0797
BEITCH, B.R. and K.E. EAKINS
The effects of prostaglandins on the intraocular pressure of the rabbit.
Birtish Journal of Pharmacology. 37: 158-167. 1969.

The effects of intracameral injections of prostaglandins E$_1$, E$_2$, F$_{1\alpha}$, F$_{2\alpha}$ and A$_1$ were studied on the intraocular pressure (IOP) of rabbits anaesthetized with urethane. All the prostaglandins studied except F$_{1\alpha}$ were found to be capable of producing a large, sustained rise in IOP, accompanied in many cases by miosis. A marked decrease in response to repeated injections was found in all the prostaglandins studied: this effect was more pronounced following a large initial response. The descending order of potency in their ability to raise IOP was E$_1$ equaled E$_2$>F$_{2\alpha}$>A$_1$>F$_{1\alpha}$. Intracameral injections of prostaglandins E$_1$ and E$_2$ resulted in an increase of the protein content of the aqueous humor which was related to the magnitude of the sustained increase in IOP. Stabilization of the blood-aqueous barrier with polyphloretin phosphate markedly reduced both the IOP response and the effect of E$_2$ on the protein content of the aqueous humor. It is concluded that the production of local vasodilation and increased permeability of the blood-aqueous barrier play an important part in the effect of prostaglandins in the IOP. Their involvement in the response of the rabbit eye to irritation is speculated. (PR) 1577

0798
BELCHER, M., N.S. MERLINO, J.T. RO'ANE, and P.D. FLYNN
Independence of the effects of epinephrine, glucagon, and adrenocorticotropin on glucose utilization from those on lipolysis in isolated rat adipose cells.
Journal of Biological Chemistry. 244: 3423-3429. 1969.

The stimulatory effects of epinephrine, glucagon, and ACTH on lipolysis and glucose uptake and oxidation were studied in rat adipose cells. It was found that PGE_1 inhibited the lypolytic effect of these hormones, but it had no effect on their stimulation of glucose uptake and oxidation. Lipolysis inhibition was never complete. (JRH) 1467

0799
BENNETT, A., and B. FLESHLER
Action of prostaglandin E_1 on the longitudinal muscle of the guinea-pig isolated colon.
British Journal of Pharmacology. 35: 351P-352P. 1969.

Tetradotoxin (0.25 or 0.5 μg/ml), a nerve blocking drug, abolished effects of nicotine and reduced the effects of PGE_1 without altering the response to acetylcholine in guinea pig colon in vitro. Hyoscine (0.1-5 μg/ml) caused reduction in the effects of PGE_1 in 2 of 11 experiments. Morphine (0.1-0.2 μg/ml) caused a reduction to PGE_1 response in 4 of 8 experiments. The results suggest PGE_1 stimulation of receptors on or in muscle cells and activation of non-cholinergic excitatory nerves. The response to tetradotoxin indicates neurogenic action of PGE_1 which was not confirmed with attempts to block intramural nerves with cocaine or lignocaine. The authors favor the idea that PGE_1 acts partly by stimulating non-cholinergic excitatory nerves, but note that the possibility of tetradotoxin acting atypically by interfering with prostaglandin receptors cannot be excluded. (RMS) 1500

0800
BERMAN, H.J., O. TANGEN, D. AUSPRUNK, and H. COLLINS
Prostaglandin E_1 inhibition of aggregation of hamsters platelets.
Bibliotheca Anatomica. 10: 507-515. 1969.

PGE_1 has been shown to inhibit the aggregation of hamster platelets in vitro as measured turbidimetrically. It was active in this respect against ADP, thrombin and collagen. On the other hand PGE_2 showed a slight potentiation of ADP-induced aggregation. Calcium ions partly inactivated PGE_1 in vitro, and the degree of inactivation was increased by increasing concentrations of calcium ions. PGE_1 was also a strong inhibitor of electrically induced platelet thrombus formation in the hamster cheek pouch. In these experiments PGE_1 was either applied topically or infused continuously. (Authors) 1749

0801
BERNIER, J.J., J.C. RAMBAUD, D. CATTAN, and A. PROST
Diarrhoea associated with medullary carcinoma of the thyroid.
Gut. 10: 980-985. 1969.

Diarrhoea, which is present in roughly one third of cases of medullary carcinoma of the thyroid, was investigated in five cases. Excessive loss of water and electrolytes in the stools was the major factor. Water and sodium diarrhoea seems to be linked to a sometimes considerable increase in the rate of transit through the small intestine and colon, and may be relieved by codeine or codethyline. The link between the tumour and the disordered motility seems definite in view of certain cases in which removal of the tumour caused the diarrhoea to disappear immediately. With regard to prostaglandins, high concentrations have been observed in the tumours and in the venous blood draining the tumours, but their presence in systemic blood is inconstant. The only hormonal substance, concentration of which seems to be definitely increased in the systemic blood of patients with a medullary carcinoma of the thyroid, is thyrocalcitonin but this hormone does not seem to have any effect on the motor activity of the digestive tract. (Authors modified) 1764

0802

BEVEGARD, S., and L. ORO

Effect of prostaglandin E_1 on forearm blood flow.

Scandinavian Journal of Clinical and Laboratory Investigation. 23: 347-353. 1969.

In 9 healthy male volunteers, the effect on forearm blood flow of infusions of prostaglandin E_1 in the brachial artery was studied by means of strain-gauge plethysmography. PGE_1 invariably caused a rapid local increase in blood flow with doses of 10^{-5} to $10^{-4}\,\mu g \cdot min^{-1}$. With $10^{-2}\,\mu g \cdot min^{-1} \cdot kg^{-1}$ of PGE_1, the forearm blood flow increased to an average of 30 $ml \cdot min^{-1} \cdot 100ml^{-1}$ (range 15-43). The PGE_1-induced vasodilatation was not blocked by atropine or propranolol, but could be abolished by simulatenous infusion of noradrenaline. (Authors) 1555

0803

BIRON, P, and J-C. BOILEAU

Fate of vasopressin and oxytocin in the pulmonary circulation.

Canadian Journal of Physiology and Pharmacology. 47: 713-717. 1969.

The only mention of prostaglandins is in a chart which shows that 90-95% of PGE_1 is inactivated by passage through the pulmonary circulation in dogs, cats and rabbits. (JRH) 1553

0804

BLAIR, E.L., S. FALKMER, C. HELLERSTROM, H. OSTBERG, and D.D. RICHARDSON

Investigation of gastrin activity in pancreatic islet tissue.

Acta Pathologica et Microbiologica Scandinavica. 75: 583-597. 1969.

The authors cite Mutt(1968) that secretin-stimulated secretion of pancreatic juice may be depressed by PG's, Ramwell and Shaw (1968) that perfusion of rat stomach mucosa with PGE_1 inhibits pentagastrin-stimulated acid secretion, and Robert et al. (1967) that i v PG's inhibit dog gastric secretion, but gives no details. (RMS) 1497

0805

BLAKELEY, A.G.H., G.L. BROWN, D.P. DEARNALEY, and V. HARRISON

The effect of nerve stimulation on the synthesis of ^{3}H-noradrenaline from ^{3}H-tyrosine in the isolated blood-perfused cat spleen.

Journal of Physiology. 200: 59P-60P. 1969.

Isolated cat spleens were perfused with blood containing PGE_1. ^{3}H-tyrosine was added to the blood. The splenic nerve was then stimulated at different frequencies and for different lengths of time. Overflowing transmitter was collected. No mention is made of the effect of the PGE_1 in the experiment. (JRH) 1659

0806

BLAKELEY, A.G.H., L. BROWN, D.P. DEARNALEY, and R.I. WOODS

Perfusion of the spleen with blood containing prostaglandin E_1: transmitter liberation and uptake.

Proceedings of the Royal Society of London; B: Biological Sciences. 174: 281-292. 1969.

The cat's spleen perfused with homologous blood containing prostaglandin E_1 provides an admirable preparation for the study of the liberation and uptake of the sympathetic transmitter. Obstruction of the circulation with platelet thrombi is eliminated; blood flow is high, and overflow of transmitter when the nerves are stimulated is higher than in other preparations. The high overflow of transmitter is attributed to adequate perfusion of a large mass of splenic tissue. There is no evidence that the amount of transmitter liberated per unit mass of tissue is increased. Uptake of liberated transmitter liberated per unit mass of tissue is increased. Uptake of liberated transmitter and of infused 1-noradrenaline is no different from that observed in other preparations of the spleen. (Authors) 1666

0807
BLATCHLEY, F.R., and B.T. DONOVAN
Luteolytic effect of prostaglandin in the guinea-pig.
Nature. 221: 1065-1066. 1969.

The authors tested the luteolytic action of pharmacologically active agents in the uterus of the guinea pig. One group of totally hysterectomized guinea pigs was injected intraperitoneally with 0.5 mg of $PGF_{2\alpha}$ twice daily for 7 days. Three other groups were treated with injections of adrenaline hydrochloride, atropine sulfate of histamine dihydrochloride with similar molar quantities. The adrenaline, atropine or histamine had no luteolytic effect on the corpora lutea. In contrast $PGF_{2\alpha}$ had a potent luteolytic effect which suggests that its action is specific. (GW) 1410

0808
BLOCK, M.A.
Familial medullary carcinoma of the thyroid.
General Practitioner. 39: 105-107. 1969.

The author briefly mentions prostaglandins being present in the tumor and blood of patients suffering from medullary carcinoma of the thyroid. (JRH) 1456

0809
BOHMAN, S-O. and A.B. MAUNSBACH
Isolation of the lipid droplets from the interstitial cells of the renal medulla.
Journal of Ultrastructure Research. 29: 569-570. 1969.

Abstract only. Lipid droplets isolated from the interstitial cells of the renal medulla may contain vasodepressor substances which the authors feel are prostaglandins. (RAP) 1790

0810
BOLDINGH, J.
Essentielle fettsauren, atherosklerose und prostaglandine. [Essential fatty acids, atherosclerosis and prostaglandins.]
Fette-Seifen-Anstrichmittel. 71: 1-10. 1969.

Chemical structures of fatty acids, and their relationships to prostaglandins are discussed with particular regard to the formation, prevention and treatment of atherosclerosis. (MEMH) 1778

0811
BORN, G.V.R.
Blood platelets as pharmacologic systems.
Abstracts, 4th International Congress on Pharmacology. Basle, 14-18 July, p. 46-47. 1969.

Abstract only. PGE$_1$ is mentioned as an inhibitor of platelet aggregation in this short description of analogies between certain pharmacological properties of mammalian blood platelets and similar properties of other cells. (JRH) 1636

0812
BOUSQUET, M.
L'AMP cyclic. [Cyclic AMP.]
Courcours Medical. 91: 7595-7601. 1969.

This review of cAMP only briefly mentions the prostaglandins. The author states that they are antilypolytic substances which seem to decrease the level of cAMP in fatty tissue. (NES) 1789

0813
BOWMAN, W.C., and M.W. NOTT
Actions of sympathomimetic amines and their antagonists on skeletal muscle.
Pharmacological Reviews. 21: 27-72. 1969.

A brief mention is made of the possible role of prostaglandins in inhibiting cAMP in denervated skeletal muscle. (JRH) 1450

0814
BOYARSKY, S., and P. LABAY
Ureteral Motility.
Annual Review of Medicine. 20: 383-394. 1969.

Only brief mention is made of prostaglandins in this investigation of ureteral muscle response. The ureter is sensitive to the prostaglandins. PGE$_1$ decelerates and stops peristalsis while PGF$_{2\alpha}$ accelerates peristalsis between the kidney pelvis and the bladder. (GW) 1465

0815
BRADLEY, P.B., and G.M.R. SAMUELS
Correlation between prostaglandin release from cerebral cortex and electrocorticogram in unanesthetized cats.
Electroencephalography and Clinical Neurophysiology. 27: 694-695. 1969.

Abstract only. A prostaglandin-like material has been found in superfusates of cerebral cortex in unanesthetized encephale isole cat preparations. The material was assayed on the rat uterus and identified by thin-layer chromatography. The level of spontaneous release of the prostaglandin-like material was higher than that which had been found previously in anesthetized preparations and it increased still further with electrical stimulation of the reticular formation at levels which induced electrocortical arousal. Administration of chlorpromazine (1.0-8.0 mg/kg) not only depressed the spontaneous release of prostaglandins, but blocked the increase evoked by stimulation, concomitantly with blocking electrocortical arousal. Increasing the stimulating voltage to restore the arousal response also restored the evoked release of prostaglandins. Most of the prostaglandin-like material released spontaneously was represented by 'E' type compounds, whereas the increase with stimulation was mainly of the 'F' compounds. It is thought that, whilst the prostaglandins are unlikely to be synaptic transmitters in the central nervous system, they must have some function connected with neuronal mechanisms. (Authors) 1752

0816

BRADLEY, P.B., G.M.R. SAMUELS, and J.E. SHAW
Correlation of prostaglandin release from the cerebral cortex of cats with the electro-corticogram, following stimulation of the reticular formation.
British Journal of Pharmacology. 37: 151-157. 1969.

Experiments were carried out on unanaesthetized "encephale isole" preparations of 15 cats. Changes in the electrocorticograms were produced by electrical stimulation of the brain stem reticular formation or by administration of drugs. Methods are described in detail. Findings provide further support for the hypothesis that there is a direct relationship between the release of prostaglandins from the cerebral cortex of the cat and neuronal activity. Prostaglandin-like material was found in superfusates of cerebral cortex in unanaesthetized "encephale isole" cat preparations. The material was assayed on the isolated rat uterus and identified by thin-layer chromatography. The level of spontaneous release of prostaglandin-like material was greater than that which had been found in anaesthetized preparations and it increased further with electrical stimulation of the reticular formation which induced cortical arousal. Chlorpromazine not only depressed the spontaneous release but blocked the increase evoked by stimulation concomitantly with blocking electrocortical arousal. Increasing the stimulating voltage to restore the arousal response also restored the evoked release of prostaglandins. Most of the prostaglandin-like material released spontaneously was represented by PGE_1 and PGE_2 but the increase with stimulation was mainly of $PGF_{1\alpha}$. (PR) 1578

0817

BU'LOCK, J.D., and P.T. CLAY
Fatty acid cyclization in the biosynthesis of brefeldin A: a new route to some fungal metabolites.
Chemical Communications. 237-238. 1969.

The author makes brief mention of the fact that prostaglandins can be shown to arise from an increasing range of structural modifications which are brought about by the assembly of enzyme-bound polydeto-acyl chains in the biosynthesis of aromatic polyketides. (GW) 1637

0818

BURKE, G.
Effects of iodide on thyroid stimulation.
Clinical Research. 17: 521. 1969.

Abstract only. The stimulatory effects of PGE_1 on adenyl cyclase activation and glucose-1-^{14}C oxidation in sheep thyroid in vitro were inhibited by excess iodide. However, PGE_1 stimulation of colloid droplet formation was unaffected. Twenty-fours hrs of KI pretreatment markedly inhibited the effects of PGE_1 on thyroid ^{131}I release in the in vivo mouse bioassay but stimulation of colloid droplet formation was unaffected. (RAP) 1758

0819

BURKE, G.
Effects of prostaglandins on basal and stimulated thyroid function.
Journal of Laboratory and Clinical Medicine. 74: 856-857. 1969.

Abstract only. Studies were made regarding the effects of prostaglandins E_1, E_2, $F_{1\alpha}$ and $F_{1\beta}$, on thyroid adenyl cyclase activity, colloid droplet formation and glucose oxidation in vitro and on radioiodine release in vivo. Each prostaglandin increased adenyl cyclase activity in sheep thyroid and stimulated endocytosis of colloid in ovine thyroid-slices but PGE_1 and PGE_2 consistently augmented thyroid-slice glucose oxidation. Adenyl cyclase activiation and endocytosis effects were additive with thyrotropin (TSH), PGE_1 and PGE_2 but not potentiated. $PGF_{1\alpha}$ and $PGF_{1\beta}$ abolished submaximal TSH effects on thyroid-slice oxidation. Comparable data were obtained in studies of prostaglandin effects of TSH and LATS-stimulated, ^{131}iodine release from mouse thyroid in vivo. In the mice each of the prostaglandins alone (50 to 200 μg) augmented thyroidal radioiodine release but less than either thyroid stimulator. However when added to submaximal concentrations of TSH or LATS significant reduction of TSH or LATS effect was found. Findings suggest: 1) The various phases of hormonogenesis in thyroid adenyl cyclase activation, endocytosis and glucose metabolism are not enhanced in concert; 2) TSH (and perhaps LATS), and prostaglandins compete for a common adenyl cyclase receptor site(s) in the thyroid; 3) hormonal stimulator caused a diminished response when these sites are occupied by less potent prostaglandins. (PR) 1562

0820
BUTCHER, R.W.
Effects of prostaglandins on cyclic AMP levels in tissues.
Abstracts, 4th International Congress on Pharmacology. Basle, 14-18 July, p. 13. 1969.

This brief review is concerned with the effects of prostaglandins on cAMP levels in tissues, especially adipose tissue. There are six references. (JRH) 1738

0821
BUTCHER, R.W. and C.E. BAIRD
The regulation of cyclic AMP in lipolysis in adipose tissue by hormones and other agents.
In: Holmes, W.L., L.A. Carlson, and R. Paoletti, eds., "Drugs Affecting Lipid Metabolism," p. 5-23. New York, Plenum Press, 1969.

The authors review their own and other published research on lipolysis in adipose tissue. Prostaglandins, insulin, and nicotinic acid are discussed as one division of a four-part presentation, the remainder of which covers the activity of hormone-sensitive triglyceride lipase, lipolytic hormones, phosphodiesterase, and phosphodiesterase inhibitors. General effects of PGE_1 on fat cell preparations are noted and the sensitivity of adipose and other tissues in terms of prostaglandin-drug-hormone interactions, is discussed briefly. The suggestion that tissue prostaglandin release is influenced by adenyl cyclase activation and that prostaglandins might act as feedback regulators upon adenyl cyclase is mentioned. (RMS) 1495

0822
BUTTRAM, V.C. and R.H. KAUFMAN
Primary dysmenorrhea: combination vs. sequential therapy.
Texas Medicine. 65: 52-55. 1969.

This article discusses briefly the possible role of prostaglandins in the etiology of dysmenorrhea. PGF elicits myometrial relaxation. Past investigations show that $PGF_{2\alpha}$ and PGE_1 are the principal human myometrial stimulants with menstrual fluid containing more PGF and PGE. Various studies suggest these compounds are formed in the endometrium, especially in the secretory phase, and released during menstruation. This release could account for increased myometrial activity during menstruation. (CM) 1446

0823

BYGDEMAN, M., K. SVANBORG, and B. SAMUELSSON
A method for determination of prostaglandins in human seminal fluid.
Clinica Chimica Acta. 26: 383-379. 1969.

Human seminal fluid has earlier been shown to contain thirteen different prostaglandins. Methods for quantitative determination of six groups of these prostaglandins, viz. PGE, PGA, PGB, 19-hydroxy-PGE, 19-hdyroxy-PGB, and PGF compounds are described. The methods included extraction with ether, group separation by thin-layer chromatography and quantitative determination by measuring a chromophore (λmax 278 mμ) which is formed by treatment with alkali or is present originally. The recovery of the various prostaglandins was determined using ^{3}H-labelled compounds and the precision of the method was studied by repetitive analyses of a pool of seminal fluid from men with suspected infertility. Evidence on the conversion of PGA derivatives to PGB derivatives and on the binding of PGA derivatives to other components of the seminal fluid during storage is discussed. (Authors) 1533

0824

BYGDEMAN, M.
Prostaglandins in human seminal fluid and their correlation to fertility.
International Journal of Fertility. 14: 228-231. 1969.

Semen samples from 137 men were analyzed for percentage content of PGE, PGA, PGB, 19-OH PGA, and 9-OG PGB. The men were grouped into samples by history of fertility into: normal, nonexamined infertility, and functional infertility. The semen from men of proven fertility differed from the other groups primarily in the concentration of PGE; 55.2 μg/ml. as compared to 36.0 and 29.4 μg/ml. Results indicated PGE concentration is important to normal fertility. (RMS) 1447

0825

CARLSON, L.A., L.-G. EKELUND, and L. ORO
Circulatory and respiratory effects of different doses of prostaglandin E_1 in man.
Acta Physiologica Scandinavica. 75: 161-169. 1969.

PGE_1 was infused i v into 8 healthy male subjects in doses from 0.032 to 0.58 μg/kg/min. With doses from 0.058 to 0.10 μg/kg/min a hyperkinetic circulation was induced, characterized by an increase in heart rate and stroke volume, decrease in peripheral resistance, and a decrease in AVO_2 difference and simultaneously a less pronounced increased in O_2 uptake. At the higher dose levels there was a further increase in heart rate and decrease in brachial artery pressure. The stroke volume decreased but the cardiac output remained above resting levels. The high doses also produced an alveolar hyperventilation, a further increase in O_2 consumption and in three cases a respiratory alkalosis. (Authors) 1508

0826

CARLSON, L.A.
Metabolic actions of prostaglandins.
Abstracts, 4th International Congress on Pharmacology. Basle, 14-18 July, p. 12-13. 1969.

Abstract only. This review is concerned with the effects of prostaglandins on metabolism especially of lipids. (JRH) 1711

0827

CARLSON, L.A., and H. MICHELI
  Paradoxical effect of prostaglandin E_1 in vitro on lipolysis in adipose tissue of diabetic rats.
  Proceedings of the 6th International Congress on Diabetes Supp. Stockholm, 1967. Amsterdam,
    Excerpta Medica, 1969. p. 189. (International Congress Series No. 1725)

Abstract only. We have added PGE_1 to adipose tissue from rats with acute insulin deficiency produced by treatment with anti-insulin serum and studied lipolysis. Unexpectedly, the release of glycerol was significantly increased by addition of PGE_1 (1 µg/ml) to adipose tissue of rats with an acute insulin deficiency of 24 hours duration, while the glycerol release was depressed when the duration of the diabetic state was of only 6 hours. These findings suggest that after an insulin deficiency of more than 24 hours the rate of lipolysis in adipose tissue may be controlled by factors other than the rate of accumulation of cyclic AMP. (Authors modified) 1773

0828
CARR, A.A.
  Renal effects of PGA_1 in man.
  Clinical Research. 17: 554. 1969.

Abstract only. PGA_1 was given by peripheral vein infusion (0.48-1.32 µg/Kg. body weight/ minute) to five men with essential hypertension to determine the effects of renal function as measured during standard renal clearances. These studies were performed during water diuresis and hydropenia with Mannitol. The changes in renal function produced by PGA_1 are expressed for water diuresis and hydropenia respectively and are as follows: inulin clearance (GFR) 6 and 8 ml/minute increase: PAH clearance (ERPF) 342 and 291 ml/minute increase: urine flow 3.8 and 4.4 ml/minute increase; sodium excretion 157 and 861 µeq/minute increase. During water diuresis free water clearance (CH_2O) increased 1.9 ml/minute and during hydropenia solute free water reabsorption (T^CH_2O) decreased 1.1 ml/minute. All these changes were significant at $P<.001$. The changes in renal function occurred during a decreased mean arterial pressure of 14 and 17 mm of Hg. Although peritubular capillary pressure was not measured it is proposed the renal effects were possibly due to a rise in peritubular capillary hydrostatic pressure as a result of the rise in effective renal plasma flow and decreased filtration fraction. The result could be decreased tubular reabsorptive capacity resulting in sodium diuresis and increased CH_2O due to increased delivery of sodium to the loop of Henle. The decreased T^CH_2O can be explained by increased medullary blood flow which lowers medullary hypertonicity. Thus all the renal effects could be due to hemodynamic changes (Author modified) 1759

0829
CARRIERE, S. and J. FRIBORG
  Correlation between the vasodilation and natriuresis produced by acetylcholine and prostaglandin.
  Clinical Research. 17: 671. 1969.

Abstract only. The correlation between the natriuresis produced by acetylcholine and PGE_1 and the intrarenal blood flow changes produced by these vasodilators has been studied using the Krypton method and autoradiograms. These substances produced significant increases of the urine volume, the natriuresis and the renal plasma flow without affecting the glomerular filtration rate. The cortical blood flow rate (ml/min./gm) significantly increased whereas the medullary blood flow did not change. However, the relative volume of the outer medulla increased in comparison to the cortex, resulting in proportional increases of the total cortical and medullary blood flow per 100 gm of kidney tissue. The increased cortical blood flow rate is probably associated with a higher velocity of the flow or an increase of the intravascular volume per gm of tissue in that region, whereas, in the outer medulla, the blood flow rate being preserved, no such changes can be observed. These results suggest that the natriuresis produced by acetylcholine and prostaglandin may be explained by the hemodynamic changes in the pertiublar capillaries of the outer cortex. The modifications of the intrarenal distribution of blood flow observed in these experiments could also be compatible with a redistribution of the glomerular filtration rate within the kidney, although we have no direct evidence to support this hypothesis. (Authors modified) 1761

0830
CHASE, L.R., S.A. FEDAK, and G.D. AURBACH
Activation of skeletal adenyl cyclase by parathyroid hormone in vitro.
Endocrinology. 84: 761-768. 1969.

The authors mention the lack of influence of PGE_1 (4.7×10^{-5}M in ethanol), cyclic AMP, phentolamine, or propanolol on skeletal adenyl cyclase in a series of studies which show that parathyroid hormone causes rapid adenyl cyclase accumulation in skeletal tissue and conclude that all effects of parathyroid hormone are mediated by cyclic AMP. (RMS) 1433

0831
CHAWLA, R.C. and M.M. EISENBERG
Effect of prostaglandin E_1 on the motility of innervated antral pouches in dogs.
Clinical Research. 17: 299. 1969.

Abstract only. An evaluation of the effect of prostaglandin E_1 on vagally innervated canine antral pouch motility was made. No change in blood sugar or electrolytes was noted and it is concluded that PGE_1 is capable of profound suppression of antral motor activities in dogs. (GW) 1619

0832
CHAWLA, R.C., and M.M. EISENBERG
Prostaglandin inhibition of innervated antral motility in dogs.
Proceedings of the Society for Experimental Biology and Medicine. 132: 1081-1086. 1969.

PGE_1 has been analyzed in terms of its effect on vagally driven gastric motor activity in dogs; profound inhibition of 2-deoxy-D-glucose stimulated motility was observed. The effect of PGE_1 on serum calcium, potassium, and blood sugar levels has been monitored; no significant change in calcium or potassium during PGE_1 infusion has been noted. Blood sugar, traditionally elevated by 2-deoxy-D-glucose, was not altered from its usual pattern after PGE_1 infusion. PGE_1 is capable, in doses previously shown to have an inhibitory effect on gastric secretion, to inhibit gastric motility as well. The precise mode of action is speculated upon but remains undisclosed. (Authors) 1440

0833
CHRIST, E.J.
Die rolle von prostaglandinen bei der hormonstimulierten lipolyse in isolierten fettgewebe. [The role of prostaglandins in homrone-stimulated lipolysis in isolated fatty tissue.]
Fette-Seifen-Anstrichmittel. 71: 932-933. 1969.

Abstract only. PGE synthesis from ^{14}C-marked unsaturated fatty acids in rat and sheep epididymal fatty tissue was demonstrated. The enzymatic effectiveness in the fatty tissue may be higher than is suggested by the quantity of the ^{14}C-marked substratum transposed. The fatty acid composition of the triglyceride, diglyceride, and free fatty acid fractions of rat epididymal fatty tissue was determined. The recorded quantities of dihomo-γ-linoleic acid or arachidonic acid probably suffice to explain PGE synthesis intact tissue. The specificity of the anti-lipolytic effect was studied with several PGE homologues. EFS-active fatty acids with 20-21 C atoms were found to be as effective as PGE_1, while PG's formed from EFS-inactive fatty acids were much less effective. PG's with fewer than 20 C atoms were only slightly effective. A direct influence of PGE on the formation and hydrolysis of cyclic AMP could not be proven with an isolated enzyme system. (Authors modified) 1779

0834
CHRISTLIEB, A.R., S.J. DOBRZINSKY, C.J. LYONS, and R.B. HICKLER
Short term PGA_1 infusions in patients with essential hypertension.
Clinical Research. 17: 234. 1969.

Abstract only. Six hypertensive patients were given a daily i v infusion of 1 or 2 mg of PGA_1 at a rate of 0.3-1.2 μg/kg/min. PGA_1 infusions lowered the blood pressure in essential hypertensives with much individual variation in response. Post infusion rebounding of the blood pressure was common, and at times was severe. No persisting or cumulative anti-hypertensive effect was observed. (GW) 1621

0835
COCEANI, F., J.J. DREIFUSS, L. PUGLISI, and L.S. WOLFE
Prostaglandins and membrane function.
In: Mantegazza, P., and E.W. Horton eds., "Prostaglandins, Peptides, and Amines," p. 73-84.
London, Academic Press, 1979.

The suggestion has been made that prostaglandins are local hormones formed by tissues in response to stimuli of different kinds, in which case they may be concerned with membrane activity as modulators of cyclic adenosine monophosphate (cAMP). The studies here reported attempt to clarify the mechanism of prostaglandin release and the role of these lipids in membrane function. A model is presented in which prostaglandins, formed from precursors within the cell membrane, control membrane excitability through an action on the distribution and availability of calcium. The formation and release of PGE_2 and $PGF_{2\alpha}$, and the influence of PGE_1 on calcium exchange, in isolated rat-stomach strips are discussed. From these studies, it was concluded that prostaglandins are formed within and released from the smooth muscle cell membrane and that this process is functionally linked to the activation of the cholinergic receptor and smooth muscle contraction. Experiments with radioactive calcium indicated that a fundamental action of PGE_1 is to cause retention of calcium during an efflux phase with half-time of about 5 minutes and to accumulate this calcium in a slowly exhangeable pool. The action of $PGF_{2\alpha}$ on synaptic transmission was studied in the cuneate nucleus of the cat brain and it was concluded that in the brain as well as in the stomach the effects of prostaglandins may be a consequence of action on the effector membrane. A simple model of the relationships between prostaglandin formation and membrane function is discussed. (ART) 1694

0836
COLLIER, H.O.J.
New light on how aspirin works.
Nature. 223: 35-37. 1969.

Aspirin seems likely to selectively antagonize certain endogenous substances that on administration elicit nociception, erythema, edema, bronchoconstriction or a fall in aterial blood pressure. Such substances include prostaglandin $F_{2\alpha}$. (GW) 1409

0837
COREY, E.J., N.M. WEINSHENKER, T.K. SHAAF, and W. HUBER
Stereo-controlled synthesis of prostaglandins $F_{2\alpha}$ and E_2 (dl).
Journal of the American Chemical Society. 91: 5675-5677. 1969.

A new approach is described to the synthesis of prostaglandins which was designed for control of stereo-chemistry, synthesis of all the primary prostaglandins and a variety of analogs from a single precursor, and optical resolution at an early stage. Methods are described in detail with structural graphic formulas of various stages. Method of obtaining the bistetrahydropyranyl ether of dl-prostaglandin $F_{2\alpha}$ from the 15α-hydroxy-11α-acetoxylactone is described. Formation of dl-prostaglandin E_2 is also described. These synthetic dl-prostaglandins exhibited the same IR and NMR and mass spectra as the natural hormones and identical chromatographic behavior. Further studies are in progress. (PR) 1548

0838
COREY, E.J., I. VLATTAS, and K. HARDING
Total synthesis of natural (levo) and enantiomeric (dextro) forms of prostaglandin E_1.
Journal of the American Chemical Society. 91: 535-536. 1969.

This letter to the editor refers to previous communications concerning total synthesis of racemic PGE_1. The readily available nitro ketal (the structural formula of which is given), an intermediate in one of the previously described approaches, has recently been found to undergo cyclization using stannic chloride in acetone to give the oily prostanoic acid derivative (the structural formula of which is given), essentially free of the undesired 11β-hydroxy epimer and readily purified by column chromatography. A recycling procedure is described. A levo amine was obtained which was converted to PGE_1 by the reaction sequence described in a previous communication. Recrystallization of the synthetic product afforded material identical in all respects with natural PGE_1. The procedure for the conversion to dl-PGE_1 when applied to a dextro-amine (obtained from a precursor nitro diol) yielded the enantiomer of natural PGE_1 showing the same infra red spectrum and chromatographic Rf values as racemic and natural forms of PGE_1. The results demonstate the first total synthesis of the natural form of PGE_1; further, together with previously accomplished transformations of PGE_1 to other prostaglandins of the first family, they constitute a formal total synthesis of the natural forms of prostaglandins $F_{1\alpha}$, $F_{1\beta}$ A_1, and B_1. (ART) 1531

0839
CROWSHAW, K.
Biosynthesis of renal prostaglandins.
Federation Proceedings. 28: 845. 1969.

Abstract only. Fresh rabbit kidney medulla (8 g/incubation) was homogenized in 0.1 M.K-phosphate buffer (pH 7.3) containing 0.001 M. EDTA. Two separate homogenates, one of which contained 15 mg added arachidonic acid, were incubated aerobically for 60 minutes and then the reaction terminated by ethanol. To a third homogenate 15 mg of arachidonic acid was added and ethanol added immediately. The lipids were extracted and separated by TLC into PGF, PGE, and carboxylic acid zones. Prostaglandin concentrations were estimated by rabbit duodenum bioassay. Both incubated homogenates had comparable prostaglandin concentrations whereas the basic levels in the non-incubated homogenates were considerably less. In identical experiments using kidney cortex, no prostaglandins were found. Results indicate that the renomedullary prostaglandins are biosynthesized only by the kidney medulla. The endogenous prostaglandin precursor may not be arachidonic acid. 1543

0840

CROWSHAW, K.
Distribution of prostaglandins in rabbit kidney.
Clinical Research. 17: 427. 1969.

Abstract only. Fatty acids, possible non-polar prostaglandin metabolites, PGA, PGE, and PGF compounds were separated from an acidic lipid extract obtained from frozen rabbit kidney cortex. Extraction and bioassay of fresh kidney cortex revealed no prostaglandin-like activity. It is concluded that rabbit kidney prostaglandins are present only in the medulla and there are no cortical mechanisms for their inactivation under normal conditions. (GW) 1620

0841

CROWSHAW, K., J.C. McGIFF, N.A.   TERRAGNO, A.J. LONIGRO, M.A. WILLIAMSON, J.C. STRAND, J.B. LEE, and K.K.F. NG
Prostaglandin-like substances present in blood during renal ischemia: patterns of release and their partial characterization.
Journal of Laboratory and Clinical Medicine. 74: 866. 1969.

Abstract only. By blood-bathed organ technique, continuous assays of E and F prostaglandins in the blood were made on 14 morphine-chloralose anesthetized dogs with unilateral renal ischemia. Results show an immediate release of prostaglandin-like substances (PLS) from the ischemic kidney in 5 experiments. Also noted was a continuous release of PLS from the ischemic (12 experiments) and the contralateral kidneys (10 experiments). The appearance of an angiotensin-like substance was noted in arterial blood in all experiments. On release of constriction a pulse of PLS appeared in the ischemic renal effluent in 7 experiments. No release of PLS was found in renal effluents of a spontaneously hypertensive dog. Angiotension II was thought to mediate release of PLS from the contratlateral kidney since renal arterial infusion of angiotensin II did the same. In 3 experiments during peak assay activity of PLS in renal effluents, blood from the renal vein was subjected to acidic lipid extraction and reassayed for PLS. Blood collected before renal ischemia had shown no PLS (assayed as PGE_2). Assayed during renal ischemia, PLS ranged from 0.002 to 0.011 μg. per mililiter as corrected for expected recoveries (30%). Chromatographic characterization suggests PLS to be a mixture of PGE_2 and $PGF_{2\alpha}$. PGA_2 would not have been detected if present. Extract of 453 gm. of dog renal medulla revealed PLS. Further purification separated this activity into 3 fractions with the properties of PGE_2 (0.09 μg per gm), $PGF_{2\alpha}$ (0.055 μg per gm) and PGA_2 (0.033 μg per gm). It is thought that an alteration of the balance between renal hormones, pressor (renin, $PGF_{2\alpha}$) and vasodepressor (PGE_2, PGA_2), may determine the development of renal hypertension. (PR) 1563

0842

CRUNKHORN, P. and A.L. WILLIS
Actions and interactions of prostaglandins administered intradermally in rat and in man.
British Journal of Pharmacology. 36: 216P-217P. 1969.

The effects of prostaglandins E_1, E_2, $F_{1\alpha}$ and $F_{2\alpha}$ on vascular permeability in the female rat skin was tested. PGE_1 and PGE_2 induced an increase in vascular permeability as shown by

214

extravasation of pontamine blue. $PGF_{1\alpha}$ and $PGF_{2\alpha}$ however, were almost without effect. Mepyramine maleate greatly reduced the response of PGE_2 and methysergide bimaleate gave complete inhibition. $PGF_{2\alpha}$ had an inhibitory effect with PGE_1 and PGE_2 when administered intradermally. Compound 48/80 was similarly inhibited. $PGF_{1\alpha}$ did not possess any of these inhibitory effects. In man PGE_1 and PGE_2 injected intradermally induced local edema and redness. (GW) 1505

0843

CURTIS, D.R., and J.M. CRAWFORD

Central synaptic transmission—microelectrophoretic studies.

Annual Review of Pharmacology. 9: 209-240. 1969.

This review makes a very brief mention of the role of prostaglandins as a possible substance that has a transmitter-like action upon certain mammalian neurons. Partially purified prostaglandins had both excitatory and depressant effects upon brain stem neurons, particularly on cells of the medullary reticular formation. (GW) 1463

0844

CUTHBERT, M.F.

The effect on airways resistance of prostaglandin E_1 administered by aerosol to healthy and asthmatic volunteers.

Abstracts, 4th International Congress on Pharmacology. Basle, 14-18 July p. 113. 1969.

Abstract only. In the present study, experiments were first carried out in six healthy volunteers. None had any history of bronchial asthma or other allergic disease and none was concurrently receiving any other drug. The forced expiratory volume in one second (FEV_1), blood pressure and pulse rate were measured at least once every ten minutes and the electrocardiogram was monitored continuously. Following the inhalation of metered aerosol doses of PGE_1 no increase in airways resistance or other untoward effect was noted in any subject. A similar study has been commenced in asthmatic volunteers, using a double blind experimental protocol to compare the effect of PGE_1, isoprenaline and a placebo. In four asthmatic volunteers in whom reversible airways resistance was demonstrated by the use of a metered isoprenlaine(sic) aerosol prostaglandin E_1 given by the same route caused an increase in the FEV_1. The relative potency of these two preparations has not yet been established but PGE_1 appears to have a brochodilator effect comparable in both degree and duration with that of isoprenaline in approximately one-tenth the dose of the latter. (Author modified) 1608

0845

CUTHBERT, M.F.

Effect on airways resistance of prostaglandin E_1 given by aerosol to healthy and asthmatic volunteers.

British Medical Journal. 4: 723-726. 1969.

The effects of the inhalation of PGE_1 and of isoprenaline, using metered aerosols, on the forced expiratory volume in one second (FEV_1) was measured in 6 healthy and 8 asthmatic volunteers. PGE_1 administered as a free acid caused throat irritation in the healthy subjects and the neutral triethanolamine salt was then used in both healthy and asthmatic subjects for comparison with the effects of isoprenaline. In healthy subjects, neither form of PGE_1 affected the FEV_1 but the triethanolamine salt form of PGE_1 produced an increase in FEV_1 in 5 of the 6 asthmatic patients with reversible airways obstruction: the increase produced by inhalation of 55 micrograms of PGE_1 was comparable in degree and duration with that produced by inhalation of 550 micrograms of isoprenalene sulphate. Two of the 8 asthmatic subjects had irreversible airways obstruction, and one of the remaining 6 exhibited coughing and bronchospasm upon inhalation of PGE_1. The effects of various prostaglandins on human bronchial smooth muscle are discussed. (ART) 1526

0846
DAIGNEAULT, E.A., and R.D. BROWN
Comparison of acetylcholine with glycine and prostaglandin E_1 on cochlear N_1.
Federation Proceedings. 28: 775. 1969.

Abstract only. Cochlear responses to injections into the left vertebral artery of cats of acetylcholine (20-25 μg), glycine (0.02-2.0 mg) and prostagnlandin E_1 (25-160 μg) were ascertained. Acetylcholine produced depression of N_1 whereas glycine and prostaglandin E_1 did not. Prostaglandin E_1 at the 25-35 μg dose range produced a small increase in N_1 and blocked the acetylcholine-induced N_1 depression for a period up to 2 hours. (PR) 1544

0847
DANIEL, E.E.
Digestion: motor function.
Annual Review of Physiology. 31: 203-226. 1969.

A summary of the latest developments in the field of prostaglandin research is presented in one section of this review. The author cites seven papers ranging in scope from chemical structure to comparative inhibition. Other papers identify prostaglandin E_1 and $F_{1\alpha}$ with the biologically active substances released by frog stomach and intestine; the finding that PGE_2 and $PGF_{2\alpha}$ are released from the rat stomach in vitro; the suggestion that contractions speed up the release of prostaglandins from smooth muscle cell membranes; the findings of PGE_2 but not $PGF_{1\alpha}$ in the mucosa of the human stomach, and the discovery that such drugs as hyoscine and methysergide have no effect on the action of prostaglandins on rat fundal smooth muscle are discussed. (GW) 1485

0848
DAVIES, B.N., and P.G. WITHRINGTON
Actions of prostaglandins A_1, A_2, E_1, E_2, $F_{1\alpha}$, and $F_{2\alpha}$ on splenic vascular and capsular smooth muscle and their interactions with sympathetic nerve stimulation, catecholamines and angiotensin.
In: Mantegassa, P. and E.W. Horton, eds., "Prostaglandins, peptides and amines," p. 53-56. London, Academic Press, 1969.

PGE_2 and $PGF_{2\alpha}$ have been shown to be released by the spleen when the postaganglionic sympathetic nerves to the spleen were stimulated. These two prostaglandins were released also when the spleen was contracted by an injection of adrenaline, and it has been suggested that, under certain conditions, stimulation of the reticuloendothelial system leads to the release of prostaglandins. A slight dilation of the splenic vascular bed was caused by infusion of PGE_2 but a similar infusion of PGE_1 (which is not one of the prostaglandins released by the spleen) resulted in a much longer and sustained dilatation. Infusion of A_1, A_2, $F_{1\alpha}$, and $F_{2\alpha}$ at rates up to 4 micrograms per minute caused some vasodilatation in the spleen but none was as effective as PGE_1. Spleen volume was slightly increased by A_1, A_2, and $F_{1\alpha}$, the cause of which was ascribed to venoconstriction rather than to contraction of the capsular smooth muscle. The possibility that prostaglandins may influence sympathetic nerve transmission and the actions of catecholamines and angiotensin was investigated. With neither PGE_1 nor PGE_2

was there any apparent differences between the responses of the spleen to the various tests made before and during an infusion. $F_{2\alpha}$ infusions potentiated a contraction of the spleen to nerve stimulation and to injected adrenaline that was 8 to 12 percent larger than the control response while the responses to noradrenaline and angiotensin were not affected. Results obtained with PGA_1 suggested that it behaves like $PGF_{2\alpha}$, while PGA_2 decreased the size of the reaction to adrenaline. $PGF_{2\alpha}$ shows the most positive interaction with the responses of the spleen to injected adrenaline and to sympathetic nerve stimulation. (ART) 1691

0849
DAVIES, B.N., B.H. ROBINSON, and P.G. WITHRINGTON
The effects of graded doses of phenoxybenzamine on the vascular and capsular responses of the isolated, blood-perfused dogs spleen to sympathetic nerve stimulation and catecholamines.
Archives Internationales de Pharmacodynamie et de Therapie. 180: 143-153. 1969.

Analysis was made of the effects of graded doses of the β-blocking agent phenoxybenzamine on the vascular and capsular responses of the isolated blood-perfused dog spleen to sympathetic nerve stimulation at various frequencies within the physiological range and to noradrenaline and adrenaline. It had previously been shown that the release of prostaglandins from the splenic nerve of the dog followed stimulation and subsequently reported that prostaglandin E_1 was a potent dilator of the splenic vascular bed. Since the vasodilation produced by nerve stimulation after phenoxybenzamine is abolished by the β-receptor antagonist propranolol, which has no effect on the vasodilation produced by the prostaglandins, it is unlikely that the prostaglandins are of functional significance in this situation. (GW) 1560

0850
DAVIS, B., U. ZOR, T. KANEKO, D.H. MINTZ, and J.B. FIELD
Effects of parathyroid extract (PTE), arginine vasopressin (AVP), and prostaglandin E_1 (PGE_1) on urinary and renal tissue cyclic 3'5' adenosine monophosphate (cAMP).
Clinical Research. 17: 458. 1969.

Abstract only. PGE_1 increased cAMP 1 to 3-fold in rat renal cortex and up to 10-fold in rat renal medulla. PGE_1 raised dog medullary cAMP 5-fold but not cortical cAMP. As PGE_1 was reported to inhibit arginine vasopressin action, both were added to rat renal medulla. Together they raised cAMP more than either one alone. Results reported provided no support for competition between PGE_1 and arginine vasopressin for cAMP generation. (GW) 1616

0851
DAVIS, H.A.
Output of prostaglandins from the rabbit kidney on renal nerve stimulation.
Journal of Physiology. 201: 76P. 1969.

Title only. 1664

0852
DeHAEN, P.
New developments in drugs.
American Journal of Pharmacy. 141: 99-103. 1969.

This review presents a general overview of the studies carried out in 39 years of pharmaceutical research (1930-1968). The prostaglandins are mentioned as being included in the 117 studies concerned with drugs that are not as yet available to the American physician. (GW) 1734

0853
von DEIMLING, O.
Eine neue Hormongruppe: prostaglandine. [A new hormone group: prostaglandin.]
Umschau. 69: 87. 1969.

Pharmacological activities which refer back to the hormone group of prostaglandins, according to the present state of knowledge, have been known since 1930. On the other hand, the substances themselves, were unknown for 6 years. Shortly thereafter, an American research group at Harvard University succeeded in synthesizing several prostaglandins. The possibility now exists that more intensive investigations will soon be made on the broader activity spectrum of these relations, that up to now were only covered to a very limited extent from natural sources. (GW) Translation of German abstract. 1675

0854
DEMIS, D.J.
Allergy and drug sensitivity of skin.
Annual Review of Pharmacology. 9: 457-482. 1969.

This review discusses types of allergic responses and drug reactions occurring in the skin, mostly as seen in human skin. In the section dealing with agents known to act as chemical liberators of stored biogenic amines and other active compounds, it is noted that histamine releasers have been considered to be of prime importance. It is further noted that a possible role for prostaglandin has been suggested by von Euler, U.S., and Eliasson, B. *Prostaglandins*. (Academic Press, New York, 1967.). (ART) 1700

0855
DEUTSCH, E.
Die konservative therapie des arterilsklerotischem gefasshadens. [Conservative therapy of arteriosclerotic vascular damage.]
Weiner Klinische Wochenschrift. 81: 421-425. 1969.

In pharmacological therapy of arteriosclerosis, PGE_1, with a number of other medications, tend to dilate blood vessels and inhibit platelet aggregation. (MEMH) 1803

0856
DeWIED, D., A. WITTER, D.H.G. VERSTEEG, and A.H. MULDER
Release of ACTH by substances of central nervous system origin.
Endocrinology. 85: 561-569. 1969.

The authors show the lack of a suitable in vivo assay system for corticotrophin releasing factor (CRF) with in vitro and in vivo rat bioassay systems pretreated to block non-specific release of ACTH. PGE_1 and PGE_2, angiotensin II, argenine, vasopressin, carbachol, and crude calf-brain CRF activated the pituitary adrenal system. No effects were observed with $PGF_{2\alpha}$. Blocking with combinations of sodium pentobarbital, chlorpromazine, atropine, morphine and dexamethasone showed CRF to be the only substance triggering ACTH release in the isolated anterior pituitary; other assay systems failed to discriminate between crude CRF and other active agents. (RMS) 1655

0857
DEYS, H.P.
Prostaglandines. [Prostaglandins.]
Pharmaceutisch Weekblad. 104: 1661-1662. 1969.

This lecture in Dutch is a brief summary in popular style of the report on prostaglandins by
V.R. Pickles, in English, published in Nature 224: 221-225. October 18, 1969. (ART) 1678

0858
van DORP, D.A.
Essentiele fettsauren und prostaglandine. [Essential fatty acids and prostaglandin].
Naturwissenschaften. 56: 124-130. 1969.

Relative activity is discussed of prostaglandins derived from linolenic acid and from arachidonic
acid. The chemical structures of PGE_1, PGE_2, PGE_3, $PGF_{1\alpha}$, PGA_1, PGB_1, 19-OH PGA_1, and
19-OH PGB_1 are illustrated and their synthesis is discussed. A multiple function for the
essential fatty acids is postulated. (ART) 1662

0859
DUBE, W.J., G.W. BELL, and M.A. ALIAPOULIOS
Thyrocalcitonin activity in metastatic medullary thyroid carcinoma. Further evidence for its
parafollicular cell origin.
Archives of Internal Medicine. 123: 423-427. 1969.

The report of Williams, E.D. et al. (1968) on prostaglandins is mentioned without comment in
discussing neoplastic growths as the site of synthesis of serotonin, prostaglandins, and
thyrocalcitonin, in a description of a patient with medullary thyroid carcinoma. (RMS) 1517

0860
DURU, S. and R.K. TURKER
Effect of prostaglandin E_1 on the strychnine-induced convulsion in the mouse.
Experientia. 25: 275. 1969.

The effect of PGE_1 on the convulsions induced by strychnine was studied in 3 groups of 12
mice each. Controls were injected with 2 mg/kg strychnine sulfate, in 0.2 ml of saline,
intraperitoneally. The second group was injected with 20 μg/kg PGE_1 and the third group with
30 μg/kg PGE_1 intraperitoneally, each 1 minute before strychnine injection. Appearance of
convulsions was delayed in those treated with 20 μg/kg PGE_1 but duration of convulsions was
increased and time elapse to last convulstion increased; no mice in the first or second group
survived the convulsions. In those receiving 30 μg/kg PGE_1 convulsions occurred after a still
larger interval, lasted longer and final convulsions were more delayed. All of this group survived.
Several possible explanations are mentioned. (PR) 1535

0861
EDMONS, J.F., E. BERRY, and J.H. WYLLIE
Release of prostaglandins caused by distension of the lungs.
British Journal of Surgery. 56: 622-623. 1969.

Abstract only. Guinea pig isolated lungs give off an effluent when perfused with Kreb's solution
and intermittently inflated. This material causes contraction of the rat stomach strip and the rat

colon. Chromatography reveals that it contains components that are indistinguishable from prostaglandin E and F_2. A similar effect could be demonstrated in the anesthetized dog. Although the substance was not formally identified, tissue reactions are consistent with its being a prostaglandin. These results suggest that positive-pressure ventilation releases a prostaglandin-like substance which may contribute to circulatory collapse. (GW) 1521

0862

EDWARDS, W.G., Jr.

A vasodrepressor lipid resembling prostaglandin E_2 in the renal venous blood of hypertensive patients.

American Journal of Cardiology. 23: 307. 1969.

Abstract only. A method is described whereby prostaglandin-like compounds in blood may be tentatively identified by thin layer chromatography. A vasodepressor lipid resembling prostaglandin E_2 was found in the venous blood from the affected or more severely affected kidney in 5 subjects. In two patients who previously had surgical repair of renal artery stenosis, the vasodepressor lipid was undetectable in blood from either renal vein. (GW) 1630

0863

EDWARDS, W.G., Jr., C.G. STRONG, and J.C. HUNT

A vasodepressor lipid resembling prostaglandin E_2 (PGE_2) in the renal venous blood of hypertensive patients.

Journal of Laboratory and Clinical Medicine. 74: 389-399. 1969.

A method is described for the extraction of prostaglandin-like material from the blood with catheterization, homogenization, homogenization and centrifugation of the withdrawn blood, extraction of the solid phase and purification by column chromatography followed by rat bioassay. The method was employed in the study of eight hypertensive patients. A vasodepressor lipid resembling prostaglandin E_2 (PGE_2) was found in 5 of these in concentrations varying from 40 to 234 ng. per milliliter. This vasodepressor lipid was found only in the venous blood from the affected or more severely damaged kidney as assessed by separated renal function studies, renal venous renin-activity, renal arteriography, exrectory urography and isotope renography. In one patient with severe bilateral dysfunction, the vasodepressor lipid was found in blood from both renal veins. In 2 postsurgical patients where renal artery stenosis had been corrected and widely patent anastomoses demonstrated, no vasodepressor lipid was detectable from either renal vein. The vasodepressor lipid lateralization correlated with lateralization of renal venous renin-activity in all but one in whom it correlated with the anatomical lesion which was shown to be cortical infarction. With the technique described a PGE_2-like compound, possibly involved in the genesis of systemic arterial hypertension, can be recovered from the venous effluent of kidneys. (PR) 1566

0864

EIK-NES, K.B.

Patterens of steroidogenesis in the vertebrate gonads.

General and Comparative Endocrinology. Supp. 2: 87-100. 1969.

This review of steroidogenesis in the gonads briefly mentions PGE_2. The author observed that the infusion of 11 mg/min of PGE_2 via the spermatic artery of dogs increased the secretion of testosterone by the infused testis. The significance of these findings was not clear although some ideas were put forth. (RAP) 1741

0865
ELIASSON, R., Z. BRZDEKIEWICZ, and B. WIKLUND
Tachphylactic response of the isolated rat myometrium to prostaglandins E.
In: Mantegazza, P. and E.W. Horton, eds., "Prostaglandins, Peptides, and Amines," p. 57-64. London, Academic Press, 1969.

The isolated oestrogen dominated rat myometrium can develop tachyphylaxis to prostaglandins E_1, E_2, and E_3. Tachyphylaxis is enhanced when the calcium concentration of the bath fluid is lowered. Variations in the magnesium and potassium concentrations have no significant effect. The progesterone dominated uteri did not show tachyphylaxis. There is no cross-desensitization between the three prostaglandins. Oxytocin can restore the reactivity of the myometrium, if it has become non-responsive by repeated applications of prostaglandins E. (Authors) 1692

0866
ELIASSON, R. and Z. BRZDEKIEWICZ
Tachyphylactic response of the isolated rat uterus to prostaglandins F.
Pharmacological Research Communications. 1: 397-402. 1969.

It has been shown that the reactivity pattern of the isolated rat uterus to PGE and PGA depends on the hormonal pretreatment and the composition of the bath fluid. This paper presents the reactivity pattern of the rat myometrium to $PGF_{1\alpha}$, $PGF_{1\beta}$, and $PGF_{2\alpha}$. In the standard salt solution, the estrogen-dominated rat uterus showed tachphylaxis to $PGF_{1\alpha}$ and $PGF_{1\beta}$ but not to $PGF_{2\alpha}$. Lowering of the calcium or potassium concentration or elevation of the magnesium concentration enhanced the tachyphylactic response and also evoked a similar response pattern for $PGF_{2\alpha}$. The tachyphylactic response was diminished when the potassium or calcium concentration was increased. the progesterone-dominated uteri did not show tachyphylaxis to the three PGF materials but the activity ratio was dependent on the ionic composition of the bath fluid. (ART) 1613

0867
ELIASSON, R. and Z. BRZDEKIEWICZ
Tachyphylactic response of the isolated rat uterus to prostaglandins A.
Pharmacological Research Communications. 1: 391-396. 1969.

The estrogen dominated uterus revealed desensitization to both PGA_1 and PGA_2 and this reactivity pattern was enhanced by lowering the calcium conentration. The tachyphylaxis was most marked for PGA_1 with a relative potency of about 1:0.7 ($A_1:A_2$). The progesterone dominated myometrium was less reactive to the PGA's than the estrogen dominated one and the activity ration was reversed ($A_1:A_2 = 1:1.5$). A positive staircase phenomenon to PGA_1 when calcium concentration was 0.27 or 0.54 mM was observed but an irregular pattern for PGA_2 was seen. The reactivity pattern was regular to both prostaglandins when calcium was 0.81 mM. The progesterone dominated myometrium was less sensitive at 0.27 mM Ca^{2+} and more sensitive to the prostaglandins at 0.81 mM Ca^{2+} than the estrogen dominated uterus. Variations in the magnesium and potassium concentrations did not significantly influence the reactivity pattern. There was no cross desensitization between the two prostaglandins, nor between the PGA's and oxytocin. (Authors modified) 1748

0868
ELKELES, R.S., J.R. HAMPTON, M.J.G. HARRISON, and J.R.A. MITCHELL
Prostaglandin E_1 and human platelets.
Lancet. 2: 111. 1969.

The authors conducted experiments to show that some aspects of the in vitro behavior of human platelets can be influenced by the intravenous administration of PGE_1. PGE_1 was injected into 9 male subjects at the rate of 0.2 μg/kg/min for up to 15 minutes. In 4 subjects mean platelet adhesiveness fell from 36% to 11% at the end of the infusion. There was no difference between the effects of adenosine diphosphate (ADP) before and after the infusion. Immediately after the PGE_1 the addition of ADP failed to produced any change in platelet electrophoretic mobility. (GW) 1444

0869
ELLIS, L.C. and M.H. BAPTISTA
A proposed mechanism for the differential radiosensitivity of the immature rat testis.
Proceedings of the 9th Annual Hanford Biology Symposium. 963-974. 1969.

Minced rat testicular tissue was used to determine the effects of PGA_1, PGE_1 and $PGF_{2\alpha}$ on lipid peroxidations and androgen biosynthesis. Three grams of testicular tissue was incubated with radiolabeled pregnenolone, progesterone and 1 mg of a prostaglandin for 1 hr. PGA_1 diminished the conversion of progesterone into androgen and the accumulation of 17α-hydroxyprogesterone. PGE_1 and $PGF_{2\alpha}$ also caused the increase in 17α-hydroxyprogesterone but only PGA_1 increased androstenedione production. Both PGE_1 and PGA_1 reduced testosterone production. The amount of lipid peroxidation was increased by PGA_1 and decreased by PGE_1. (JRH) 1799

0870
EMBREY, M.P.
The effect of prostaglandins on the human pregnant uterus.
Journal of Obstetrics and Gynaecology of the British Commonwealth. 76: 783-789. 1969.

A study of the effects of intravenous infusion of pure prostaglandins on the mobility of the human pregnant uterus is described. Using both the E prostaglandins, and those of the F series, stimulation of myometrial contractility was observed, the oxytocic properties of the E prostaglandins being particularly striking and reliable. In late pregnancy, and at term, the response was characterized by an increase in the frequency and amplitude of contractions, without—in the dosage used—any appreciable increase in tone. In the early months of pregnancy, hypertonus was a feature of the response. Slow in action, the induced level of uterine activity tended to persist or recur long after the infusion was discontinued and, especially in the case of PGE_2, was frequently followed by the successful induction of labour. (Author) 1403

0871
von EULER, U.S. and P. HEDQVIST
Inhibitory action of prostaglandins E_1 and E_2 on the neuromuscular transmission in the guinea pig vas deferens.
Acta Physiologica Scandinavica. 77: 510-512. 1969.

The observation that low doses of PGE_1 and PGE_2 inhibit contractile response of isolated guinea pig vasa differentia to transmural nerve stimulation was tested and the present results agree with earlier conclusions. Findings suggest that oligodynamic concentrations inhibit transmitter release from adrenergic nerve terminals by negative feedback rather than by blocking conduction. PGE_2 was about half as active as PGE_1. The direct effect of NA produced a contractile response which was enhanced by PGE_1. Larger doses of PG enhanced the response to electrical stimulation; inhibitory responses were abolished by atropine which was without effect on nerve stimulation or the direct effects of NA or PGE_1. (RMS) 1516

0872
FAIN, J.N. and S.C. LOKEN
Response of trypsin-treated brown and white fat cells to hormones.
Journal of Biological Chemistry. 244: 3500-3506. 1969.

Rat white and brown fat cells were prepared by enzymatic digestion. Some fat cells of each
type were prepared by digestion with an enzyme mixture containing trypsin and some were
prepared by a trypsin-free method. It was found that trypsin treatment did not block the
antilipolytic effect of PGE_1 as it did that of insulin. (JRH) 1452

0873
FASSINA, G., F. CARPENEDO, and R. SANTI
Effect of prostaglandin E_1 on isolated short-circuited frog skin.
Life Sciences. 8: 181-187. 1969.

Frog abdominal skin was clamped between two lucite chambers containing frog Ringer solution
and short circuit current (S.C.C.) across the skin was measured at room temperature for five
months. PGE_1 was added to the solution facing the internal surface of the skin, increasing the
S.C.C. in proportion to its concentration with maximum activity reached in 20 minutes.
Removal of the PGE_1 reestablished the original condition. Pretreatment of the skin with
digitoxin or ouabain inhibited the effect of PGE_1 on the S.C.C. Introduction of digitoxin after
addition of PGE_1 evidenced the same inhibitory effect. Additional experiments show that the
activity of PGE_1 is inhibited by an excess of calcium and by the removal of sodium in the
Ringer solution. The results indicate that the action of PGE_1 on frog skin is dependent on the
normal rate on Na^+ transport. (ICM) 1472

0874
FASSINA, G. and P. DORIGO
Transport-inducing antibiotics (gramicidin and valinomycin) as inhibitors of catecholamine-
stimulated lipolysis in vitro.
In: Holmes, W.L., L.A. Carlson, and R. Paoletti, eds., "Drugs Affecting Lipid Metabolism, p.
117-124. New York, Plenum Press, 1969.

The suggestion that the lipolytic process requires a continuous supply of energy was
investigated using valinomycin and gramicidin, toxic antibiotics which induce extensive energy
dependent cation transport into mitochondria, to effect a sustained increase in respiration in
rats and determine if lipolysis could be depressed by the lowering of available ATP. An
antagonistic effect was demonstrated ascribed to 3 postulated mechanisms including an affect
on membrane permeability. In this context, PGE_1 inhibition of lipolysis is mentioned as possibly
due to its stimulating of ion transport, and it is noted that an important control factor in lipid
metabolism may be energy-dependent cation transport. (RMS) 1496

0875
FICHMAN, M.P.
Natriuretic effect of prostaglandin (PGA_1) in man.
Clinical Research. 17: 429. 1969.

Abstract only. The effect of prostaglandin A_1 on H_2O and Na^+ excretion in 12 patients (3
normals, 4 cirrhotics, 2 hypertensives, and 4 hyponatremics) was determined. The results
suggest that while PGA_1 failed to alter the effect of ADH on H_2O excretion, PGA_1 inhibited
proximal tabular reabsorption of Na^+ and that its natriuretic effect in man was potentiated by
ADH infusion. (GW) 1618

0876
FINCH, N. and J.J. FITT
  The synthesis of dl-prostaglandin E_1 methoxime.
  Tetrahedron Letters. 4639-4642. 1969.

The synthesis of the dl-PGE_1 methoxime from 2'-carboxy-5-oxocyclopent-1-eneptanoic acid is discussed with respect to technique. Some chemical and physical properties are given at each reaction as well as diagrams. The authors reported that when the methoxime was removed from the complex, very little dl-PGE_1 was obtained. Therefore, a new oxime reagent was developed which can be removed under conditions mild enough for PGE_1 to survive. The use of this new reagent in the synthesis scheme was not explained. (RAP) 1780

0877
FLACK, J.D., and P.W. RAMWELL
  Prostaglandin receptors of smooth muscle and hormone activated tissues.
  Abstracts, 4th International Congress of Pharmacology. Basle, 14-18 July, p. 92. 1969.

Abstract only. The nature of the tissue receptors of a variety of isolated preparations for different prostaglandins has been systematically studied. Drugs used include i) naturally occurring prostaglandins, ii) modified prostaglandins, iii) isomers and analogues of prostaglandins, and iv) smooth muscle agonists and hormones. The effect of a) ring and b) side chain substitution, c) configuration changes of different groups, as well as d) absolute configurational changes has been evaluated. Of special interest were a group of compounds which had antagonist properties; different degrees of specificity for individual prostaglandins and marked tissue differences were observed. (Authors) 1635

0878
FLACK, J.D., R. JESSUP, and P.W. RAMWELL
  Prostaglandin stimulation of rat corticosteroidogenesis.
  Science. 163: 691-692. 1969.

Prostaglandins and their $C20:_\omega 6$ fatty acid precursors are present in rat adrenal glands. Small doses of prostaglandins (PGE_1, PGE_2, or $PGF_{1\alpha}$ 1.4 to 2.4 micromolar) increased steroidogenesis in the superfused adrenal glands obtained from hypophysectomized rats. This effect was mimicked in part by both adrenocorticotropin and its postulated intracellular intermediate adenosine 3',5'-cyclic monophosphate; all three responses were inhibited by cycloheximide. (Authors) 1625

0879
FLESHLER, B. and A. BENNETT
  Responses of hyman, guinea pig, and rat colonic circular muscle to prostaglandins.
  Journal of Laboratory and Clinical Medicine. 74: 872-873. 1969.

Abstract only. The effects of prostaglandins E_1, E_2, $F_{1\alpha}$, and $F_{2\alpha}$ on isolated preparations of human, guinea pig and rat distal colon were studied. Strips of circular colon muscle were suspended in an organ bath at $37^\circ C$ containing Krebs solution bubbled with 5% carbon dioxide in oxygen. Movements of the muscle were recorded on a kymograph. Results show sensitivity varied to a low of 0.01 μg per milliliter; E_1 and E_2 prostaglandins caused relaxation or inhibited contractions induced by acetylcholine, while $F_{1\alpha}$ and $F_{2\alpha}$ prostaglandins produced contractions or increased responses to acetylcholine. Nerve block by tetrodotoxin did not alter the effects of the prostaglandins, and hyoscine did not reduce contractions produced by the F type prostaglandins. Also, adrenergic block did not reduce the inhibitory effect of E type compounds. Therefore, prostaglandins appear to act directly on the muscle cells and not on the intrinsic nerves which supply the circular layer. Results are consistent with the view that prostaglandins may play a role in gastro-intestinal motility. Elevated concentrations may produce abnormal motility. (PR) 1564

0880

FORD, S.H. and J. FRIED

Smooth muscle activity of $RAC.$-7-oxaprostaglandin $F_{1\alpha}$ and related substances.
Life Sciences. 8: 983-987. 1969.

This study assayed six synthetic preparations of 7-oxaprostaglandin derivatives on gerbil colon. All of the variants were active with the most potent possessing activity of the order of 1/500 that of $PGF_{1\alpha}$. Results show that substances possessing the degree of hydroxylation of $PGF_{1\alpha}$ are more active than those lacking the 15-hydroxyl group. Also, a 13,14-trans-double bond or triple bond is preferred for activity over the cis-double bond or no double bond. The fact that all the substances assayed are racemates and are highly potent must be considered in evaluating the data. (ICM) 1470

0881

FOX, C.A. and B. FOX

Blood pressure and respiratory patterns during human coitus.
Journal of Reproduction and Fertility. 19: 405-415. 1969.

It has been suggested that oxytocin is released during human coitus which could account for such circulatory manifestations as a fall in blood pressure, flushing and an increase in limb blood flow. Prostaglandins produce similar physiological changes, and their release into the blood stream during coitus and orgasm must be considered. (GW) 1484

0882

FOX, H. and H.N. JACOBSON

Nerve fibers in the umbilical cord and placenta of two species of subhuman primates. (*Macaca phillipensis* and *Galago crassicaudatus*)
Acta Anatomica. 73: 48-55. 1969.

The author briefly mentions that, besides in subhuman primates, prostaglandins have also been demonstrated in the human umbilical cord and that there is increasing evidence that such substances play an important role in neural transmission mechanisms. (GW) 1644

0883

FREDHOLM, B. and K. STRANDBERG

Release of histamine and formation of smooth-muscle stimulating principles in guinea-pig lung tissue induced by antigen and bee venom phosphatidase A.
Acta Physiologica Scandinavica. 76: 446-457. 1969.

This investigation showed that bee venom phosphatidase A releases histamine and spasmogenic lipids from guinea-pig lung tissue in vitro. The phosphatidase A containing fraction (FI) of bee venom was found to be as active as bee venom in inducing the formation of spasmogenic lipids.

In order to characterize these spasmogenic lipids, lung tissue extracts were subjected to biological assay. The rabbit duodenum and the rat uterus activities are expressed in terms of prostaglandin $F_{1\alpha}$. (GW) 1507

0884

FRIED, J., T.S. SANTHANAKRISHNAN, J. HIMIZU, C.H. LIN, S.H. FORD, B. RUBIN, and E.O. GRIGAS
Prostaglandin antagonists: synthesis and smooth muscle activity.
Nature. 223: 208-210. 1969.

The authors report the synthesis of 9,11-di and 9,11,15-trideoxy PG anaolgs which are capable of selectively antagonizing the smooth muscle effects of both prostaglandins E_1 and $F_{1\alpha}$. Smooth muscle assays were performed with three tissues, guinea pig ileum, rabbit duodenum and gerbil colon. The concentration required for a 50% inhibition was determined. Acetylenic diol V was the only compound that achieved a 50% inhibition of PGE_1 within a 10-25 μg/ml concentration range. The close similarity in structure between agonists and antagonists in this series of compounds suggests the possibility that the observed antagonism towards PGE_1 and $PGF_{1\alpha}$ might involve competition for the same receptor site. (GW) 1405

0885

FUJIMOTO, S., and M.F. LOCKETT
The intrarenal release of prostaglandin E by noradrenaline.
Abstracts, 4th International Congress on Pharmacology. Basle, 14-18 July, p. 122. 1969.

Abstract only. Noradrenaline, infused intravenously into cats under chloralose anaesthesia at rates ranging from 0.1 to 0.18 μg per kg per min. more than doubles the flow of renal lymph and more than doubles the concentration of prostaglandin E in the renal lymph. The diuresis and natriuresis produced by noradrenaline after complete α-adrenergic block (caused by phentolamine 0.2 mg/kg, or by dihydroergocristine 0.8 mg/kg, i.v.) with β-adrenergic block (propranolol, 2 mg/kg, i.v.) resembles the diuresis and natriuresis caused by the intraaortic infusion of prostaglandin E_1, 0.1 μg per kg min. The effects of prostaglandin E_1 on renal function are unaffected by α- and β-adrenergic blocking agents. Angiotensin II amide, 0.1 μg per kg per min., although very markedly pressor, enhances the renal lymph flow and increases the prostaglandin content of renal lymph only to a small degree as compared with noradrenaline. The diuretic natriuretic effects of noradrenaline are found in large part attributable to the intrarenal release of prostaglandin E by noradrenaline. (Authors) 1607

0886

FUNAKI, T.
[Effects of prostaglandin E_1 on the isolated intestine, tracheal preparation and the stomach fundus preparation.]
Folia Pharmacologica Japonica. 65: 106-107. 1969.

Article in Japanese. Abstract not available at present. 1674.

0887

FUNDER, J.W., J.R. BLAIR-WEST, J.P. COGHLAN, D.A. DENTON, B.A. SCOGGINS, and R.D. WRIGHT
The effects of prostaglandin E_1 upon corticosteroid secretion.
Australian Journal of Experimental Biology and Medical Science. 47: P11-P12. 1969.

Abstract only. PGE_1 was infused into the autotransplanted adrenal gland of the Na depleted sheep. Blood flow was consistently increased when the adrenal arterial PGE_1 concentration was 0.2 to 2.0 µg/100 ml. The effects upon steroid secretion fell into two groups. One group showed a regular and equivalent fall in the secretion rate of cortisol, corticosterone and aldosterone. The other group showed no change in aldosterone secretion but modest increases in the rate of secretion of cortisol and corticosterone. The finding that PGE_1 can inhibit adrenocortical secretion is consistent with the possibility of a physiological role in the modulation of steroidogenesis. (GW) 1511

0888
GARBARSCH, Ch.
Histochemical studies on the lipids in the epithelium of the fetal colon.
Histochemie. 18: 168-173. 1969.

The author mentions that lipids discovered in the colon of fetuses might be prostaglandins and might be involved in water and salt balance in the colon. (JRH) 1453

0889
GESSA, G.L., G.A. CLAY, and B.B. BRODIE
Evidence that hyperthermia produced by d-amphetamine is caused by a peripheral action of the drug.
Life Sciences. 8: 135-141. 1969.

Experiments were performed to determine the effects of various antilipolytic compounds on the hypothermia and release of free fatty acids (FFA) produced by injection of d-amphetamine. It was found that PGE_1 given 500µg/kg subcutaneously prevented the hyperthermia and the release of FFA and the release of norepinephrine from fat tissue. (JRH) 1473

0890
GESSA, G.L., G.A. CLAY, and B.B. BRODIE
Evidence that the rise in temperature produced by d-amphetamine is caused by a peripheral action of the drug.
In: Holmes, W.L., L.A. Carlson, and R. Paoletti, eds., "Drugs Affecting Lipid Metabolism," p. 35-43. New York, Plenum Press, 1969.

The possibility that those effects of d-amphetamine to which tolerance is readily acquired might result from a depletion of catecholamine at the locus of action was investigated in rats. Evidence was found that amphetamine-produced hyperthermia is a peripheral effect that results from the calorigenic action of norepinephrine released from sympathetic nerve endings in adipose tissue. Treatment of the animals with nicotinic acid or with PGE_1 blocked the rise in FFA and body temperature induced by amphetamine. This is termed significant in that these substances do not act at the receptor level. (RMS) 1493

0891
GILES, T.D., A.C. QUIROZ, and G.E. BURCH
The effects of prostaglandin E_1 on the systemic and pulmonary circulations of intact dogs. The influence of urethane and pentobarbital anesthesia.
Experientia. 25: 1056-1058. 1969.

The effects of pentobarbital and urethane anesthesia upon the hemodynamic responses to PGE_1 in dogs are reported. Measurements of cardiac index (CI) were influenced by the anesthetic, increasing under pentobarbital and decreasing under urethane. Heart rate (HR) decreased with i a infusion of PGE_1 in urethane-anesthetized dogs whereas both i a and i v administration produced increases under pentobarbital. With regard to the effect of anesthetics, some other differences are noted and discussed, and it is concluded that any evaulation of the hemodynamics of PG must include a consideration of the anesthetic. Pulmonary artery pressure, small pulmonary vein pressure, and pulmonary blood volume changes were observed to be very small. PGE_1 in the doses used was found to have a probable negative chronotropic effect on the heart. (RMS) 1537

0892

GILLESPIE, I.E.
British Society of gastroenterology meetings.
Gastroenterology 56: 989-1003. 1969.

Two short papers mentioned prostaglandins. One paper cited the possibility that prostaglandins play a role in the occurrence of diarrhea in patients with medullary carcinoma of the thyroid. Removal of the tumor usually cured the diarrhea. The other paper discussed the role of prostaglandin E_1 and E_2 in gastrointestinal motility. Prostaglandins in both humans and guinea pigs contracted longitudinal muscle and relaxed circular muscle of the ileum. The rapid clearance of prostaglandin by the liver and lungs make it unlikely that they are involved normally in the regulation of gastrointestinal motility. (GW) 1594

0893

GILMORE, N., J.R. VANE, and J.H. WYLLIE
Prostaglandin release by the spleen in response to infusion of particles.
In: Mantegazza, P., and E.W. Horton, eds., "Prostaglandins, Peptides, and Amines," p. 21-29. London, Academic Press, 1969.

The discovery is reported that the injection of several colloidal materials into the dog spleen in vitro causes the release of PGE_2 and $PGF_{2\alpha}$ without any associated muscle contraction. The colloidal materials used were iron dextran complex, iron saccharate, amorphous insulin, and crystalline insulin suspension. These substances were perfused intraarterially into the isolated dog spleen and the venous effluent from the spleen was used to superfuse a series of isolated assay tissues, usually a rat stomach strip and a rat colon. Contractions in these tissues indicated the release of prostaglandins. The same concentrations of the colloid materials applied directly (intravenously) to the assay tissues usually had no effect on them. It is speculated that because the spleen contains an abundance of reticuloendothelial cells, it is reasonable to suppose that these cells will try to remove various colloids from the perfusion medium by phagocytosis. It is concluded that the results therefore suggest that stimulation of the reticuloendothelial cells in this way involves the release of prostaglandins. (ART) 1688

0894

GONDA, A., N. WONG, J.F. SEELY, and J.H. DIRKS
The role of hemodynamic factors of urinary calcium and magnesium excretion.
Canadian Journal of Physiology and Pharmacology. 47: 619-626. 1969.

A brief mention is made of the work of other researchers who found that prostaglandin causes a natriuresis which is independent of the filtered load in the kidney. (JRH) 1551

0895

GOODSON, J.M. and V. DISTEFANO

Isolation of non-histamine hypotensive substances from porcine thyroid extracts.

Pharmacologist. 11: 270. 1969.

Abstract only. A first stage ethanol tissue extraction followed by acidic lipid isolation indicates the existance of prostaglandin-like substances. A second stage weak acid extraction followed by gel chromatography resolved four distinct hypotensive peaks when assayed in the rat: I, (molecular weight ca. 1000 to 200) evoked a hypotension unaffected by diphenhydramine. This peak which may contain adenosine diphosphate has been further resolved into one neutral and two acidic hypotensive substances by electrophoresis. II, (molecular weight ca. 200 to 60) contains predominantly histamine but possibly low concentrations of other hypotensive substances. III, (molecular weight ca. 40) represents hypotensive inorganic salts. IV, exhibits delayed elution by gell interaction. It absorbs U.V. strongly in the region of 260 m μ but has elution characteristics which differentiate it from adenine derivatives. Neither bradykinin nor acetylcholine has been demonstrated to exist in these extracts; serotonin has been eliminated by its strong gel binding properties. (Authors modified) 1794

0896

GOTTENBOS, J.J. and G. HORNSTRA

The influence of prostaglandin E_1 on experimental platelet aggregation in rats.

Abstracts, 2nd International Symposium of Athersclerosis. Chicago, 2-5 November, p. 130-133. 1969.

The first part of this symposium article dealt with the development of a system for investigating the influence of PGE_1 on the hemostatic process in rats in vivo. 0.06 mg ADP/30 seconds was infused into this system with increasing doses of PGE_1 $(0.16 \times 10^{-2} \mu g/30$ sec to $5.00 \times 10^{-2} \mu g/30$ sec). The results showed a positive rectilinear relationship between the log dose PGE_1 and its aggregation inhibiting effect. The method of aggregation measurement developed was then used to investigate the clinical application of PGE_1 for the prevention of intravascular thrombosis. A single subcutaneous, intramuscular or intraperitoneal administration of 50 μg PGE_1 was made and measurements taken every 10 minutes. After subcutaenous PG administration, the aggregation inhibition was no longer significant within 21 minutes: after IM administration the significant inhibition period was <31 minutes. The aggregation inhibition after IP administration remained significant for 70 minutes. The authors suggested that PGE_1 might be useful for the prevention of intravascular thrombosis and for the maintenance of thrombocyte lost in hemodialysis, heart-lung machines, etc. (RAP) 1589

0897

GRANSTROM, E., and B. SAMUELSSON

The structure of a urinary metabolite of prostaglandin $F_{2\alpha}$ in man.

Journal of the American Chemical Society. 91: 3398-3400. 1969.

The structure of a urinary metabolite of $PGF_{2\alpha}$ in man is illustrated graphically. The method of obtaining the urinary metabolite is described, including intravenous injection of tritium-labeled $PGF_{2\alpha}$ into female subjects, and processing of pooled urine (extraction of the urinary radioactivity, purification of extract by reversed-phase partition chromatography, esterification, purification of the metabolite by silicic acid chromatography and mass spectrometry of 4 derivatives: acetate 6, trimethylsilyl ether 7, 0-methyloxime (methoxime) acetate 8, and methoxime trimethylsilyl ether 9). The difference between the retention times of the acetates and the trimethylsilyl ethers indicated the presence of two hydroxyl groups in the metabolite. Metabolite 4 of $PGF_{2\alpha}$ corresponds to metabolite 3 formed from PGE_2 and differs only in the functional group at C-5. (ART) 1527

0898

GRANSTROM, E., and B. SAMUELSSON

The structure of the main urinary metabolite of prostaglandin $F_{2\alpha}$ in the guinea pig.
European Journal of Biochemistry. 10: 411-418. 1968.

Two metabolites were isolated from the urine of male guinea pigs which had been given subcutaneously administered tritium-labeled $PGF_{2\alpha}$. Extraction methods, radioactivity assay, chromatographic methods, mass spectrometry, and analysis by gas-liquid chromatography are described. The main metabolite was identified by comparison with reference compounds. Its structure is 5α,7α-dihydroxy-11-oxotetranor-prostanoic acid. The second, less polar product was tentatively shown to be the delta-lactone form of the main metabolite. (ART) 1567

0899

GREEN, K.

Gas chromatography: mass spectrometry of O-methyloxime derivatives of prostaglandins.
Chemistry and Physics of Lipids. 3: 254-272. 1969.

This report describes gas chromatographic and mass spectrometric studies of O-methyloxime derivatives of E-type prostaglandins as trimethylsilyl ethers (TMS) or acetates (AC). Preparations of materials and methods used are described. Appropriate figures and tables showing findings are prepared. O-mthyloxime trimethylsilyl ether derivatives and O-methyloxime acetyl derivatives of prostaglandins containing a β-ketol system in the five-membered ring and in some cases an additional keto group in the side chain have been found to be stable during gas chromatography. GLC data obtained under different conditions are presented. The mass spectrometric fragmentation of the compounds has been investigated using derivatives deuterated in the methyl ester group, the methoxime group, the trimethylsilyl ether groups or the acetyl groups. The molecular ion was generally seen. Eliminations characteristic for trimethylsilyl ether and acetyl derivatives were observed, also β-cleavage at the methoxime group. The latter reaction was in some cases accompanied by transfer of hydrogen to the charge retaining ion and loss of CH_3O from the methoxime group. Findings are considered valuable in determining location of keto groups and of the nature of substitutents. (PR) 1572

0900

GREEN, K.

Structures of urinary metabolites of prostaglandin $F_{2\alpha}$ in the rat.
Acta Chemica Scandinavica. 23: 1453-1455. 1969.

Tritium labelled $PGF_{2\alpha}$ was administered intravenously to female rats. The urine was collected for 24 hours, acidified and butanol extracted. Gas liquid chromatography of O-methyloxime derivatives followed by mass spectral analysis gave data indicative of the structures of 5 metabolites, four are C-16 and one C-18 which are given along with probable structures. (RMS) 1420

0901

GREENBERG, R.A. and H.V. SPARKS

Prostaglandins and consecutive vascular segments of the canine hindlimb.
American Journal of Physiology. 216: 567-571. 1969.

Intra-arterial administration of prostaglandin E_1 or A_1 causes decreased resistance, increased capillary filtration coefficient (K_F), and increased vascular capacity of the isolated canine hindlimb. $PGF_{2\alpha}$ causes first decreased and then increased resistance, decreased vascular capacity, and has variable effects on K_F. Constant-flow studies and experiments on isolated smooth muscle indicate that PGE_1 and PGA_1 cause relaxation of veins and arteries whereas, at the same dose, $PGF_{2\alpha}$ causes contraction. The increased K_F during PGE_1 and PGA_1 infusion is probably due to decreased precapillary sphincter tone, since 1) there is no net filtration associated with the increased K_F, and 2) the relative changes in resistance and K_F are similar in magnitude to those during exercise when there is no increase in capillary permeability. (Authors) 1457

0902

GREENOUGH, W.B., N F. PIERCE, Q. AL AWQATI, and C.C.J. CARPENTER
  Stimulation of gut electrolyte secretion by prostaglandins, theophylline, and cholera exotoxin.
  Journal of Clinical Investigation. 48: 32a-33a. 1969.

Abstract only. The effects of theophylline, PGA_1, PGE_1, cAMP, and *Vibrio cholerae* exotoxin were measured in dogs with chronic jejunal loop preparations and in vitro with rabbit ileal mucosa stripped of serosa and muscularis. PGE_1 PGA_1 2 μg/min for 90 minutes infused into the superior mesenteric arteries with theophylline reduced absorption of the dog jejunum significantly. Infusion rates of 8, 24, and 100 μg/min caused successively greater fluid loss. 100 μg/min caused net isotonic fluid secretion (mean±SE) of 21.1±8.2 (PGE_1) and 24.0±11.0 ml/loop/hr. (PGA_1) with altered electrolyte transport persisting 3 hr. Theophylline at 22.2 mg/min caused net isotonic fluid secretion of 24.0±11.2 ml/loop/hr. cholera toxin resulted in net secretion of 57.4±8.1 ml/loop/hr. Fluid composition of secretions was identical in each case. No histological evidence of gut damage was noted. Rabbit ileum preparation showed similar changes in the short circuit current with all agents. (RMS) 1481

0903

GRIMLEY, P.M., L.J. DEFTOS, J.R. WEEKS, and A.S. RABSON
  Growth in vitro and ultrastructure of cells from a medullary carcinoma of the human thyroid
    gland: transformation by simian virus 40 and evidence of thyrocalcitonin and prostaglandins.
  Journal of the National Cancer Institute. 42: 663-680. 1969.

Continuous monolayer tissue cultures were established with explants from a metastatic nodule of medullary carcinoma primary in the thyroid gland. Some cultures were infected with simian virus 40 and the cells acquired morphologic and growth characteristics associated with viral transformation. Thyrocalcitonin activity was detected by radioimmunoassay of culture fluids as late as 7 months after initiation of the monolayers. Substances with the chemical and biological properties of PGE_2 and $PGF_{2\alpha}$ were extracted from culture fluids after 4 months. (Authors) 1486

0904

GROLLMAN, A.
  Pathogenesis of hypertension and implications for its therapeutic management.
  Clinical Pharmacology and Therapeutics. 10: 755-764. 1969.

PGE is mentioned briefly in a review of some 49 papers on arterial hypertension with emphasis on that of renal origin. The antihypertensive action of renal extracts is discussed, and it is noted that the relationship of the various preparations termed: lipid, phospholipid inhibitor of renin, and in the case of Lee (1967), PGE, remains to be established. (RMS) 1534

0905
GUTNECHT, G.D., J.C. CORNETTE, and B.B. PHARRISS
Antifertility properties of prostaglandin $F_{2\alpha}$.
Biology of Reproduction. 1: 367-371. 1969.

$PGF_{2\alpha}$ was found to be 100% effective in preventing or terminating pregnancy in the rat when administered subcutaenously at a dose of 2 mg/day over any consecutive 3-day period from day 4 through day 13 after coitus. Orally administered $PGF_{2\alpha}$ at a 10-fold increase in dosage was ineffective for interrupting pregnancy. Prostaglandin $F_{1\alpha}$, at a subcutaneous dose approximately 10-fold above that of $PGF_{2\alpha}$, was 100% active in rats. In the rabbit, $PGF_{2\alpha}$ was 100% effective in blocking nidation when administered by subcutaenous injection at a dose of 5 mg/kg/day on days 4 through 8 of pregnancy. (GW) 1510

0906
HAMBERG, M.
Biosynthesis of prostaglandins in the renal medulla of rabbit.
FEBS Letters. 5: 127-130. 1969.

Efficient conversion of arachidonic acid into prostaglandins by the renal medulla of rabbit is reported. Materials and methods of study are described; 5,6,8,9,11,12,14,15-octa-tritio-arachidonic acid, prepared by reduction of 5,8,11,14-eicosatetraynoic acid with tritium gas and Lindlar catalyst and purified by silicic chromatography, was used in the synthesis. This work demonstrated the formation of PGE_2 and $PGF_{2\alpha}$ from the aracidonic acid by homogenates of rabbit renomedullary tissue. Only traces of a labeled compound tentatively identified as PGA_2 could be isolated after incubations of octatritio-arachidonic acid. Studies on the biosynthesis of prostaglandins in renomedullary tissue of other species and the further metabolism of prostaglandins in renal tissue are reported in progress. (PR) 1573

0907
HAMBERG, M. and B. SAMUELSSON
The structure of a urinary metabolite of prostaglandin E_2 in the guinea pig.
Biochemical and Biophysical Research Communications. 34: 22-27. 1969.

Identification of the major urinary metabolite of PGE_2 in the guinea pig as 5β ,7α-dihydroxy-11-ketotetranor-prostanoic acid is described. Radioactive PGE_2, 1.5-1000 μg, was injected subcutaenously into male guinea pigs. Of the injected radioactivity, 35-40% was recovered in the urine in 24 hours of which 85-90% could be extracted by diethyl ether after acidification. Comparative chromatographic (GLC) methods and mass spectra analyses of derivatives were used. The formation of this compound from PGE_2 involves oxidation of the secondary alcohol group in the side chain, reduction of the *trans* double bond, β-oxidation of the carboxyl side chain and reduction of the keto group in the five-membered ring. The first two reactions were originally discovered in the guinea pig lung and were demonstrated in other tissues and species by other observers. Also degradation of the carboxyl side chain of prostaglandins by β-oxidation had been demonstrated previously both in vivo and in vitro by others. However, reduction of the keto group in the five-membered ring had not been previously observed. The configuration (β) of the hydroxyl group formed is opposite to that found in "primary" prostaglandins. Studies on the enzyme system catalyzing this reaction as well as the sequence of the reactions involved in the transformation of PGE_2 into its urinary metabolite in the guinea pig are in progress. (PR) 1519

0908

HAMBERG, M., and B. SAMUELSSON
The structure of the major urinary metabolite of prostaglandin E_2 in man.
Journal of the American Chemical Society. 91: 2177-2178. 1969.

This letter to the editor reports the determination of the structure of the major urinary metabolite of PGE_2 in humans. Radiolabeled PGE_2 was injected intravenously into human subjects. The urine was collected and analyzed. The major urinary metabolite was a 16-carbon compound. (JRH) 1657

0909

HAMPTON, J.R.
The potential of thrombolysis in the treatment of thrombotic coronary occlusion.
Circulation. (supp. 4): IV 231-239. 1969.

The author discusses the potential use of thrombolytic agents for improved myocardial function in cases of coronary occlusion. Results of 3 papers on the effects of PGE_1 on platelet behavior are reviewed. The author notes the inhibition of experimental thrombus formation in rabbits, and in vitro using human platelets, as well as dissolution of ADP-induced aggregation of human platelets by doses of PGE_1 of 1 mg/ml or less. Prelimnary studies in man are mentioned showing reduced platelet adhesiveness with infusion of PGE_1 at 0.1 μg/kg/min. (RMS) 1430

0910

HATCH, F.E., and J.G. JOHNSON
Intrarenal blood flow.
Annual Review of Medicine. 20: 395-408. 1969.

Brief mention is made of the effect of prostaglandins on intrarenal blood flow. The infusion of prostaglandin in non-depressor doses into the renal artery of dogs shows an increased natriuresis independent of changes in glomerular filtration rate, an increase in renal blood flow and decrease in p-amino hippurate (PAH) extraction. An increase in free water clearance and a decrease in free water reabsorption is also seen to occur. (GW) 1464

0911

HAYAISHI, O.
Enzymic hydroxylation.
Annual Review of Biochemistry. 38: 21-44. 1969.

In a review of monooxygenase catalyzed reactions, the author cites two papers by Samuelsson et al. (1965 and 1967) noting that the conversion of 8,11,14 eicosatrienoic acid by sheep vesicular gland to PGE_1 involves one dioxygenase and one monooxygenase reaction and that the 9 and 11 hydroxyl oxygens of $PGF_{1\alpha}$ are derived from the same oxygen molecule. (RMS) 1466

0912

HAYAISHI, O. and M. NOZAKI
Nature and mechanisms of oxygenases.
Science. 164: 389-396. 1969.

Brief mention is made of the chemical composition of prostaglandins. Prior experiments provided evidence that the biosynthesis of prostaglandins from unsaturated fatty acids is catalyzed by the action of two successive reactions involving a dioxygenase and a monooxygenase. Experiments with oxygen-18 indicated that the two oxygen atoms at the 9- and 11- positions of prostaglandins are derived from a single molecule of oxygen and the hydroxyl group at the 15- position seems to be incorporated after the pentacyclic ring is formed by the action of the dioxygenase. (GW) 1627

0913
HEDQVIST, P.
   Antagonism between prostaglandin E_2 and phenoxybenzamine on noradrenaline release from the cat spleen.
   Acta Physiologica Scandinavica. 76: 383-384. 1969.

PGE_2 might exert possible braking action on the release of NA from the nerves. Antagonism was indicated between PGE_2 and PBA (phenoxybenzamine) on transmitter release from the splenic nerves. Part of the increase of NA outflow caused by PBA is based on inhibition of local release of endogenous PGE_2, besides other known effects of PBA. This might explain the striking difference between the effect of PBA and certain other drugs blocking NA uptake, such as cocaine and desmethylimipramine, on outflow of NA in response to nerve stimulation. (BA) 1506

0914
HEDQVIST, P. and J. BRUNDIN
   Inhibition by prostaglandin E_1 on noradrenaline release and of effector response to nerve stimulation in the cat spleen.
   Life Sciences. 8: 389-395. 1969.

The authors studied the action of PGE_1 on the release of noradrenaline and on the mechanical effector response to nerve stimulation in the isolated, perfused cat spleen. Electrical nerve stimulation produced an immediate large rise in perfusion pressure. Intraarterial infusion of PGE_1 at final concentration of $3.8 \times 10^{-8} M$ to $3.8 \times 10^{-6} M$ markedly reduced the pressor response. PGE_1 also inhibited to the same extent the pressor response caused by intraarterial injection of NA. In four experiments PGE_1 markedly and reversibly reduced the NA overflow response initiated by stimulation of the splenic nerves. (GW) 1471

0915
HEDQVIST, P.
   Modulating effect of prostaglandin E_2 on noradrenaline release from the isolated cat spleen.
   Acta Physiologica Scandinavica. 75: 511-512. 1969.

Isolated cat spleens arranged for electrical stimulation were used to measure the effect of PGE_2 infusion on norepinephrine (NE) release by sympathetic nerves. The amount of NE in the pooled effluent from the perfused spleen was measured before, during and after nerve stimulation. PGE_2 $(0.8-1.6 \times 10^{-6} M)$ was infused for 5 min., starting 3 min before nerve stimulation. Stimulation of the splenic nerves (10 cycles/sec for 20 sec) caused a large increase in NE release. PGE_2 infusion reduced the response to nerve stimulation, often more than 50%. There was a complete recovery of responsiveness 20-30 min after termination of PGE_2 infusion. PGE_2 may function as a local negative feedback mechanism for controlling NE release. (JRH) 1509

234

0916
HEDQVIST, P.
Prostaglandin E as a modulator of noradrenaline release from sympathetic nerves.
Acta Physiologica Scandinavica. Supp. 330: 59. 1969.

Abstract only. Experiments were performed to study whether PGE may modify the release of NA from sympathetic nerves. Close arterial infusion of PGE_1, 10^{-6} to 4×10^{-6} M into isolated cat spleens perfused at a constant rate with salinic medium, was found to occasionally reduce the outflow of NA in response to stimulation of the splenic nerves. PGE_2 at the same concentration caused the NA overflow response to consistently fall. Phenoxybenzamine (PBA) in the perfusion medium (a PGE release inhibitor) resulted in a several-fold increase of the NA overflow response to nerve stimulation. It is suggested that PGE may counteract further release of NA from the nerves by a negative feed back mechanism. (GW) 1515

0917
HEINEMANN, H.O. and A.P. FISHMAN
Nonrespiratory functions of mammalian lung.
Physiological Reviews. 4I: 1-47. 1969.

The author reports that prostaglandins suppress cyclic AMP production, and thereby inhibit indirectly lipolysis in adipose tissue. Prostaglandins raise the heart rate and lower the systematic vascular resistance. They are removed from the circulation by the lung and are actively metabolized by lung tissue. But the biological meaning of the prostaglandins and what they do in the lungs remains to be unraveled. (GW) 1441

0918
HELLER, J. and A. NOVAKOVA
Proximal tubular reabsorption during renal vasodilation and increased arterial blood pressure in saline loaded rats.
Pflugers Archiv. 309: 250-265.   1969.

Author briefly mentions that PGE_1 and PGE_2 among many other vasodilators produce a natriuresis and diuresis without altering the glomerular filtration rate or the perfusion pressure. It was also mentioned that PGE_1 did not alter the ratio of inulin tubular fluid concentration to plasma concentration in the dog proximal tubule. (JRH) 1451

0919
HEPP, D., L.A. MENAHAN and O. WIELAND
On the role of 3',5'-cyclic AMP phosphodiesterase in insulin action in rat liver and adipose tissue.
Hormone and Metabolic Research. 1: 93-94. 1969.

The authors briefly mention that when preparations of rat liver and fat cells were perfused with PGE_1 at 2.9×10^{-7} or 2.9×10^{-6} molar concentration, no appreciable change of PGE activity was observed. (RAP) 1753

0920
HISSEN, W., J.S. FLEMING, M.E. BIERWAGEN and M.H. PINDELL
Effect of prostaglandin E_1 on platelet aggregation in vitro and in hemorrhagic shock.
Microvascular Research. 1: 374-378. 1969.

PGE$_1$ was investigated for its inhibition of platelet aggregation both in vitro and in vivo. In vitro, PGE$_1$ inhibited ADP/5-HT-induced platelet aggregation in both rabbit platelet rich plasma and dog whole blood concentrations as low as 2.8×10^{-8}M (10ng/ml). The duration of PGE$_1$ activity in either system was markedly influenced by the nature of the anticoagulant used. Rapid loss of PGE$_1$ activity occurred in the presence of heparin but not with sodium citrate alone or in combination with heparin. PGE$_1$ was inactiviated rapidly in heparinized blood at 33°C, while no inactivation was seen in heparinized blood at 5°C over a period of 60 min. In vivo, PGE$_1$ was without activity when infused into a normal dog. However, the increased screen filtration pressure, representing platelet aggregation associated with hemorrhagic shock was reduced by PGE$_1$ infusion. An infusion of 5 μg/kg/min, which was without effect of ADP/5-HT-induced platelet aggregation in the normal animal, had a profound influence on platelet aggregation induced by hemorrhagic shock. The significance of these findings is discussed with respect to the shock process and the reported short biological half-life of PGE$_1$. (Authors modified) 1776

0921

HOFFER, B.J., G.R. SIGGINS and F.E. BLOOM
  Cyclic 3',5' adenosine monophosphate (C-AMP) mediation of the response of rat cerebellar
    Purkinje cells to norepinephrine (NE): blockade with prostaglandins.
  Pharmacologist 11: 238. 1969.

  Abstract only. Evidence of possible cAMP mediation of NE response is that micro-electrophoresis of prostaglandins E$_1$ or E$_2$ selectively block the Purkinje cell response to NE, but do not affect responses to cAMP or to non-specific depression with gamma-amino-butyric acid. (RAP) 1791

0922

HOFFER, B.J., G.R. SIGGINS, and F.E. BLOOM
  Prostaglandins E$_1$ and E$_2$ antagonize norepinephrine effects on cerebellar Purkinje cells:
    microelectrophoretic study.
  Science. 166: 1418-1420. 1969.

  In microelectrophoretic experiments, prostaglandins E$_1$ and E$_2$ antagonize the reduction in discharge rate of cerebellar Purkinje cells produced by norepinephrine. Slowing of discharge evoked by 3',5'-adenosine monophosphate or gamma aminobutyric acid is not antagonized. These data provide the first indication that endogenous prostaglandins may physiologically function to modulate central noradrenergic junctions. (Authors) 1629

0923

HOLT, P.G. and I.T. OLIVER
  Studies on the mechanism of induction of tyrosine aminotransferase in neonatal rat liver.
  Biochemistry. 8: 1429-1437. 1969.

  Postnatal synthesis of tyrosine aminotransferase in rat liver is slightly repressed by ergotamine tartrate, prostaglandin, insulin and pyridoxine and is increased by repeated injection of glucagon or 3',5'-cyclic adenosine monophosphate. It is suggested that apparent multiple induction pathways for tyrosine aminotransferase in both fetal and adult animals may be explained in terms of the discrete induction of different forms of the enzyme. (GW) 1522

0924

HOROWITZ, J.D. and M.L. MASHFORD

A perfused vein preparation sensitive to plasma kinins.

Naunyn-Schmiedebergs Archiv fur Pharmakologie und Experimentelle Pathologie 263: 332-339. 1969.

Plasma kinins were found to be the most potent of several substances in eliciting constriction from the isolated, perfused central vein of the rabbit ear. Catecholamines and histamine also caused response, but to a lesser degree. PGE_1 caused no response at 10^{-5} g/0.05-0.2 ml. (RMS) 1591

0925

HORTON, E.W. and I.H.M. MAIN

Actions of prostaglandin E_1 on spinal reflexes in the cat.

In: Mantegazza, P. and E.W. Horton, eds., "Prostaglandins, Peptides, and Amines," p. 121-122. London, Academic Press, 1969.

It has been shown that in the cat with a spinal transection at C_2, intravenous injection of PGE_1 potentiates the crossed extensor reflex but that in the presence of the intact brain, PGE_1 may have a predominantly inhibitory effect on spinal reflexes. A study was undertaken to determine whether this inhibition results from a supraspinal site of action. Adult cats were anesthetized with chloralose. Mid-collicular decerebration and spinalization at the cervical level were carried out under ether anesthesia. The actions of PGE_1 on the patellar reflex in chloralosed cats, on the crossed extensor reflex in chloralosed cats, and on decerebrate and spinal cats are described. PGE_1 inhibited the crossed extensor reflex, and usually the monosynaptic patellar reflex also, in chloralosed and decerebrate cats. In the spinal cat, PGE_1 facilitated crossed extensor reflex responses. It is concluded that PGE_1 has an action at the spinal level facilitating the crossed extensor reflex and in addition a second spinal action inhibiting the reflex. Since facilitation was never observed with PGE_1 when either the brain or brain stem was present, it seems likely that it also exerts an inhibitory effect via higher centers which overshadow the facilitatory action at the spinal level. (ART) 1697

0926

HORTON, E.W. and R.L. JONES

The biological assay of prostaglandins A_1 and A_2.

Journal of Physiology. 200: 56P-57P. 1969.

Findings are briefly reported on the response of several isolated smooth muscle preparations to PGA_1 and A_2. As little as 5-10 ng A_2 and 25-50 ng A_1 were detected by recording systemic arterial blood pressure changes in the anesthetized spinal cat, which was found to be more sensitive than the rat or rabbit. The rat fundus was found to be the most sensitive smooth muscle preparation, detecting 100-200 ng A_2 and 1-2 ng A_1. It is noted that rat fundus used in conjunction with cat blood pressure can be used to distinguish PGE and PGA. (RMS) 1437

0927

HORTON, E.W. and P.B. MARLEY

An investigation of the possible effects of prostaglandins E_1, $F_{2\alpha}$, and $F_{2\beta}$ on pregnancy in mice and rabbits.

British Journal of Pharmacology. 36: 188P-189P. 1969.

The effects of PGE_1, $PGF_{2\alpha}$, and $PGF_{2\beta}$ on fertility in mice and as oxytocic substances in mice and rabbits was investigated. Subcutaneous PGE_1, 1 mg/kg, on days 1, 1 and 2, 3, 3 and 4, 4, 5, or 9 had no effect on gamete transport rate and caused a 22% increase in mean number of fetuses. PGE_1 injected subcutaneously, intravenously, and intraperitoneally on day 19 was without oxytocic effects at doses of 0.5 and 2.0 mg/kg, but was sedative at the higher doses. $PGF_{2\alpha}$, 1 and 0.25 mg/kg, and $PGF_{2\beta}$, 10 mg/kg intraperitoneally induced parturition in 7/16, 1/8, and 1/12 of mice respectively. Neither PGE_1 at 100-400 μg nor $PGF_{2\alpha}$ at 100-250 μg given intravenously or intraperitoneally to rabbits altered the time of parturition. (RMS) 1503

0928
HORTON, E.W.
Hypotheses on physiological roles of prostaglandins.
Pharmacological Reviews. 49: 122-161. 1969.

Literature covering the possible role of prostaglandins in sperm transport, menstruation, parturition, placental blood flow and central nervous system activity is reviewed as is material on adenyl cyclase-cAMP interactions, ion transport and platelet aggregation. Cited are 296 references. (GW) 1401

0929
HORTON, E.W. and R.L. JONES
Prostaglandins A_1, A_2 and 19-hydroxy A_1; their actions on smooth muscle and their inactivation on passage through the pulmonary and hepatic portal vascular beds.
British Journal of Pharmacology. 37: 705-722. 1969.

This study has determined the activity of prostaglandins A_1, A_2, and 19-hydroxy A_1 relative to prostaglandin E_1 on a selection of non-vascular smooth muscle and cardiovascular preparations. Also investigated are the role of the lung and the liver in terminating the vasodilator actions of these prostaglandins. Tissues tested were proximal jejunum of rabbit, rat fundal stomach muscle, guinea-pig seminal vesicles, guinea-pig ileum, chicken crop muscle, cat and guinea-pig tracheal chains, rat uterine horns, and guinea-pig uterine horns. Each tissue was suspended in a suitable bath solution and contractions measured. Rabbit oviduct, cats with a spinal section at C_2 level, and cats and dogs under pentobarbitone anesthesia were also used in a series of in vivo investigations. Both cats and dogs were used for intravenous and intra-arterial injections through the pulmonary circulation. Cats were used for intraportal infusion for passage of prostaglandins through the liver. Findings are as follows: 1) Prostaglandins A_1, A_2, and 19-hydroxy A_1, have qualitatively similar actions to prostaglandin E_1 on smooth muscle. 2) The prostaglandins A have little activity on gastrointestinal, respiratory and reproductive smooth muscle but are potent depressors of the systemic arterial blood pressure of the dog, cat and rabbit. 3) The depressor action of the prostaglandins E and A is due to a direct dilator action on many peripheral vascular beds and not due to changes in nervous tone of these beds. 4) A single passage through the pulmonary circulation of the cat or dog causes substantial loss of the vasodilator activity of prostaglandin E_1 but little if any loss of the vascular activity of the prostaglandin A. 5) A single passage trhough the hepatic portal circulation of the cat causes substantial loss of the vasodilator activity of prostaglandins E_1, A_1, and A_2. 6) The cat blood pressure and rat fundal strip would be a suitable combination for the parallel biological assay of prostaglandins A_1 and A_2. (PR) 1580

0930
HUMES, J.L., M. ROUNBEHLER and F.A. KUEHL, Jr.
A new assay for measuring adenyl cyclase activity in intact cells.
Analytical Biochemistry. 32: 210-217. 1969.

Intact fat cells from epididymal fat pads of rats were assayed for cAMP-C^{14}. In testing the assay, norepinephrine (1×10^{-6} M) and norepinephrine + PGE_1 (1 mg/ml) were used. The amount of cAMP-C^{14} measured when using NE was about twice that of NE + PGE_1; thereby demonstrating the role of PGE_1 in inhibiting lipolysis. The authors point out that only newly synthesized nucleotide is measured by their assay and therefore, it should be particularly useful for determining the role of agents affecting the rate of formation and breakdown of cAMP. (RAP) 1742

0931
HYMAN, A.L.
The active responses of pulmonary veins in intact dogs to prostaglandins $F_{2\alpha}$ and E_1.
Journal of Pharmacology and Experimental Therapeutics. 165: 267-273. 1969.

Dogs were prepared so that blood flow to one lobe of the lung was controlled by a constant infusion pump. Blood pressure was recorded in a systemic artery, the main pulmonary artery, a small lobar artery and vein. Similar measurements were made in a normally infused lobe of the same dog. Dye dilution techniques were used to determine cardiac output and blood flow rates in the normal and controlled lobe of the lung. $PGF_{2\alpha}$ was infused into the lobar artery at a rate of 4.0 to 5.1 μg/kg/min. PGE_1 was infused at 0.8 to 1.3 μg/kg/min. PGE_1 produced a reduced pressure in lobar veins and arteries, the main pulmonary artery and the systemic artery. This decrease was slow but progressive during the infusion. Artrial and pleural pressures and respiratory and heart rates remained unchanged. Blood flow and volume in the normally infused lobe significantly decreased but remained unchanged in the pump infused lobe. $PGF_{2\alpha}$ produced an increase in pressure in the small lobar artery and vein. The arterial response was much more abrupt than venous change. In most of the dogs there was an abrupt increase in respiratory rate. These results indicate that PGE_1 causes an active dilation of pulmonary vessels while $PGF_{2\alpha}$ causes a constriction of the same vessels. (JRH) 1468

0932
HYMAN, A.L.
The direct effects of vasoactive agents on pulmonary veins, studies of responses to acetyl-
    choline, serotonin, histamine and isoproterenol in intact dogs.
Journal of Pharmacology and Experimental Therapeutics. 168: 96-105. 1969.

This paper deals with a method of measuring the effects of vasoactive substances on the pulmonary vessels in intact dogs. The author briefly mentions a previous paper in which segmented redistribution of pulmonary blood was suggested as an explanation for the response of the lungs' $PGF_{2\alpha}$. (JRH) 1469

0933
HYNIE, S.
Prostaglandiny. [Prostaglandins.]
Ceskoslovenska Fysiologie. 18: 43-71. 1969.

This general review in Czech covers the structure, classification, biosynthesis, metabolism and biological activity of prostaglandins, including pharmacology and mechanisms of action. The literature cited includes 455 references. (ART) 1683

0934

IRION, E. and H. BLOMBACK

Prostaglandins in platelet aggregation.

Scandinavian Journal of Clinical and Laboratory Investigation. 24: 141-144. 1969.

Prostaglandins were studied with respect to their influence on human platelet aggregation. The inhibitory effect of several prostaglandins on ADP-induced platelet aggregation was investigated using a technique similar to that of Born (1962) and O'Brien (1962). Methods are described. It was demonstrated that PGE_1 has a high inhibiting capacity while PGE_2 shows about one-fifth of the capacity. The activity of PGE_1 217, $PGF_{2\alpha}$, $PGF_{1\alpha}$ and $PGF_{1\beta}$ is much lower. These findings are in agreement with results reported by others for PGE_1. PGE_2 was shown to have a pronounced inhibiting effect. No stimulating effect could be observed, as has been reported by Kloeze (1967). (PR) 1571

0935

ISHII, M. and L. TOBIAN

Interstitial cell granules in renal papilla and the solute composition of renal tissue in rats with Goldblatt hypertension.

Journal of Laboratory and Clinical Medicine. 74: 47-52. 1969.

Thirty-eight male Wistar rats were made hypertensive by unilateral left renal artery constriction with the right contralateral kidney intact ("untouched"). Nineteen normotensive sham operated control rats had an average osmiophilic renal papillary granule count of 133 per 100 squares for the right and 136 for the left kidney. The hypertensive rats had an average granule count of 99 for the untouched kidneys and 115 for the "clipped" kidneys; also the granules were smaller than in the controls. Untouched kidneys of the hypertensive rats showed marked decreases in the concentration of papillary sodium as well as decreases in the papilla per cortex gradients for sodium and urea. The clipped kidneys of the hypertensive rats also had a decrease in the sodium gradient. The number of granules in the untouched kidneys and in the controls correlated with the sodium gradient, the papillary sodium concentration and the urea gradient. In the clipped kidneys the granule count correlated with the sodium gradient. These alterations indicate abnormal function of the renal papilla in hypertension. The number of granules correlated with the level of the blood pressure. Plasma chemical analyses revealed no differences in sodium or potassium concentration between the two groups but plasma urea nitrogen and creatinine levels were increased in the hypertensive rats. The authors cite 4 papers indicating a relationship between prostaglandins and the known antihypertensive function of the kidney and note that vasodepressor prostaglandins may be present in the interstitial cell granules and their reduction in both kidneys of the Goldblatt hypertensive rats might contribute to maintaining their chronically elevated blood pressure. (PR) 1565

0936

ISRAELSSON, U., M. HAMBERG, and B. SAMUELSSON

Biosynthesis of 19-hydroxy-prostaglandin A_1.

European Journal of Biochemistry. 11: 390-394. 1969.

Prostaglandin A_1, prostaglandin E_1, and 11α-hydroxy-9,15-diketoprostanoic acid have been incubated with guinea pig liver microsomes plus boiled high speed supernatant supplemented with NADPH. Prostaglandin A_1 was converted in 10-18% yield into 19-hydroxy-prostaglandin A_1 (35%) and 20-hydroxy-prostaglandin A_1 (65%). Prostaglandin E_1 gave a very small yield (less than 3%) of a hydroxylated derivative and 11α-hydroxy-9,15-diketoprostanoic acid was not hydroxylated at all. In the latter case, however, a reduction of the side chain keto group took place, affording 11α, 15-dihydroxy-9-ketoprostanoic acid. (Authors) 1656

0937
JAMES, G.W.L.
The use of the in vivo trachea preparation of the guinea-pig to assess drug action on lung.
Journal of Pharmacy and Pharmacology. 21: 379-386. 1969.

A method for recording the effect of drugs on an isolated in vivo segment of trachea and on lung pressure of a deeply anaesthetized animal is described. Acetylcholine, angiotensin, histamine, 5-hydroxytryptamine, prostaglandin (PG)$F_{2\alpha}$ and slow-reacting substance of anaphylaxis, but not bradykinin, caused an increase in tracheal segment pressure and an increase in lung pressure. Bradykinin caused an increase in lung pressure, but a fall in tracheal segment pressure. It was concluded that bradykinin acts mainly on the smaller airways and $PGF_{2\alpha}$ mainly on the large airways. The tracheal segment responded to adrenaline, aminophylline, ephedrine, isoprenaline, papaverine, PGE_1 and PGE_2. Propranolol reduced or abolished responses to all these brochodilators except PGE_1 and PGE_2. (Author) 1714

0938
JAMES, R.C. and T.W. BURNS
Effects of insulin, prostaglandin E, and cations on the lipolytic activity of catecholamines, theophylline, and dibutyryl cyclic AMP in rat and human adipose tissue cells.
Diabetes. 18(supp. 1): 360. 1969.

Abstract only. It was previously shown that cells prepared from rat epididymal fat pads were responsive to the lipolytic activity of glucagon, growth hormone plus dexamethasone, and ACTH, while cells from human subcutaneous adipose tissue were not. This paper reports the effect of inhibitors and cations on the lipolytic activity of norepinephrine, epinephrine, theophylline, and dibutyryl cyclic AMP. Lipolysis was stimulated in both rat and human cells by norepinephrine, theophylline, and dibutyryl cyclic AMP. Lipolysis was inhibited by PGE in both rat and human cells. Insulin or PGE inhibited the lipolytic effect of theophylline. PGE had little or no effect on dibutyryl cyclic AMP stimulation while insulin tended to inhibit its effect. The influence of the presence or absence of potassium, magnesium, or calcium ions is mentioned. (ART) 1532

0939
JANUSZEWICZ, W. and M. SZNAJDERMAN
Patogeneza nadcisnienia teniczego—wspolczesne kierunki i perspeltywy. [Pathogenesis of arterial hypertension—current trends and perspectives']
Polskie Archiwum Medycyny Wewnetrznej. 43: 1221-1228. 1969.

Article in Polish. Abstract not available at present. 1681

0940
JAQUES, R.
Morphine as inhibitor of prostaglandin E_1 in the isolated guinea-pig intestine.
Experientia. 25: 1059-1060. 1969.

Some narcotic and non-narcotic analgesics, anti-inflammatory drugs and other compounds were tested for possible antagonistic action on PGE_1-induced contractions of isolated guinea-pig intestine. The inhibitory action of the narcotics on PGE_1-induced contractions was compared to contractions induced by arachidonic acid peroxide (AAP). It was found that doses of PGE_1 as low as 0.1 μg/ml were effective in including contractions in isolated guinea-pig ileum and jejunum. 10-20 times more AAP was required to produce an equal effect. The analgesic narcotics tested were morphine and etonitazene. With morphine, the ED_{50} for producing inhibition to PGE_1-induced contractions was 0.1 mg/ml. The ED_{50} for etonitazene was 0.000023 mg/ml. Other compounds which produced inhibition of PGE_1 and the ED_{50} were nalorphine 0.014 mg/ml, ethyl-3,5,6,-tri-O-benzyl-glucofuranoside (glyuenol$^{®}$) 5mg/ml, atropine 0.34 mg/ml, tripelennamine 6.2 mg/ml, and cyanide (sodium) 22mg/ml. Compound tested which produced no inhibition were aminopyrine, azamethonium and sodium fluouride. The inhibitors worked equally well on contraction induced by AAP. (JRH) 1538

0941

JELKS, G.W., T.E. EMERSON, Jr., R.M. DAUGHERTY, Jr., and R.E. HODGMAN
Cardiovascular effects of prostaglandin $F_{2\alpha}$ in dogs.
Clinical Research. 17: 513. 1969.

Abstract only. The effects of a 5 min IA infusion of $PGF_{2\alpha}$ at 58 μg/min on venous return (VR), aortic pressure (Pa), peripheral (TPR) and pulmonary (PVR) resistances, heart rate (HR) and myocardial contractile force (MCF) were studied in anesthetized dogs (pentobarbital) weighing an average of 19 kg. VR was diverted from the venae cavae into a reservoir and returned to the right atrium with a pump. Cardiac inflow was either held constant and reservoir volume (RV) allowed to vary (KF; N-11) or changed continually to match VR (NF; N=10). In KF studies, $PGF_{2\alpha}$ caused an initial increase in VR of 75 ml/min which fell below control after 2 min of infusion. Pa and TPR rose initially by 10 mmHg and 0.01 PRU, respectively, but returned to near control level by the 3rd min. RV rose 32 ml, reflecting translocation of blood centrally, but fell below control from 3-5 min. Pulmonary artery pressure, PVR, HR and MCF rose throughout the infusion, while left atrial pressure (LAP) was unaffected. In NF studies, $PGF_{2\alpha}$ caused a marginal increase in VR which quickly fell below control. Pa and TPR were elevated throughout the infusion. Other changes were similar to those of the KF study except for LAP which fell. Occasionally VR fell from the onset of infusion. These data suggest that $PGF_{2\alpha}$ stimulates both cardiac and peripheral arteriolar smooth muscle but in the steady state, the cardiac stimulation may not raise cardiac output because of opposing changes in Pa and venous filling pressure. With respect to the latter, perhaps venous constriction initially increases VR but then later decreases it. (Authors modified) 1762

0942

JELKS, G.W., R.M. DAUGHERTY and T.E. EMERSON, Jr.
Effects of prostaglandin E_1 on venous return and peripheral resistance in the dog.
Physiologist. 12: 261. 1969.

Transient and steady state effects of a 10 minute prostaglandin E_1 infusion on venous return, arterial blood, pulmonary artery and left atrial pressures were studied in anesthetized dogs. The results indicated that the dominant effect of PGE_1 is a marked fall in the total peripheral resistance apparently resulting from active vasodilation. (GW) 1602

0943

JESKE, W.
Prostaglandyny i ich rola w ukladzie rozrodczym. [The prostaglandins and their role in the procreative system.]
Ginekologia Polska. 40: 935-938. 1969.

Suggestions concerning the role of prostaglandins are presented in the physiology of conception, delivery and menstruation cycle phenomenons. The possible complications due to the excess of deficiency of those substances are discussed. (Authors) 1671

0944
JONES, R.L.
  The metabolism by the lung of prostaglandins E and A.
  Journal of Physiology. 201: 76P. 1969.

  Title only. 1436

0945
JUHLIN, L. and G. MICHAELSSON
  Cutaneous vascular reactions to prostaglandins in healthy subjects and in patients with urticaria
    and atopic dermatitis.
  Acta Dermato-Venereologica. 49: 251-261. 1969.

  This investigation compares the effects of intradermally injected PGE_1, PGE_2, $PGF_{1\alpha}$ and
  $PGF_{2\alpha}$. The reaction of PGE_1 was studied in healthy subjects and in patients with various
  types of skin disorders, including urticaria, atopic dermatitis, eczematous dermatitis and
  psoriasis. Also the effects of several drugs on the reaction of PGE_1 were studied. Intradermal
  injections of 0.001-5 μg of prostaglandins in 0.1 ml saline were given on the volar surface of
  the forearm. Measurements of the areas of erythema resulting were made at 20 minutes, 1,2,5
  and 24 hours later. Intravenous retrograde injection of drugs in a forearm vein was made with
  arterial occulsion for 20 minutes. PGE_1 was injected intradermally 3 to 5 minutes later. In all
  subjects intradermal injection of 1 ng-5 μg of PGE_1 gave a characteristic erythematous reaction
  by 20 minutes and lasting for 1-10 hours. Hyperalgesia and erthematous streaks from the test
  site were induced with doses above 10 μg. PGE_2 had the same effects as PGE_1. $PGF_{1\alpha}$ and
  $PGF_{2\alpha}$ produced less erythema and no erythematous streaks or hyperalgesia. The relationship in
  response between the 4 prostaglandins was the same in all patients tested. None of the
  prostaglandins produced an increase in sweating. Patients with chronic urticaria showed greater
  and more prolonged reactions than others. In patients with atopic dermatitis 5 μg of PGE_1
  causes an erythematous reaction which contrasted with the blanching seen after the other
  vasodilators as histamine and bradykinin. Regional blockage of the α-adrenergic receptors with
  dibenzylene was without certain effect on the PGE_1-induced erythema. After intradermal
  infiltration of the skin with a β-adrenergic blocking agent, PGE_1 reactions were reduced. Similar
  pretreatment with lidocaine 1% influenced the PGE_1 reaction at 20 minutes only. In skin
  pretreated with 2% gluocinolone cream under an occlusive dressing the reaction to PGE_1 was
  reduced by 50%. Large doses of epinephrine are required to inhibit the vasodilatory effects of
  PGE_1. The reaction to PGE_1 was not influenced by histamine depletion, antihistamines or
  atropine. Intravenous infusion of kallikrein inhibitor reduced the effects of high but not of low
  doses of PGE_1. Findings suggest that PGE_1 might be one of the main mediators of antidromic
  vasodilatation and an important factor in inflammatory response. (PR) 1583

0946
JUST, G., C. SIMONOVITCH, F.H. LINCOLN, W.P. SCHNEIDER, U. AXEN, G.B. SPERO, and
J.E. PIKE
  A synthesis of prostaglandin $F_{1\alpha}$ and related substances.
  Journal of the American Chemical Society. 91: 5364-5371. 1969.

  Details are given of a synthesis of $PGF_{1\alpha}$, $PGF_{1\beta}$, and several other prostaglandins isomeric
  with these. The key step of the synthesis involves the acid-catalyzed opening and rearrangement
  of epoxybicyclo (3.1.0) hexanes. Structures of intermediate compounds are shown. (PR) 1550

0947
KADAR, and F.A. SUNAHARA
Inhibition of prostaglandin effects by ouabain in the canine vascular tissue.
Canadian Journal of Physiology and Pharmacology. 47: 871-879. 1969.

The effects of prostaglandins on the isolated mesenteric vein and artery of the dog were investigated. PGE_1 inhibited spontaneous contractions of the tissue whereas $PGF_{1\alpha}$ and $PGF_{2\alpha}$ stimulated them. The effects of prostaglandins were not influenced by pretreatment with atropine, phenoxybenzamine, propranolol, or tetrodotoxin. The norepinephrine-induced contractions were inhibited by PGE_1 and enhanced by $PGF_{1\alpha}$ and $PGF_{2\alpha}$. Both these contrasting effects were enhanced in a low concentration of K^+ (1.2mM) and diminished when the media contained K^+ in high concentration (23.2mM). Pretreatment of the tissue with ouabain in sufficiently high concentration (1.5×10^{-5} M) produced an initial contracture followed by relaxation. PGE_1 and $PGF_{1\alpha}$ had no effect on the ouabain-treated tissue by $PGF_{2\alpha}$ still induced dose-dependent contractions. In the Ouabain-treated tissue, the effects of PGE_1 and $PGF_{1\alpha}$ on the norepinephrine-induced contraction were also absent. From those experiments it is concluded that the transport enzyme (Na^+ K^+)-dependent ATPase is necessary for PGE_1 and $PGF_{1\alpha}$ to elicit their action on vascular tissue. The $PGF_{2\alpha}$ effect is probably mediated by an enzyme which is not sensitive to ouabain. (Authors) 1554

0948
KANEKO, T., U. ZOR, and J.B. FIELD
Thyroid-stimulating hormone and prostaglandin E_1 stimulation of cyclic 3',5'-adenosine
   monophosphate in thyroid slices.
Sciences. 163: 1062-1063. 1969.

Thyroid-stimulating hormone increased the cyclic 3',5'-adenosine monophosphate concentration in dog thyroid slices during a 1-minute incubation period and produced a maximum effect soon thereafter. The elevation persisted for at least 30 minutes. The concentrations of the cyclic 3',5'-adenosine monophosphate increased as the TSH concentration was increased from 0.125 to 50 milliunits per milliliter. PGE_1, which increases glucose oxidation in dog thyroid slices, also increased the concentration of cyclic 3',5'-adenosine monophosphate. Although sodium fluoride stimulates thyroid adenyl cyclase, it did not increase concentration of cyclic 3',5'-adenosine monophosphate. Carbamylcholine and menadiol sodium diphosphate augment glucose oxidation in dog thyroid slices but do not change concentrations of cyclic 3',5'-adenosine monophosphate. (Authors) 1624

0949
KAPLAN, E.L., G.W. PESKIN and C.D. ARNAUD
Nonsteroid, calcitonin-like factor from the adrenal gland.
Surgery. 66: 167-174. 1969.

In the discussion section, the authors briefly state that prostaglandins are found in medullary carcinomas and in adrenal tissue. They imply that the PG's may have some connection with the cells producing calcitonin in the thyroid and the cells producing the calcitonin-like factor in the adrenals. (RAP) 1795

0950
KAPLAN, H.R., G.J. GREGA, G.P. SHERMAN, and J.P. BUCKLEY
Central and reflexogenic cardiovascular actions of prostaglandin E_1.
International Journal of Neuropharmacology. 8: 15-24. 1969.

The centrally mediated cardiovascular effects of PGE_1 were investigated using cross-circulation procedures. PGE_1 was injected into the arterial inflow (IA-R) of vascularly isolated, neurally intact heads of anesthetized recipient dogs. Following the administration of 5 and 10 $\mu g/kg$ of PGE_1 (IA-R) consistent depressor responses occurred in the donor dogs and in the recipient's isolated trunk. The administration of 5 and 10 $\mu g/kg$ of PGE_1 (IA-R) to debuffered recipients (carotid sinus-body areas bilaterally denervated) elicited centrally mediated pressor responses in the recipient's trunk paralleled by depressor effects in the donor and a decline in the recipient's perfusion pressure. Centrally mediated pressor responses of angiotensin II were inhibited by PGE_1 only in the non-debuffered recipients. Pressor responses to PGE_1 in debuffered recipients were blocked by the ganglionica blocking agent, hexamethonium. This study provides evidence for the involvement of the carotid sinus-body structures and the central nervous system as additional loci for the cardiovascular effects of PGE_1. (Authors) 1575

0951

KARIM, S.M.M., K. SOMERS, and K. HILLIER

Cardiovascular actions of prostaglandin $F_{2\alpha}$ infusion in man.

European Journal of Pharmacology. 5: 117-120. 1969.

The effect of $PGF_{2\alpha}$ on the cardiovascular system in six normal volunteers was studied by intravenous infusion of 0.01-2.0 $\mu g/kg/min$. This prostaglandin had no effect on the systolic and diastolic blood pressures, heart rate, ECG, or the respiration rate. Similarly rapid single intravenous injections of 1-40 μg of $PGF_{2\alpha}$ did not alter any of these parameters. One volunteer was also infused with $0.2\mu g/kg/min$ of prostaglandin E_1 for 30 min. This produced an increase in heart rate and a fall in systolic and diastolic pressures. (Authors) 1425

0952

KARIM, S.M.M., R.R. TRUSSELL, K. HILLIER, and R.C. PATEL

Induction of labour with prostaglandin $F_{2\alpha}$.

Journal of Obstetrics and Gynaecology of the British Commonwealth. 76: 769-782. 1969.

The effect of $PGF_{2\alpha}$ on the activity of the pregnant human uterus in vivo has been studied in 35 women at or near term. Labour was successfully induced with continuous infusion of 0.05 $\mu g/kg/min$ of $PGF_{2\alpha}$ in 29 women. The pattern of uterine activity produced by $PGF_{2\alpha}$ was similar to that of normal labour with increase in frequency and amplitude of contractions. No increase in the resting tone of the uterus was recorded. The possible physiological role of prostaglandins in the process of parturition and the potential value of $PGF_{2\alpha}$ in obstetrics is discussed. (Authors) 1402

0953

KARIM, S.M.M.

The role of prostaglandin $F_{2\alpha}$ in parturition.

In: Mantegazza, P. and E.W. Horton, eds., "Prostaglandins, Peptides, and Amines," p. 65-72. London, Academic Press. 1969.

This study of the effects of $PGF_{2\alpha}$ in 5 healthy male volunteers and one non-pregnant female volunteer, aged between 18 and 31 years, showed that, in a concentration up to 2.0 micrograms/kg/min, there was no alteration of blood pressure, heart rate, electrocardiogram, or respiration. However, an infusion of 0.05 microgram/kg/min into 10 pregnant women at or near term caused the uterus to contract in a rhythmic and regular fashion, the pattern of which was not unlike that of normal labor. The author states that this "successful induction of labour with $PGF_{2\alpha}$ represents the first ever clinical application of a substance belonging to the prostaglandin group", (ART) 1693

0954
KATSUBE, J. and M. MATSUI
  Synthetic studies on cyclopentane derivatives. Part I. Alternative routes to dl-prostaglandin B_1
  and dihydrojasmone.
  Agricultural and Biological Chemistry. 33: 1078-1086. 1969.

  An alternative route to dl-prostaglandin B_1 using the Grignard reaction of 2-(6'-
  tertbutyloxycarbonylhexyl)-3-methoxy-2-cyclopenten-1-one with 3-tetra-hydropyranyloxy-
  1-octyne was developed. An easy synthesis of dihydrojasmone was also described.
  (Authors) 1680

0955
KESSLER, E., C. JIMENEZ, R. MANAKIL and R.C. HUGHES
  A possible new humoral factor in shock.
  Circulation. 40(supp. 3): 121. 1969.

  Abstract only. Lipid extracts were made of normal, endotoxin shock and hemorrhagic shock
  dog plasma with ethyl acetate at pH 3. After treatment with alcoholic KOH and incubation at
  $50°$ shock plasma showed an absorption peak at 278 ± 1.6 mμ. Normal plasma had no peak.
  Various fatty acids, histamine, bradykinin, serotonin, acetylcholine, pyruvic acid, and lactic acid
  showed no peaks. Prostaglandins E_1, E_2 and $F_{1\alpha}$ showed absorption peaks suggesting the
  material extracted from palsma may be a prostaglandin. Peaks were absent in 25 normal
  subjects, but were present in 16 of 48 patients with various forms of shock. Where present the
  mortality rate was 75%, compared to 56% when absent ($X^2=19.0$, P<.005). In one patient with
  associated renal failure, peritoneal dialysis fluid and plasma showed peaks, indicating the
  material was dialyzable. The results suggest a prostaglandin-like substance(s) may contribute to
  the disturbed physiology in human and experimental shock. (Authors modified) 1751

0956
KLOEZE, J.
  Influence of prostaglandins on ADP-induced platelet aggregation.
  Acta Physiologica and Pharmalogica Neerlandica. 15: 50-51. 1969.

  Abstract only. None of the prostaglandins examined induced platelet aggregation in citrated
  platelet-rich plasma (cPRP), as was measured by a turbidometric technique. However, it
  appeared that PGE_1 inhibited platelet aggregation induced by addition of ADP to cPRP of rats,
  pigs and men. PGE_2 stimulated platelet aggregation induced by addition of ADP in cPRP of
  rats and pigs, but it was inactive in human cPRP. Dose-response curves showed that PGE_1 and
  PGE_2 were already active in doses as low as 10^{-3} to 10^{-2} μg/ml. The effect of PGE_1 and
  PGE_2 was found to be specific, as a great number of prostaglandins with very small differences
  in chemical structure appeared to be inactive. Only three related substances, viz. PGE_1-217,
  dinor-PGE_1 and an iso-PGE_1 were found to cause PGE_1-like effects on ADP-induced platelet
  aggregation. In a roughly quantitative assay and expressed on a dose-level, these substances
  showed 5,1.25 and 1% PGE_1-activity respectively. (Author modified) 1744

0957
KLOEZE, J.
  Relationship between chemical structure and platelet-aggregation activity of prostaglandins.
  Biochimica et Biophysica Acta. 187: 285-292. 1969.

  Small amounts of prostaglandins, are found widely distributed in different organs. Most of them
  have potent biological activity of some kind. One of these activities is the effect on platelet

aggregation in vitro induced by adenosine diphosphate (ADP). PGE_1 was found to have an inhibiting effect, whereas PGE_2 displayed a stimulating effect. These findings led to experiments to determine the potencies of the inhibiting and stimulating effects in relation to each other by examining mixtures of PGE_1 and PGE_2 at different concentrations. In order to gain information on the relationship between chemical structure and biological activity, a number of analogues and unnatural PG's related to PGE_1 and PGE_2 were examined for their effect on ADP-induced platelet aggregation in citrated platelet-rich rat blood plasma. The potency of PGE_1 to inhibit platelet aggregation was found to be many times gteater than the stimulating potency of PGE_2. The E-type structure of the cyclopentane ring is required for either an inhibiting or a stimulating effect of prostaglandins on platelet aggregation. The oxo-groups at the ring and in the carboxyl group (C-9 and C-1 respectively), as well as the distance between them, are essential as is the hydroxyl group at C-15. Changing the length of the a-side chain containing the carboxyl group inactivates the prostaglandin. (ART) 1416

0958
KISCHER, C.W.
   Accelerated maturation of chick embryo skin treated with prostaglandin (PGB_1): an electron
      microscopic study.
   American Journal of Anatomy. 124: 491-512. 1969.

PGB_1 completely blocks feather organ development, but stimulates proliferation and keratinization of the epidermis when applied to organ cultures of developing chick skin with feather organ loci. Electron microscopy of control and PGB_1 treated explants indicated the following. 1) Tonofilaments appear much sooner in treated explants than in controls. 2) The cristae of mitochondria in epidermal cells of treated skins are long, closely apposed, and oriented parallel to the longest axis of the organelle as contrasted with a transverse orientation for them in control epidermis. 3) There is increased deposition of collagen in the mesenchyme of treated skins as contrasted to that of the controls. 4) Unusual structures occur throughout the stratified and superficial epidermis of treated skins. 5) Small but numerous vacuoles appear in the epidermal cells of treated skins. (GW) 1514

0959
KISCHER, C.W.
   Fibrous elements of mesenchyme during development of a skin derivative.
   Journal of Cell Biology. 43: 69a. 1969.

Abstract only. This brief paper reports results of electron microscope studies on the differences observed in chick embryo skin development when grown in vitro with and without PGB_1. PGB_1 causes rapid maturation of the epidermis but completely inhibits feather development. The electron microscope revealed increase deposition of collagen or collagen related elements. Large anchor filaments which normally disappear during maturation persist in PGB_1 treated skin. (JRH) 1488

0960
KISCHER, C.W.
   The role of collagen in developing skin and a skin derivative: a biochemical, light and electron
      microscopical study.
   Anatomical Record. 163: 212. 1969.

Abstract only. It had previously been shown that if dorsal skin with presumptive down feather organ sites from the back of chick embryo is explanted to grow in a medium containing prostaglandin B_1, morphogenesis of feather organs is completely blocked while the entire skin is stimulated to precociously mature. To determine the significance of collagenous substrate in the mesenchyme three techniques were used to compare the distribution, alignment and amount of collagen in PGB_1-treated explants and controls. Observation by electron microscopy shows more collagen in the mesenchyme of PGB_1-treated explants than in developing feather organs. The basal lamina of the former is thicker than in the latter and is often infiltrated with collagen and anchoring filaments. The basal epidermis of treated skins shows a scattering of structures similar to hemidesmosomes. Examined by polarized light whole tissue mounts demonstrate a lattice-like birefringence in control tissue which is predominantly located in the interfollicular areas. The pattern observed in PGB_1-treated tissues is one of uniform dispersion. Assay for hydroxyproline indicates no essential differences in amount of collagen, on a dry weight basis, between control and treated explants. (PR) 1559

0961
KOPP, E.
Prostaglandine und magensekretion. [Prostaglandins and gastric secretion.]
Schweizerische Medizinische Wochenschrift. 99: 1369-1370. 1969.

Abstract only. Experiments were perfor,ed to see if prostaglandins could inhibit ulcers induced by pyloric ligature (Shay-ulcers) or injection of pregnisolone (steroid-ulcers) in rats. Subcutaneous or intravenous prostaglandin reduced the volume and concentration of acid and pepsin secreted by the stomach. The formation of ulcers was also inhibited in a dose dependent manner. While the mechanics of PG action are unknown, they are not due to anticholinergic effects. Since there is significant secretion of PG's in the stomach, it is likely that they have a physiological role in the stomach. (JRH) 1590

0962
KORVER, O.
Optical rotatory disperison and circular dichroism of prostaglandins.
Recueil des Travaux Chimiques des Pays-Bas et Belgique. 88: 1070-1079. 1969.

Optical rotatory dispersion and circular dichroism data are given for eight prostaglandins. An interpretation of the data is given in terms of the conformation of the cyclopentanone ring. (Author) 1724

0963
KOTTEGODA, S.R.
An analysis of possible nervous mechanisms involved in the peristaltic reflex.
Journal of Physiology. 200: 687-712. 1969.

The effect of drugs on peristalsis and on the contractions of the two muscle coats of the isolated guinea pig ileum in response to co-axial electrical stimulation have been studied. The response of the circular muscle to 23 drugs was investigated. Prostaglandin E_1 and E_2 in a dose range between 10^{-7} and 10^{-6} g/ml caused a maintained longitudinal muscle contraction and a suppression of the contractile response of the circular muscle to electrical stimulation. The prostaglandins were the only substances used which antagonized the circular muscle contraction. (GW) 1415

0964

KUNZE, H.

Formation of ^{14}C-prostaglandin E_2 from (1-^{14}C) arachidonic acid during perfusion through the frog intestine.

Abstracts, 4th International Congress on Pharmacology. Basle, 14-18 July, p. 373. 1969.

Abstract only. Frog intestine preparations were perfused through the vascular bed with Ringer solution at a rate of 0.5 ml/min. After a single injection of 50 μg(1-^{14}C— arachidonic acid (specific activity 2.12 μC/mmol) the venous effluents were collected over 20 min or in fractions of 1 min. Non-labelled prostaglandins E_2 and $F_{2\alpha}$ were added to each fraction and lipids were extracted with ethyl acetate. The lipid extracts were chromatographed on columns of silicic acid and on thin layer plates. The results indicate that ^{14}C-PGE$_2$ is formed from (1-^{14}C) arachidonic acid during perfusion through the frog intestine. After a single injection about 3% of the arachidonic acid was detected converted product in the perfusate collected during 20 min. The conversion begins rather quickly, ^{14}C-PGE$_2$ appearing already 1-2 min after injection of the precursor. (Author modified) 1599

0965

KUNZE, H. and R. BOHN

Formation of prostaglandin in bovine seminal vesicles.

Naunyn-Schmiedebergs Archiv fur Pharmakologie und Experimentelle Pathologie. 264: 263-264. 1969.

In this letter to the editor, the author notes that bovine seminal vesicles contain much less prostaglandin than sheep vesicular tissue does. Homogenized vesicular tissue was incubated with O_2, arachidonic acid or phospholipase A. All of these substances increased the yield of prostaglandin. Experimental methods are not given. (JRH) 1584

0966

KUO, J.F. and E.C. DeRENZO

A comparison of the effects of lipolytic and antilipolytic agents on adenosine 3'5'-mono-phosphate levels in adipose cells as determined by prior labeling with adenine-8-^{14}C.

Journal of Biological Chemistry. 244: 2252-2260. 1969.

A direct method for assaying cAMP in isolated adipose cells is described. The method involves measuring the formation and accumulation of cAMP-8-^{14}C by cells previously treated with adenine-8-^{14}C. Experiments with this method confirm that PGE$_1$ reduces the norepinephrine elevated cAMP levels. (JRH) 1487

0967

LADINSKY, H. and K. STRANDBERG

Formation of spasmogenic lipids in cat lung tissue induced by bee venom.

In: Mantegazza, P. and E.W. Horton, eds., "Prostaglandins, Peptides, and Amines," p. 39-49. London, Academic Press, 1969.

Spasmogenic lipids were formed in minced cat lung tissue on incubation with bee venom at 37°C and pH 7. A concentration of bee venom of 50 μg/ml/100mg lung tissue and an incubation period of 20 minutes gave the highest amounts of spasmogenic lipids. Fractionation of the spasmogenic lipids by solvent partition and silicic acid chromatography revealed the presence of at least two different biologically active principles. Pure prostaglandin E$_1$ and F$_1$ were mainly found in the ether fraction and produced a rapid, transient contraction of the guinea-pig ileum followed by tachyphylaxis. (GW) 1690

0968
LAITY, J.L.H.
The release of prostaglandin E_1 from the rat phrenic nerve-diaphragm preparation.
British Journal of Pharmacology. 37: 698-704. 1969.

Detailed evidence is presented that the prostaglandin released from the diaphragm on nerve stimulation is mainly PGE_1 and that not only (+)-tubocurarine but also hemicholinium, bretylium and phenoxybenzamine fail to block the prostaglandin output in response to nerve stimulation. Furthermore (+)-tubocurarine and phenoxybenzamine themselves cause prostaglandin release in the absence of nerve stimulation. Rat phrenic nerve-diaphragm preparations were studied during tetanic contraction of the muscle in response to nerve stimulation. Tentative identification of the prostaglandin depended on solvent extraction column and thin-layer chromatography and parallel biological assay. Polar lipid substances were released from the preparation, in the absence of nerve stimulation, by (+)-tubocurarine and noradrenaline. It is concluded that release of prostaglandin does not result from muscle contraction nor solely from adrenergic nerve stimulation. The detection of more polar lipids with biological activity suggests that PGE_1 is only one of a group of substances (including PGE_2) released under the conditions of the experiment. (PR) 1579

0969
LARGE, B.J., P.F. LESWELL, and D.R. MAXWELL
Bronchodilator activity of an aerosol of prostaglandin E_1 in experimental animals.
Nature. 224: 78-80. 1969.

The authors compared the bronchodilator activities of aerosols of PGE_1 and isoprenaline in the anesthetized guinea pig. Intravenous administration of PGE_1 and isoprenaline caused tachycardia and a fall in blood pressure, isoprenaline being 5 times more potent as a bronchodilator. Aerosol administration of PGE_1 caused no cardiovascular changes while isoprenaline produced marked tachycardia. PGE_1 was 10 to 100 times more potent in inhibiting bronchoconstriction. The suprisingly high bronchodilator potency noted with aerosols of PGE_1 may be related to the observation that up to 95% of PGE_1 and PGE_2 infused intravenously in the dog, cat and rabbit was removed during a single pulmonary circuit, suggesting that in the lungs there is a system for the inactivation of prostaglandin. (GW) 1407

250

0970
LAVERY, H., R.D. LOWE, and G.C. SCROOP
Cardiovascular effects of prostaglandins infused into extra—cranial arteries of the dog.
Journal of Physiology. 204: 109P-110P. 1969.

To investigate their actions on the cardiovascular centers in the brain, PGA_1, PGE_1, $PGF_{1\alpha}$ and $PGF_{2\alpha}$ were infused into one vertebral or one internal carotid artery of greyhound dogs premedicated with morphine and anesthetized with α-chloralose. Effects were compared with those obtained during intravenous infusions at the same rate. Other arteries to the brain were ligated. $PGF_{1\alpha}$ infused into the vertebral artery at rates of 1-30μg/minute caused rise in arterial pressure and tachycardia. Propranolol 10 mg intravenously did not affect the tachycardia; vagotomy abolished the tachycardia. Residual pressor effect was abolished by bethanidine (10 mg/kg intravenously). $PGF_{1\alpha}$ infused into the internal carotid artery or intravenously had no cardiovascular effects at this dose range. Confirmation is assumed that effects were due to activation of a centre within the area of distribution of the vertebral artery. $PGF_{2\alpha}$ was 10 times as potent in producing similar effects. PGE_1 had less effects, with variations between animals. PGA_1 had no demonstrable central effect with varying doses when given by all three routes. Findings show that $PGF_{1\alpha}$ and $PGF_{2\alpha}$ can activate autonomic center in the area of distribution of the vertebral artery. Effects appear to be due to inhibition of vagal tone to the heart resembling response to angiotensin. (PR) 1556

0971
LEAKE, C.D.
Review of reviews.
Annual Review of Pharmacology. 9: 521-530. 1969.

This review mentions, inter alia, five reviews dealing with work on prostaglandins as follows: Euler, U.S. von, Clin. Pharmacol. Ther. 9: 228-239. (1968); Hinman, J.W., Bioscience 17: 779-785 (1967); Bergstrom, S., Science 157: 382-391 (1967); Bergstrom, S., Carlson, L.A., and Weeks, J.R., Pharmacol. Rev 20: 1-48. (1968); and Euler, U.S. von, and Eliasson, R., *Prostaglandins* (Academic Press, New York, 164 p. 1968 (sic) listed elsewhere as 1967.) (ART) 1699

0972
LEAR, J., S. KOHATSU, and H.A. OBERHELMAN, Jr.
Gastric secretion of acid in response to right and left atrial infusion of gastrin.
American Journal of Digestive Diseases. 14: 870-874. 1969.

The authors describe previous work by S.M. Ferreira and J.R. Vane (1968) and E. Anggard and B. Samuelsson (1967) who performed in vivo and in vitro experiments in mammals and revealed that the lungs are capable of removing approximately 80% of prostaglandins. (GW) 1647

0973
LEE, J.B., J.C. McGIFF, H. KANNEGIESSER, J.G. MUDD, Y. AYKENT, and T.F. FRAWLEY
Antihypertensive and natriuretic activity of prostaglandin A_1 in human hypertension.
Clinical Research. 17: 456. 1969.

Abstract only. PGA_1 infused iv at low rates (0.1-0.3 μg/kg/min) into patients with essential hypertension resulted in a rise in renal blood flow and urinary sodium excretion but the arterial blood pressure was not affected. At higher infusion rates (2-3 μg/kg/min) the blood pressure fell in association with a reduction in renal blood flow. (GW) 1617

0974

LEE, J.B.

Hypertension, natriuresis, and the renal prostaglandins.

Annals of Internal Medicine. 70: 1033-1038. 1969.

The author discusses current knowledge on the subject of prostaglandins as of May 1969 citing 24 references on structure, physiological roles, sites of synthesis and relation to cyclic AMP. (RMS) 1576

0975

LEE, J.B. and J.F. FERGUSON

Prostaglandins and natriuresis: The effect of renal prostaglandins on PAH uptake by kidney cortex.

Nature. 222: 1185-1186. 1969.

The effect of prostaglandins on para-aminohippurate (PAH) accumulation by rabbit kidney cortex was studied in vitro. Slices of rabbit renal cortex were incubated in Krebs-Ringer buffer with sodium acetate and PAH. Incubation with PGA_1 and PGA_2 resulted in a 50% and 40% fall. PGE_1 and PGE_2 showed a 38% and 40% decrease. PGF_2 was significantly lower than the control. Because the plasma of saline loaded animals seems to contain a factor which inhibits PAH uptake and promotes salt excretion, suggests a close relationship between "natriuretic hormone" and the prostaglandin. PGA_2 escapes pulmonary degradation making it possible that during saline infusion, PGA_2 acts on the renal cortex and enhances sodium excretion. The evidence thus points to the fact that PGA_2 may be the "natriuretic hormone" of the kidney. (GW) 1428

0976

LEFEBVRE, P.J., J.C. SODOYEZ, A.S. LUYCKX, and P.P. FOA

Metabolism of adipopose tissue of the golden hamster with chronic hypoglycaemia and hyperinsulinemia due to a transplantable islet-cell tumour.

Diabetologia. 5: 201-202. 1969.

Abstract only. Prostaglandin PGE_1 reduces basal glycerol release in the normal, fasted or fed hamsters and in the fed tumor-bearing animals. This compound had no effect on basal glycerol release in the fasted insuloma-bearing animals. (Authors modified) 1798

0977

LEWIS, G.P. and N.J.P. WINSEY

The action of pharmacologically active substances on the flow and composition of cat hind-limb lymph.

British Journal of Pharmacology. 35: 377P-378P. 1969.

Arterial infusion of histamine, acetylcholine, bradykinin, 5-hydroxytryptamine and PGE_1 and $PGF_{2\alpha}$ were observed for their effects on femoral venous outflow and flow and composition of lymph from a cannulated femoral lymphatic. No increase in the concentration of intracellular enzymes was observed. Vasodilation and increase in lymph flow usually accompanied by increased protein concentration followed histamine, acetylcholine, and bradykinin infusion. (RMS) 1501

0978

LEWIS, G.P. and J. MATTHEWS

The cause of the vasodilatation accompanying free fatty acid release in rabbit adipose tissue.
Journal of Physiology. 202: 95P-96P. 1969.

Author states that the vasodilatation which follows fat mobilization induced by ACTH, β MSH, and growth hormone is probably caused by PGE_1 released during lipolysis. This assumption is based on similarities in chemical properties of this vasodilatator and PGE_1. It also behaves similarly biologically and chromatographically. (JRH) 1421

0979

LIPPMANN, W.

Inhibition of gastric acid secretion in the rat by synthetic prostaglandins.
Journal of Pharmacy and Pharmacology. 21: 335-336. 1969.

The author examined the effects of four synthetic PG's on gastric secretion in the rat. Basal gastric acid secretory activity was measured in control rats. The synthetic PGE analogue, AY-20,524 was found to be a more potent antisecretory agent than the corresponding synthetic PGF, AY-16,809. The synthetic PGF analogues, AY-16,809, AY-21,669 and AY-21,670, were about one-eighth as active as AY-20,524, which in turn inhibited the gastric acid secretion at an activity about one-tenth that of PGE_1. (GW) 1478

0980

MAIN, I.H.M., and P.M. WRIGHT

The abolition of tremor by prostaglandin E_1.
In: Mantegazza, P. and E.W. Horton, eds., "Prostaglandins, Peptides, and Amines," p. 125-127.
    London, Academic Press, 1969.

The object of this investigation was to elucidate the site of action of PGE_1 in inhibiting tremor observed when cats and chicks are lightly anesthetized. Topical application of PGE_1 (10-100μg/ml) to the cerebral hemispheres of chicks abolished tremor. In cats, E_1 administered by a single intravenous injection reduced or abolished tremor and produced a fall in mean arterial blood pressure. Other vasodepressor drugs such as oxytocin in chicks and hexamethonium in cats and chicks also inhibited tremor. $PGF_{2\alpha}$ in chicks sometimes inhibited tremor while causing only a rise in blood pressure. Hexamethonium in cats not only reduced blood pressure but also caused a 7°C rise in skin temperature which was accompanied by abolition of tremor. PGE_1 administered by slow intravenous injection had similar effects. (GW) 1695

0981

MAIN, I.H.M.

Effects of prostaglandin E_2 (PGE_2) on the output of histamine and acid in rat gastric secretion
    induced by pentagastrin or histamine.
British Journal of Pharmacology. 36: 214P-215P. 1969.

Preliminary results in an in vivo study of the relationship of pentagastrin and histamine infused intravenously into rats, and secretion of histamine and acid from the stomach suggest that inhibition of acid secretion by 0.2-2μg/min intravenous PGE_2 is not accompanied by reduction in histamine output. PGE_2 inhibited acid secretion during histamine infusion but histamine output was increased. This perhaps indicates increased mucosal blood flow or permeability. (RMS) 1504

0982
MAIN, I.H.M.
The effects of prostaglandins on tremor.
Journal of Physiology. 201: 77P. 1969.

Title only. 1663

0983
MAREK, J., O. KUCHEL, and K. HORKY
Nektere nove pohledy na fysiologicke mechanismy posturalni adaptage. [A few current thoughts
on the physiological mechanisms of postural adaptation.]
Ceskoslovenska Fysiologie. 18: 159-174. 1969.

Article in Czech. Abstract not available at present. 1672

0984
MARK, A.L., P.G. SCHMID, J.W. ECKSTEIN, and M.G. WENDLING
Venous responses to prostaglandin $F_{2\alpha}$ ($PGF_{2\alpha}$) and norepinephrine (NE).
Clinical Research. 17: 252. 1969.

Abstract only. Experiments were done to determine effects of $PGF_{2\alpha}$ on the lateral saphenous
vein and to compare its effects with those of norepinephrine in anesthetized dogs. Parallel line
bioassay indicated that the dose of $PGF_{2\alpha}$ required to produce a given response was 19.2 times
the dose of norepinephrine required to produce the same response. This suggests that relatively
large concentrations of $PGF_{2\alpha}$ may be required to exert a physiological influence on venous
tone. (GW) 1622

0985
MARMO, E., F. DIMEZZA, A. IMPERATORE, and S. DIGIACOMO.
Metoclopramide e muscolatura esofagea, gastrica, enterica, splenica, tracheale, della cistifellea e
della vescica urinaria: richerce in vitro. [Metoclopramide and esophageal, gastric, enteric,
splenic, and tracheal musculature, and that of the gall bladder and of the urinary bladder:
research in vitro.]
Gazzetta Internazionale di Medicina e Chirurgia. 73: 1459-1478. 1969.

1) Investigations in vitro with metochlopramide on the oesophageal, gastric, duodenal, ileal,
cholic, appendicular, splenic and tracheal muscle and that of the gall bladder and urinary
bladder have been carried out. Humans organs and those from various species of animals (frog,
rat, guinea pigs, cat, rabbit, dog) were used. 2) Metochlopramide in vitro produced effects
which were mainly directed at the various smooth muscles. Those for low concentrations were
of a stimulating type and for high concentrations of a depressing type. These effects should not
occur at the receptor level. 3) For high concentrations metochlopramide in vitro had an
aspecific antagonistic effect of the papaverine type on various kinds of smooth muscle, for
acetylcholine, acetyl-β-methylcholine, histamine, 5-hydroxy-tryptamin, noradrenalin, bradykinin
and prostaglandin E_1. (Authors) 1717

0986
MARQUIS, N.R., R.L. VIGDAHL, and P.A. TAVORMINA
Platelet aggregation. I. Regulation by the cyclic AMP and prostaglandin E_1.
Biochemical and Biophysical Research Communications. 36: 965-972. 1969.

An investigation into the possibility that PGE_1 inhibition of platelet aggregation is mediated by increased cAMP is reported. Observations on human platelets showed adenyl cyclase stimulation by PG in the order of potency of PGE_1, PGA_1, and $PGF_{1\alpha}$ which is the same order of inhibition of aggregation. PGE_1 activity was potentiated by caffeine which in contrast to PG increased cAMP by inhibition of platelet phosphodiesterase. Exogenous cAMP mimicked PGE_1 in effects, and cAMP levels in whole platelets changed simultaneously with PGE_1-induced inhibition of aggregation. It is suggested that intracellular cAMP regulation is of major importance in platelet aggregation. (RMS) 1525

0987
MAXWELL, G.M.
The effect of prostaglandin $F_{2\alpha}$ upon coronary venous flow and myocardial metabolism.
Australian Journal of Experimental Biology and Medical Science. 47: 713-721. 1969.

General hemodynamics, coronary sinus flow and myocardial extraction of glucose, lactate, pyruvate and nonesterified fatty acid were measured before and after the injection of $PGF_{2\alpha}$ in anaesthetized dogs. Systemic and pulmonary arterial pressures increased; cardiac output decreased transiently, coronary venous flow increased. Little change was found in myocardial O_2 and CO_2 metabolism. Cardiac efficiency decreased. Arterial lactate, pyruvate and non-esterified fatty acid values increased; glucose was unchanged. Myocardial extraction values for these substrates did not change significantly. Some of the changes noted did not appear when stoichiometric amounts of $PGF_{2\alpha}$ and PGE_1 were injected together. (Author modified) 1745

0988
MAY, B. and E. BOHLE
Metabolishche effekte von prostaglandin E_1 (PGE_1) in tierversuch. [Metabolic effect of prostaglandin E_1 in animal experimentation.]
Fette-Seifen-Anstrichmittel. 71: 933. 1969.

Title only. 1777

0989
MAY, B., D. HELMSTAEDT, K. JEUCK and E. BOHLE
Uber metabolische wirkungen verschiedener prostaglandine im tierversuch. [On the metabolic effects of certain prostaglandins in animal experiments.]
Verhandlungen der Deutschen Gesellschaft fur Innere Medizin. 75: 806-808. 1969.

Intraperitoneal administration of PGE_1 to rats (0.1 mg/kg, or about 10-12 µg/rat) repeated three times was found to result in a marked decrease in the glycogen content of the liver, to about 50% control values twelve hours afterwards. In mice a similar hyperglycemia could be induced with similar dosages. A rise in blood sugar correlated well with loss of glycogen in the liver. The direct glycogen effect of PGE_1 was demonstrated on human and rabbit thrombocytes (25-100 µ/ml, 60 min incubation) by a decrease of 20-45% in platelet glycogen. In the authors' opinion, the in vivo hypoglycemic effect of PGE_1 is the result of synergistic direct (glycogenolytic) and indirect (pyro-catecholamine-releasing) properties of this substance. (MEMH) 1768

0990

MAY, B., K. JEUCK, D. HEMLSTAEDT, and E. BOHLE
Primare und sekundare prostaglandineffekte auf den fett-und kohlenhydratstoffwechsel.
[Primary and secondary effects of prostaglandin on fat and carbohydrate metabolism.]
Medizin und Ernahrung. 10: 172-174. 1969.

This article in German points out that PGE_1, in addition to its antilipolytic effect and insulin-like property also has a weak direct glycogenolytic effect in vitro (on liver tissue and thrombocytes). In vivo, PGE_1, depending upon the dosage, can initiate a secondary reaction, e.g., release of catecholamines from the adrenal cortex—so that there is an increase (in vivo) in blood sugar level and decrease of liver glycogen content with the result that there is a direct as well as indirect effect. It is also tentatively suggested that the inhibition of gluocose incorporation in the glycogen of the diaphragm in vivo may play a role in the mechanism involved. (ART) 1722

0991

McGIFF, J.C., N.A. TERRAGNO, A.J. LONIGRO, and K.K.F. NG
Patterns of release and identification of renal antihypertensive substances produced by renal ischemia.
Journal of Clinical Investigation. 48: 57a. 1969.

Abstract only. Interactions of angiotensins and prostaglandins in the development of renal hypertension are suggested by experiments on dogs in which several patterns of response to unilateral renal ischemia were detected by superfused organ banks containing rat stomach, rat colon and chick rectum. Upon renal arterial constriction, patterns of response included immediate release of prostaglandin-like substance(s) (PLS) from ischemic kidneys, continuous release of angiotensin-like substance(s) (ALS) and PLS from ischemic kidneys, and release of PLS from contralateral kidneys. A pulse of PLS appeared in ischemic renal effluent following release of constriction. PGE's, and less frequently, $PGF_{2\alpha}$ were tentatively identified only in renal venous blood. Contralateral PLS was probably mediated by angiotensin II derived from release of angiotensin I from ischemic kidney. Ischemic effects on PLS were reproduced by arterial infusion of angiotensin II. (RMS) 1435

0992

McGIFF, J.C., K. CROWSHAW, N.A. TERRAGNO, A.J. LONIGRO, J.C. STRAND, M.A. WILLIAMSON, K.K.F. NG, and J.B. LEE
Release of prostaglandin-like substances during acute renal ischemia.
Circulation 40 (supp. 3): III-144. 1969.

Abstract only. Chromatographic characterization of prostaglandin-like substances obtained from renal venous blood obtained during peak activity of bioassay organs in a study of renal ischemia in dogs showed PGE_2 to be the major constituent. (RMS) 1432

0993

McGIFF, J.C., N.A. TERRAGNO, J.C. STRAND, J.B. LEE, A.J. LONIGRO, and K.K.F. NG
Selective passage of prostaglandins across lung.
Nature. 223: 742-745. 1969.

Prostaglandin solutions were infused into the left carotid artery of an anesthetized male dog. Prostaglandin A_1 and A_2 produced increases in renal blood and urine flow on intra-aortic infusion and intravenous infusion. PGE_1 and PGE_2 increased renal blood flow only on intra-aortic infusion. The effects of $PGF_{2\alpha}$ on renal blood flow were variable. The fact that PGA_1 and PGA_2 are able to pass through pulmonary circulation without loss of activity while PGE_1 and PGE_2 are oxidized by 15-hydroxy PGDH in the lung merits consideration of the former as circulating hormones. (GW) 1424

0994

McGIFF, J.C., N.A. TERRAGNO, K.K.F. NG, and J.B. LEE
Selective passage of prostaglandins across the lungs.
Federation Proceedings. 28: 286. 1969.

Prostaglandins (PG's) are considered to be unimportant as circulating hormones since the lungs remove more than 95% pf PGE's infused intravenously (i.v.). Renal blood flow (RBF), which is extremely sensitive to infusion of PGE's and PGA's (0.0001 μg/ml renal blood) into the renal artery, was used as an index of the activity and fate of PG's. The effects on RBF of PG's infused i.v. were compared to the same dose infused into the aorta (i.a.). In chloralose anesthetized dogs, the effects of PGA's (0.005 μg/kg/min) on RBF (electromagnetic flowmeter) were independent of the route of administration. On i.v. and i.a. administration PGA_1 increased RBF by 40±12% and 39±17% respectively, from a control of 190±13 ml/min. The effects of PGA_2 were indistinguishable from PGA_1 whereas $PGE's_{1,2}$ increased RBF only on i.a. infusion. The threshold dose (0.001 μg/kg/min) for PGA's on RBF was identical i.v. or i.a. The increased RBF produced by PGA's occurred at subpressor doses and could not be dissociated from their diuretic action (photoelectric drop counter). Since PGA's escape removal by the lungs and increase RBF and urine flow at concentrations which had no effect on blood pressure, the PGA's may function as circulating hormones after their release from an organ (e.g., kidney). (Authors) 1542

0995

McKINNON, E.L., O. TANGEN, and H.J. BERMAN
Effects of prostaglandin E_1 on platelet aggregation: an electron microscopic study.
Anatomical Record. 163: 315. 1969.

Abstract only. No structural evidence was found that 0.83 μg/ml PGE_1 altered hamster platelets in citrated platelet-rich plasma. ADP-, thrombin-, and collagen-induced sphering, formation of dendrites, appearance of extracellular granules and aggregation were eliminated or reduced by PGE_1 with the least effects seen on thrombin induced aggregation and membrane and organelle damage. (RMS) 1558

0996

McQUEEN, D.S. and A. UNGAR
The modification by prostaglandin E_1 of central nervous interaction between respiratory and cardio-inhibitor pathways.
In: Mantegazza, P. and E.W. Horton, eds., "Prostaglandins, Peptides and Amines," p. 123-124. London, Academic Press, 1969.

Dogs showing marked sinus arrhythmia in which the heart accelerated during inspiration and slowed in expiration were selected for study. Any reflex component of the arrhythmia was excluded by surgical denervation of the lungs in open-chested animals or by applying positive pressure ventilation at a frequency unrelated to that of the arrhythmia to animals paralyzed by decamethonium. PGE_1 was injected into the common carotid artery in doses of 5 to 30 nMoles/kg body weight. The arrhythmia was abolished or markedly reduced in every experiment, the cardiac period (duration of the cardiac cycle, beat by beat) becoming equal to that in the inspiratory phase of the arrhythmia. PGE_1 thus appears to inhibit an expiratory slowing rather than an inspiratory speeding. The effect was not altered by denervation of the carotid sinus region; this establishes a central site of action. While the response to PGE_1 was at its height, stimulation of the carotid body chemoreceptors gave rise to a bradycardia similar to one evoked before injection of PGE_1, indicating that the pathways for reflex cardiac inhibition are not blocked. If doses of PGE were reported at 20 minute intervals, tachyphylaxis was seen after two or three injections. The results suggest that PGE_1 may have a selective action on a group of neurones within the brain the effect of which is to inhibit the spread of activity from respiratory to cardio-inhibitor and vasomotor pathways. (ART) 1698

0997
MERLEVEDE, W.
Het werkingsmechanisme van hormonen. [The mechanism of action of hormones.]
Tijdschrift voor Geneeskunde. 19: 967-974. 1969.

This textbook type article in Dutch briefly mentions that prostaglandins affect the level of cAMP in some organs. It is also mentioned that PGE_1 blocks the effect of epinephrine on adenyl cyclase. (JRH) 1721

0998
MICHAL, F., and B.G. FIRKEN
Physiological and pharmacological aspects of the platelet.
Annual Review of Pharmacology. 9: 95-118. 1969.

It is noted that prostaglandin E is an inhibitor of platelet clumping caused by adenosine diphosphate (ADP). (ART) 1703

0999
MICHELI, H., L.A. CARLSON, and D. HALLBERG
Comparison of lipolysis in human subcutaneous and omental adipose tissue with regard to effects of noradrenaline, theophylline, prostaglandin E_1 and age.
Acta Chirurgica Scandinavica. 135: 663-670. 1969.

Subcutaneous and omental adipose tissue from 13 male and 26 female patients, 13 to 76 years, was incubated in vitro and the glycerol release followed as a measured of lipolysis under various conditions. The major purpose of the study was to compare subcutaneous and omental tissues. The phospholipid content was lower in subcutaneous than omental tissue but there was a strong correlation ($r=0.90$) between the content of the two tissues. Glycerol release under various conditions was correlated between tissues from the two locations. PGE_1 was added giving a final concentration of 1 μg/ml and incubated for 1 hour. There was a moderate correlation for basal ($r=0.43$), noradrenaline stimulated ($r=0.44$) and PGE_1 inhibited ($r=0.66$) lipolysis. There was a stronger correlation for lipolysis in the two tissues after stimulation with theophylline ($r=0.90$) and with both theophylline and noradrenaline ($r=0.85$). In general, the omental tissue had a greater lipolytic response than subcuatneous. It is evident that certain lipolytic parameters may vary considerably from tissue site to tissue site and extrapolations from subcutaneous tissue to the major part of adipose tissue may be erroneous. (Authors modified) 1756

1000

MICHELI, H.

Factors modifying the effect of prostaglandin E_1 on lipolysis in adipose tissue.

In: Holmes, W.L., L.A. Carlson, and R. Paoletti, eds., "Drugs Affecting Lipid Metabolism, p. 75-84. New York, Plenum Press, 1969.

The effect of PGE_1 on lipolysis was studied on adipose tissue in a basal state and under stimulation with noradrenaline. Rats were tested in both fed and fasting states; rabbits had been fed; dogs were fasting. Tissue was treated also with nicotinic acid, anti-insulin serum or glucose in the rat series. Stimulation was seen in fasting rat tissue treated with anti-insulin serum or nicotinic acid in the basal state and in some fasting dogs in the basal state. Other dogs showed inhibition in stimulated or basal states. Rabbits showed inhibition in the noradrenaline stimulated state. Fed rats, untreated, were inhibited in both basal and noradrenaline stimulated states; fasting rats, untreated, were inhibited in the noradrenaline state. Fasting rats, treated with glucose, were inhibited in the basal state. It is concluded that altering the in vivo conditions or the use of different species can cause various and opposite effects in vitro. (RMS) 1491

1001

MICHELI, H.

Lipolysis in rabbit, dog and human adipose tissue.

Dissertation, Department of Geriatrics, Uppsala, 4-20. 1969.

This dissertation is a summary of six papers published by the author and collaborators. The presentation is a study of lipolysis in rabbit, dog and human adipose tissue with in vivo and in vitro results concerning PGE_1. (RAP) 1767

1002

MICHIBAYASHI, T.

Studies on the inhibitory action of prostaglandin E_1 on the isolated guinea-pig ureter: I. Effect of prostaglandin E_1 on the excitability of smooth muscle cell membrane.

Sapporo Medical Journal. 35: 361-371. 1969.

The mechanism of the inhibitory action of prostaglandin E_1 on the electrical and mechanical activities of the isolated guinea pig ureter was studied. PGE_1 decreased or abolished contractions of the ureter, and caused an increase of its membrane potential. Increases in potassium ion concentration increased the threshold concentration of PGE_1 required to abolish the contraction. Increases in PGE_1 also raised the threshold voltage required to evoke the action potential of this preparation. The presence of tetrodotoxin, atropine, phentolamine and propranolol did not affect the inhibitory action of PGE_1. The application of PGE_1 decreased the number of spike oscillations, increased amplitude, shortened duration and prolonged rising phase of the action potential of the preparation evoked by a single electrical stimulation. From the results obtained, it may be concluded that PGE_1 decreases the excitability of the smooth muscle membrane of the isolated guinea pig ureter and shows an inhibitory action by increasing the membrane potential and the threshold voltage, by prolonging the relative refractory period, and presumably by decreasing the cell membrane permeability to calcium ions. (GW) 1718

1003

MICHIBAYASHI, T.

Studies on the inhibitory action of prostaglandin E_1 on the isolated guinea-pig ureter. II. Inhibitory action of prostaglandin E_1 on the isolated guinea-pig ureter and calcium ions.

Sapporo Medical Journal. 35: 440-448. 1969.

The guinea pig ureter was used to investigate the relationship between the inhibitory action of prostaglandin E_1 and calcium ions. Calcium-induced contractions of isolated guinea pig ureter depolarized by calcium-free Krebs solution were prevented or abolished by PGE_1. The height of contraction evoked by a single electrical stimulation is dependent upon the extracellular calcium concentration but this height of contraction is inhibited by PGE_1 in concentration of 2×10^{-8} g/ml. This same concentration of PGE_1 also increased the threshold voltage for evoking contraction. The effect of PGE_1 which prolongs the refractory period of the preparation was enhanced with the decrease in extracelluler calcium concentration. These effects of PGE_1 were thus surmised to be enhanced with the decrease in extracellular calcium concentration. (GW) 1676

1004

MIELE, E.

Lack of effect of prostaglandins E_1 and $F_{1\alpha}$ on adreno-medullary catecholamine secretion evoked by various agents.

In: Mantegazza, P. and E.W. Horton, eds., "Prostaglandins, Peptides and Amines," p. 85-93. London, Academic Press, 1969.

A study was made of the effects of PGE_1 and $PGF_{1\alpha}$ on the catecholamine secretion from the perfused cat adrenal gland evoked by calcium reintroduction (into the perfusion medium) barium, acetylcholine (ACh), stimulation of the splanchnic nerve, and an excess of potassium. Cat adrenal glands were perfused in situ. The total catecholamine (adrenaline plus noradrenaline base) output per minute of perfusion was determined by analysis of the pertusate. The addition of PGE_1 or $PGF_{1\alpha}$ to the perfusion fluid did not modify either the direct secretory response to calcium and barium, or the calcium-dependent secretory response to splanchnic nerve stimulation, ACh, and excess potassium. Thus, it appears that PGE_1 and $PGE_{2\alpha}$ are pharmacologically inactive on the calcium-dependent secretory process of the perfused cat adrenal medulla. (ART) 1696

1005

MILHAUD, G., G. CORTRIS, and M. TUBIANA

Le cancer medullaire de la thyroide et le syndrome d'hyperthyrocalcitonine. [Medullary carcinoma of the thyroid and hypersecretion of thyrocalcitonin.]

Presse Medicale. 77: 2129-2123. 1969.

The authors mention without detail the work of (Williams, 1966) and (Williams, Karim et al) who established the biochemical aspects of medullary carcinoma of the thyroid with amyloid stroma as being the production and secretion of important quantities of thyrocalcitonin (TCT) in the blood, although the production of 5-hydroxytryptamine or prostaglandin is inconsistent. (GW) 1665

1006

MILLER, J.W.nd J.E. LEWIS

Drugs affecting smooth muscle.

Annual Review of Pharmacology. 9: 147-172. 1969.

This general review cites the work of five investigatiors and their collaborators who were concerned with prostaglandins. The mechnaism of the contractile effect of PGE_1 on the isolated rat uterus was studed by D.M. Paton and E.E. Daniel, Can. J. Physiol. Pharmacol. 45: 795-804. (1967); the inhibitory effect of PGE_1 on the isolated human uterus was studied by R. Eliasson, Biochem. Pharmacol. 15: 755. (1966); and the relative inhibitory potencies of a series

260

of prostaglandins present in human semen were estimated on the isolated human uterus by M. Bydgeman and M. Hamberg, Acta Physiol. Scan. 69: 320-325. (1967). The effects of $PGF_{2\alpha}$ and PGE_2, which have been found in human lung, were investigated on isolated human bronchus by W.J.F. Sweatman and H.O.J., Collier Nature 217: 69 (1968), while other studies of the actions of the prostaglandins on the lung and tracheal muscle have been thoroughly reviewed by S. Bergstrom, L.A. Carlson and J.R. Weeks, Pharmacol. Rev. 20: 1-48. (1968). (ART) 1702

1007

MILLER, R.P., B.D. POLIS, A.M. PAKOSKEY, E. POLIS, H.P. SCHWARZ, and L. DREISBACH
Prostaglandin E_1 and LSD induced changes of the plasma phospholipids which vary with physical and psychic stress.
Physiologist. 12: 303. 1969.

Plasma phospholipid changes found in stress reflect energetic control mechanisms mobilized by brain excitation which are mediated by prostaglandin and pituitary hormones. Injection of prostaglandin E_1 and LSD into normal rats showed an elevation in both plasma phosphatidyl glycerol and phosphatidyl ethanolamine and a decrease in lecithin. The brains of these rats showed elevations in phosphatidyl glycerol. Hypophysectomized rats failed to significantly alter the plasma phospholipid pattern; their brains showed a rise in phosphatidyl glycerol of lesser magnitude. (GW) 1605

1008

MIRONNEAU, J., V. SMEJKAL and C. OJEDA
Influence des substances adrenergiques et de la prostaglandine E_1 sur l'activite electrique spontanee de l'oreillette de lapin. [Influence of adrenergic substances and prostaglandin E_1 on the spontaneous electric activity of the rabbit auricle.]
Comptes Rendus des Seances de la Societe de Biologie et de ses Filiales. 163: 1414-1419. 1969.

The influence of several adrenergic agents and PGE_1 on the amplitude and duration of action potentials and heart rate was studied in isolated rabbit auricles. PGE_1 (10^{-8} g/ml) caused increase in amplitude and duration of the action potential and an increase in heart rate. Both noradrenaline and PGE_1 could cause heart arythmias. The changes caused by PGE_1 are similar to those caused by adrenaline and isoprenaline in these experiments. PG's seem to be less important than the adrenergic agents in controlling heart function. (JRH) 1765

1009

MISIEWICZ, J.J., S.L. WALLER, N. KILEY, and E.W. HORTON
Effect of oral prostaglandin E_1 on intestinal transit in man.
Lancet. 1: 648-651. 1969.

The effects of ingestion of 2 mg. prostaglandin E_1 on intestinal transit and motor activity was studied in four healthy volunteers. Prostaglandin E_1 notably increased transit-rate through the small intestine and the colon, with the production of abdominal colic and the passage of fluid and faeces per rectum. The significance of these findings in relation to the diarrhea associated with certain tumours is discussed. (Authors) 1445

1010

MIYANO, M.
Synthetic studies on prostaglandins II. A novel synthesis of methyl esters of 15-dehydro-PGB$_1$ and PGE 237.
Tetrahedron Letters. 2771-2774. 1969.

3-*exo*-n-hexanoyl-2-*endo*-carboxylic acid was converted to methyl esters of 15-dehydro-PGB$_1$ and PGE 237. Each step of the synthesis is explained complete with diagrams. Some chemical and physical properties of the products of each reaction are given. (RAP) 1781

1011

MIYANO, M. and C.R. DORN
Total synthesis of 15-dehydro-prostaglandin E$_1$.
Tetrahedron Letters. 1615-1618. 1969.

The complete synthesis of 15-dehydro-PGE$_1$ from 3-keto-undecan-1,11-dioic acid is discussed. Diagrams of each step are included and some chemical and physical properties are given at each step. The authors point out that their synthesis involves only a few steps and that this approach can readily be adpated for large scale preparation of 15-dehydro-PGE$_1$. (RAP) 1783

1012

MOGHISSI, K.S.
Sperm migration in the human female genital tract.
Journal of Reproductive Medicine. 3: 156-168. 1969.

This review has one section on PG's in conjunction with the migration of sperm in the female genital tract. Several papers are cited concerning the PGEs and PGFs found in the seminal plasma and the uterus and the response of the female reproductive tract to these prostaglandins. (RAP) 1754

1013

MOSKOWITZ, J. and J.N. FAIN
Hormonal regulation of lipolysis and phosphorylase activity in human fat cells.
Journal of Clinical Investigation. 48: 1802-1808. 1969.

PGE$_1$ at a concentration of 1 ng/ml antagonized theophylline, and norepinephrine-induced release of glycerol and free fatty acids (FFA) in human fat cell preparations. Insulin at higher doses also inhibited theophylline-stimulated lipolysis. The N$_6$-2-0'dibutyryl derivative of cyclic adenosine monophosphate (DCAMP) stimulated lipolysis. PGE$_1$ did not significantly inhibit the lipid mobilizing effects of DCAMP. Changes in glycogen phosphorylase activity after treatment with theophylline, norepinephrine, DCAMP, and PGE$_1$ paralleled those of lipolysis. These results suggest that in man as in experimental animals lipolysis and phosphorylase activity are regulated through processes involving cyclic AMP and that PGE$_1$ appears to exert its antilipolytic effect in human fat cells, as in rat fat cells, by interfering at the level of adenyl cyclase with the accumulation of cyclic AMP. (Authors) 1406

1014

MUEHRCKE, R.C., A.K. MANDAL, M. EPSTEIN, and F.I. VOLINI
Cytoplasmic granularity of the renal medullary interstitial cells in experimental hypertension.
Journal of Laboratory and Clinical Medicine. 73: 299-308. 1969.

The authors report the investigation of previous workers who extracted an acidic lipid from the porcine renal medulla. This lipid caused acute vasodepression in the anesthetized, vagotomized and pentolinium-treated rat. It was subsequently termed "medullin" and identified as prostaglandin E_2-217. It is now termed PGA_2. (GW) 1641

1015
MULLER, M.
Etude clinique at anatomo-pathologique de 31 carcinomes medullaires a stroma amyloide de la thyroide. [Clinical and anatomo-pathological study of 31 cases of medullary carcinoma with amyloid stroma of the thyroid.]
Schweizerische Medizinische Wochenschrift. 99: 433-439. 1969.

Only brief mention is made of prostaglandins as the author describes the origin of diarrhea in medullary carcinoma. He mentions the work of Williams (1968) who recently gave evidence of the fact that prostaglandins had action on smooth muscle in tumoral tissue and blood of two patients with medullary carcinoma. (GW) 1723

1016
MURAD, F., H.B. BREWER, and M. VAUGHAN
Effect of thyrocalcitonin (TCT) on cyclic AMP formation by rat kidney.
Clinical Research. 17: 591. 1969.

Abstract only. Subcellular fractions of rat kidney were used to synthesize cAMP. The effects of various hormones and compounds on this synthesis were measured. It was found that PGE_1 (5 µg/ml) had a small stimulatory effect on the accumulation of cAMP. (JRH) 1705

1017
MUSTARD, J.F., F.A. McELROY, R.L. KINLOUGH, M. GUCCIONE and M.A. PACKHAM
Changes in platelet metabolism during ADP-induced aggregation.
Clinical Research. 17: 652. 1969.

Abstract only. The authors briefly mention that PGE_1 does not block the initial platelet shape change but inhibits the burst in CO_2 production which is essential for ADP-induced aggregation. (RAP) 1760

1018
MUSTARD, J.F. and M.A. PACKHAM
Platelet function and myocardial infarction.
Circulation. 40 (supp. 4): IV 20-30. 1969.

The authors review the subject citing 80 papers. Prostaglandins are mentioned very briefly in that PGE_1 and adenosine produce impairment of hemostatic plug formation and alteration of thrombus formation. (RMS) 1652

1019
MUSTARD, J.F.
Thromboembolism—a manifestation of the response of blood to injury.
Circulation. 40 (supp. 3): III 31-32. 1969.

Abstract only. The author briefly mentions that PGE_1 inhibits the release of certain compounds from blood platelets during the formation of a thrombus. (RMS) 1653

1020

NAIMZADA, M.K.

Effects of some naturally-occurring prostaglandins (PGE_1, PGE_2, PGA_1, and $PGF_{1\alpha}$) on the hypogastric nerve vas deferens and seminal vesicle preparations of the guinea pig.

Chimica Therapeutica. 4: 34. 1969.

The activity of some naturally-occurring prostaglandins (PGE_2, PGA_1 and $PGF_{1\alpha}$) has been studied in vitro on the hypogastric nerve vas deferens and seminal vesicle preparations of the guinea pig and compared with that of prostaglandin E_1 (PGE_1). It has been shown that PGE_2, PGA_1 and $PGF_{1\alpha}$ like prostaglandin E_1 (PGE_1), are able to increase the responses of the sympathetic stimulation and of sympathetic amines in both preparations. PGE_2 has however only 1/10 PGA_1 1/50-1/100 and $PGF_{1\alpha}$ 1/100 of the potency of PGE_1. No correlation has been observed between the ability to stimulate the smooth muscle and that of potentiating the responses of guinea pig vas deferens and seminal vesicle to adrenergic stimulation. (Author) 1643

1021

NAIMZADA, M.K.

Effects of some naturally occurring prostaglandins on the isolated hypogastric nerve seminal vesicle preparation of the guinea pig.

Life Sciences. 8: 49-55. 1969.

The activity of some naturally occurring prostaglandins (PGE_2, PGA_1, and $PGF_{2\alpha}$) has been studied in vitro on the hypogastric nerve-seminal vesicle preparation of the guinea pig and compared with that of prostaglandin E_1. It has been shown that PGE_2, PGA_1 and $PGF_{1\alpha}$ are capable of increasing the responses of sympathetic stimulation and of epinephrine and norepinephrine. PGE_2 has however only 1/10, PGA_1 1/50 and $PGF_{1\alpha}$ 1/100 the potency of PGE_1. No correlation has been observed between the ability to stimulate the smooth muscle and that of potentiating the responses of guinea pig seminal vesicle to adrenergic stimulation. (Author) 1474

1022

NAKANO, J.

Cardiovascular effect of a prostaglandin isolated from a gorgonian *Plexaura homomalla*.

Journal of Pharmacy and Pharmacology. 21: 782-783. 1969.

Study comparing the cardiovascular effects of a prostaglandin A_2-like compound from a sea animal with those of PGE_2 and PGA_2 anesthetized dogs is presented. Intravenous administration of PGE_2 and PGA_2 increased heart rate and myocardial contractile force as mean systemic arterial pressure decreased. Chemical analysis of the PGA_2-like compound isolated from the sea animal showed it to be a 15-epimer of PGA_2 (15-S-PGA_2), 15-R-PGA_2. This compound caused no essential change in the three hemodynamic parameters. (GW) 1482

1023

NAKANO, J. and B. COLE

Effects of prostaglandins E_1 and $F_{2\alpha}$ on systemic, pulmonary, and splanchnic circulations in dogs.

American Journal of Physiology. 217: 222-227. 1969.

The effect of PGE_1 and $PGF_{2\alpha}$ on the systemic, pulmonary and splanchnic circulations were studied in anesthetized dogs. It was confirmed that PGE_1 is a potent hypotensive agent, whereas $PGF_{2\alpha}$ is a hypertensive agent. Both prostaglandins produced multiple hemodynamic changes due to their direct actions on the peripheral vascular beds and on myocardial contractility. PGE_1 caused a biphasic response in systemic venous return and portal venous pressure, an initial increase being followed by a decrease. On the other hand, $PGF_{2\alpha}$ decreased significantly systemic venous return and increased portal venous pressure. The hemodynamic effects of the injection of PGE_1 and $PGF_{2\alpha}$ into the portal vein were qualitatively and quantitatively different from those of the same doses of PGE_1 and $PGF_{2\alpha}$ into a femoral vein or the left atrium. It is concluded that both PGE_1 and $PGF_{2\alpha}$ directly influence the splanchnic circulation and also appear to be metabolized significantly in the liver and slightly in the lungs. (Authors) 1454

1024

NAKANO, J., E. ANGGARD, and B. SAMUELSSON
15-hydroxy-prostanoate dehydrogenase. Prostaglandins as substrates and inhibitors.
European Journal of Biochemistry. 11: 386-389. 1969.

This study was undertaken to determine the Michaelis constants and maximum reaction rates for a number of the prostaglandin substrates of prostaglandin dehydrogenase (15 hydroxyprostanoate dehydrogenase). Also the inhibitory effect of some prostaglandins were studied on the enzymatic oxidation of PGE_1. Sources of chemicals and methods of preparing enzymes and substrates are described, and the methods of the investigation are outlined. Results show that the enzyme is stereospecific with regard to the configuration at C-15. The nature of the substitutents of the cyclopentane ring did not markedly affect the properties as a substrate, whereas the nature of the carboxyl side chain was important. Prostaglandin B compounds and a synthetic epimer of prostaglandin E_1, 15-R-prostaglandin E_1, produced non-competitive inhibition. (PR) 1569

1025

NAKANO, J., E. ANGGARD and B. SAMUELSSON
Substrate specificity and inhibition of 15-hdyroxy-prostaglandin dehydrogenase (PGDH).
Pharmacologist. 11: 238. 1969.

We examined the substrate specificity for purified PGDH and its inhibition by PG metabolites isolated from liver and by synthetic PG analogues. PGDH activity was assayed fluorometrically by measuring NADH formed. Km and Vmax for substrates and Ki for inhibitors were determined from Lineweaver-Burk plots. The comparison of Km values of PG shows that PGE_1 (15-S-PGE_1) is the best substrate whereas 15-R-PGE_1 is not a substrate for PGDH. This indicates that PGDH is stereospecific at the c-15 configuration of PG. Certain modifications on the cyclopentane ring of PGE_1 (i.e.), PGA and PGF compounds do not markedly reduce the substrate specificity. On the other hand, if one alters the planar configuration of the carboxyl side chain relative to the ring (PGB compounds and 8-iso-PGE_1) or shortens the side chain (dinor-$PGF_{1\alpha}$, and tetranor-PGE_1, and -$PGF_{1\alpha}$) the resulting PG compounds are poor substrates. Using PGE_1 as a substrate, PGB compounds inhibit competitively PGDH, while 15-R-PGE_1 produces non-competitive inhibition. (Authors modified) 1792

1026

NEVILLE, E. and E.S. HOLDSWORTH
A "second messenger" for vitamin D.
FEBS letters. 2: 313-316. 1969.

White crossbred cockerels were reared on a diet deficient in vitamin D from hatching to age 4 months. They were then given vitamin D 400 I.U. intramuscularly 16 hours or 5,000 I.U. in 0.1 ml propylene glycol by mouth 3 hours before experiments. Adenyl cyclase activity was measured in mucosal cells scraped from the lower half of the small intestine. The method is described. Calcium absorption in vivo was measured from closed sacs of the ileum by injecting labeled $CaCl_2$ and killing the bird after 20 minutes. To test the effect of cAMP on absorption of calcium, the N^6,2-0-dibutyryl-3'5'cyclic monophosphate derivative was used intraperitoneally before placing the labeled calcium in the intestine. Calcium translocation was measured by everted sacs of ileum and determining the amount of labeled calcium which accumulated in the serosal fluid. The "membrane fraction" of vitamin D deficient chick intestine catalysed the formation of small amounts of cyclic AMP from ATP in the presence of caffeine. The amount of cAMP was increased by pretreatment with vitamin D_3. Results suggest that the amount of or the activity of adenyl cyclase is increased by vitamin D_3 treatment. The addition of prostaglandin E_1 to the incubation mixture increased the yield of cyclic AMP from the vitamin D_3 deficient preparation and from the preparation from the birds given vitamin D three hours previously but little when 400 I.U. of vitamin D_3 was given 16 hours previously. Further experiments investigated the effect of dibutyryl cAMP on calcium absorption in vivo and its effect on calcium translocation in everted sacs. (N.K.) 1574

1027

NISSEN, H.M. and I. BOJESSEN

On lipid droplets in renal interstitial cells. IV. Isolation and identification.
Zeitschrift fur Zellforschung und Mikroskopische Anatomie. 97: 274-284. 1969.

It is suggested that the lipid fat droplets in the cytoplasm of renal interstitial cells contain compounds that play a physiological role in the functioning kidney. Lipid material extracted from rat renal papillae consisted mainly of triglycerides, cholesterol esters and free fatty acids. The prostaglandin precursor arachidonic acid constitutes a significant fraction of the triglyceride long chain fatty acid. Only small quantities of prostaglandin were present. 16 μg PGE_2 and 14 μg PGE_2-217 have been determined from 5 rats while PGE_1 and PGE_2 were found in 20 experiments. PGF_1 was found only 3 times. (GW) 1732

1028

NISSEN, H.M. and H. ANDERSEN

On the activity of a prostaglandin-dehydrogenase system in the kidney. A histochemical study during hydration/dehydration and salt-repletion/salt-depletion.
Histochemie. 17: 241-247. 1969.

Following a brief period of acute hydration in dehydrated rats, a decrease in a prostaglandin-dehydrogenase activity was observed in the loop of Henle and the collecting tubule. Salt-depletion and salt-repletion were of no influence on the enzyme activity. (Authors) 1462

1029

NORMANN, T. and B. OTNES

Intestinal ganglioneuromatosis, diarrhoea and medullary thyroid carcinoma.
Scandinavian Journal of Gastroenterology. 4: 553-559. 1969.

The authors briefly mention the work of Williams et al. (1968) in which high levels of prostaglandins were reported in medullary carcinomas of the thyroid. (JRH) 1707

1030
NOVER, L.

Chemische und biologische aspekte der wirkungsweise mischfunktioneller oxygenasen. [Chemical and biological aspects of the sphere of action of the fixed functioning oxygenase.]
Die Pharmazie. 24: 361-378. 1969.

The author mentions in this article in German the work of Bergstrom 1967 who incorporated the dioxygenase reaction in the biosynthesis of the prostaglandins. (GW) 1725

1031
NUTTER, L.J., O.S. PRIVETT and W.O. LUNDBERG

Isolation of 22:5ω6 and 22:5ω3 acids and their conversion to prostaglandins via sheep vesicular glands.
Abstracts, 60th Meeting of the American Oil Chemists' Society. San Francisco, 20-24 April, abs. 146. 1969.

Docosapentaenoic acids were isolated in highly purified form from tuna oil and testicular lipids of rats fed a corn oil diet via a combination of liquid-liquid partition and argentation chromatography. Reductive ozonolysis of the fish oil 22:5 showed that the double bonds resided in the 7,10,13,16,19 positions; the acid isolated from rat testicular lipid had the double bonds in the 4,7,10,13,16 positions. In contrast to arachidonic acid and 5,8,11,14,17-20:5 which gave single homogenous prostaglandins PGE_2 and PGE_3, respectively, when incubated with sheep vesicular glands, both 22:5 acids gave a mixture of prostaglandins. These prostaglandins were isolated by argentation chromatography. Studies on their formation, isolation and structure are described. (Authors) 1804

1032
NUTTING, E.F. and P.S. CAMMARATA

Effects of prostaglandins on fertility in female rats.
Nature. 222: 287-288. 1969.

Large quantities of readily available prostaglandins enabled the authors to investigate the effects of prostaglandins on fertility. Female rats were caged with males. Biosynthetic prostaglandins E_2 and $F_{2\alpha}$ solutions were injected subcutaneously either once, twice or three times daily during the first 6 or 7 days of pregnancy. Injection of 0.5 mg twice daily of PGE_2 or $PGF_{2\alpha}$ during the first 7 days of pregnancy resulted in a significantly smaller number of implantation sites. Disturbances in the central nervous system within 5 minutes of an injection of 0.5 mg PGE_2 may cause a delay of nidation. 3 of 5 unimplated rats given 0.5 mg of PGE_2 twice daily from day 1-6 of pregnancy yielded only one normal blastocyte. Several modes of action must thus be postulated to explain the antifertility effect of prostaglandins. (GW) 1422

1033
O'CONNELL, J.M.B. and R.J. SHALHOUB

Renal vasodilator drugs and P-aminohippuric acid transport: in vitro studies.
Clinical Research. 17: 622. 1969.

Prostaglandin E_1 is shown to only slightly inhibit the transport of PAH which seems to cast serious doubt on the concept that this drug lowers the renal extraction of PAH by redistribution of renal blood flow. (GW) 1615

1034

ORANGE, R.P., D.J. STECHSCHULTE, and K.F. AUSTEN
Cellular mechanisms involved in the release of slow reacting substance of anaphylaxis.
Federation Proceedings. 28: 1710-1715. 1969.

In a study of the mechanisms of release, inhibition, and isolation by thin layer chromatography of SRS-A, the pharmacological characteristics of SRS-A from guinea pigs, rats, and monkeys are compared to those of bradykinin, PGE_1, and $PGF_{2\alpha}$ as to effects on guinea pig ileum, estrous rat uterus, and ascending gerbil colon. The three preparations of SRS-A were also compared to histamine, bradykinin and PGE_1 and $PGF_{2\alpha}$ in producing vascular permeability changes in rat and guinea pig skin pretreated with high doses of histamine and serotonin. Smooth muscle bioassay comparisons may be summarized by noting that the prostaglandins and bradykinins were more active on gerbil colon that SRS-A; bradykinins were more active on rat uterus than SRS-A or prostaglandins, and SRS-A was more active on guinea pig ileum than bradykinin or the prostaglandins. Permeability changes induced in the skin of guinea pigs were significant with the SRS-A preparations, but not with histamine, bradykinin, or the prostaglandins. (RMS) 1541

1035

ORANGE, R.P., T. ISHIZAKA, K. ISHIZAKA, M.L. KARNOVSKY, and K.F. AUSTEN
Pharmacologic characterization of slow reacting substance of anaphylaxis (SRS-A).
Federation Proceedings. 28: 678. 1969.

Abstract only. In the course of characterizing SRS-A from several sources, the biological activity on smooth muscle preparations and the skin permeability enhancement effect of this substance was compared to bradykinin, histamine, serotonin, and PGE_1 and $PGF_{2\alpha}$. (RMS) 1545

1036

ORR, D.E. and F.B. JOHNSON
Synthesis of 3-(1,1-ethylenedioxyoctyl)-cyclopentanone.
Canadian Journal of Chemistry. 47: 47-50. 1969.

A synthesis of the naturally occurring prostaglandins is described using a readily available six-membered carbocyclic ring compound as the starting material. In the first step, a seven carbon atom side chain is incorporated by direct alkylation or by rearranging the appropriately substituted cyclo-hexane ring. At a later stage in the reaction sequence, the length of the seven carbon atom chain is increased by one carbon atom by the contraction of a six-membered ring. The ring contraction not only gives a side chain of the correct length but also generates a ketone function in the eight carbon side chain, in a position suitable for later introduction of the allyl alcohol function. (GW) 1633

1037

PALMER, E.C., F. SULSER and G.A. ROBISON
The effects of neurohumoral agents on the level of cyclic AMP in different brain areas in vitro.
Pharmacologist. 11: 258. 1969.

Abstract only. The use of chopped tissue from various areas of the rat brain allows consistent cyclic AMP responses to be observed. Incubation with norepinephrine (NE) (10^{-5} M) for 6 min causes a marked increase in cAMP in all areas tested. Histamine (10^{-5} M) acts synergistically with NE. Serotonin (5HT), dopamine, amphetamine (all 10^{-5} M) and prostaglandin E_1 (1 µg/ml) do not change cAMP. Moreover, preincubation of chopped tissue with 5HT for 3 min does not appreciably modify the response to NE. (Authors modified) 1793

1038
PARKINSON, T.M. and J.C. SCHNEIDER
Absorption and metabolism of prostaglandin E_1 by perfused rat jejunum in vitro.
Biochemica et Biophysica Acta. 176: 78-85. 1969.

Segments of rat jejunum were perfused in vitro with Krebs-Ringer biocarbonate buffer containing [I-^{14}C] prostaglandin E_1 or [$5,6$-$^3H_2$] prostaglandin E_1. When segments were perfused with [I-^{14}C] prostaglandin E_1, approximately 15% of the radioactivity disappearing from the mucosal fluid in 60 min was translocated to the serosal fluid, 345 was taken up by the perfused tissue and 49% was lost from the system, presumably as $^{14}CO_2$. ^{14}C activity was distributed throughout tissue lipids: 17% was in the total nonsaponifiable fraction, 6% in the digitonin precipitable sterols, 475 in ethyl ether extracts of the saponifiable fraction, and 36% in the ether extracted aqueous phase. When segments were perfused with [$5,6$-$^3H_2$] prostaglandin E_1, only about 0.1% of the total radioactivity translocated to the serosal fluid was identified as intact prostaglandin E_1 by reverse isotope dilution. [$5,6$-$^3H_2$] prostaglandin E_1 was not incorporated into intestinal tissue glycerol or cholesterol esters under conditions where [$9,10$-$^3H_2$] palmitic acid was esterified. These data support the hypothesis that the intestine may be a primary site of oxidative metabolism of orally administered PGE_1. (Authors) 1426

1039
PENG, T.-C., K.M. SIX, and P.L. MUNSON
Effect of prostaglandin E_1 on the hypothalamo-hypophyseal-adrenal cortical axis of rats.
Federation Proceedings. 28: 437. 1969.

Abstract only. The possible identity of hypothalamic corticotropin releasing factor (CRF) was investigated. PGE_1, found in the hypothalamus, stimulated ACTH release in intact anaesthetized rats as shown by increased adrenal and plasma corticosterone and decreased adrenal ascorbic acid (AAA) and cholesterol. There is a linear log dose response relationship between 0.5-2 μg PGE_1 (iv) and 30-100% maximum depletion of AAA. Since this effect was blocked or inhibited by morphine, PGE_1 like other known constituents of the hypothalamus, is probably not CRF. PGE_1 (5 μg) had no effect on AAA in hypophysectomized rats indicating that it does not have an ACTH-like effect on the adrenal cortex. It probably causes ACTH release indirectly by stimulating release of CRF from the hypothalamus. Other prostaglandins, PGA_1 and $PGF_{2\alpha}$, at 5 μg had no effect on AAA in intact rats indicating some degree of specificity for the effect of PGE_1. (PR) 1547

1040
PERRIER, C.V. and L. LASTER
Adenyl cyclase activity of guinea-pig gastric mucosa.
Clinical Research. 17: 596. 1969.

Abstract only. The effects of histamine, a histamine analogue and an antihistaminic on a preparation unresponsive to choline esters and gastrin might support the theory that histamine is, or is related to, the final stimulus for gastric acid secretion. However, the stimulation by prostaglandins, which inhibit gastric secretion in vivo, necessitates further study before physiological conclusions can be drawn. (ART) 1704

1041
PERSSON, B.
Lipid metabolism.
Acta Obstetricia et Gynecologica Scandinavica. 48 (supp. 3): 92-96. 1969.

The author briefly mentions that prostaglandins are antilipolytic substances that lower the intracellular cyclic AMP-level in the new born lamb. (GW) 1645

1042
PHARRISS, B.B. and L.J. WYNGARDEN
The effect of prostaglandin $F_{2\alpha}$ on the progestogen content of ovaries from pseudopregnant rats.
Proceedings of the Society for Experimental Biology and Medicine. 130: 92-94. 1969.

Prostaglandin $F_{2\alpha}$ was infused into pseudopregnant rats for 2 days (days 5 and 6) at 1 mg/kg/day. The infusion tubing was placed either in the uterine lumen or the right heart. The progestogen content of the ovaries of these animals was compared to that of animals receiving only saline; progesterone levels were decreased and 20_{α}-dihydroprogesterone concentrations were increased. Vaginal smear records of pseudopregnant rats receiving $PGF_{2\alpha}$ subcutaneously showed a shortening of pseudopregnancy to 7 days from a normal of 14 days. This information supports an indirect mechanism for local luteal control by uterine tissue. (Authors) 1448

1043
PICARELLI, Z.P. and M. DOLNIKOFF
Kininolytic activity of human amniotic fluid.
Pharmacological Research Communications. 1: 183-184. 1969.

The authors briefly mention that some researchers have theorized that the oxytocic principles in human amniotic fluid may be prostaglandins. (JRH) 1623

1044
PICCININI, F., P. POMARELLI, and A. CHIARRA
Further investigations on the mechanism of the inotropic action of prostaglandin E_1 in relation to the ion balance in frog heart.
Pharmacological Research Communications. 1: 381-389. 1969.

It has been shown the PGE_1 can antagonize the effect of potassium ions on the frog heart only when an adequate amount of calcium ions are present in the perfusing fluid, and that no direct antagonistic action on potassium can be demonstrated. In order to clarify which mechanism is primarily involved in the cardiac action of PGE_1, the relationship between PGE_1 and intracellular exchangeable calcium has been investigated. PGE_1 brings about a significant increase in the rate of uptake of Ca-45, whereas the total myocardial calcium content and the net amount of cellular exchangeable calcium is not affected. These results could be explained by an increase of the membrane permeability to Ca^{++}. (Authors) 1612

1045
PICKLES, V.R.
Prostaglandins.
Nature. 224: 221-225. 1969.

Prostaglandins are widely distributed in mammalian tissues and have a high order of biological activity, but their functions are not fully known. This article is intended to give the non-biologist some insight into the events leading to the present minor publication explosion and the current state of knowledge of these substances. (Author) 1404

1046

PIPER, P.J. and J.R. VANE

Release of additional factors in anaphylaxis and its antagonism by anti-inflammatory drugs.
Nature. 223: 29-35. 1969.

The authors describe the use of new methods to detect the release of active substances during anaphylaxis in guinea pig isolated lung. Prostaglandin E_2 and $F_{2\alpha}$ as well as histamine and SRS-A were found to be released into the perfusate. The presence of prostaglandins was indicated by the contraction of the rat stomach strip, rat colon and the chick rectum by the effluent from the shocked lungs. Isolated lungs from guinea pigs also inactivated prostaglandins. PGE_2 and $PGF_{2\alpha}$ infused intra-arterially into sensitized or unsensitized lungs was removed by 95-98%. PGE_1 and $PGF_{2\alpha}$ contracted the isolated tracheal muscle whose tone had not been artificially raised. (GW) 1411

1047

PIPER, P.J. and J.R. VANE

The release of prostaglandins during anaphylaxis in guinea pig isolated lungs.

In: Mantegazza, P. and E.W. horton, eds., "Prostaglandins, Peptides and Amines," p. 15-19. London, Academic Press, 1969.

When isolated lungs from ovalbumen sensitized guinea pigs are challenged, histamine, slow reacting substance in anaphylaxis, and kallikrein have been found to be released into the perfusate. This paper reports that prostaglandins are also released. Contraction of all the assay tissues (guinea pig ileum, rat stomach strip, rat colon, chick rectum) in the presence of antagonists (mepyramine, hyoscine, phenoxybenzamine combination, propranolol, and methysergide) which eliminated responses to histamine, acetylcholine, catecholamines, and serotonin, strongly suggested the release of prostaglandins. Extracts of perfusate were assayed for prostaglandin content. Chromatograms showed that $PGF_{2\alpha}$ and one or more prostaglandins of the E type were released during anaphylaxis. The latter were found to be mainly PGE_2. Although neither histamine nor bradykinin infusions into unsensitized isolated lungs caused a release of prostaglandin, such release did occur upon infusion with slow-reacting substance in anaphylaxis (SRS-A). Thus it may be that the release of prostaglandins into the pulmonary circulation when sensitized lungs are challenged is mediated, in part or in whole, by the initial release of SRS-A. (ART) 1686

1048

POLIS, B.D., A.M. PAKOSKEY, and H.W. SHMUKLER

Regeneration of oxidative phosphorylation in aged mitochondria by prostaglandin B_1.
Proceedings of the National Academy of Sciences. 63: 229. 1969.

The possible role of prostaglandins and phosphatidyl glycerol in the phosphorylation mechanism was studied. Dephosphorylation occurs when aged mitochondria further "uncoupled" with Triton X-100 and reacted with adenosine diphosphate and ^{32}P under conditions for oxidative phosphorylation. Addition of prostaglandin E_1 and phosphatidyl glycerol to the reaction reversed the dephosphorylation. In the presence of Triton X- 100, both prostaglandins E_1 and B_1 were equally effective in reactivating phosphorylation. In the absence of Triton X-100, only PGB_1 was effective. The action of prostaglandin B_1 as a possible intermediate in mitochondrial phosphorylation is implied. (GW) 1476

1049
POISNER, A.M.
Inhibition of ATP-induced effects on chromaffin granules.
Federation Proceedings. 28: 287. 1969.

Abstract only. In the present study, the effect of ATP on the light scattering of chromaffin granules was studied in the absence or presence of various enzyme inhibitors. The effect of ATP was not inhibited by prostaglandin (10 µg/ml). (GW) 1713

1050
PROTIVA, M.
Entwicklung von neuen strukturen im gebiet der arzenistoffe, die krieslauf und herzfunktion beeinflussen. [Development of new structures in drugs that influence blood circulation and heart function.]
Pure and Applied Chemistry. 19: 131-151. 1969.

Prostaglandins are briefly mentioned in this review of new drugs which affect heart and circulatory functions. Thirteen papers on prostaglandins are cited in the references. (JRH) 1631

1051
PURO, K.
Prostaglandiinit [Prostaglandins.]
Duodecim. 85: 673-675. 1969.

Article in Finnish. Abstract not available at present. 1679

1052
RAMWELL, P.W., J.E. SHAW, E.J. COREY and N. ANDERSEN
Biological activity of synthetic prostaglandins.
Nature: 221: 1251-1252. 1969.

This communication is a preliminary account of the results obtained from the effect of synthetic prostaglandin on intestinal and reproductive smooth muscle preparation and on the arterial blood pressure of the rat. Unexpected biological potency of certain unnatural synthetic steroisomers was found. The concentrations of prostaglandin E_1, A_1, and $F_{1\alpha}$ necessary to contract isolated smooth muscle preparation to 50% of maximum and to elicit a 10% change in blood pressure is indicated. The activity of the racemic synthetic prostaglandin E_1 and $F_{1\alpha}$ is approximately one-half that of the naturally occurring compounds but the synthetic racemates of prostaglandin A_1 in certain assays have been found to be more active than the normal isomer. 15-epi prostaglandin E_1 has been found to be more active than race-prostaglandin E_1. (GW) 1423

1053
RAMWELL, P.W. and E.G. DANIELS
Chromatography of the prostaglandins.
In; Marinetti, G.V., ed., "Lipid Chromatographic Analysis," p. 213-344. New York, Marcel Dekker, 1969.

This paper provides a guide to the many chromatographic techniques that may be employed for the separation of prostaglandins. Eight topics with eight tables and four figures are presented. Included in the discussion is the structure of the naturally occurring prostaglandins, their nature and distribution, extraction from tissues and biological fluids, bioassay procedures, paper, gas and thin layer chromatography and column chromatography to purify and separate the prostaglandins. The authors refer to 67 other papers. (GW) 1460

1054

RAMWELL, P.W., J.E. SHAW and S.J. JESSUP
Follicular fluid kinin and its action on fallopian tube.
Endocrinology. 84; 931-936. 1969.

Only brief mention is made of prostaglandin in this paper. The authors report that prostaglandin E_1 was found to antagonize the effect of both nervous and hormonal stimulation on the smooth musculature of the oviduct in vivo. Whether the prostaglandins modify the patency of the isthmus following coitus remains to be determined. (GW) 1429

1055

RAMWELL, P.W.
Release of prostaglandins.
Abstracts, 4th International Congress on Pharmacology. Basle, 14-18 July, p. 12. 1969.

Abstract only. This article reviews the biochemistry and cell physiology of prostaglandins. (JRH) 1727

1056

ROBISON, G.A., A. ARNOLD and R.C. HARTMANN
Divergent effects of epinephrine and prostaglandin E_1 on the level of cyclic AMP in human blood platelets.
Pharmacological Research Communications. 1: 325-332. 1969.

PGE_1 increased the level of cAMP in human blood platelets while epinephrine caused a decrease. The cAMP-lowering effect of epinephrine was prevented by the α-adrenergic blocking agent phentolamine but not by the β-adrenergic blocking agent propranolol. Platelets were incubated with 2 mm theophylline for 30 minutes, and for the final 20 minutes the authors added PGE_1 or PGE_2 at a concentration of 1, 10, 100, or 1000 ng/ml. PGE_1 increased the level of cAMP at concentrations of 10 ng/ml and above(i.e. the level of cAMP increased with the concentration of PGE_1). PGE_2 was effective only at a concentration 1,000 ng/ml. These results are of interest in that they provide further evidence that α-adrenergic responses in general may be mediated by decreased levels of cAMP and that the level of cAMP in blood platelets is an important determinant of their tendency to aggregate. (JRH) 1747

1057

RODESCH, F., P. NEVE, C. WILLEMS, and J.E. DUMONT
Stimulation of thyroid metabolism by thyrotropin, cyclic 3':5'-AMP, dibutyryl cyclic 3':5'-AMP and prostaglandin E_1.
European Journal of Biochemistry. 8: 26-32. 1969.

This article is mainly concerned with the stimulation of thyroid metabolism as shown by the binding of $[^{131}I]$ and the stimulation of glucose oxidation at C-1, by several compounds including PGE_1. It was found that PGE_1 did increase the metabolism of thyroid tissue in vitro. No effect of fluoride and PGE_1 on intracellular colloid droplet formation was observed. (JRH) 1568

1058
ROSELL, S.
Nervous and pharmacological regulation of vascular reactions in adipose tissue.
In: Holmes, W.L., L.A. Carlson and R. Paoletti, eds., "Drugs Affecting Lipid Metabolism," p. 25-34. New York, Plenum Press, 1969.

A segment of this review, which is adapted from a lecture, is devoted to those substances, stored or formed at the site of action, vasoactive on adipose tissue. Five references discussing PGE_1 activity as to potency, mode of action, and physiological function are cited and briefly discussed. (RMS) 1494

1059
RUCINSKA, E.
Prostaglandin und die myotrope wirkung der brenzkatechinamine. [Prostaglandin and the myotropic effect of the pyrocatcholamines.]
In: Pfister, Cl., ed., "Beitrage zur klinischen und biochemischen Pharmkologie," p. 240-245. Berlin, Volk und Gesundiheit. 1969.

Rat diaphragm and phrenic nerve preparation were placed in Tyrode baths. The diaphragm was stimulated directly or indirectly via the phrenic nerve. Pyrocatecholamines (noradrenaline, adrenaline, isoproterenol) added to the bath had a positive myotropic effect under direct and indirect stimulation: the muscle contraction was both stronger and longer. When PGE_1 was added to the bath, the contraction was shorter. When one of the pyrocatecholamines was added to the bath before PGE_1, the normal pyrocatecholamine effect was obtained, but when added after PGE_1, the effect was significantly weakened. (MEMH) 1772

1060
RYAN, T.J. and P.W.M. COPEMAN
Microvascular pattern and blood stasis in skin disease.
British Journal of Dermatology. 81: 563-571. 1969.

The authors briefly mention the work of O'Brien (1966, 1969) who reviewed the factors involved in platelet stickiness and aggregation. Many factors had to be considered and little is known about the precise part played by the constituents of the plasma, adenosine diphosphate (ADP), noradrenalin, adrenalin, the naturally occurring vasodepressor prostaglandin E and the turbulence of blood flow. (GW) 1649

1061
RYAN, W.L., D.M CORONEL, and R.J. JOHNSON
A vasodepressor substance of the human placenta.
American Journal of Obstetrics and Gynecology. 105: 1201-1206. 1969.

Acidic extraction of human mature placentas yielded a vasodepressor material resembling PGE_1 in chemical and rat bioassay properties. Extraction at neutral pH with water, acetone, ether, or ethanol yielded no vasoactive extract. Assay of 4 normal and 4 toxemic patients suggests that there is less vasodepressor material in toxemic placentas. (RMS) 1499

1062
SADOWSKI, J., A.L. MORRISON, and E.E. SELKURT
Examination of possible renal origin of the humoral factor responsible for saline diuresis in the dog.
Pflugers Archiv. 312: 99-109. 1969.

Johnson et al. (1966) are quoted as saying that prostaglandins might be leached from the kidney medulla. The author hypothesized a renal origin of the humoral factor responsible for the natriuresis which follows saline infusion in the dog. (GW) 1733

1063
SAID, S.I. and V. MUTT
Long acting vasodilator peptide from lung tissue.
Nature. 224: 699-700. 1969.

The authors report the occurrence of a vasoactive peptide in lung tissue from normal hogs. A vasoactive peptide extracted from lung tissue was freed of prostaglandin by precipitation. Injection of the lung preparation into the hind limb of anaesthetized dogs caused an increase in femoral arterial flow. Doses of prostaglandin E_1 and E_2 produced greater peak increases in blood flow. (GW)' 1609

1064
SAID, S.I. and V. MUTT
A peptide fraction from lung tissue with prolonged peripheral vasodilator activity.
Scandinavian Journal of Clinical and Laboratory Investigation. 107: 51-57. 1969.

Extracts of hog lung tissue were tested for vasodilator response by intraarterial injection in the hind limb of anesthetized dogs. Anesthesia was achieved by intravenous sodium pentobarbital, and the mean femoral arterial flow was measured directly by an electromagnetic probe and flow meter. The preparations were injected into a small branch of the same femoral artery while the mean systemic arterial blood pressure was monitored by a catheter in a carotid artery. An active peptide fraction from lung elicited a vasodilator response that was slower to develop and longer-lasting than that resulting from PGE_1. PGE_2, however, gave a response comparable to that of lung preparations, even those which had been freed of prostaglandins. 2mMg of lung extract was the smallest dose showing a measurable vasoactivity. (RAP) 1770

1065
SALZMAN, E.W. and L.L. NERI
Cyclic 3',5'-adenosine monophosphate in human blood platelets.
Nature. 224: 609-610. 1969.

Author briefly mentions that other researchers have reported PGE_1 to be a potent inhibitor of platelet clumping. (JRH) 1660

1066
SAMUELSSON, B.
Biosynthesis and metabolism of prostaglandins.
Abstracts, 4th International Congress on Pharmacology. Basle, 14-18 July, p. 11. 1969.

Abstract only. The mechanism of the conversion of essential fatty acids into prostaglandins will be discussed as well as in vitro and in vivo studies on the metabolsim of prostaglandins in several species including man. (Author) 1588

1067
SAMUELSSON, B.
Biosynthesis of prostaglandins.
Progress in Biochemical Pharmacology. 5: 109-128. 1969.

While biosynthesis of prostaglandins was first accomplished with vesicular glands from rams and bulls, in conjunction with the essential fatty acid precursors, arachidonic acid, dihomo-gamma-linolenic acid, and 5,8,11,14,17-eicosapentaenoic acid, it was later shown that such biosynthesis was not restricted to the vesicular gland. PGE_2 and $PGF_{2\alpha}$ were synthesized from arachidonic acid using homogenates of guinea pig lung. The finding that these two prostaglandins were not interconvertible was important for the elucidation of the mechanism of the biosynthesis, which is described in detail in this paper. The enzymatic synthesis of prostaglandins involves cyclization of unesterified precursor acids. In the vesicular gland, the capacity of the cyclizing enzyme exceeds by several orders of magnitude the amount of unesterified precursor acids available. It is hypothesized that the hydrolytic reaction is the rate-limiting step and that prostaglandins may control their own synthesis by a feed-back mechanism. Also, the control of prostaglandins by hormonal and nervous mechanisms could use the hydrolytic reaction as a regulatory step. (ART) 1667

1068

SANDLER, M., E.D. WILLIAMS, and S.M.M. KARIM

The occurrence of prostaglandins in amine-peptide-secreting tumours.

In: Mantegazza, P. and E.W. Horton, eds., "Prostaglandins, Peptides and Amines," p. 3-7. London, Academic Press, 1969.

High levels of prostaglandins $F_{2\alpha}$ and E_2 were found in a wide variety of amine-peptide-secreting tumors. The first group of tumors were derived from the neural crest. Ganglioneuroma contained high levels of $PGF_{2\alpha}$ and was associated with severe diarrhea. The second group of tumors investigated were derived from the foregut. Tissue from the 5-hydroxytryptamine-kinin-producing bronchial carcinoid contained a high level of $PGF_{2\alpha}$. Diarrhea was an important clinical feature in the bronchial tumor with high PG levels. The third group consisted of two 5-hydroxytryptamine-kinin-producing midgut carcinoids neither of which contained identifiable amounts of PGE_1, PGE_2, $PGF_{1\alpha}$ or $PGF_{2\alpha}$, but diarrhea was associated with both. This paper makes a case for a closer examination of prostaglandin-peptide-amine relationship. (GW) 1687

1069

SANNER, J.H.

Antagonism of prostaglandin E_2 by 1-acetyl-2-(8-chloro-10,11-dihydrodibenz [b,f] [1,4] oxazepine-10-carbonyl) hydrazine (SC- 19220).

Archives Internationales de Pharmacodynamie et de Therapie. 180: 46-56. 1969.

This report concerns the specific inhibition of prostaglandin PGE_2-induced contractions of the guinea pig ileum in vitro by 1-acetyl-2-(8-chloro-10,11-dihydrodibenz [b,f] [1,4] oxazepine-10 carbonyl) hydrazine (SC-19220). Thirty-one female albino guinea pigs were used. In preliminary trials approximately equal submaximal contractions were obtained by adjusting the doses of PGE_2, bradykinin and acetylcholine added to the bath. Two control contractions were obtained at 3.5 minute intervals. A suspension of SC-19220 in the bathing solution was then kept throughout the experiment while 3 more contractions were elicited. To allow time for the tissue to become equilibrated with the antagonist the first of the 3 sets of contractions were not used. There were reductions in PGE_2-induced contractions with SC-19220 concentrations of 7.5×10^{-6} M and above. N responses to PGE_2 were elicited when the highest concentration of SC-19220 was 1.5×10^{-4} M or lower. SC-19220 was thus shown to be a specific inhibitor of contractions induced on the guinea pig ileum of PGE_2. PGE_2 contractions could be completely blocked without a significant effect on contractions produced by bradykinin or acetylcholine. Log concentration-response curves made with cumulative doses of PGE_2 were shifted toward higher concentrations of agonist, a characteristic of competitive inhibition. Higher concentrations of SC-19220 caused a depression of the PGE_2 log concentration-response curves, indicating a non-competive component at high inhibition concentrations. (PR) 1518

1070
SCHLENK, H., T. GERSON, and D.M. SAND
Conversions of non-biological polyunsaturated fatty acids in rat liver
Biochimica et Biophysica Acta. 176; 740-747. 1969.

It is concluded that the odd numbered $\omega 3$ and $\omega 6$ and even numbered $\omega 5$ fatty acids with four or more methylene interrupted double bonds when administered to the rat, behave metabolically in several respects like the prostaglandins. (GW) 1648

1071
SCHNATZ, J.D. and T.C. CUMMISKEY
Neutral and alkaline lipolytic activities in biopsies of human adipose tissue.
Life Sciences. 8; 1273-1279. 1969.

Authors briefly report the work of other researchers who found the low lipolytic activity in homogenates of adipose tissue was due to an inhibitor, possibly prostaglandin. (JRH) 1661

1072
SCHNEIDER, W.P.
The synthesis of ($\pm$)-prostaglandins E_2, $F_{2\alpha}$, and $F_{2\beta}$.
Chemical Communications. 304-305. 1969.

This report details a synthesis of ($\pm$)-PGE$_2$, and PGF$_{2\alpha}$ and PGF$_{2\beta}$ based on 6 end-substituted bicyclo [3,1,0] hexane intermediates. The synthetic ($\pm$)-PGE$_2$ showed at least 50% of the biological activity of natural PGE$_2$ in its effects on blood pressure in rats and on contraction of smooth muscle. Reduction of ($\pm$)-PGE$_2$ with sodium borohydride gave ($\pm$)-PGF$_{2\alpha}$ and its 9-epimer, ($\pm$)-PGF$_{2\beta}$, in about a 45:55 ratio. The ($\pm$)-PGF$_{2\alpha}$ had a least 50% the activity of natural PGF$_{2\alpha}$ in its affect on blood pressure and smooth muscle. (GW) 1639

1073
SCHNEIDER, W.P., U. AXEN, F.H. LINCOLN, J.E. PIKE, and J.L. THOMPSON
The synthesis of prostaglandin E$_1$ and related substances.
Journal of the American Chemical Society. 91: 5372-5378. 1969.

The total synthesis of crystalline dl-prostaglandin E$_1$ and its methyl ester by way of bicyclo (3.1.0) hexamone intermediates is described in detail. The same reaction sequence also produces dl-8-iso-prostaglandins and also dl-PGA$_1$ and dl-PGB$_1$ methyl esters. This is a continuation of a previously reported study. (PR) 1549

1074
SCHREIBMAN, P.H., D.E. WILSON, and R.A. ARKY
Inhibiton of cyclic 3',5'-adenosine monophosphate-activated lipase.
Journal of Clinical Investigation. 48: 75abs. 1969.

Abstract only. The authors report that inhibition of lipolysis by β-adrenergic blocking does not require the adipocyte membrane. It was also shown that a cAMP-activated monoglyceridase is present in a cell free extract of rat and human adipose tissue, and that antilipolytics, propanolol (10^{-8}M), phentolamine (10^{-8} M), nicotinic acid (10^{-7}M), and PGE$_1$ and PGF$_{1\beta}$ (10^{-7}M), completely inhibit the monoglyceridase stimulation at a site distinct from α- and β-adrenergic receptors. (RMS) 1434

1075

SHAW, J.E. and P.W. RAMWELL
Direct effect of prostaglandin E_1 on the frog gastric mucosa.
Abstracts, 4th International Congress on Pharmacology. Basle, 14-18 July, p. 109-110. 1969.

The mechanism by which PGE_1 modifies acid secretion has been examined using the isolated frog gastric mucosa. After removing the serosa, the nutrient surface of the mucosa was superfused with a bicarbonate buffered Ringer solution. The mucosal secretory surface was similarly superfused but with an unbuffered ringer solution. After 2 hrs the effect of PGE_1 and the pentapeptide ICI 50.123 was determined following their addition to either the secretory of nutrient surface. In contrast to the results obtained with rats and dogs in vivo PGE_1 $(1.4\text{-}5.6 \times 10^{-5}\,M)$ in this preparation increased H^+ secretion 2-3 fold, when added to either the nutrient or secretory surface. Pentapeptide added to the nutrient surface, also increased H^+ secretion. Simultaneous perfusion of PGE_1 and pentapeptide increased H^+ secretion, but the response was less than that obtained with either drug alone. PGE_1 stimulation of H^+ secretion has now been correlated with a significant increase in oxygen consumption, suggesting for the first time an interdependence of PGE_1 action and oxidative metabolism. However, further studies indicate that the primary site of PGE_1 action in this tissue is on the cell membrane. (Authors modified) 1610

1076

SHAW, J.E. and P.W. RAMWELL
Separation, identification and estimation of prostaglandins.
In: Glick, D., ed., "Methods of Biochemical Analysis," Vol. 17. p. 325-371. New York, Interscience, 1969.

The authors review 108 papers in an extremely comprehensive review of the subject covering definitions, occurrence, properties, extraction, and identification. (RMS) 1490

1077

SHEHADEH, Z., W.E. PRICE and E.D. JACOBSON
Effects of vasoactive agents on intestinal blood flow and motility in the dog.
American Journal of Physiology. 216: 386-392. 1969.

Observations were made of the effects of prolonged intraarterial infusion of various doses of seven vasoactive substances including $PGF_{2\alpha}$ and PGE_1, on mesenteric hemodynamics and intestinal motility in anesthetized dogs. Both blood flow and pressure were measured in the mesenteric bed. $PGF_{2\alpha}$ produced a variable reaction, but in general, vasoconstriction. Arterial pressure was not affected while venous pressure rose as did intestinal motility. Blood flow was reduced. The dose levels ranged from 0.10 to 1.00 $\mu g/kg/min$. PGE_1 infused at the same dose levels produced a dramatic vasodilation at the lowest dose levels. There was no difference in effect at the higher levels. Intestinal motility was abolished, venous pressure rose slightly and blood flow was greatly increased. (JRH) 1458

1078

SHIMIZU, K., T. KURASAWA, T. MAEDA, and Y. YOSHITOSHI
Free water excretion and washout of renal medullary urea by prostaglandin E_1.
Japanese Heart Journal. 10: 437-455. 1969.

An experiment is described, the purpose of which was to determine whether PGE_1 has a direct antagonistic effect on the action of antidiuretic hormone (ADH) in vivo in dogs. PGE_1 was infused directly into the isolated left renal artery in a subdepressor dosage of about 0.2 microgram/min/kg. Plasma level was maintained at approximately 3 mg/100 ml by continuous intravenous infusion of para-aminohippurate (PAH) in normal saline at a rate of 0.2 to 0.4 ml/min. Bilateral urine collections for clearance studies and chemical determinations were made at 10 to 20 minute intervals and blood samples were taken at the midpoint of each period. Results are shown in tables, graphs and figures. In the kidney infused with PGE_1, free water excretion and washout of renal medullary urea and sodium occurred without marked natriuresis. Increases in sodium and water excretion with increased renal plasma flow and decreased extraction ratio of PAH seem to indicate support for the presence of a direct anti-ADH action of PGE_1. (ART) 1557

1079
SHIO, H., N.H. ANDERSEN, E.J. COREY and P.W. RAMWELL
Stereospecificity of 15-Hydroxyprostaglandin dehydrogenase.
Abstracts 4th International Congress on Pharmacology. Basle, 14-18 July p. 100. 1969.

Abstract only. Dehydrogenation of the 15-hydroxy group is one of the main metabolic pathways of prostaglandins in the mammalian lung. The swine lung enzyme appears to be so specific for prostaglandins and NAD dependent, that it could be used in an enzymatic assay. The availability of a series of recently synthesized epimers of PGE_1 has made it possible to study the stereoisometric requirements of this enzyme. The reaction was estimated by two methods—OD 340 increase due to NADH production and OD 500 increase in alkaline solution due to 15-keto PGE_1 itself. The relative reaction rate of prostaglandin epimers and optical rotary dispersion studies suggested, i) one-half of racemic PGE_1 and racemic 11-epi PGE_1 was converted into the keto form, which had the optically natural configuration, ii) the reaction rate of racemic 11, 15-epi PGE_1 showed a slower but steady conversion up to 50% and ORD indicated that the unnatural antipode was attacked by the enzyme, iii) 15-epi PGE_1 was not converted into the keto form at all and furthermore, it competitively inhibited the reaction of natural PGE_1. These findings indicate that the enzyme requires 15(S)-OH configuration for binding as well as for dehydrogenation. (Authors modified) 1634

1080
SIGGINS, G.R., B.J. HOFFER, and F.E. BLOOM
Cyclic adenosine monophosphate: possible mediator for norepinephrine effects on cerebellar Purkinje cells.
Science. 165: 1018-1020. 1969.

Only brief mention is made of the prostaglandins in the study of the effect of norepinephrine on the responsivenss of Purkinje cells when administered electrophoretically. Microelectro- phoretic administration of the prostaglandins E_1 and E_2 reduces the slowing of spontaneous discharge obtained by application of norepinephrine. The authors propose that the action of norepinephrine on Purkinje cells may be specifically mediated by the formation of cyclic AMP. (GW) 1628

1081
SIH, C.J., G. AMBRUS, P. FOSS, and C.J. LAI
A general biochemical synthesis of oxygenated prostaglandin E.
Journal of the American Chemical Society. 91: 3685-3687. 1969.

Until now, no oxygenated prostaglandin E (PGE) has been synthesized or isolated from natural sources. This paper describes the synthesis of 11α-15-(S)-dihydroxy-9,18-dioxo-5-*cis* 13-*trans*-prostadienoic acid (18-oxo-PGE$_2$), and 11α-15-(S)-dihydroxy-9,19-dioxo-5-*cis*, 13-*trans*-prostadienoic acid (19-oxo-PGE$_2$). The principle of this method involves microbiological hydroxylation of arachidonic acid by exposure to a fungus, *Ophiobolus graminis*, which gives rise to 2 polar products, characterized as 18E-hydroxy-arachidonic acid and 19E-hydroxy-arachidonic acid. These unsaturated fatty acids or their derivatives can then be cyclized by exposure to bull seminal vesicle microsomes to yield the desired oxygenated prostadienoic acid derivatives, 18-oxo-PGE$_2$ and 19-oxo-PGE$_2$. It is claimed that the method described provides a relatively convenient route for the synthesis of oxygenated PGE derivatives, which would be exceedingly difficult to prepare by conventional partial or total chemical synthesis. (ART) 1528

1082
SINGLETON, J.W.
Humoral effects of the pancreas upon the gastrointestinal tract.
Gastroenterology. 50: 342-362. 1969.

In this review, the author briefly mentions that prostaglandins may be the "diarrhea hormone" that has been hypothesized in conjunction with pancreatic neoplasms. (RAP) 1739

1083
SIRCUS, W.
Peptide-secreting tumours with special reference to the pancreas.
Gut, 10: 506-515. 1969.

Only brief mention is made of prostaglandins in this review in conjunction with a series of case reports. Of two diarrhea problems, one was due to a neural crest tumor in which the tissue concentration of prostaglandin F$_{2\alpha}$ was grossly excessive, and the other to an alpha-cell pancreatic tumor, in the blood of which prostaglandin E$_2$ was circulating in excess. (GW) 1709

1084
SJOSTRAND, N.O. and G. SWEDIN
Potentiation by smooth muscle stimulants of an hypogastric-nerve-seminal vesicle preparation from the guinea-pig.
Acta Physiologica Scandinavica. Supp. 330: 60. 1969.

The contractions of the seminal vesicle were recorded with a balloon inserted in the lumen of the seminal vesicle. Adrenaline, noradrenaline, acetylcholine, histamine, serotonine, bradykinin, prostaglandin E$_1$ and barium chloride were found to potentiate the responses of the seminal vesicle to hypogastric nerve stimulation. This potentiation occurred with doses much smaller than those giving a direct contraction of the organ. It is suggested that in vivo such a potentiation can occur and be of importance for the emission. (Authors) 1646

1085
SMEJKAL, V., J. MIRONNEAU, C. OJEDA and Y.N. GARGOUIL
The influence of adrenergic drugs on the frog and rabbit heart artium.
Abstracts, 4th International Congress on Pharmacology. Basle, 14-18 July, p. 281. 1969.

Abstract only. In a 'double-sucrose gap' method with use of frog isolated auricles trabecles a series of adrenergic and cholinergic drugs was tested. Prostaglandin E_1 10^{-5} and 10^{-6} mimics some adrenaline and isoprenaline effects, but its influence on the threshold interferes with the influence of corresponding pharmacologic solvent e.g. ethanol. Ethanol increases threshold. (Authors modified) 1603

1086
SMEJKAL, V.

Influence de substances cardioactives sur les proprietes electrophysiologiques de l'oreillette de grenouille. [Influence of cardioactive substances on the electrophysiological properties of the frog auricle.]
Journal de Physiologie. 61(supp. 2): 406-407. 1969.

The sucrose gap technique was used on isolated frog auricles to measure the effects of several α- and β-adrenergic agents on the action potential and threshold levels. PGE_1 (up to 10^{-5} M) produced effects similar to the β-adrenergic agents which lower threshold value and increase the action potential. (JRH) 1755

1087
SMEJKAL, V.

[On the action of prostaglandin E_1 on smooth muscle of the digestive system in experimental animals.]
Ceskoslovenska Gastroenterologie a Vyziva. 23: 32-35. 1969.

Prostaglandin E_1 is an active lipid, a higher long-chain fatty acid. In vitro in doses of 215 μg/ml, it causes after short-term relaxation contraction of the isolated rat duodenum. This contraction is inhibited by the anticholinergic atropine, antihistaminic bromadryl, the antiserotonin substance lysenyl but only in concentrations above 10^{-5} M/20 ml Tyrode's solution. Therefore we cannot consider the specific effect of prostaglandin on acetylcholine, histamine or serotonin receptors but rather the direct effect of prostaglandin on smooth muscle of the duodenum and on a special receptor. In situ prostaglandin stimulates the motor activity of the rabbit stomach, the effect being manifested after a dose of 500 μg/kg iv and it persists for 10 min. When evaluating the effect of prostaglandin, it is important to differentiate the effect of the pharmacological solvent, ethanol. The author draws attention to the possible importance of prostaglandin in the regulation of the digestive tract of animals and man. (Author) 1632

1088
SMELIK, P.G.

The effect of a CRF preparation on ACTH release in rats bearing hypothalamic dexamethasone implants: A study on the "implantation paradox."
Neuroendocrinology. 5: 193-204. 1969.

Prostaglandin E_1 was one of several substances used to test the capacity of the pituitary gland system of the rat to respond to stressful stimuli after implantation of dexamethasone into the hypothalamus. The activation of the pituitary-adrenal axis by PGE_1, adrenaline, histamine, carbachol or angiotensin II was prevented by dexamethasone implanted 18 hours previously. It is the author's opinion that the dexamethasone implants block the effect of stressful stimuli by interfering with the production of corticotrophin releasing factors (CRF). (GW) 1677

1089
SMITH, J.W., A.L. STEINER, W.M. NEWBERRY, Jr. and C.W. PARKER
Cyclic nucleotide inhibition of lymphocyte transformation.
Clinical Research. 17: 549. 1969.

Abstract only. The role of cyclic adenosine monophosphate (cAMP) was investigated in the transformation of human blood lymphocytes. It was demonstrated that intracellular levels in lymphocytes were raised by isoproterenol, aminophylline, and PGA_1, among other effects. (ART) 1706

1090
SOBEL, B.E. and A.K. ROBISON
Activation of guinea pig myocardial adenyl cyclase by prostaglandins.
Circulation. (supp. 3): III-189. 1969.

Abstract only. Measurements of adenyl cyclase by precipitation and chromatography of ^{3}H-cAMP and of phosphodiesterase by spectrophotometry revealed marked stimulation of guinea pig myocardial adenyl cyclase by PGE_1 and $PGF_{1\alpha}$ at 10^{-9} M but no effect on phosphodiesterase with PG's at 10^{-7} to 10^{-9} M. These findings suggest that the inotropic action of PG's may be due to enhanced cAMP synthesis due to adenyl cyclase activation. (RMS) 1431

1091
SOMOVA, L.
Dynamic tracing of the activity of vasodepressor lipid isolated from kidneys of hypertensive
    animals.
Comptes Rendus de l'Academie Bulgare des Sciences. 22: 1189-1192. 1969.

The author mentions without detail the work of Hickler (1966) who showed that a purified vasodepressor lipid (VDL) from the renal medulla is close in its chemical, biochemical and biological properties to the biologically active hormone-like substances known as prostaglandins and more particularly to those of the E group. (GW) 1719

1092
STAUB, N.C.
Respiration.
Annual Review of Physiology. 31: 173-226. 1969.

The 1966-1968 literature on respiration is critically reviewed and the effects of prostaglandins on respiration are discussed. 126 citations are included. (MT) 1746

1093
STOCK, K. and E. WESTERMANN
Interactions between ACTH, adrenolytic drugs and prostaglandin E_1 in the lipolytic system.
In: Margoulies, M., ed., "International Symposium on Protein and Polypeptide Hormones,
    Liege, 1968," p. 159-161. Amersterdam, Excerpta Medica Foundation, 1969.

Rat epididymal adipose tissue was incubated in vitro with lipolytic stimulators ACTH, theophylline, N^6-2'-o-dibutyryl-3',5'-AMP and norepinephrine and with various concentrations of the inhibitors phentolamine, a β adrenolytic drug, Kö 592, and PGE_1. Intrinsic activities and affinities for the lipolytic system of the stimulators employed were calculated from dose response curves, and the inhibitor constants were determined. In contrast to the adrenolytics, PGE_1 inhibited lipolysis induced by ACTH, theophylline, and norepinephrine competitively without competition for binding sites. It is proposed that PGE_1 inhibits the binding of ATP to adenyl cyclase without reducing the rate of breakdown of the ATP-adenyl cyclase complex once formed. Total inhibition was never observed and no inhibitor constant for PGE_1 could be calculated. (RMS) 1561

1094
SUNAHARA, F.A. and M.G. VIGUERA
Interaction of prostaglandin and other vasoactive agents on the microcirculation.
Abstracts, 4th International Congress on Pharmacology. Basle, 14-18 July, p. 386-387. 1969.

Abstract only. In vivo and in vitro experimental procedures were used to investigate the interaction of prostaglandins and other vasoactive agents on the circulation. Changes in dimension of microvessels were monitored using a microscope, image-splitting eyepiece, TV camera, videoscreen and recorder. Intravenous administration of 1.5 to 15 $\mu g/kg/min$ of PGE_1 resulted in a dose-dependent decrease in the systemic blood pressure, a concomitant constriction in cremasteric blood vessels and a dilatation in the mesenteric blood vessels. During and after infusion of 1.5 $\mu g/kg/min$ of PGE_1, the effects of topically applied norepinephrine were significantly inhibited. Constrictor effects of topically applied angiotensin were not altered. $PGF_{2\alpha}$ did not modify norepinephrine effects. In in vitro experiments, in which electrical stimulation of arterial vessels of a perfused isolated rabbit ear was employed, perfusion of prostaglandin caused a dose-dependent decrease in the sensitivity to the electrical stimulation. Our data supports the view that the site of action of PGE_1 is different from those of phentolamine, tetrodotoxin or papaverine. (Authors modified) 1592

1095
SUTHERLAND, E.W. and G.A. ROBISON
The role of cyclic AMP in the control of carbohydrate metabolism.
Diabetes. 18: 797-819. 1969.

Prostaglandins (which suppress the accumulation of cAMP in rat adipose tissue), reduced levels of stimulation of adenyl cyclase, glucagon and the catecholamines, as well as thyroxine, growth hormone and the glucocorticoids, can prevent insulin deficiency which leads to the excessive production of cAMP. (GW) 1654

1096
SUZUKI, T., Y. ABIKO, and T. FUNAKI
[Effects of prostaglandin E_1 on the cardiovascular system in dogs.]
Folia Pharmacologica Japonica. 65: 1-6. 1969.

Article in Japanese. Abstract not available at present. 1682

1097
SZABO, M.M. and M.S. GOLDSTEIN
The effect of guanine derivatives on palmitate-1-^{14}C incorporation into rat epidydymal adipose tissue.
Proceedings of the Society for Experimental Biology and Medicine. 131: 1055-1059. 1969.

The corporation of palmitate-I-^{14}C into rat epidydymal adipose tissue lipids was remarkedly enhanced by the in vitro addition of guanosine and its 5'-phosphate ester and to a lesser extent by dGMP, G2', 3'cP, adenosine, uridine and cytidine. Such antilipolytic agents as nicotinic acid or PGE$_1$ could not duplicate this stimulation of palmitate incorporation. (GW) 1443

1098

TAKANO, S. and T. SUZUKI
[Effect of prostaglandin E$_1$ on the responses to noradrenaline, adrenaline, acetylcholine and serotonin of isolated spleen strips of dog and isolated uterus of rabbit, guinea pig and rat.]
Folia Pharmacologica Japonica. 65: 105-106. 1969.

Article in Japanese. Abstract not available at present. 1673

1099

TAUNTON, O.D., J. ROTH, and I. PASTAN
Studies on the adrenocorticotropic hormone-activated adenyl cyclase of a functional adrenal tumor.
Journal of Biological Chemistry. 224: 247-253. 1969.

The effect of various compounds and hormones on the ACTH and fluoride stimulated adenyl cyclase activity in mouse adrenal tumor cells grown in vitro and in vivo were studied. It was found that PGE$_1$, (5-10 μg/ml), had no effect on the increase. (JRH) 1477

1100

TEMPLE, D.M.
Isolation techniques for pharmacologically active substances (animal).
Annual Review of Pharmacology. 9: 407-418. 1969.

This review deals with methods of isolation which have been used for hormones, neurohumors, and autacoids. In discussing lipid-soluble smooth muscle stimulants, consideration is given to the prostaglandins, a group of lipid-soluble carboxylic acids with the ability to contract smooth muscle. Six references are cited dealing with extraction of prostaglandins from blood, brain, seminal fluid, and other tissue homogentes. (ART) 1701

1101

THOMASSON, H.J.
Prostaglandins and cardiovascular diseases.
Nutritio et Dieta. 11: 228-240. 1969.

Generally, the favourable influence of poly-unsaturated fatty acids on the atherosclerotic process is explained by means of the lipid-filtration theory (consequently, via the lipoprotein level of the blood). In the present article, an additional theory has been proposed, according to which the prostaglandins (E$_1$ as well as F), synthesized in the organism from poly-unsaturated fatty acids, play a part in the prevention of both atherosclerosis and thrombosis. This effect of prostaglandins would occur in the presence of divergent pathogenic factors, e.g. platelet adhesion and aggregation, vasoconstriction and dilatation, hypertension, lack of physical exercise, mental stress, smoking, poor condition of the heart muscle as well as diminished coronary flow. In addition, it is postulated that the synthesis of certain prostaglandins is promoted, if the dietary supply of the precursors (e.g. linoleic acid) is increased. (Authors) 1669

1102

THOMASSON, H.J.
   Prostaglandins and cardiovascular diseases.
   Nutrition Reviews. 27: 67-69. 1969.

This review discusses experimental results on prostaglandins relative to cardiovascular diseases. PGE_1 in very low concentration prevents or counteracts both in vitro and in vivo platelet adhesion to vessel walls as well as the reversible platelet aggregation. Therefore, PGE_1 may be involved in the formation of the initial thrombus and may thus represent a link between dietary fat and atherosclerosis. A tentative theory is that the vessel wall is premeated by and/or coated with PGE_1. Local concentration of PGE_1 may then determine development or disappearance of the initial thrombus. Also PGE_1 may induce the release of heparin from mast cells and thus retard fibrin formation. PGE's decrease blood pressure by vasodilatation; therefore, catecholamines which cause hypertension and which are released by smoking or emotional stress may be counteracted by PGE_1. $PGF_{1\alpha}$ and $PGF_{2\alpha}$ synthesized in the lung, affected only the contractile force of the heart and aided improvement in experimental perfused hearts. PGE_1 and PGE_2 caused an increased flow in coronary vessels in the isolated perfused rat heart. PGE_1 has an insulin-like reaction by stimulating the uptake and oxidation of glucose and the synthesis of glycogen and triglycerides and may therefore be correlated with its effects on blood pressure and blood lipid levels. (PR) 1581

1103

THOMPSON, J.H. and M. ANGULO
   Prostaglandin-induced serotonin release.
   Experientia. 25: 721-722. 1969.

Reported here are the effects of PGE_1 on gastrointestinal serotonin levels following amine depletion by p-chloro-phenylalanine (PCPA). The total amount of serotonin in the stomach fundus, pyloris antrum and mid-jejunum in rats was determined. Prostaglandin E_1 (200 µg/kg sc or iv) reduced the serotonin levels whether in the control or in the p-chlorophylalanine (150 or 300 mg/kg) or with pretreated reserpine (5.0 mg/kg) animals. (GW) 1536

1104

TOBIAN, L., M. ISHII, and M. DUKE
   Relationship of cytoplasmic granules in renal papillary interstitial cells to "postsalt"
      hypertension.
   Journal of Laboratory and Clinical Medicine. 73: 309-319. 1969.

Authors report that other workers have isolated prostaglandins with vasodepressor action from the renal medulla and these prostaglandins have been shown to reduce blood pressure in experimentally hypertensive rats and rabbits. (JRH) 1658

1105

TURCHETTO, E., H. WEISS, G. BOCCHINI, and E. FORMIGGINI
   Die gewebslipoide der ratte in verschiedenen altersstufen. [Tissue lipids in the rat at different
      ages.]
   Nutritio et Dieta. 11: 109-114. 1969.

In the past experiment we have studied the behaviour of fatty acids of liver, heart, and voluntary muscle lipids, in Wistar male rats, fed after weaning a standard diet and killed at different ages, i.e. at weaning, in the 3rd, 4th, 6th, 12th and 18th months of life. We have thus studied systematically the alterations which accompany aging through the variations of the fatty acid patterns in the total lipids, neutral fats, and phospholipids. We think [these] phenomena are worth noting: first C 16:1 and C 20:3 acids progressively increase from low values (or traces) to appreciable percentages; also, there is a decrease of arachidonic and linoleic acids between the third and 12th month, with the tendency to go back to the initial values after one year of life. The first observation might be related to a progressive aging pattern; the second observation might be related to prostaglandin production, particularly evident during the middle period of life; prostaglandins are believed to derive from some C 20 fatty acids as precursors. (Authors modified) 1668

1106
TURKER, R.K. and B.K. KIRAN
Interaction of prostaglandin E_1 with oxytocin on mammary gland of the lactating rabbit.
European Journal of Pharmacology. 8: 377-379. 1969.

Prostaglandin E_1 when given by arterial or intravenous injection did not produce any change in the milk-ejection pressure of the lactating rabbit but it decreased the mild-ejection activity of oxytocin. Adrenergic blockade did not influence this inhibition. It is concluded that the inhibitory action of prostaglandin E_1 on oxytocin-induced milk-ejection activity of the lactating rabbit is a consequence of a direct action of the lipid on myoepithelial cells. (Authors) 1712

1107
TURKER, R.K. and P.A. KHAIRALLAH
Prostaglandin E_1 action on canine isolated tracheal muscle.
Journal of Pharmacy and Pharmacology. 21: 498-501. 1969.

PGE_1 inhibits contractions of dog isolated tracheal muscle stimulated by different agents, but the degree of inhibition varies with the agent used. Low concentrations of PGE_1 completely block the stimulant effect of 5-hydroxytryptamine, but even large concentrations of PGE_1 do not completely antagonize the contractions caused by acetylcholine. The inhibitory effect of PGE_1 is blocked by methysergide and not by propranolol, morphine or dihydroergotamine. PGE_1 does not relax depolarized smooth muscle, although bradykinin and isoprenaline do. It is concluded that in tracheal smooth muscle, PGE_1 interacts with cell membranes close to the 5-hydroxytryptamine D receptors. This causes activation of the smooth sarcoplasmic reticulum, leading to accumulation of calcium ions and relaxation. (Authors) 1483

1108
TURKER, R.K., Ph. A. KHAIRALLAH, S.O. KAYAALP, and S. KAYMAKCALAN
Response of the nictitating membrane to prostaglandin E_1 and angiotensin.
European Journal of Pharmacology. 5: 173-179. 1969.

Prostaglandin E_1 has previously been shown to potentiate contractile responses of smooth muscle to various stimuli. To study this phenomenon further, responses of cat nictitating membranes to angiotensin and preganglionic electrical stimulation were measured, and so were the changes in these responses following intraarterial administration of PGE_1. In adrenalectomized animals PGE_1 contracted the nictitating membrane independent of the superior cervical ganglion, but response was decreased following adrenergic or cholinergic blockade, and abolished following both. Angiotensin also contracted the nictitating membrane in adrenalectomized animals independent of the superior cervical ganglion, and the response was inhibited by cholinergic but not by adrenergic blockade. PGE_1 potentiated responses to acetylcholine, norepinephrine and angiotensin. It also antagonized ganglionic blockade by hexamethonium. (Authors) 1408

1109
VANE, F. and M.G. HORNING
   Separation and characterization of the prostaglandins by gas chromatography and mass
      spectrometry.
   Analytical Letters. 2: 357-371. 1969.

   Prostaglandins of the A,B,E and F series together with 8-iso-E_1, have been separated and
   characterized by gas chromatography and mass spectrometry through the use of TMSi and
   MO-TMSi derivatives. These derivatives are suitable for work with GLC systems with flame
   ionization detectors. Additional derivatives must be sought for detection by electron capture
   techniques. (Authors) 1788

1110
VANE, J.R.
   The release and fate of vaso-active hormones in the circulation.
   British Journal of Pharmacology. 35: 209-242. 1969.

   The author briefly mentions prostaglandins several times without particular details in a general
   discussion of superfusion and the blood-bathed organ technique. Subsequently, in a more
   detailed discussion, the release and systemic site(s) of deactivation of bradykinin, 5-hydroxy-
   tryptamine, prostaglandins, angiotensins I and II, and adrenaline and noradrenaline are discussed
   in a detailed literature review. (RMS) 1502

1111
VARGAFTIG, B.B., E.O. DeMIRANDA, B. LACOUME
   Inhibition by non-steroidal anti-inflammatory agents of in vivo effects of "slow reacting
      substance C."
   Nature. 222: 883-885. 1969.

   SRS-A (slow reacting substance A) is a liposoluble spasmogen which is released by guinea pig
   lung during in vitro anaphylaxis. It increases the resistance of lungs to inflation. Non-steroidal
   anti-inflammatory agents block this as well as prostaglandin $F_{2\alpha}$ activity. The activity of $PGF_{2\alpha}$
   on the isolated human trachea is also blocked by fenamates. (GW) 1735

1112
VAUGHAN, M. and F. MURAD
   Adenyl cyclase activity in particles from fat cells.
   Biochemistry. 8: 3092-3099. 1969.

   A study of adenyl cyclase activity in a particle fraction prepared from epididymal fat pads of
   rats is reported. Prostaglandin E_1 at 7.7-38.5 μg/ml is noted to have had no effect on basal or
   epinephrine stimulated cyclase activity. (RMS) 1524

1113
VERGROESEN, A.J., J. DE BOER and J.J. GOTTENBOS
   Effects of prostaglandins on perfused isolated rat hearts.
   Acta Physiologica and Pharmacologica Neerlandica. 15: 72. 1969.

Abstract only. To elucidate the role of the heart in the cardiovascular reactions known to occur after the administration of several prostaglandins, isolated rat hearts were perfused according to Langendorff with different concentrations of a range of PG's. The criteria used were the contractile force of the myocard, the frequency of the heart beat and the amount of fluid perfused through the coronary vascular system. PGE_1, PGE_2, PGE_1-217, $PGF_{2\alpha}$ and $PGF_{1\alpha}$ appeared to be active in concentrations of 0.01 μg/ml and higher. The three PGE-compounds acted mainly as vasodilators, PGE_1 being the most active, and had little or no effect on the frequency of the heart beat. The effect of PGE_1 on the contractile force was variable; mostly a moderate increase was observed, but sometimes a definite decrease occurred. On the other hand, the perfusion with $PGF_{1\alpha}$ or $PGF_{2\alpha}$ had practically no vasodilating effect but resulted in a sharp increase in mechanical activity of the myocard without causing a change in the frequency of the heart beat. $PGF_{1\beta}$ and $PGF_{2\beta}$ seem to have hardly any activity in concentrations of I μg/ml and lower. Several isomers and homologues of PGE_1 and PGE_2 the homologues with 18, 19, 20, 21 and 22 carbon atoms were inactive in this experimental set-up. (Authors modified) 1743

1114

VERGROESEN, A.J. and J. DeBOER
Effects of prostaglandins E_1 and $F_{1\alpha}$ on potassium-depressed frog, rat and guinea-pig hearts.
Abstracts, 4th International Congress on Pharmacology. Basle, 14-18 July, p. 284. 1969.

Abstract only. The inotropic and chronotropic effects of PGE_1 and $PGF_{1\alpha}$ were studied in isolated frog and rat hearts. Strong positive inotropic and chronotropic effects of PGE_1 were observed in the potassium intoxicated frog and rat heart contrary to the absence of these effects in normally functioning hearts. $PGF_{1\alpha}$ showed hardly any effect on the frog heart but appeared to be more active than PGE_1 on the rat heart. The effects of PGE_1 and $F_{1\alpha}$ were also studied in potassium-depressed Starling heart-lung preparations of rats and guinea-pigs. Heart rate, cardiac output and electrocardiographic activity were decreased by the elevated $[K^+]$ of the perfused blood-saline mixture but could be restored to normal values by the addition of either PGE_1 or $PGF_{1\alpha}$. On account of these results it is assumed that PGE_1 and $F_{1\alpha}$ antagonize the effects of K+ rather than enhancing the effects of Ca^{2+} as is suggested by several investigators. (Authors modified) 1600

1115

VIGDAHL, R.L., N.R. MARQUIS, and P.A. TAVORMINA
Platelet aggregation. II. Adenyl cyclase, prostaglndin E_1, and calcium.
Biochemical and Biophysical Research Communications. 37: 409-415. 1969.

Intracellular cyclic AMP synthesis from ATP generated by incubation of intact human platelets with adenosine-^{14}C is increased by PGE_1. The level of cyclic AMP synthesized is temporally related to the degree of inhibition of platelet aggregation effected by PGE_1. Aggregation is dependent upon the presence of calcium ions, which inhibit both the stimulation of adenyl cyclase and the inhibition of aggregation of PGE_1. (Authors) 1523

1116

VIGUERA, M.G. and F.A. SUNAHARA
Microcirculatory effects of prostaglandins.
Canadian Journal of Physiology and Pharmacology. 47: 627-634. 1969.

In vivo microscopy of Wistar rats was used to evaluate the vasomotor action of prostaglandins E_1 and $F_{2\alpha}$ in the mesocecal and cremasteric muscle circulation. Minute to minute measurements of vessel diameters were monitored. Intravenous PGE_1 decreased systemic blood pressure with an increase in diameter of the metarterioles of the cremasteric muscle and decrease in diameter of the metarterioles of the mesocecum. It effectively inhibited the constrictor effect of topically applied norepinephrine and epinephrine on the mesocecum metarterioles. Constrictor effects of topically applied angiotensin II were not altered. Intravenous $PGF_{2\alpha}$ increased systemic blood pressure with decrease in metarteriolar diameter of both mesocecum and cremasteric muscles. Vascular effects of PGE_1 could be explained by its apparent ability to block at the postganglionic sympathetic neuroeffector site where it acts on the vascular tissue, modulating its response to adrenergic stimulants. $PGF_{2\alpha}$ acts as a direct stimulant on the vascular smooth muscle. (PR) 1552

1117
VISWANATHAN, C.V.
   Chromatographic analysis of molecular species of lipids: a general survey.
   Chromatographic Reviews. 11: 153-201. 1969.

   A general survey of the chromatographic fractionation techniques of the 6 closely related prostaglandin compounds is reported. The two main groups of naturally occurring prostaglandins PGE and PGF were separated from one another as a class by silicic acid column chromatography. (GW) 1611

1118
VOGT, W., U. MEYER, H. JUNZE, E. LUFFT, and S. BABILLI
   Entstchung von SRS-C in der durchstromten meerschweinchenlunge durch phospholipase A: identifizierung mit prostaglandin. [Formation by phospholipase A of SRS-C in perfused guinea-pig lung: identification with prostaglandins.]
   Naunyn-Schmiedebergs Archiv fur Pharmakologie und Experimentelle Pathologie. 262: 124-134. 1969.

   SRS-C appears in perfusates of guinea-pig lungs when the lungs are treated with phospholipase A or venoms containing this enzyme. SRS-C has been concentrated and purified from such perfusates. The biological activity of SRS-C depends mainly on the presence of prostaglandins and to a minor degree also on peroxides. Both compounds originate from unsaturated fatty acids. The liberated prostaglandins are only partially preformed. The bulk of them are formed after cleavage of the precursor acids from tissue phosphatides. Prostaglandins have not been detected as constituents of lung phosphatides. (Authors) 1586

1119
VOGT, W., J. BARTELS, H. KUNZE and U. MEYER
   A possible physiological role of phospholipase A for the formation and release of prostaglandin.
   Abstracts, 4th International Congress on Pharmacology. Basle, 14-18 July, p. 378. 1969.

   Abstract only. This study of factors controlling the formation and release of prostaglandins suggests that mediators of prostaglandin formations and release function by activating phospholipase A, thus releasing substrate. (JRH) 1597

1120
VOGT, J.
Release from brain tissue of compounds with possible transmitter function: interaction of drugs with these substances.
British Journal of Pharmacology. 37: 325-337. 1969.

The author makes reference to the work of Ramwell and Shaw (1966) who showed that prostaglandins were released into a cup applied to the surface of the somatosensory cortex of the cat when afferent nerves or the contralateral cortex are stimulated. The author suggests prostaglandins perform a general metabolic role in cell activity since they are also set free when many peripheral nerves are stimulated and might be transformed breakdown products of phospholipid membranes. (GW) 1650

1121
VOHRA, M.M.
Effect of prostaglandin E_1 on the nictitating membrane of the cat and its influence on the response to some agonists.
Abstracts, 4th International Congress on Pharmacology. Basle, 14-18 July, p. 369. 1969.

Abstract only. The effect of intraarterial (external carotid) administration of PGE_1 was studied on the response of the nicitiating membrane of cats anaesthetized with chloralose. The contractile response of the n.m. to PGE_1 varied widely from animal to animal and often persisted for 30-60 min. Injections of 1-3 μg of PGE_1 caused only contraction of the ipsilateral n.m., however contractions of the contralateral n.m. occurred with 7-10 μg doses of PGE_1. PGE_1 showed highly significant ($P < 0.01$) dose dependent effects. Neither devervation of the n.m. nor pretreatment with reserpine altered significantly the cumulative D-R curve for PGE_1. Contractile responses to PGE_1 were not blocked by atropine, phenoxybenzamine or promethazine. No effect of 1-100 mg doses of PGE_1 on the D-R curves for noradrenaline, adrenaline, acetycholine and serotonin was observed on either normal or devervated n.m. It is concluded that PGE_1 caused dose-dependent contraction of n.m. of cats by a direct-action of the smooth muscle. (Author modified) 1598

1122
VONKEMAN, H., D.H. NUGTEREN, and D.A. van DORP
The action of prostaglandin 15-hydroxydehydrogenase on various prostaglandins.
Biochimica et Biophsica Acta. 187: 581-583. 1969.

A study was made of a variety of prostaglandins derived from the newly discovered essential fatty acids. Prostaglandin E was prepared biosynthetically from the corresponding fatty acid. Prostaglandin A_1 and A_2 were prepared from prostaglandin E_1 and E_2. An enzyme preparation derived from swine lung was used. Two series of incubations were performed with separately prepared batches of enzyme. Each prostaglandin tested, the amount of 15-keto prostaglandin E produced was plotted against the time of incubation. As the amounts of 15-keto-PGE formed are proportional to the incubation times, the reaction rates are constant for periods up to one hour. The prostaglandins tested are oxidized by the 15-hydroxy dehydrogenase at rates which are of the same order of magnitude. The results of this investigation indicate that "unnatural" but biologically active prostaglandins administered to an animal will be inactivated as rapidly as the ordinary prostaglandins. (GW) 1414

1123
WAITZMAN, M.B.
Effects of prostaglandin and α-adrenergic drugs on ocular pressure and pupil size.
American Journal of Physiology. 217: 1593-1598. 1969.

Ocular administration in the rabbit eye of isoproterenol or norepinephrine causes significant antagonism of prostaglandin-induced miosis. Significant antagonism of PG-induced elevation of intraocular pressure occurs with much lower doses of isoproterenol than with norepinephrine. Ocular administration of low doses of phenoxybenzamine or propranolol does not interfere with PG effects. In the rabbit, phenoxybenzamine (as reported earlier) blocks the antagonizing action of norepinephrine, but propranolol is much less effective in blocking isoproterenol. Physiological antagonism between PG and the catecholamines in the cat eye appears similar relative to pupil effects, although the cat eye responds to much lower doses of isoproterenol. Phenoxybenzamine blocks more readily than propranolol the isoproterenol action in the cat eye, indicating some α-adrenergic action of isoproterenol in this animal. (Author) 1461

1124
WALASZEK, E.J., D.D. SUMNER, D.C. DYER and R.A. WILEY
A pharmacologically active lipid in kidney tissue.
Abstracts, 4th International Congress on Pharmacology. Basle, 14-18 July, p. 126. 1969.

Abstract only. A pharmacologically active lipid isolated from rabbit kidney was a potent stimulator of smooth muscle preparations like the guinea pig ileum, rabbit duodenum, hamster colon and the guinea pig seminal vesicle. The experimentors were able to distinguish it from prostaglandins. (JRH) 1604

1125
WARD, C.O.
Prostaglandins—a new class of hormones.
Journal of the American Pharmaceutical Association. 9: 127-128. 1969.

The author poses the question as to what is the normal physiologic or pathologic role for the new series of prostaglandins. Their exogenous administration would make them likely mediators of physiological activity, at least in the tissues where prostaglandins are found. Their release upon nerve stimulation suggests a possible function as an inhibitory mediator of neuronal activity. The presence of prostaglandins in the gastrointestinal tract and their ability to inhibit gastric secretion and to stimulate intestinal smooth muscle suggests a function as local hormones aiding in regulation of gastrointestinal activity. The lack of prostaglandins then could conceivably contribute to many disorders. (GW) 1730

1126
WARD, J.P. and D.A. van DORP
Synthesis of methyl dl-13-hdyroxy-9cis-octadecenoate and methyl dl-19-hydroxy-all cis-8,11,14-eicosatrienoate.
Recueil des Travaux Chimiques des Pays-bas et Belgique. 88: 1345-1357. 1969.

The authors report that other workers have isolated 19-hydroxy prostaglandins from human seminal plasma and that these compounds were formed by gamma-l-oxydation of prostaglandins by enzymes from human liver. The authors had synthesized dl-19-hydroxy-all cis-8,11,14-eicosatrienoic acid from prostaglandins with enzymes from vesicular glands. It was found that, although hydroxylation of this hydroxytrienoic acid occurred it was not a precursor of 19-hydroxy prostaglandins in the enzyme systems from both human and sheep glands. (JRH) 1737

1127
WARNER, W.A.
Release of free fatty acids following trauma.
Journal of Trauma. 9: 692-699. 1969.

It is noted that free fatty acid levels in blood plasma are depressed by a variety of agents, including prostaglandin E. (ART) 1708

1128
WAY, L. and R.P. DURBIN
Inhibition of gastric acid secretion in vitro by prostaglandin E_1.
Nature. 221: 874-875. 1969.

The action of prostaglandin E_1 on isolated gastric mucosa was studied in vitro. Strips of frog gastric mucosa were incubated 3-4 hours in a nutrient solution free of histamine. Synthetic human gastrin was added to the solution. Gastrin produced a sharp rise in the rate of secretion in the control strip while PGE_1 almost completely blocked it in the experimental. PGE_1 $10^{-7}M$ blocked acid secretion when much smaller concentrations of histamine were used. This same concentration of PGE_1 had no effect on cAMP stimulation. It seems likely the PGE_1 inhibits acid secretion by decreasing the mucosal levels of cAMP. (GW) 1412

1129
WEEKS, J.R.
Book review: Ramwell, P.W. and J.E. Shaw, eds. "Prostaglandin Symposium of the Worcester Foundation for Experimental Biology," 402 pp. New York, John Wiley & Sons, Inc., 1968.
American Journal of Pharmaceutical Education. 33: 172. 1969.

Many aspects of the prostaglandins are discussed in the 33 papers in this book. Although their physiological roles are yet undefined, roles are implied in reproduction, kidney function, control of fat metabolism, ion transport in cell membranes, and the nervous, gastrointestinal, and cardiovascular systems. Coverage related to fat metabolism, the cardiovascular system, and renal function is especially comprehensive. Additional information which suggests an association of low seminal prostaglandin with male subfertility is presented. Other new activities of prostaglandins reported relate to inhibition of gastric secretion and ulcer formation. For the first time, details of the large-scale biosynthesis of prostaglandins are described. (GW) 1728

1130
WEEKS, J.R.
The prostaglandins: biologically active lipids with implications in circulatory physiology.
Circulation Research. 24/25 (supp. 1): I-123-129. 1969.

The author reviews experimental observations regarding prostaglandins, especially cardiovascular and renal effects on dogs. He emphasizes that the cardiovascular actions of the prostaglandins represent only a small part of the many actions of these powerful and versatile lipids and that their mechanism of action involves some fundamental aspect of the function of many types of cells. There follows a discussion by other researchers. Dr. E.E. Murihead of Memphis, Tenn. stated that in his laboratory PGE_2 does have some influence on the blood pressure of the animal with reno-vascular hypertension but only in extremely high doses, 1000 μg/kg. Another lipid extracted from the renal medulla, termed "antihypertensive neutral renomedullary lipid," lowers the blood pressure of the hypertensive rabbit in doses of 10-20 μg/kg. By gas-chromatography this lipid has at least 10 peaks, one of which representing about 6% of the lipid is believed to contain the active principle. (PR) 1529

1131

WEEKS, J.R.

Pharmacology of prostaglandin $F_{2\alpha}$ on the uterus and corpus luteum.

Abstracts, 4th International Congress on Pharmacology. Basle, 14-18 July, p. 12. 1969.

Abstract only. This article reviews the possible role of $PGF_{2\alpha}$ as a luteolytic agent. (JRH) 1593

1132

WEEKS, J.R., N.C. SEKHAR, and D.W. DUCHARME

Relative activity of prostaglandins E_1, A_1, E_2, and A_2 on lipolysis, platelet aggregation, smooth muscle and the cardiovascular system.

Journal of Pharmacy and Pharmacology. 21: 103-108. 1969.

The relative activities of four prostaglandins (PGE_1, PGA_1, PGE_2, and PGA_2) were determined in several biological tests. They were compared as intestinal muscle stimulants on rabbit duodenum and guinea-pig ileum, as inhibitors of platelet aggregation in rabbit plasma, as inhibitors of adrenaline-induced lipolysis in rat isolated epididymal fat, as vasodepressor agents in anaesthetized rats and dogs, and on both blood pressure and cardiac output in unanaesthetized dogs. Formation of PGA's by dehydration and introduction of one additional double bond virtually abolished activity in all of these systems except the cardiovascular system. PGE_2 was more active than PGE_1 on isolated rabbit duodenum and as an antilipolytic agent, but less active in the other systems. Only PGE_1 had high potency as an inhibitor of platelet aggregation. (Authors) 1480

1133

WEINER, R. and G. KALEY

Influence of prostaglandin E_1 on the terminal vascular bed.

American Journal of Physiology. 217: 563-566. 1969.

Experiments were performed to explore the influence of PGE_1 on the local regulation of blood flow in the terminal vascular bed of the rat mesoappendix. The vasomotor activity of PGE_1 on specific vascular components of the microcirculation as well as its effects on vascular responsiveness to locally applied vasoconstrictor and vasodilator agents were documented by direct in vivo microscopic observation. Locally administered PGE_1 transiently dilated all muscular microvessels. Unlike other naturally occurring vasodilator agents such as bradykinin and histamine, PGE_1 antagonized the constrictor action of angiotensin, epinephrine, norepinephrine, and vasopressin long after its vasodilator activity had vanished. However, PGE_1 did not interfere with the constrictor activity of serotonin, nor did it alter vascular responsiveness to the dilators, bradykinin and histamine. Our findings suggest that prostaglandin E_1 may serve as a local hormonal regulator of blood flow in the microcirculation by virtue of its vasodilator properties and its ability to suppress vascular responsiveness to endogenous constrictor amines and polypeptides. (Authors) 1455

1134

WEINHEIMER, A.J. and R.L. SPRAGGINS

The occurrence of two new prostaglandin derivatives (15-epi-PGA_2 and its acetate, methyl ester) in the gorgonian plexaura homomalla. Chemistry of coelentrates. XV.

Tetrahedron Letters. 5185-5188. 1969.

Two new prostaglandin derivatives found in a gorgonian were isolated and their structures determined. One of these derivatives was found to have a structure very similar to mammalian PGA_2. The authors suggest that the compounds they identified might possibly be used as synthetic precursors to currently useful prostaglandins. (RAP) 1782

1135
WESTERMANN, E. and K. STOCK
  Effects of adrenergic blocking agents of FFA mobilization.
  In: Holmes, W.L., L.A. Carlson, and R. Paoletti, eds., "Drugs Affecting Lipid Metabolism," p.
    45-61. New York, Plenum Press, 1969.

The site and mode of action of methoxamine derivatives, various adrenolytics, and PGE_1 on the inhibition of catecholamine-induced lipolysis are reviewed in addition to the presentation of new experimental data. Results with PGE_1 show inhibition of norepinephrine, ACTH and theophylline-induced lipolysis in a competitive manner, but no inhibition of dibutyryl cyclic AMP-stimulated lipolysis. It is concluded that PGE_1 interferes with the binding of the substrate ATP to adenyl cyclase rather than interfering with the triglyceride lipase activation. Thus the picture produced is one of apparent competition with various lipolysis activators. (RMS) 1492

1136
WESTERMANN, E., K. STOCK and P. BIECK
  Phenlisopropyl-adenosin (PIA): Ein poteneter hemmstoff der lipolyse in vivo und in vitro.
    [Phenylisopropyl adenosine (PIA): A powerful inhibitor of lipolysis in vivo and in vitro.]
  In: Berg, G., ed., "Fettstoffwechsel 5. Lipolyse und Lipolyseblocker," p. 68-73. Munich, Pallas
    Verlag. 1969.

The inhibitory effects of PIA on lipolysis are reported, and are compared with those reported for a number of other substances including prostaglandins. (MEMH) 1769

1137
WESTERMANN, E., K. STOCK, and P. BIECK
  Phenlisopropyl-adenosin (PIA): Ein poteneter hemmstoff der lipolyse in vivo und in vitro.
    [Phenlisopropyl-adenosine (PIA): a potential inhibitor of lipolysis in vivo and in vitro.]
  Medizin und Ernahrung. 10: 143-147. 1969.

Phenylisopropyl adenosine (PIA) in vivo is about a 20,000 times stronger inhibitor of lipolysis than adenosine. Numerous substances of diverse chemical structure, including the prostaglandins, are capable of the in vitro inhibition of lipolysis. These substances can influence the reaction in various places. They can inhibit the formation of cAMP as, for example, prostaglandin E_1 does; they can initiate the decomposition of cyclic 3',5'-AMP, and they can block the reaction of cyclic 3',5'-AMP with triglyceridlipase. As a result of their investigations, the authors assume that PIA attacks the adenyl cyclase system of the fat cell, possibly like PGE_1, interfering with combination of ATP to the enzyme adenyl cyclase. (GW) 1530

1138
WINDHAGER, E.E.
  Kidney, water, and electrolytes.
  Annual Review of Physiology. 31: 117-172. 1969.

The author reviews several recent studies that have demonstrated that prostaglandins exert significant effects on the secretory function of the kidney. Prostaglandin E_1 infused into the renal artery of dogs increased urine volume, urinary Na excretion, free water clearance and renal plasma flow. A similar study determined that the extraction ratio of PAH was significantly reduced but the filtration rate did not change. Perfusion experiments on isolated collecting tubules of rabbit kidneys suggest that PGE_1 interferes with vasopressin-induced osmotic water flow. Another study established that contractions of helical strips from small renal arteries induced by catecholamines are relaxed by low concentrations of prostaglandins. The identification of prostaglandins E_2, $F_{2\alpha}$, and A_2 from the medulla of rabbit kidneys is also reported. (GW) 1710

1139
WILLIAMS, C.R. and D.B. BREWER
Medullary carcinoma of the thyroid.
British Journal of Surgery. 56: 437-443. 1969.

The author briefly mentions a paper by E.D. Williams in which the medullary carcinoma from a patient was found to contain large quantities of prostaglandins. (JRH) 1651

1140
WILLIS, A.L.
Parallel assay of prostaglandin-like activity in rat inflammatory exudate by means of cascade superfusion.
Journal of Pharmacy and Pharmacology. 21: 126-128. 1969.

Features of experimental methods used in showing the pharmacological activity of inflammatory exudates and reported elsewhere (Willis, In "Prostaglandins, Peptides and Amines." Symposium, Florence, Italy, 1968) are described in detail. The author suggests that the pharmacological activity in inflammatory exudates from rats is mainly attributable to the presence of E-type prostaglandins. The first source of prostaglandin-like activity was the edema fluid from rat feet inflamed by carrageenin. Rat stomach strips suspended in a modified cascade apparatus showed a high sensitivity to prostaglandins (0.5 to 1 ng of PGE_2). The thyrode superfused rectum from chicks of 150-200 g responded in a selectively sensitive manner to E-type prostaglandin. This tissue was equisensitive to PG's E_1 and E_2 but at least 100 times less sensitive to $PGF_{1\alpha}$ and 20 times less sensitive to $PGF_{2\alpha}$. Other tissues used have been the gerbil colon, the guinea-pig proximal colon and the rat colon. (GW) 1479

1141
WILLIS, A.L.
Release of histamine, kinin and prostaglandins during carrageenin-induced inflammation in the rat.
In: Mantegazza, P. and E.W. Horton, eds., "Prostaglandins, Peptides and Amines," p. 31-38. London, Academic Press, 1969.

The release of pharmacologically active substances during carrageenin inflammation has been studied with a view to establishing their role as mediators of inflammation. Inflammatory exudate, obtained by use of the carrageenin air bleb technique, produced (in the absence of antagonists) responses similar to those of histamine; these responses were blocked by mepyramine. Groups of rats were killed at intervals up to 24 hours after injection of carrageenin. The exudate from the air blebs was examined for its prostaglandin-like activity. Extracts were assayed on selective tissues such as rat stomach, gerbil colon, rat colon, guinea pig proximal colon, and chick rectum, which show different sensitivities to prostaglandins of the E and F series. The time-course for prostaglandin-like activity in the air bleb exudate showed little activity at one hour (less than 5 ng/ml) but rising values after 3 hours that leveled off somewhat between 18 and 24 hours. Mean maximum values were on the order of 80 ng/ml. The time course of histamine and kinin values is discussed. Prostaglandins were identified as PGE_1 and PGE_2. It is concluded that they must now be considered to play a significant role as mediators of inflammation. (ART) 1689

1142

WILSON, D.E. and R.A. LEVINE

Decreased canine gastric mucosal blood flow induced by prostaglandin E_1: a mechanism for its inhibitory effect on gastric secretion.

Gastroenterology. 56: 1268. 1969.

Abstract only. Dogs with Heidenhain pouches were infused iv with histamine to produce maximum gastric secretion. PGE_1 was then infused iv at a rate of 1 μg/kg/min or 2 μg/kg/min for 30 min. Gastric secretions were collected every 15 min, starting 30 min before PG infusion and continuing until 30 min after infusion. The secretions were analyzed for volume, pH, and titratable acidity. Mucosal blood flow was also measured. PGE_1 caused a reduction in all measured factors at both dose levels. The only difference in the two dose levels was that the larger dose caused a significantly greater reduction in mucosal blood flow. The authors report that they have also found a decrease in gastric secretion after administering cAMP to intact dogs. (JRH) 1459

1143

WOLFE, L.S.

Features of chemical structure of synaptic membranes.

In: Jasper, H.H., A.A. Ward and A. Pope eds., "Basic Mechanisms of the Epilepsies," p. 782-790. Boston, Little, Brown and Co., 1969.

Particular attention should be given the possibility that abnormal behavior of epileptogenic tissue is due to response of normal neurons to an abnormal chemical environment. In the brain, the commonest essential polyunsaturated fatty acid is arachidonic acid from which the most common abundant brain prostaglandin $PGF_{2\alpha}$ is derived. Two general hypothesis of the primary site of action of prostaglandins have been advanced: (1) prostaglandins modulate membrane-bound adenyl cyclase activity by feedback inhibition of hormonally stimulated cyclic 3',5'-adenosinemonophosphate formation, (2) prostaglandins regulate the calcium distribution and fixation in cellular compartments after stimulation of receptor sites by neurohormones. Prostaglandins may affect calcium distribution in effector cells. The brain prostaglandins formed and released upon stimulation may control neuronal responses by their action on distribution of calcium between membrane and cytoplasmic phases. (GW) 1715

1144

WOLFE, S.M. and N.R. SHULMAN

Adenyl cyclase activity in human platelets.

Biochemical and Biophysical Research Communications. 35: 265-272. 1969.

Adenyl cyclase activity has been demonstrated in washed particles from ultrasonically disrupted platelets. NaF increased activity 10-fold and PGE_1 increased activity 18-fold. This is the first time a substance other than NaF has caused maximal stimulation of adenyl cyclase. Activation was found with PGE_1 concentrations below 10^{-8}M. The inhibitory effect of PGE_1 on platelet aggregation may be related to activation of adenyl cyclase. (Authors) 1512

1145
WOOSTER, M.J.
    Prostaglandins, membrane permeability and renal function.
    Nephron. 6: 691-692. 1969.

    Abstract only. Prostaglandin E_1, applied in the serosal medium to the isolated urinary bladder of the Dominican toad, produced a sustained, dose-dependent increase in short-circuit current and net sodium flux. PGE_1 also induced a small, transient increase in hydro-osmotic permeability in bladders from these toads, but was without effect on chloride permeability. PGE_1 inhibited the increased hydro-osmotic permeability response to vasopressin. There was a lesser inhibition of the increased SCC response to either low or high concentrations of vasopressin. The increased SCC response to 1×10^{-8} g/ml PGE_1 was potentiated by 1 mM theophylline, and may have been mediated via stimulation of adenyl cyclase. The pathway for PGE_1-stimulated sodium transport was more sensitive to removal of sodium from the mucosal solution than was that for basal or vasopressin-stimulated sodium transport. (Author modified) 1763

1146
WYLLIE, J.H., P.J. PIPER, and J.R. VANE
    Fate of prostaglandins in the lungs.
    British Journal of Surgery. 56: 623. 1969.

    Abstract only. The removal mechanism of prostaglandin E_2 and $F_{2\alpha}$ in the lungs was studied. Prostaglandins were infused intra-arterially into the lungs of guinea pigs and were assayed in the effluent either continuously or intermittently on isolated tissues. Chromatography showed that prostaglandin E_1, E_2, $F_{2\alpha}$ and A_2 were changed to less polar derivatives during passage through the lungs. The average percentage conversion in a single transit was estimated to be 75% for E_1, 82% for E_2, 58% for $F_{2\alpha}$ and 38% for A_2. For prostaglandin E_2 this conversion was much slower at $10^{\circ}C$ and was undetectable at $6^{\circ}C$, indicating that it was due to metabolism. (GW) 1520

1147
YAMAMURA, Y. and T. KOKUBU
    [Prostaglandins.]
    Naika. 22: 891-899. 1969.

    This review article in Japanese covers all aspects of prostaglandin research with sections on synthesis, structure, biological effects, prostaglandin metabolites, chemical identification, possible modes of action on an enzymatic level, and interaction with other biologically active compounds. The bibliography has 32 entries. (JRH) 1684

1148
YOSHIDA, H.
    Endocrinology.
    Naika. 23: 40-46. 1969.

    Article in Japanese. Abstract not available at present. 1670.

1149
YURA, Y. and J. IDE
A total synthesis of a dl-prostaglandin B_1.
Chemical and Pharmaceutical Bulletin. 17: 408-410. 1969.

The authors report the total synthesis of the racemic prostaglandin B_1. A keto acid (β-oxo acid) was cyclized to a cyclopentone-1,3-dione derivative. Treatment of the dione with diazomethane in ether gave the melthylenol ether. This product was then reacted with magnesium bromide by the Grignard reaction to yield acetylene ester. Selective hydrogenation in the presence of Lindlar's catalyst produced a cis-olefinic ester which was subject to prolonged catalytic hydrogenation to produce the ethyl ester of PGE_1-237. Treatment of the cis-olefinic ester with 10% KOH in 50% methanol at 50% for 10 hours followed by acidification furnished pure dl-prostaglandin B_1. (GW) 1716

1150
ZETLER, G., D. MONKEMEIER, and H. WIECHELL
Peptid-receptoren fur tachykinine in der tuba uterina des menschen. [Peptide receptors for tachykinins in the human fallopian tube.]
Naunyn-Schmiedebergs Archiv fur Pharmakologie und Experimentelle Pathologie. 262: 97-111. 1969.

Article in German with an English summary that only briefly mentions prostaglandins. (JRH) 1585

1151
ZETLER, G. and H. WIECHELL
Pharmakologisch aktive lipide in extrakten aus tube und ovar des menschen. [Pharmacologically active lipids extracted from fallopian tube and ovary of humans.]
Naunyn-Schmiedebergs Archiv fur Pharmakologie und Experimentelle Pathologie. 265: 101-111. 1969.

Polypeptides with smooth-muscle stimulating activity are not present in extracts made from isthmus and ampulla of human Fallopian tube, and from human ovary. Prostaglandins-like activity exists in the ampulla but not in the isthmus of the tube or in the ovary. In the ampulla, there are probably two prostaglandin-like compounds which do not seem to be identical with prostaglandins E_1 and $F_{2\alpha}$. (Authors) 1587

1152
ZETLER, G., D. MONKEMEIR, and H. WIECHELL
Stimulation of fallopian tubes by prostaglandin $F_{2\alpha}$, biogenic amines and peptides.
Journal of Reproduction and Fertility. 18: 147-149. 1969.

The quantitative pharmacological investigation of the ability of various biogenic compounds to stimulate human fallopian tubes in vitro was undertaken. Responses were calculated relative to eledoisin, which was more than twice as potent as the next most active substance. $PGF_{2\alpha}$ was slightly less active than physalaemin and substance P, and these 4 substances were vastly more potent than 5-hydroxytryptamine, acetylcholine, hydrochloride and histamine dihydrochloride. Isolated tubes of guinea pig, rabbit, pig, and sheep were also used in a series of experiments briefly mentioned. A qualitative and quantitative similarity to human preparations was found only in the rabbit. It is noted that the tachykinin peptides and $PGF_{2\alpha}$ may have a physiological role in ovulation. (RMS) 1489

1153
ZIEVE, P.D. and W.B. GREENOUGH III
   Adenyl cyclase in human platelets: activity and responsiveness.
   Biochemical and Biophysical Research Communications. 35: 462-466. 1969.

Hormonally responsive adenyl cyclase activity was measured in lysates of human platelets. Activity was stimulated by fluoride, prostaglandin E_1, and glucagon; and was inhibited by thrombin, epinephrine, norepinephrine and serotonin. The results suggest that adenosine-3',5' monophospate (cAMP) may be important in the regulation of platelet adhesiveness. (Authors) 1513

1154
ZOR, U., T. KANEKO, I.P. LOWE, G. BLOOD, and J.B. FIELD
   Effect of thyroid-stimulating hormone and prostaglandins on thyroid adenyl cyclase activation
      and cyclic adenosine 3',5'-monophosphate.
   Journal of Biological Chemistry. 244: 5189-5195. 1969.

Thyroid-stimulating hormone (TSH) rapidly increased cyclic adenosine 3',5'-monophosphate (cAMP) concentration and increased ^{14}C-l-glucose oxidation which was still present 60 minutes after TSH injection in canine thyroid slices. Prostaglandin E_2 reproduced effects of TSH on both ^{14}C-l-glucose oxidation and cyclic AMP concentration. Prostaglandin $F_{1\alpha}$ stimulated ^{14}C-l-glucose oxidation, but it did not increase cAMP levels. Prostaglandin B_1 did not modify either cyclic AMP concentrations or ^{14}C-l-glucose oxidation. (GW) 1442

1155
ZOR, U., G. BLOOM, I.P. LOWE, and J.B. FIELD
   Effects of theophylline, prostaglandin E_1 and adrenergic blocking agents on TSH stimulation of
      thyroid intermediary metabolism.
   Endocrinology. 84: 1082-1088. 1969.

Prostaglandin E_1 $(0.95 \times 10^{-6}\,M)$ significantly increased glucose-1-^{14}C oxidation in dog thyroid slices but had no effect on ^{32}P incorporation into phospholipid. Prostaglandin E_1 neither inhibited nor potentiated the effects of TSH on $^{14}CO_2$ production. Phenoxybenzamine $(2.9 \times 10^{-5}M)$, an α-adrenergic blocking agent, did not modify basal or TSH stimulated glucose oxidation. DCI $(1.15 \times 10^{-4}M)$, a β-adrenergic blocking agent, occasionally stimulated glucose oxidation but did not consistently reduce the effect of TSH. $3 \times 10^{-5}M$ Propranolol, another β-adrenergic blocking agent, did not influence basal glucose oxidation or the increase caused by TSH. At higher concentrations $(3 \times 10^{-4}M)$ basal glucose oxidation was sometimes increased and TSH stimulation was consistently inhibited. This higher dose also abolished dibutyryl 3',5'-cyclic AMP (DBC) enhanced glucose oxidation but had less of an effect on carbamylcholine stimulation. Glucose oxidation was not increased by ATP, ADP, AMP, or α-MSH. These results are discussed in relation to the hypothesis that TSH controls thyroid gland function via generation of 3',5'-cyclic adenosine monophosphate. (Authors modified) 1419

1156
ZOR, U., T. KANEKO, H.P.G. SCHNEIDER, S.M. McCANN, I.P. LOWE, G. BLOOM, B.
BORLAND, and J.B. FIELD
   Stimulation of anterior pituitary adenyl cyclase activity and adenosine 3':5'-cyclic phosphate by
      hypothalamic extract and prostaglandin E_1.
   Proceedings of the National Academy of Sciences. 63: 918-925. 1969.

Hypothalamic extract, containing the releasing factors for anterior pituitary hormones, within minutes stimulated adenyl cyclase activity and adenosine 3':5'-cyclic phosphate (cAMP) concentrations in rat anterior pituitary in vitro. Cerebral cortical extract was ineffective and hypothalamic extract had no effect on these parameters in posterior pituitary or thyroid. Prostaglandin E_1 also increased adenyl cyclase activity and cAMP levels in anterior pituitary tissue. (Authors) 1475

1157
ZOR, U., T. KANEKO, H.P.G. SCHNEIDER, S.M. McCANN and J.B. FIELD
Stimulation of anterior pituitary (AP) adenyl cyclase activity (ACA) and cyclic 3',5'-adenosine-monophosphate (cAMP) by hypothalamic extract (HE) and prostaglandin E_1.
Abstracts, 4th International Congress on Pharmacology. Basle, 14-18 July, p. 381. 1969.

Abstract only. The role of ACA and CAMP in release of AP hormones mediated by HE was studied in rat AP in vitro. PGE_1 (0.6 μg/ml) augmented ACA (150%) and increased CAMP from 15.5 to 169 mμmoles/gm in AP while PGF was inactive. (Authors modified)

1158

ABDEL-SAYED, W., F.M. ABBOUD, P.R. HEDWALL, and P.G. SCHMID
Vascular responses to prostagladin E_1 (PGE_1) in muscular and cutaneous beds.
Circulation. 42 (supp. 3): III, 126. 1970.

In this study, responses to intraarterial infusions of PGE_1 were examined simultaneously in the isolated, perfused gracilis muscle and hind paw of the dog. At low concentrations (10^{-9} g/ml) consistent arterial dilation was seen in the muscle, but not in the paw. At higher concentrations 7×10^{-9} to 10^{-8} g/ml) significant and equivalent dilation was seen in both beds. Venous responses were negligible. Acute denervation reduced the effect of PGE_1. The results indicate that the threshold of the vasodilator effect of PGE_1 is lower in skeletal muscle than in the cutaneous bed, that the major site of action is on the arterial segment, and that the dilation is not dependent on intact sympathetic innervation in a specific way. (GW) 2261

1159

ABE, K., M. SUZUKI, K. YOSHINAGA and T. TORIGAI
Chemistry and physiology of prostaglandins.
Clinical Endocrinology (Tokyo). 18: 567-578. 1970.

Article in Japanese. Abstract not available at present. 2518

1160

ABIKO, Y. and A. MINAMIDATE
Coronary vasodilatation produced by prostagladin E_1 (PGE_1), dipyridamole, Mg-345, acetylcholine (ACh), papaverine and sodium nitrate in the dog.
Folia Pharmacologica Japonica. 66: 95-96. 1970.

Article in Japanese. Abstract not available at present. 2515

1161

ADOLPHSON, R.L. and R.G. TOWNLEY
A comparison of the bronchodilator activities of isoproterenol and prostaglandin E_1 aerosols.
Journal of Allergy. 45: 119-120. 1970.

Abstract only. PGE_1 is shown to be inhibitory to bronchoconstriction induced by methacholine and histamine in guinea pigs with a five-fold greater potency than isoproterenol; with induced β-adrenergic blockade, PGE_1 continued more effective. In vitro, β-adrenergic blockade prevented isoproterenol-induced relaxation of methacholine-treated respiratory smooth muscle. No cardiac stimulating activity was noted for PGE_1 in the doses used, unlike proterenol. (RMS) 2300

1162

AHERN, D.G. and D.T. DOWNING
Inhibition of prostaglandin biosynthesis by eicosa-5,8,11,14-tetraynoic acid.
Biochimica et Biophysica Acta. 210: 456-461. 1970.

Conversion of arachidonic acid to prostaglandin E_2 by acetone powders of sheep seminal vesicles was irreversibly inhibited by low concentrations of eicosa-5,8,11,14-tetraynoic acid. This compound also inhibited the hydroxylation of linoleic and linolenic acids by the tissue preparation, thereby preventing the elimination of the inhibitory effects of these polyethylenic acids. The tetraacetylenic acid can therefore limit prostaglandin biosynthesis both by its direct effect on prostaglandin synthetase and by preventing the inactivation of endogenous inhibitors. (Authors) 2042

1163

AHERN, D.G. and D.T. DOWNING

Inhibition of prostaglandin biosynthesis in sheep seminal vesicular tissue by eicosa-5:8:11:14-tetraynoic acid.

Federation Proceedings. 29: 854 abs. 1970.

Abstract only. Orally administered eicosa-5,8,11,14-tetraynoic acid has been reported to inhibit lipid biosynthesis by human sebaceous glands in vivo. In seeking the mechanism for this effect the authors considered that the inhibitor might function by interference with the metabolism of its ethylenic analogue, arachidonic acid (eicosa-5,8,11,14-tetraenoic acid), the only known biologically active metabolites of which are the prostaglandins. This hypothesis was supported by an observation that the tetraacetylenic acid inhibits soy bean lipoxidase, which, like the prostaglandin synthetase, mediates an attack on specific polyunsaturated fatty acids by molecular oxygen. It is now reported that the tetraacetylenic acid is indeed a powerful and irreversible inhibitor of prostaglandin biosynthesis in sheep seminal vesicular tissue. In addition to inhibiting prostaglandin synthesis, the tetraacetylenic acid prevents the conversion of linoleic and linolenic acids to hydroxy acids. It is therefore apparent that the acetylenic compound can inhibit prostaglandin biosynthesis both by its direct effect on prostaglandin synthetase and by preventing inactivation of endogenous linoleic and linolenic acids. (ART) 2113

1164

AHN, C.S. and I.N. RESENBERG

Iodine metabolism in thyroid slices: effects of TSH, dibutyryl cyclic $3',5'$- AMP, NaF and prostaglandin E_1.

Endocrinology. 86: 396-405. 1970.

Dog thyroid slices were used to determine the effect on iodine metabolism of thyroid stimulating hormone (TSH) dibutyryl cAMP, NaF and PGE_1. All the thyroid slices were preincubated in Krebs-Ringer phosphate (KRP) with a slightly modified Ca^{++} concentration for one hour before being used in an experiment. They were then placed in KRP with $Na^{131}I$ added to follow organic binding and uptake of iodine. In some experiments adenyl cyclase activity and cAMP formation were followed with a labeled ^{32}P-ATP and 8-^{14}C-adenine respectively. The preincubation procedure significantly increased organic binding of iodine even in control slices. It was found that TSH, dibutyryl cAMP, NaF and PGE_1 (0.03 mM) produced a 2-3 fold increase in binding of iodine and it was independent of iodide transport. In thyroid slices in which organic binding was blocked with methimazole, the uptake of radio iodine was lower with TSH, dibutyryl cAMP, NaF, and PGE_1 than control slices. Adenyl cyclase activity was stimulated with TSH, NAF and PGE_1 while accumulation of labeled cAMP was stimulated only by TSH and PGE_1. The authors feel that increase in organic binding of iodine induced by TSH and PGE_1 are likely due to increased formation of cAMP. (JRH) 2223

1165

AHN, C.S. and I.N. ROSENBERG

Protoeolysis in thryoid slices: Effects of TSH, dibutyryl cyclic $3',5'$-AMP and prostaglandin E_1.

Endocrinology. 86: 870-873. 1970.

Proteolysis of labeled iodoprotein in dog thyroid slices was enhanced by the presence in the medium of dibutyryl cyclic $3',5'$-AMP (dbcAMP) and prostaglandin E_1 (PGE_1) as well as TSH.

Inorganic iodide, iodotyrosines and iodothyronines were the major labeled products found. The stimulating effects of maximally effective concentrations of TSH, dbcAMP, and PGE_1 on proteolysis in thryoid slices was not additive, suggesting a common pathway for their actions. Theophylline, caffeine, epinephrine and NaF had no effect on proteolysis. (Authors) 2222

1166

AKANUMA, M.

Modes of stimulating action of prostaglandin E_1 on the gastrointestinal tract from the guinea-pig.

Sapporo Medical Journal. 38:41-52. 1970.

The present experiments were performed to clarify the mechanism of stimulating action of prostaglandin E_1 (PGE_1) on the isolated gastrointestinal smooth muscle from the guinea-pig as compared with the action of acetylcholine (ACh). The results obtained were as follows: 1) Dose-response curves of stomach colon strips to PGE_1 were similar to those of ACh. The threshold concentration of PGE_1 for contraction was lower than that of ACh. 2) Contractile responses to PGE_1 and ACh were strong in the stomach and colon whereas they were weak in the esophagus, duodenum, jejunum, and ileum. 3) Atropine (10^{-6}-10^{-5}g/ml) had no effect on the response of the stomach and colon to PGE_1 (10^{-9}-10^{-7}g/ml), but showed an inhibitory effect on the response of the duodenum, jejunum and ileum by as much as 29.6, 26.6, and 34.8%, respectively. 4) Eserine (10^{-7}g/ml) did not potentiate the response of the stomach and colon to PGE_1, but always potentiated the response of the duodenum, jejunum, and ileum by about 200-300%. 5) Tetrodotoxin (10^{-7}-10^{-6}g/ml) had no effect on the response of the stomach and colon, but some effects on the ileum to PGE_1 (10^{-8}-10^{-7}g/ml). 6) The isolated gastrointestinal strips from reserpinized guinea-pigs (5mg/kg $\times$ 2 for 24 hr) were similarly sensitive to PGE_1 and ACh as compared with normal strips. 7) Dibenamine (10^{-5}g/ml) and propranolol (2×10^{-6}g/ml) did not pontentiate the response to PGE_1 of the stomach or ileum. 8) PGE_1 had no potentiating effect on the stimulating action of the stomach, duodenum, jejunum, ileum and colon to ACh. 9) Additive contraction was observed by the application of PGE_1 (10^{-7}g/ml) and ACh (10^{-5}g/ml) in the depolarized stomach strips. From these results obtained, it is postulated that, in the gastrointestinal strips from the guinea-pig, PGE_1 appears to have a direct action on the muscle cells and a partial neuronal action which may act on intrinsic nerves therefore bring about a release of acetylcholine-like substances. The nervous component, in the action of PGE_1, may be different in various portions of the digestive tract and in different species. (Author) 2530

1167

AKANUMA, M.

Relationship between stimulating action of prostaglandin E_1 and calcium on the gastrointestinal smooth muscle from the guinea-pig.

Sapporo Medical Journal. 38: 53-59. 1970.

The stimulating effects of prostaglandin E_1 (PGE_1) on the depolarized muscle preparation of the gastrointestinal tract from guinea-pig were studied with regard to external calcium concentrations. The results obtained were as follows: 1) The stomach and colon preparations depolarized by the KCL-Krebs solution produced tensions by the addition of ACH (10^{-5}g/ml), PGE_1 (10^{-7}g/ml) and 2.5mM Ca, respectively. Only the tension produced by ACh was abolished in the presence of atropine (10^{-6}g/ml). 2) The colon preparation relaxed by the Ca-free KCl-Krebs solution was contracted with increasingly added Ca. The response reached its maximum at 2.5mM Ca addition. 3) The Ca response of the colon preparation relaxed by the

Ca-free KCl-Krebs solution was potentiated by the addition of PGE_1 (10^{-7}g/ml) with decreasingly added Ca concentrations. ACh (10^{-6}g/ml) also showed a similar effect to that of PGE_1. The effect of ACh was abolished by atropine (10^{-6}g/ml). From these results obtained, it is strongly suggested that PGE_1 acts on different sites of the cell membrane as compared against ACh, and that PGE_1 increases cell membrane permeability to Ca in a depolarized preparation which in turn increases the entry of external Ca which causes the development of tension. (Author) 2544

1168

ALABASTER, V.A. and Y.S. BAKHLE

The release of biologically active substances from isolated lungs by 5-hydroxytryptamine and tryptamine.
British Journal of Pharmacology. 40: 582P-583P. 1970.

These experiments were performed to determine if large doses of serotonin would cause the release of biologically active substances from isolated perfused lungs from guinea-pigs, rats and dogs. The effluent from the pulmonary veins were bioassayed on up to six assay tissues. The assay tissues were pretreated with specific inhibitors to allow the detection of serotonin, histamine, kinins, prostaglandins, slow reaction substance and rabbit aorta contraction substance. It was found that prostaglandins were released in response to high serotonin infusion (0.05-1 μg/ml). (JRH) 2050

1169

AL-AWQATI, Q., M. FIELD, N.F. PIERCE, and W.B. GREENOUGH III

Effect of prostaglandin E_1 on electrolyte transport in rabbit ileal mucosa.
Journal of Clinical Investigation. 49: 2a. 1970.

Abstract only. Rabbit ileal mucosa in vitro was used to determine the effect of PGE_1, cholera exotoxin (CE), and theophylline on sodium absorption and chloride secretion. It was found that PGE_1 produces the same directional changes in ion movement as do CE and theophylline. The authors feel this indicates a common mechanism of action possibly involving cAMP. (JRH) 2221

1170

ALDRIDGE, R.R., S. BARRETT, J.B. BROWN, J.W. FUNDER, J.R. GODING, C.C. KALTENBACH and B.J. MOLE

The effect of prostaglandins on ovarian steroidogenesis in vivo.
Journal of Reproduction and Fertility 21: 369-371. 1970.

The effect of PGE_1 and $PGF_{1\alpha}$ on the progesterone secretion of superovoulated ovaries was investigated in 4 sheep with cervical ovarian autotransplants. Control secretion rates of 13, 17, 19, and 35 fell to 10, 9, 14, and 20 mg/min within 10 minutes of the start of PGE_1 infusion despite a two- to three-fold increase in blood flow. $PGF_{1\alpha}$ caused a dramatic drop in progesterone secretion rates from 7, 10, 15, and 21 mg/min to 5,5,5, and 7 mg/min, though there was no consistent change in blood flow. Progesterone secretion remained depressed even 60 minutes after the end of PG infusion. (RMS) 2071

1171

ALEXANDRE, C. and F. ROUESSAC

La dicyclohexylcarbodiimide, agent de deshydratation intramoleculaires des cetols. [Dicyclo-hexylcarbodiimide, an intramolecular, dehydrating agent of ketols.]
Tetrahedron Letters. 1011-1012. 1970.

The authors mention that during the systhesis of prostaglandins Corey et al. used dicyclohexyl-carbodiimide to dehydrate a β ketol intermediary. (NES) 2497

1172

AMBACHE, N. and M.A. ZAR

An inhibitory action of histamine on the guinea-pig ileum.

British Journal of Pharmacology. 38: 229-240. 1970.

In atropinized, plexus-containing preparation of the longitudinal muscle from the guinea-pig ileum, in which histamine contractions were abolished by mepyramine or diphenhydramine, an inhibitory action of histamine was revealed on the "tetanic spasms" produced by field stimulation. Histamine did not reduce contractions elicited by prostaglandin E_2 bradykinin, 5-hydroxytryptamine, nicotine or dimethylphenylpiperazinium. (GW) 2232

1173

AMBACHE, N. and M.A. ZAR

An inhibitory effect of prostaglandin E_2 on neuromuscular transmission in the guinea pig vas deferens.

Journal of Physiology. 208: 30P-32P. 1970.

The authors discuss the high activity of PGE_2 which seems to indicate a regulatory role for prostaglandins in some types of humoral transmission. PGE_2 but not $PGF_2\alpha$ is able to inhibit contraction of the vas deferens which have been electronically stimulated by trains pf 5-10 pulses of supramaximal voltage per minute. This effect of PGE_2 is not attributed to a non-specific depression of the smooth muscle or to a specific antagonism against noradrenaline or acetylcholine. (GW) 2005

1174

AMBACHE, N., J. VERNEY, and M.A. ZAR

Evidence for the release of two atropine-resistant spasmogens from Auerbach's plexus.

Journal of Physiology. 207: 761-782. 1970.

Two atropine-resistant response-components of nervous origin have been detected in the plexus-containing preparations of the longitudinal muscle from the guinea-pig ileum, by alternate field stimulation with equal numbers of pulses at 50 Hz (response A) and a 5 Hz (response B). Neither response is due to a release of histamine, serotonin, or prostaglandins, since both A and B persist in the presence of mepyramine, methysergide, and the prostaglandin-antagonist Sc-19220 (Searle & Co.), 1-acetyl-2-(8-chloro-10,11-dihydrodibenz [b,f] [1,4] oxazepine-10-carbonyl) hydrazine. In the present experiments on atropinized preparations, it was found that the A and B responses persisted after the extinction of contractions, induced by 5 to 10 ng/ml of PGE_2, by the presence of 1 to 5 $\mu g/ml$ of Sc-19220 in the bath fluid. When the Sc-19220 was washed out, the PGE_2 contractions recovered fully but the bradykinin contractions and the A and B responses (which had been slightly reduced) failed to recover from the slight depression. These results exclude prostaglandins as possible mediators of the A or B responses. (ART) 2236

1175

AMBACHE, N., J. VERNEY, and M.A. ZAR

Evidence for the release of two non-cholinergic spasmogens from the plexus-containing longitudinal muscle of the guinea-pig ileum.

Journal of Physiology. 207: 8P-10P. 1970.

Auerbach's plexus-containing longitudinal muscle preparations from the guinea-pig ileum, bathed in Krebs solution containing atropine, mepyramine and methysergide were stimulated every minute by trains of 10 or 20 pulses at frequencies of, alternately, 5 and 50 Hz. Two non-cholinergic spasmogens and acetylcholine were released. Neither spasmogen is a prosta-glandin, as shown by tests with the specific antagonist SC-19220. (GW) 2059

1176

AMOUROUX, J., J.J. BERNIER, J.C. RAMBAUD, A.M. DUPUY-COIN and S.R. KALIFAT

Etude optique et ultrastructurale d'un cas de carcinome medullaire de la thyroide. [Optical and ultrastructural study of a case of medullary carcinoma of the thyroid.]

Zeitschrift fur Krebsforschung. 74: 122-130. 1970.

The authors briefly cite Karim and Sandler, 1968 on the secretion of prostaglandins by medullary carcinoma of the thyroid. (RMS) 2498

1177

ANGGARD, E., G. ARTURSON, and C.-E. JONSSON.

Efflux of prostaglandins in lymph from scalded tissues.

Acta Physiologica Scandinavica. 80: 46A-47A. 1970.

PGE_2 has been found in paw lymph after a dog sustained scalding injury. The lymph was extracted with ethanol and the acidic lipid fraction was chromatographed on silicic acid along with tritium tagged PGE_2 for group separation. Separation into individual components was effected on hydrophobic kieselguhr by reversed phase partition chromatography. Biological activity assays and radioactivity measurements were carried out on individual fractions. The methylated and silylated derivatives obtained from the prostaglandin separated by silicic acid and reversed phase chromatography had identical retention time on a gas-liquid chromatograph with an electron capture detector as the same derivative prepared from authentic PGE_2. (SR) 2037

1178

ANGGARD, E. and H. BERGKVIST

Group separation of prostaglandins on Sephadex LH-20.

Journal of Chromatography. 48: 542-544. 1970.

This report shows that a lipophilic dextran derivative, Sephadex LH-20, may be used for group separation of the methyl esters of prostaglandins E, F and A or B. The prostaglandin E and A compounds were detected by their absorption at 280 nm after conversion into the respective prostaglandin B by dilute NaOH. The prostaglandins of the F group were analyzed by quantitative gas-liquid chromatography after their conversion into the trimethylsilyl ether derivatives. The recovery of 50 ng of the methyl ester of prostaglandin E_2 was studied using the tritium-labeled compound added to extracts of acidic lipids from human plasma followed by chromatography. (GW) 2310

1179

ANGGARD, E.

Pharmacology of prostaglandins. Introductory survey.

In. Eigenmann, R., ed., "Proceedings of the Fourth International Congress on Pharmacology," July, 1969. P. 1-11,54-69. Basle, Schwabe, 1970.

This article is the chairman's report on a symposium on the pharmacology of prostaglandins. It contains a summary of all the theories presented at the symposium. (JRH) 2435

1180

ANONYMOUS

Abortion without surgery?

Time Magazine. Feb. 9: 39-40. 1970.

This article very briefly reviews the use of prostaglandins for abortion. (M.T.) 2448

1181
ANONYMOUS
  Abortions made safe.
  Nature. 225: 415-416. 1970.

  In this editorial review the author discusses the use of $PGF_{2\alpha}$ by Karim and Filshie to produce abortions in humans. The author generally favors the use of $PGF_{2\alpha}$ as an abortifacient. (JRH) 2487

1182
ANONYMOUS
  ALZA describes its prostaglandins research.
  Pharmaceutical Journal. 205: 426. 1970.

  The ALZA Corporation reports its commitment to develop products on naturally occurring prostaglandin. A prostaglandin contraceptive product will probably be developed within 3-4 years; an interim product intended to be administered once or twice a year to induce menses will probably appear before then. Bronchodilator and ocular therapeutic products are expected within a short time. Development of a prostaglandin product for causing an abortion will likely take 2-4 years. (GW) 2368

1183
ANONYMOUS
  Clinical Use of Prostaglandins.
  British Medical Journal. 4: 253-254. 1970.

  This editorial discusses prostaglandins as new and powerful oxytocic agents. Research is cited on the termination of pregnancy at 9 to 28 weeks. The effectiveness of $PGF_{2\alpha}$ over oxytocin without the antidiuretic effect of the latter is discussed and it is noted that PGE_2 has proven to be 10 times more potent that $PGF_{2\alpha}$, enabling the effective usage of prostaglandins in dosages free from side effects. The use of prostaglandins or an inhibitor, polyphloretin phosphate, as fertility control agents is speculated upon. (RMS) 2035

1184
ANONYMOUS
  Drug is held safe in late abortions.
  New York Times. Jan. 28: 26. 1970.

  This newspaper article briefly reports the use of prostaglandins to induce abortion in the first and second trimesters. (JRH) 2533

1185
ANONYMOUS
  Grant awarded for prostaglandin study.
  The Southwest Foundation Reporter. 4: 1, 5. 1970.

  A historical background of the discovery of prostaglandin in the semen of man and sheep is presented. One of the first medical applications likely will be that of reproduction regulation. The objectives of the research program will be to find a convenient method of prostaglandin administration and to look for possible side effects on the developing fetus. Another phase of the study will relate the use of prostaglandins as a contraceptive agent where it is hoped that two methods of reproduction control will be found. (GW) 2028

1186
ANONYMOUS
Pill progress.
Nature. 228: 807-808. 1970.

Recent developments in prostaglandin research may lead to the development of a once-a-month female contraceptive pill. Vaginally administered, prostaglandin tablets induce abortions at one-tenth of the dose required for successful intravenous injection. During the past three years, chemical synthesis of racemic prostaglandin E_1 with properties identical with the naturally occurring levorotatory enantiomer but half of its biological activity has been described. (GW) 2043

1187
ANONYMOUS
Pregnancy just melts away.
Medical World News. 11(41): 20-21. 1970.

This newspaper-type article discusses the use of prostaglandins as birth control substances and as abortifacients. The major clinical workers are listed with their methods and amount of success. The two prostaglandins discussed are PGE_2 and $PGF_{2\alpha}$. (JRH) 2391

1188
ANONYMOUS
Prostaglandin research intensifies.
Chemical and Engineering News. 48: 42-44. 1970.

At present, 16 prostaglandins have been discovered. It is hoped that prostaglandins can be used to regulate menstruation and fertility, prevent conception, induce child birth, lower blood pressure, prevent blood clots, treat asthmatics, and act as long-lasting nasal decongestants. These modulators of cellular metabolism are separated into 4 categories—E,F,A, and B. Although prostaglandin activity was first discovered in human semen in 1930, the first crystalline pure compounds were not produced until 1957. Scientific papers in all phases of prostaglandin research are currently appearing at the rate of more than one a day. (GW) 2373

1189
ANONYMOUS
Prostaglandins.
Lancet. 1: 223-226. 1970.

A brief history of the prostaglandins is presented with a statement about their complex chemical composition and diverse actions. They are exciting because of their extreme potency: effective concentrations are in the order of nanogrammes per 100 ml. The prostaglandins also show similarities to such humoral transmitters as acetylcholine and the catecholamines. The answer to what the prostaglandins mean in broad biological terms is at the present not know. (GW) 2299

1190
ANONYMOUS
Prostaglandins.
Lancet. 2: 1236. 1970.

Possible physiological, pathological and therapeutic roles of E and F type prostaglandins as discussed at the Society for Drug Research symposium in London on November 25, 1970 are mentioned very briefly and with very few details. (RMS) 2001

1191
ANONYMOUS
[Prostaglandins.]
Policlinico; Sezione Pratica. 77: 724-725. 1970.

After a brief history of prostaglandin research, the author summarizes some of their physiological effects on smooth muscle fiber, on the blood vessels, and on the metabolism. In his view, too little is yet known of these substances even to suggest possible therapeutic uses but he foresees great progress in the near future. (MEMH) 2406

1192
ANONYMOUS
Prostaglandins and the induction of labour or abortion.
Lancet. 1: 927-928. 1970.

The article reviews the use of prostaglandins for the induction of labor and for therapeutic abortion. They have a definite advantage over oxytocin to induce these processes with only trivial side effects. Extracts of human menstrual fluid are now known to contain PGF_α which has a spasmogenic action on the human uterus. Further study is needed to confirm the presence of prostaglandins in all tissues. (GW) 2046

1193
ANONYMOUS
Prostaglandins and therapeutic abortion.
Research in Reproduction. 2: 1. 1970.

Two papers, Karim and Filshie and Roth-Brandel et al., Lancet I. 1970, p. 157 and 190 respectively, are reviewed in a news-type article on the oxytocic effects of prostaglandins $F_{2\alpha}$ and E_1. In the first study, 14 of 15 patients, 9-22 weeks pregnant, aborted successfully after receiving $PGF_{2\alpha}$ 50 mg/min intravenously. Side effects were confined to diarrhea in 7 patients and vomiting in 3. In the second study, lower doses of both $PGF_{2\alpha}$ and PGE_1 were used, but abortion was induced in only 3 of 11 women. The suggestion is that long-term treatment should be more effective. The results reported for both papers are deemed most encouraging as being simple, minimal risk, non-traumatic and effective in both the first and second trimesters of pregnancy. (RMS) 2090

1194
ANONYMOUS
Prostaglandins: Talents of ubiquitous intracellular hormone regulators may range from abortion to treating peptic ulcer.
Medical World News. 11(35): 27-35. 1970.

An extremely wide-ranging news-type review on prostaglandins research is presented. Outlines of historical and ongoing research including the quoted personal remarks of many key investigators serve to connect and put into perspective the findings from the diverse fields of interest within prostaglandins research. A spearate report within the body of the review indicates the focus of research at 10 major pharmaceutical companies. (RMS) 2112

1195
ANONYMOUS
   Prostaglandins terminate pregnancy.
   Family Planning Perspectives. 2: 51. 1970.

This news item briefly reports the work of Karim and Filshie in inducing abortions with $PGF_{2\alpha}$. Some of the advantages and problems in therapeutic abortion with prostaglandins are discussed. (JRH) 2531

1196
ANTONACCIO, M.J. and B.R. LUCCHESI
   The interaction of glucagon with theophylline, PGE_1, isoproterenol, ouabain, and $CaCl_2$ on the dog isolated papillary muscle.
   Life Sciences 9: 1081-1089. 1970.

The effects of isoproterenol, PGE_1, theophylline, ouabain, $CaCl_2$, and glucagon on the isolated ventricular papillary muscle of dogs was measured. Some of these drugs were used in combination. PGE_1 (2.5×10^{-7}M) had no effect on the muscle, nor did it affect the response to glucagon. PGE_1 at 10^{-5}M increased the isometric tension of the muscle but still did not affect its response to glucagon. (JRH) 2012

1197
APPLEMAN, M.M. and C.L. SEVILLA
   Cyclic 3',5'-adenosine monophosphate and the effects of hormones on hormone sensitive lipase from adipose tissue.

   In: Greengard, P. and E. Costa, eds., "Role of Cyclic Amp in Cell Function: Advances in Biochemical Psychopharmacology 3." p. 209-216. New York, Raven Press. 1970.

Brief mention is made of the fact that prostaglandins are antilipolytic because they can partially reverse the action of the lipolytic hormones on adipose tissue. They also tend to reduce the amount of nucleotide present in adipose cells. (GW) 2364

1198
ARAKAWA, K.
   [Role of hormone in blood pressure control.]
   Clinical Endocrinology (Tokyo). 18: 617-622. 1970.

Article in Japanese. Abstract not available at present. 2076

1199
ARMSTRONG, D.T.
  Reproduction.
  Annual Review of Physiology. 32: 439-470. 1970.

  This review on selected aspects of reproduction briefly mentions that the luteolytic effects $PGF_{2\alpha}$ and LH may be secondary to their effects on the follicle. It is also mentioned that prostaglandins may have a role in sperm transport and parturition. (JRH) 2466

1200
ARORA, S., P.K. LAHIRI, and R.K. SANYAL
  The role of prostaglandin E_1 in inflammatory process in the rat.
  International Archives of Allergy and Applied Immunology. 39: 186-191. 1970.

  The injection of prostaglandin E_1 produced rupture of subcutaneous mast cells in the albino rat. There was also an increase in oedema formation and in capillary permeability. Fibrous tissue formation around implanted cotton wool pellets was augmented as well. On a dosage basis, prostaglandin E_1 was about a thousand times more effective in all these actions than was either histamine or 5HT. Depletion of histamine and 5HT from the tissues only slightly affected the inflammatory process. A possible role of prostaglandin E_1 as a mediator of inflammation is discussed. (Authors) 2241

1201
ARTURSON, G. and C.-E. JONSSON
  Tankbara mediatorer i brannskadad vavnad. [Imaginable mediators in burn damaged tissue.]
  Nordisk Medicin. 84: 959. 1970.

  Article in Norwegian. Abstract not available at present. 2111

1202
ASAKAWA, T.
  Release of prostaglandins.
  Shinryo. 23: 1339-1346. 1970.

  Article in Japanese. Abstract not available at present. 2519

1203
ASSAN, R.
  La Cetogenese: mecanism, regulation, retentissements metaboliques.
  [Ketogenesis: mechanism, regulation and metabolic repercussions.]
  Pathologie et Biologie. 18: 507-520. 1970.

  The author reviews current concepts of the biochemical aspects of ketogenesis stressing metabolic and hormonal regulation. The action of PGE as a powerful anti-lipolytic with an unknown role in ketogenesis is mentioned very briefly. (RMS) 2075

1204
AURBACH, G.D. and L.R. CHASE
  Cyclic $3',5'$-adenylic acid in bone and the mechanism of action of parathyroid hormone.
  Federation Proceedings. 29: 1179-1182. 1970.

  The authors review 23 studies pertaining to the action of parathyroid hormone and the adenyl cyclase-cyclic AMP system in bone and kidney. The increase in concentration of $3',5'$-AMP in fetal rat calvaria in response to prostaglandin E_1 is briefly mentioned. The physiological significance of PGE_1 in skeletal metabolism is said to be uninvestigated. (RMS) 2060

1205

AXEN, U., J.L. THOMPSON, and J.E. PIKE
A total synthesis of (±)-prostaglandin E_3 methyl ester via endo-bicyclo-hexane intermediates.
Chemical Communications. 602. 1970.

(±)-prostaglandin E_3 methyl ester has been synthesized via endo-bicyclo-hexane intermediates. (Authors) 2297

1206

BAGLI, J.F.
Photochemical synthesis of prostanoic acids.
In: Abstracts, 5th Mid Atlantic Regional Meeting of the American Chemical Society, 1-3 April. p. 49. Newark. 1970.

Abstract only. Experiments are discussed leading to the cleavage of the bicyclo [3,2,0] -heptane system to yield the prostanoic acid derivatives. Transformation of these to the analogs of prostaglandins are reported. Biological activity of some of these components are discussed. (GW) 2372

1207

BAGLI, J.F.
Prostaglandins and related compounds.
In: Cain, C.K., ed., "Annual Reports in Medicinal Chemistry," p. 170-179. New York, Academic Press. 1970.

This report dealing with recent chemical advances of the prostaglandins is divided into two parts: 1) synthesis of natural prostaglandins and 2) synthesis of unnatural prostanoic acids and related analogs. Since PGE's can be transformed to PGF's and PGA's, any synthesis of PGE also constitutes a synthesis of the corresponding PGF and PGA. (GW) 2181

1208

BALL, G., G.G. BRERETON, M. FULWOOD, D.M. IRELAND, and P. YATES
Effect of prostaglandin E_1 alone and in combination with theophylline or aspirin on collagen-induced platelet aggregation and on platelet nucleotides including adenosine $3':5'$-cyclic monophosphate.
Biochemical Journal. 120: 709-718. 1970.

The effect of PGE_1, theophylline, or aspirin, alone or in combinations, on collagen induced platelet aggregation was tested. By means of radio tracers the release of platelet nucleotides, the breakdown of platelet ATP and cAMP formation was followed. PGE_1 (0.11 μM to 0.56 μM), or aspirin alone inhibited collagen induced platelet aggregation, release of ATP and ADP, and the decay of ATP in a dose dependent manner. High doses of theophylline also inhibited platelet aggregation but its effect on nucleotides was unclear perhaps due to a direct effect of theophylline on the chromatographic analysis system. Only PGE_1 produced a significant increase in cAMP levels. PGE_1 and aspirin together and PGE_1 and theophylline together produced a synergistic effect on the inhibition of platelet aggregation even though they caused little or no change in cAMP. The authors concluded that it seems possible to inhibit platelet aggregation with combinations of drugs which produce little or no increase in cAMP. (JRH) 2393

1209

BAR, H.-P., O. HECHTER, I.L. SCHWARTZ, and R. WALTER
Neurohypophyseal hormone-sensitive adenyl cyclase of toad urinary bladder.
Proceedings of the National Academy of Sciences. 67: 7-12. 1970.

An adenyl cyclase preparation derived from epithelial cells of the urinary bladder of the toad, *Bufo marinus*, is described. This cyclase preparation is specifically stimulated by neuro-hypophyseal hormones and various synthetic analogs which evoke a hydro-osmotic response in the intact bladder. The relative stimulatory effects of these compounds have been compared on the cyclase preparation and in the intact bladder. The peptide concentrations required for half-maximal stimulation (affinity) in the cell-free and intact systems were parallel; however the magnitude of stimulation produced by saturating concentrations of peptides did not correlate. Futhermore, it was found that peptide analogs which inhibit the hydro-osmotic effect of [8-arginine]-vasopressin on the intact bladder also inhibit the stimulation of the toad bladder cyclase preparation by vasopressin. Prostaglandin E_1, mercaptans, and disulfides, which inhibit the hormone-induced hydro-osmotic response of the intact bladder, did not antagonize the stimulation of the toad bladder cyclase preparation by vasopressin. (Authors). 2040

1210

BARDEN, T.P.
The role of prostaglandins in reproductive physiology.
Ohio State Medical Journal. 66: 1009-1012. 1970.

The biochemical nature and physiopathologic roles of prostaglandins have been reviewed as related to reproductive medicine. These compounds have been implicated in uterine motility related to sperm transport, menstruation, and labor; lysis of the corpus luteum; and fallopian tube motility. From a therapeutic viewpoint, potential applications of prostaglandins have been suggested in the induction of labor, treatment of infertility, "morning-after" contraception, treatment of dysmenorrhea, and contraception by alteration of fallopian tube motility. (Author) 2423

1211

BARTELS, J., H. KUNZE, W. VOGT, and G. WILLE
Prostaglandin: Liberation from and formation in perfused frog intestine.
Naunyn-Schmiedebergs Archiv fur Pharmakologie. 266: 199-207. 1970.

Frog intestines were perfused through their vascular system and the content of prostaglandin (PG) in the effluent was estimated biologically. The spontaneous release of PG ranged from 0.55 to 29 ng PGE_1 equivalents/min. Thin-layer chromatography indicated that the liberated compound was a PGE compound. Acetylcholine and dimethyl-phenyl-piperazinium iodide raised the output of PG significantly. Adrenaline had no definite effect. Lysolecithin weakly stimulated the release of PG, but phospholipase A caused an extraordinary rise which was apparently due to formation of new PGE_1. A considerable formation of PG was also seen when arachidonic acid was infused. Being PGE_2 it obviously originated from the infused fatty acid. This fact demonstrates directly that a PG-forming enzyme system is active in the intact tissue. It is suggested that physiological stimuli can increase the formation of PG by providing free precursor acids. This could be effected by the activation of an endogenous phospholipase A. (Authors) 2228

1212
BAYLIN, S.B., M.A. BEAVEN, K. ENGELMAN, and A. SJOERDSMA
Elevated histaminase activity in medullary carcinoma of the thyroid gland.
The New England Journal of Medicine. 283: 1239-1244. 1970.

Williams et al., 1968 are cited without discussion in this report covering 9 cases of medullary carcinoma of the thyroid gland. (RMS) 2473

1213
BEAZLEY, J.M., C.J. DEWHURST and A. GILLESPIE
The induction of labour with prostaglandin E_2.
Journal of Obstetrics and Gynaecology of the British Commonwealth. 77: 193-199. 1970.

Prostaglandin E_2 was used to induce labour in 40 patients, between 29 and 42 weeks gestation. Vaginal delivery occurred in 37 patients; 3 other patients required Caesarean section. No fetal complications attributable to PGE_2 were observed, nor any maternal cardiovascular or alimentary effects. One patient demonstrated transient uterine hypertonus when infused with 40 ng/kg/min. It is suggested that as the uterine threshold for PGE_2 is variable, a slow rate of PGE_2 infusion should be used initially, but increased steadily until an effective infusion rate is obtained. (Authors) 2317

1214
BECK, N.P., J.B. FIELD, and B. DAVIS
Effect of prostaglandin E_1 (PGE_1) chlorpropamide (CPM) and vasopressin (VP) on cyclic 3'5'-adenosine monophosphate (cAMP) in renal medulla of rats.
Clinical Research. 18: 494. 1970.

Abstract only. PGE_1 was incubated with slices of rat renal medulla (10^{-8} to 10^{-3} M) and 10^{-2} M theophylline. Control cAMP levels were 5.71 $\mu\mu$Moles/g. PGE_1 per se had no effect on cAMP levels in renal cortex and inner medulla, but PGE_1 increased cAMP level in outer medulla. Dose responses were found starting at 10^{-3} M PGE_1. Vasopressin also increased cAMP levels in outer and inner medulla, but not in cortex. When vasopressin was tested in combination with 10^{-8} M PGE_1, vasopressin activity was decreased. These data indicate that PGE_1 inhibits the effect of vasopressin at the level of cAMP generation in the kidney. (Authors modified) 2157

1215
BECK, N.P., F. DeRUBERTIS, M.F. MICHELIS, R.D. FUSCO, J.B. FIELD, and B.B. DAVIS
Prostaglandin E, inhibition of parathyroid hormone-induced increases in renal cortical adenyl cyclase activity, cyclic adenosine monophosphate concentration, ^{32}P incorporation into phospholipid and in phosphate excretion.
Journal of Laboratory and Clinical Medicine. 76: 1005. 1970.

Abstract only. The interaction of PGE_1 and parathyroid hormone (PTH) on the renal cortex was measured. It was found that PGE_1 inhibits PTH induced activation of adenyl cyclase, increase in intracellular cAMP levels, and ^{32}P incorporation into phospholipids. It also inhibited the increase in renal phosphate excretion induced by PTH. However, it did not inhibit dibutyryl cAMP-induced increases in ^{32}P incorporation into phospholipids. It was suggested that PGE_1 interacts with PTH by inhibiting adenyl cyclase. (JRH) 2054

1216
BELL, N.H.
  Regulation of calcitonin secretion in vitro.
  Clinical Research. 18: 599. 1970.

  Abstract only. It was found that the secretion of calcitonin from pig thyroid slices in vitro was increased by PGE_1. Calcitonin secretion was inhibited by norepinephrine. (JRH) 2151

1217
BENNETT, A.
  Control of gastrointestinal motility by substances occurring in the gut wall.
  Rendiconti Romani di Gastroentrologia. 2: 133-142 1970.

  Several of the substances in the gut wall which may affect gastrointestinal muscle are discussed. This article briefly reviews some amines, polypeptides, prostaglandins and other uncharacterized substances and their possible involvement in gut motility under physiologic or pathologic conditions. Particular reference is made to the human gastrointestinal tract. (Author) 2443

1218
BENNETT, A. and B. FLESHLER
  Prostaglandins and the gastrointestinal tract.
  Gastroenterology. 59: 790-800. 1970.

  Literature on the place of prostaglandins in the physiological and pathological control of gastrointestinal activity is reviewed. Compounds PGE and PGF have been found in all gastrointestinal tissues studied so far and have been identified by physical, chemical, and pharmacological properties as well as by chromatography, mass spectrometry, and bioassay. They can be synthesized in the isolated stomach and intestinal mucosa of animals or released by the rat stomach at rest. Location and storage in the gut are not fully known but autoradiographic studies on the mouse pinpoint the connective tissue of the gut and the lumen. In vitro studies on animal tissue (muscle strips and intestine segments) showed contraction upon exposure to millimicrogram concentrations of PGE and PGF. In vivo studies carried out by injecting them in guinea pigs, rats, and dogs showed various degrees of gut motility. Different reactions are obtained in vitro and in vivo from longitudinal and circular muscles. In vitro studies of human gut muscle replicate the behavior of animal muscle when exposed to small amounts of prostaglandins. In vivo studies in man by ingestion of 0.8 to 3.2 mg of PGE_1 caused bile reflux into the stomach and mild colicky pains with loose watery stools. The clear role of prostaglandins in gut motility has yet to be studied. (SR) 2045

1219
BIRNBAUMER, L., S.L. POHL, H. MICHIEL, J. KRANS, and M. RODBELL
  The actions of hormones on the adenyl cyclase system.
  In: Greengard, P. and E. Costa, eds., "Role of Cyclic AMP in Cell Function. Advances in Biochemical Psychopharmacology 3." p. 185-208. New York, Raven Press. 1970.

  This review mentions without comment that trypsin does not cause a loss of cellular integrity; neither does it damage the metabolism response of fat cells to prostaglandin, oxytocin, epinephrine, theophylline, and growth hormone plus glucocorticoids. (GW) 2367

1220
BIRON, P.
Prostaglandins: Action ocytocique, luteolytique et bronchodilatatrice.
[Prostaglandins: oxytocic, luteolytic and bronchodilatory action.]
L'Union Medicale du Canada. 99: 246-247. 1970.

In this editorial, the author briefly discusses the structure of prostaglandins as well as several of their physiological and pharmacological roles in humans. (NES) 2074

1221
BLAIR-WEST, J.R., M.C. CAIN, K.J. CATT, J.P. COGHLAN, D.A. DENTON, J.W. FUNDER, J.F. NELSON, B.A. SCOGGINS and R.D. WRIGHT
Role of the kidney in experimental renovascular hypertension.
Circulation Research. 27 (supp. 2): 11-149-158, 1970.

The authors observed the effects of chronic experimental renovascular hypertension in sheep. Mention of prostaglandins is brief and restricted to the discussion section where several reports on the release of a prostaglandin-like material into renal venous blood from manipulated and/or contralateral kidneys after a variety of renal manipulations are noted. (RMS) 2492

1222
BLAND, K.P.
Uterine autotransplantation to the abdominal wall in the guinea-pig.
Journal of Endocrinology. 48: 615-620. 1970.

It is suggested indirectly that $PGF_{2\alpha}$ may be the uterine leutolysin since it was found that transplantation of the uterus to a site where its venous blood had to pass through the lungs before reaching the ovary produced ovarian effects similar to hysterectomy. (JRH) 2489

1223
BLOOR, C.M., and B.E, SOBEL
Enhanced coronary blood flow following prostaglandin infusion in the conscious dog.
Circulation. 42 (supp. 3): III 123. 1970.

Abstract only. Prostaglandins E, A, and $F_{2\alpha}$, 0.3-3.0 $\mu g/kg/min$, were administered through a left atrial catheter into 5 intact, conscious dogs to study the effect of these substances on coronary hemodynamics. PGE and PGA immediately increased coronary blood flow from 60 to 81 ml/min along with characteristic systemic hemodynamic changes including decreased aortic pressure and peripheral vascular resistance and augmentation of cardiac output and heart rate. PGE and PGA infused via an intracoronary tube, 3 $\mu g/min$, increased coronary blood flow prior to any changes in aortic pressure or heart rate. PGA and PGE infusion diminished or abolished myocardial reclusion of the left circumflex artery. PGF exhibited no direct effect on the coronary vascular bed. (GW) 2263

1224
BOBERG, J., H. MICHELI, and L. RANNER
Effect of nicotinic acid on ACTH and noradrenaline stimulated lipolysis in the rabbit. II. In vitro studies including comparison with prostaglandin E_1.
Acta Physiologica Scandinavica. 79: 299-304. 1970.

Rabbit adipose tissue was incubated in vitro and the glycerol release followed as an indicator of the amount of lipolysis. Several compounds known to stimulate and inhibit lipolysis were tested in combination. It was found that PGE_1 had no effect on lipolysis in the basal state, but it did lower the lipolysis stimulated by low submaximal doses of ACTH or noradrenaline. PGE_1 did not inhibit higher submaximal doses of ACTH. PGE_1 also almost completely blocked the synergistic stimulation of ACTH plus theophylline. (JRH) 2055

1225
BORN, G.V.R.
  Observation on the rapid morphological reaction of platelets to aggregating agents.
  Series Haematologica. 3: 114-120. 1970.

  When platelets are exposed in vitro to certain aggregating agents, a change in shape and aggregation are inhibited by prostaglandin E_1. PGE_1 also increases the adenyl cyclase activity of human platelets. (Author modified) 2538

1226
BORN, G.V.R. and J.B. SMITH
  Uptake, metabolism and release of [^{3}H]-adrenaline by human platelets.
  British Journal of Pharmacology. 39: 765-778. 1970.

  Only brief mention is made of prostaglandin in the uptake, metabolism and release of [^{3}H]-adrenaline by human platelets. Prostaglandin E_1 (2.8×10^{-7}M) had no effect on adrenaline uptake by human platelets. It was added to the platelet-rich plasma to inhibit aggregation. (GW) 2203

1227
BORTIGNON, C., F. CARPENEDO, I. MARAGNO, E. SANTI SONCIN, G.D. STELLA, and M. FERRARI
  Cardiac effects of strychnine and their mechanism.
  Journal of Pharmacy and Pharmacology. 22: 380-381. 1970.

  The in vivo and in vitro effects of strychnine on heart rate are reported. Strychnine exerts a clear, dose-related negative chronotropic activity on the frog heart. The decrease of heart rate ranges from 20 to 60% depending on the drug concentration. Prostaglandin E_1 (10^{-8} to 10^{-7}) failed to affect the negative chronotropic effects of strychnine. (GW) 2286

1228
BOURNE, H.R., R.I. LEHRER, K.L. MELMON, and M.J. CLINE
  Cyclic adenosine-3',5'-monophosphate and the regulation of human granulocyte function.
  Journal of Clinical Investigation. 49: 11a-12a. 1970.

  Abstract only. The possible role of cAMP in phagocytosis by human granulocytes was investigated. PGE_1 was used because of its known ability to stimulate adenyl cyclase activity in other cells. It was found that PGE_1 inhibited degranulation which normally follows phago-cytosis and the burst of O_2 consumption. It was concluded that factors which effect cAMP in granulocytes can control the ability of granulocytes to kill phagocytized micro-organisms. (JRH) 2220

1229
BOWERY, N.G., G.P. LEWIS, and J. MATTHEWS
  The relationship between functional vasodilation in adipose tissue and prostaglandin.
  British Journal of Pharmacology. 40: 437-445. 1970.

  Earlier it had been found that during fat mobilization there was an increased blood flow in the adipose tissue and the tissue contained a vasodilator substance. Extract of an activated fat pad contained 3 to 25 times as much activity as the contralateral resting fat pad. The following findings suggest that the vasodilator substance is prostaglandin E_2: 1) It caused contractions of the guinea-pig ileum which were not reduced by mepyramine, but were reduced by atropine. 2) It caused a prolonged vasodilator response when injected close-arterially to the epigastric fat pad. 3) It was eluted from a silicic acid column by a solvent system which is known to elute prostaglandins of the E series but not those of the F series. 4) Its indices of discrimination were similiar to those of prostaglandin E's when assayed on three different pharmacological preparations. 5) On thin-layer chromatography it behaved more like prostaglandin E_2 and E_1. Neither prostaglandin E_1 nor prostaglandin E_2 inhibited the release of free fatty acids from the rabbit epigastric fat pad by ACTH. It seems likely that prostaglandin E_2 is responsible for the vasodilation accompanying fat mobilization from adipose tissue. (Authors) 2023

1230
BRODY, J.E.
Greater emphasis on research held key to birth control gains.
New York Times. April 10: 26. 1970.

Prostaglandins are briefly mentioned in this newspaper article as showing promise of being a morning-after pill. (JRH) 2537

1231
BRODY, J.E.
Hormone-like chemicals seen as ideal birth control method.
New York Times. Sept. 20: 30. 1970.

This newspaper article reports the promising results of clinical trials with prostaglandins as the "ideal birth control method." (JRH) 2534

1232
BRODY, M.J., C.S. SWEET and P.J. KADOWITZ.
Neurogenic and humoral contributions to the etiology of renal hypertension.
University of Michigan Medical Center Journal. 36: 235-236. 1970.

Abstract only. The studies to be presented were undertaken in an effort to determine whether heretofore unrecognized neurogenic and humoral mechanisms might be operative in the production of the renal hypertensive state. The possibility that a defect in the reflex compensation for acute elevations of arterial pressure might be involved in renal hypertension was examined in acute experiments performed on chronic renal hypertensive dogs. In other experiments, a new humoral factor with actions like those of angiotensin was uncovered. Renal venous blood was demonstrated to contain material which facilitated adrenergic vasoconstrictor responses produced by nerve stimulation beyond the level obtainable by the infusion of angiotensin alone. This factor could be a prostaglandin since several of these agents produce similar effects on adrenergic transmission. (Authors modified) 2335

1233
BRUDENELL, J.M.
The problem of uterine inertia.
Practitioner. 204: 377-382. 1970.

Prostaglandins E_2 and $F_{2\alpha}$ are briefly mentioned as stimulants to uterine contractions with possible future use in the management of prolonged labor. (RMS) 2067

1234

BUCCI, M.G., J.P. GIRALDI and P. LISCHETTI

Azione dell'umore acqueo plasmoide omologo sulla motilita spontanea in vitro dell'ileo di coniglio. [Action of the plasmoid aqueous humor on the spontaneous motility of the rabbit ileum in vitro.]

Pubblicita del Bolletino di Oculistaica. 49: 598-603. 1970.

The aqueous humor from rabbits subjected to paracentisis (both with and without atropine) were tested for their effect on the rabbit ileum in vitro. Homologous plasma and aqueous humor from untreated eyes were used as controls. The aqueous humor from all the animals subjected to paracentisis caused contractions of the ileum. The authors suggest that this activity was due to the production of one or more prostaglandins by the traumatized eyes. (MEMH) 2502

1235

BUCKALEW, V.M. Jr., F.J. MARTINEZ and W.E. GREEN

The effects of dialysates and ultrafiltrates of plasma of saline-loaded dogs on toad bladder sodium transport.

Journal of Clinical Investigation. 49: 926-935. 1970.

Bioassay with toad bladder of dialysates and ultra filtrates of blood from normal and saline expanded dogs revealed a substance which inhibits sodium transport in the bioassay system. It is mentioned that this substance is not likely to be a prostaglandin since prostaglandins have been shown to stimulate sodium transport in the toad bladder. (JRH) 2478

1236

BURKE, G.

Effects of iodide on thyroid stimulation.

Journal of Clinical Endocrinology and Metabolism. 30: 76-84. 1970.

To investigate the mechanism of iodide inhibition of TSH and LATS-stimulated thyroidal radioiodine release in vivo, the effects of iodide on TSH, LATS, prostaglandin and DBC stimulation of thyroidal radioiodine release and colloid droplet formation in the McKenzie mouse bioassay were compared. The stimulatory effects of TSH and PGE_1 on adenyl cyclase and glucose oxidation and of dibutyryl cyclic-AMP (DBC) on glucose oxidation were inhibited by excess iodide but TSH, PGE_1 and DBC stimulation of colloidal drop formation was unaffected. (GW) 2336

1237

BURKE, G.

Effects of prostaglandins on basal and stimulated thyroid function.

American Journal of Physiology. 218: 1445-1452. 1970.

The effects of prostaglandins E_1, E_2, $F_{1\alpha}$, and $F_{1\beta}$, (PGE_1, PGE_2, $PGF_{1\alpha}$ $PGF_{1\beta}$) on thyroid adenyl cyclase, endocytosis and glucose oxidation in vitro and on radioiodine release in vivo were studied. Each prostaglandin (2.5×10^{-6} M to 5×10^{-5} M) increased adenyl cyclase activity in sheep thyroid mitochondria and stimulated endocytosis of colloid in ovine thyroid slices, but only PGE_1 and PGE_2 consistently augmented thyroid-slice glucose oxidation. Although the effects of prostaglandins on adenyl cyclase and endocytosis were additive with TSH, PGE_1 and PGE_2 did not potentiate, and $PGF_{1\alpha}$ and $PGF_{1\beta}$ abolished submaximal TSH effects on thyroid-slice gulcose oxidation. Each prostaglandin alone (50-200 μg) augmented mouse thyroid radioiodine release in vivo but when added to a submaximal dose of thyrotropin (TSH) or long-acting thyroid stimulator (LATS) led to a significant reduction in hormonal effect. These findings suggest that: 1) the various phases of hormonogenesis in thyroid tissue (adenyl cyclase activation, endocytosis, and glucose metabolism) are not necessarily enhanced in concert, and 2) TSH (and perhaps LATS) and prostaglandins compete for a common adenyl cyclase receptor site(s) in thyroid. (Author) 2185

1238
BURKE, G.
On the role of adenyl cyclase activation and endocytosis in thyroid slice metabolism.
Endocrinology. 86: 353-359. 1970.

The main purpose of this experiment was to determine the interrelationship if any of thyroid stimulating hormone (TSH) effect on glucose oxidization and phospholipogenesis in thyroid tissue from sheep. PGE_1 was used as an adenyl cyclase stimulator to compare to TSH. It was found that low doses of TSH which stimulated phospholipogenesis had no effect on adenyl cyclase or on endocytosis. Higher doses of TSH produced a stimulation of glucose oxidization, adenyl cyclase activity and endocytosis which were parallel to those produced by PGE_1 $(1.5 \times 10^{-5} M)$. These results indicate that TSH can produce an increase in phospholipogenesis by mechanisms that are independent of the adenyl cyclase system. (JRH) 2226

1239
BURNS, T.W. and P.E. LANGLEY
Lipolysis by human adipose tissue: the role of cyclic 3',5'-adenosine monophosphate and adrenergic receptor sites.
Journal of Laboratory and Clinical Medicine. 75: 983-997. 1970.

The effects of cAMP, its dibutyryl derivative (DAMP), and the xanthine derivative, theophylline, singly and in combination with insulin on lipolysis by human and rat adipose tissue cells have been observed. The stimulating substances used were isoproterenol, a pure beta site agonist, and epinephrine, a mixed agonist with both alpha and beta stimulating capabilities. Prostaglandin E_1 (1.0 μg/ml) significantly reduced basal lipolysis but not the lipolysis stimulated by norepinephrine. (GW) 2348

1240
BUTCHER, R.W.
Prostaglandins and cyclic AMP.
In: Greengard, P., and E. Costa, eds., "Role of Cyclic AMP in Cell Function. Advances in Biochemical Psychopharmacology 3. p. 173-183. New York, Raven Press. 1970.

The first suggestion that prostaglandins might exert some of their effects through altered levels of cyclic AMP (cAMP) came from Steinberg et al. (J. Clin. Invest. 43: 1533-1540, 1964) who reported that PGE_1 inhibited the effects of catecholamines, ACTH, glucagon, and thyroid stimulating hormone (TSH) on both lipolysis and phosphorylase activation in the rat epididymal fat pad. Initial results from experiments to determine if prostaglandins exerted their anti-lipolytic effect on the fat pad through lowering cAMP levels were paradoxical. However, when the fat cells were separated from the other components of the fat pad (i.e., the stromovascular cells), the stimulatory effect was lost and PGE_1 did not cause increased cAMP levels in this essentially homogeneous cell population but still strongly antagonized the stimulatory effect of epinephrine. PGE_1 antagonized the effects of ACTH, glucagon, TSH and epinephrine on cAMP levels in isolated fat cells. Extremely low concentrations of prostaglandins lowered cAMP levels in fat cells stimulated by 5.5 μM epinephrine, the half-maximal concentrations of PGE_1 and PGE_2 being around 4×10^{-9}M. PGA_1 and $PGF_{1\alpha}$ were somewhat less effective and the synthetic $PGF_{1\beta}$ was without effect on cAMP levels. These same orders of potency had been found previously when the anti-lipolytic activities of the prostaglandins were measured. The effects of prostaglandins on cAMP levels or on adenyl cyclase have also been detected in several systems, including lung, diaphragm, spleen, brown fat, adenohypophysis, corpus luteum, and fetal bone. Two aspects of prostaglandin-cAMP relationships are selected for speculation; the first is the mechanism by which prostaglandins change cAMP levels, and the second is the possibility that in some cases the prostaglandins may be part of a feedback control system on cAMP levels. (ART) 2362

1241

BUTCHER, R.W. and C.E. BAIRD

The relationships of prostaglandins and cyclic AMP levels.

In: Eigenmann, R., ed., "Proceedings of the Fourth International Congress on Pharmacology," July, 1969. p. 42-48. Basle, Schwabe, 1970.

In this paper presented at a symposium, the authors review research about the effects of prostaglandins on cAMP levels in adipocytes. In isolated rat adipocytes, prostaglandins lower cAMP levels no matter which lipolytic agents are used to stimulate cAMP synthesis. However, prostaglandins are ineffective in adipocyte homogenates. Prostaglandins have been found to be lipolytic in some species other than rats. It is known that prostaglandins stimulate cAMP synthesis in many tissues. It seems likely that prostaglandins lower cAMP levels by inhibiting adenyl cyclase. (JRH) 2439

1242

BUTCHER, R.W.

The role of cyclic AMP in the actions of some lipolytic and antilipolytic agents.

Hormone and Metabolic Research. 2: 5-10. 1970.

This review on the role of cAMP in lipolysis contains a section on prostaglandins. Special attention is given to the difference in the effects of prostaglandins on fat tissues and isolated adipose cells. (JRH) 2446

1243

BYGDEMAN, M., S.U. KWON, T. MUKHERJEE, U. ROTH-BRANDEL, and N. WIQVIST

The effect of the prostaglandin F compounds on the contractility of the pregnant human uterus.

American Journal of Obstetrics and Gynecology. 106: 567-572. 1970.

Recordings of uterine motility were made of 13 women in the 14th-26th week and 1 woman in the 36th week of pregnancy in a series of experiments to determine the effect of low intravenous doses of $PGF_{1\alpha}$ and $PGF_{2\alpha}$ upon uterine motility as compared to dosages of PGE_1. The threshold dose of $PGF_{1\alpha}$ was 200-500 μg; the threshold dose of $PGF_{2\alpha}$ was around 100 μg which was roughly 8 times the threshold dose of PGE_1. In the doses employed, $PGF_{1\alpha}$ had no effect on pulse or blood pressure and no side effects. This was also true of less than 200 μg of $PGF_{2\alpha}$. Slight increases in pulse rate in 5 of 7 experiments, chest discomfort, and occasional vomiting, but no alteration of blood pressures were noted with 500 or more micrograms of $PGF_{2\alpha}$. A number of experimental limitations prevented exact calculation of dose response curves; however, mean values approach linearity and slopes are similar. (RMS) 2103

1244
BYGDEMAN, M., U. ROTH-BRANDEL, and N. WIQVIST.
Induction of labour by intravenous infusion of prostaglandin.
International Journal of Gynaecology and Obstetrics. 8: 162. 1970.

Abstract only. This report represents a study where labour was induced in term pregnant women by prostaglandin E_1 and $F_{2\alpha}$. Uterine contractility was recorded by measuring the amniotic pressure during 5-8 hours. As a comparison oxytocin was administered during approximately 1 hour at the end of each experiment. The contractility pattern was evaluated as far as uterine tone, intensity and frequency of the contractions is concerned. The results reveal that both prostaglandin E_1 and $F_{2\alpha}$ stimulated to effective labour-like contractions without causing significant side effects. However, the preliminary impression is that oxytocin infusion results in somewhat more coordinated contraction cycles. An interesting characteristic of the prostaglandins is that the sensitivity of the uterus to these compounds is practically independent of the stage of pregnancy. This is in distinctive contrast of oxytocin, where the sensitivity of the uterus is very low during early stages of pregnancy. (Authors modified) 2420

1245
BYGDEMAN, M., B. FREDRICSSON, K. SVANBORG, and B. SAMUELSSON
The relation between fertility and prostaglandin content of seminal fluid in man.
Fertility and Sterility. 21: 622-629. 1970.

Three groups, 29 fertile men, 100 men from non-investigated infertile marriages, and 21 men in infertile marriages with no abnormal clinical or laboratory findings provided semen samples which were investigated by ether extraction, thin layer chromatography and ultraviolet absorption at 278 mμ before and after alkali treatment. Prostaglandin E compounds in the fertile group were never below 11 μg/ml. In the non-investigated infertile group, 17% had less than 11 μg PGE/ml, and in the final group 41% had less than 11 μg/ml PGE. Concentrations of 19-OH and dehydrated prostaglandins remained constant. Since the highest incidence of low prostaglandin E concentration occurred in an infertile group from which most accepted causes of infertility had been excluded, it is felt that PGE compounds are likely of importance in clinically unexplained infertility and should be investigated as treatment. (RMS) 2176

1246
CALLINGHAM, B.A.
Prostaglandins.
Pharmaceutical Journal. 204: 433-436. 1970.

A comprehensive review of the history and biological activity of the prostaglandins is presented. Under suitable conditions a great number of tissues and organs in the body can be induced to release prostaglandins into the circulation. They are not themselves neurotransmitters but rather they exert their effects by modifying neuronal activity by an action on the cell membrane. They have widespread and diverse effects upon the organs of the female reproductive tract, and have potent effects on certain metabolic pathways in the body. (GW) 2369

1247
CARLSON, L.A. and H. MICHELI
Characteristics of lipolysis in dog adipose tissue. Effects of noradrenaline PGE_1 and nicotinic acid.
Acta Physiologica Scandinavica. 79: 145-152. 1970.

Omental adipose tissue from twelve fasted dogs of both sexes was incubated in vitro and the glycerol release determined as a measure of lipolysis. The effect of addition of noradrenaline, prostaglandin E_1 (PGE$_1$) and nicotinic acid into the medium was studied. On basal lipolysis the following effects were obtained with PGE$_1$. 1.0 μg PGE$_1$: Inhibition 4 dogs, no effect 6 dogs. 1 μg PGE$_1$: Inhibition 1 dog, no effect 5 dogs, stimulation 6 dogs, 10 μg PGE$_1$: Inhibition 1 dog no effect 3 dogs, stimulation 8 dogs. A dualistic action of PGE$_1$ was seen in some of the dogs. PGE$_1$, 1 μg, inhibited lipolysis stimulated by noradrenaline only in those dogs where the same amount of PGE$_1$ either inhibited or did not modify basal lipolysis. In those dogs where 1 μg of PGE$_1$ stimulated basal lipolysis this dose did not inhibit the stimulation caused by noradrenaline. The -adrenergic blocker Ko 592 inhibited the stimulation of lipolysis caused by PGE$_1$ only slightly, while Ko 592 completely inhibited the stimulation elicited by noradrenaline. Nicotinic acid inhibited the basal lipolysis in five of ten dogs and the stimulation induced by noradrenaline in three of ten. In those dogs where nicotinic acid inhibited lipolysis PGE$_1$ was in general also inhibitory. (Authors) 2172

1248
CARLSON, L.A., L-G. EKELUND and L. ORO
Clinical, metabolic and cardiovascular effects of different prostaglandins in man.
Acta Medica Scandinavica. 188: 553-559. 1970.

Prostaglandin A$_1$ (PGA$_1$), PGE$_2$, PGF$_{1\alpha}$, PGF$_{1\beta}$, and PGF$_{2\alpha}$ have been infused intravenously into nine healthy male subjects. PGA$_1$ was given to three subjects in doses from 0.056 to 0.56 μg/kg/min. In one of these subjects a slight flush was seen. The heart rate tended to increase without significant change in blood pressure. The levels of free fatty acids (FFA) in plasma increased slightly. PGE$_2$ was given to two subjects in doses from 0.056 to 0.56 μg/kg/min. A flush was seen in all four subjects given PGE$_2$, and the other effects were also similar to those previously seen with PGE$_1$. Thus an increase in heart rate and of plasma FFA levels occurred. With the prostaglandins of the F-series in doses up to 0.32 or 0.56 μg/kg/min, no clinical effects were noted. There were no significant changes in heart rate, blood pressure or plasma FFA levels. (Authors) 2140

1249
CARLSON, L.A., L-G. EKELUND, and L. ORO
Effect of intravenous prostaglandin E$_1$ on noradrenaline-stimulated mobilization of plasma free fatty acids in man.
Acta Medica Scandinavica. 188: 379-383. 1970.

Five male volunteers were constantly infused intravenously with noradrenaline (0.1μg/kg/min) for 210 min. Starting at 80 min, PGE$_1$ was also infused in increasing rates 0.10, 0.18 and 0.32 μg/kg/min, each rate during a 30 min period. After a few weeks 3 of the subjects were infused for 210 min with noradrenaline only as controls. Samples of blood were collected often and analyzed for glucose and free fatty acid content. Heart rate and blood pressure were constantly monitored. Clinical effects were similar to those caused by PGE$_1$ alone, i.e. flushing, headache, abdominal pain, and respiratory distress. PGE$_1$ reduced the increase in blood pressure caused by the noradrenaline, but it did not reduce the increase in free fatty acids. This unexpected failure to inhibit free fatty acid release in vivo in man leads the authors to believe that PGE$_1$ is not transported by the blood to adipose tissue. (JRH) 2139

1250
CARLSON, L.A. and H. MICHELI
Effect of prostaglandin E$_1$ on fat mobilizing lipolysis in rat adipose tissue in relation to the nutritional condition.
Acta Physiologica Scandinavica. 80: 289-294. 1970.

The effect of PGE_1 on fat mobilizing lipolysis in rat adipose tissue was studied in relation to the nutritional condition by following the release of glycerol during incubation in vitro. In fed tissues PGE_1 at concentration of 0.01, 0.1, 1 and 10 mg/ml incubation medium significantly lowered the lipolysis to 64, 41, 48, and 68% of the basal rate (average of 6 experiments.) The lipolytic rate was significantly higher with 10 than with 0.1 mg/ml of PGE_1. The corresponding figures in adipose tissue from fasted rats were 90, 91, 86, and 95% respectively (average of 7 experiments). These average decreases were not statistically significant. Refeeding fasted rats with glucose one hour before the study restored the sensitivity of adipose tissue to the antilipolytic effect of PGE_1. In rat fat the nutritional condition is thus of importance for the appearance of the antilipolytic effect of PGE_1. The possible role of the tissue sensitivity to PGE_1 in regulating fat mobilizing lipolysis during fasting was discussed. (Authors) 2029

1251
CARLSON, L.A.
Mobilization and utilization of lipids after trauma: Relation to caloric homeostasis.
In: "Energy Metabolism in Trauma, CIBA Foundation Symposium," p. 155-171. 1970.

Prostaglandins are briefly mentioned as having antilipolytic properties and having physiological importance in regulating human free fatty acids. (JRH) 2449

1252
CARPENTER, M.P. and B. WISEMAN
Prostaglandins of rat testis.
Federation Proceedings. 29: 248 abs. 1970.

Abstract only. The presence of prostaglandins in testicular tissue has not previously been reported. Chromatography of acid-soluble lipid fractions of extracts from rat testis indicated the presence of PGE_1 and PGF_1 and suggested that PGE_2 and PGF_2 were also present. That these prostaglandins are synthesized in this tissue is indicated in experiments in which linolenic acid tagged with radioactive carbon was injected into rat testis and acid-soluble lipid extracted. Chromatography and subsequent scintillation counting of prostaglandin spots scraped from the plates showed that the radioactivity was incorporated into material corresponding to PGE_1, PGE_2, $PGF_{1\alpha}$, and $PGF_{2\alpha}$. (ART) 2119

1253
CARR, A.A.
Effects of PGA_1 on plasma renin activity in man.
Clinical Research. 18: 60. 1970.

Abstract only. The effect of PGA_1 on plasma renin activity (PRA) during water and mannitol diuresis with either 10 mEq/day or 250 mEq/day sodium intake on 5 men was measured. No change in PRA was noted, but glomerular filtration, renal plasma flow and sodium excretion increased indicating either altered renal hemodynamics or tubular sodium transport or the inhibition of some intrarenal or extrarenal receptor for renin release. (GW) 2168

1254
CARR, A.A.
Hemodynamic and renal effects of a prostaglandin, PGA_1, in subjects with essential hypertension.
American Journal of the Medical Sciences. 259: 21-26. 1970.

Prostaglandin A_1 was given to each of 5 adult males at dosages of 0.48 to 1.32 mg/kg/min. in a test of the effects of PGA_1 on hemodynamics and renal function in human essential hypertension. Striking decreases in peripheral vascular resistance were shown; the renal fraction of cardiac output increased dramatically. Enhancement of free water clearance (CH_2O) and decreased solute free water clearance (T^CH_2O) occurred in association with a sodium diuresis. The experiment was designed so that each man served as his own control. Dizziness, abdominal cramps, voluntary defecation and flushing of the face with headache were experienced by one subject receiving 1.3 mg/kg/min. The side effects disappeared when the infusion rate was halved. (RMS) 2102

1255
CARTER, L.J.
Contraceptive technology: advances needed in fundamental research.
Science. 168: 805-809. 1970.

The activities of the Agency for International Development and National Institutes of Health in promoting research on prostaglandins as a "once a month pill" are briefly mentioned in a general discussion of birth control technology. (RMS) 2070

1256
CHASE, L.R. and G.D. AURBACH
The effect of parathyroid hormone on the concentration of adenosine. 3',5'-monophosphate in skeletal tissue in vitro.
The Journal of Biological Chemistry. 245: 1520-1526. 1970.

Adenosine 3',5'-monophosphate (3',5'-AMP) was assayed in calvaria isolated from fetal rats at term. A marked increase in 3',5'-AMP was detected in vitro within 1 minute after adding purified parathyroid hormone to calvaria incubated in Krebs-Ringer-bicarbonate of Krebs-Ringer-Tris buffer. Prostaglandin E_1 and E_2 also caused a significant increase in skeletal 3',5'-AMP. Prostaglandin E_1, E_2, $F_{1\alpha}$, and $F_{2\alpha}$ also stimulated adenyl cyclase activity. (GW) 2312

1257
CHIANG, T.S. and F.E. LEADERS
Antagonism of formaldehyde-induced ocular hypertension by phenylethylamines.
Proceedings of the Society for Experimental Biology and Medicine. 135: 249-252. 1970.

It is briefly mentioned in the discussion section that the ocular hypertension in rabbit eye caused by intracameral injection of formaldehyde could involve prostaglandins. (JRH). 2477

1258
CHRIST, E.J. and D.H. NUGTEREN
The biosynthesis and possible function of prostaglandins in adipose tissue.
Biochemica et Biophysica Acta. 218: 296-307. 1970.

The possibility of a physiological function for prostaglandins as expressed in the feedback theory of the regulation of lipolysis by prostaglandin formation is compatible with results of this investigation using homogenized epididymal fat pads and fat cells of rats in vitro. The release of arachidonic acid as well as PGE_2 during lipolysis was observed. Low concentrations of PGE_1, and PGE_2, and ω-homo PGE_1 effectively inhibited stimulated lipolysis. A correlation was observed in the biological activity and ability to inhibit lipolysis as inactive homologues did not show this inhibition ability. Decreased release of PGE_2, increased lipolysis, and pronounced inhibition of stimulated lipolysis by added PGE was seen in adipose tissue deficient in essential fatty acids. (RMS) 2009

1259
CHRIST, E.J.V.J.
Die rolle von prostaglandinen bei der hormonal stimulierten lipolyse in isolierten fettegeweben. [The role of prostaglandins in hormone stimulated fatty tissues.]
Fette-Seifen-Anstrichmittel]. 72: 849-853. 1970.

Biosynthesis of PGE from ^{14}C labelled unsaturated fatty acids by enzyme preparations from epididymal fatty tissues of rats and rams is reported. Di-homo-γ-linolenic acid and arachidonic acid were detected in di- and triglyceride fractions of the fatty tissues of rats. The amounts found are perhaps sufficient for explaining quantitatively the synthesis of PGE in intact fatty tissues. The specificity of anti-lipolytic action of PGE homologues was studied. Prostaglandins from EFA-active fatty acids with 20 and 21 C atoms are as active as PGE_1, whereas prostaglandins from EFA-inactive fatty acids are much less active. The possible action of PGE_1 on the formation and degradation of cycle adenosine-3',5' monophosphate was studied with isolated enzyme systems. A direct action of PGE_1 could not be detected. (Author) 2541

1260
CLASSEN, M., H. KOCH, P. DEYHLE, S WEIDENHILLER, and L. DEMLING
Wirkung von prostaglandin E_1 auf die basale magensekretion des menschen. [Reaction of prostaglandin E_1 on the basal stomach secretion of humans.]
Klinishche Wochenschrift. 48: 876-878. 1970.

Six volunteers received 300 μg prostaglandin E_1 (4-5 μg/kg body weight) in 250 cc solution intravenously within 30 minutes following measurement of basal acid secretion. For the first time it could be demonstrated that PGE_1 inhibits acid secretion significantly in man. The side effects in these persons were tolerable. In another case a controllable fall in blood pressure below 90 mm Hg systolic was noted. (Authors) 2344

1261
CLIFTON, T.
A drug for the 70's.
Sunday Times (London). Sept. 13: 9. 1970.

This newspaper article reports some of the potential uses of prostaglandins and some of the problems yet to be solved. (JRH) 2535

1262
COLE, B., G.A. ROBISON, and R.C. HARTMANN
Effects of prostaglandin E_1 and theophylline on aggregation and cyclic AMP levels of human blood platelets.
Federation Proceedings. 29: 316 abs. 1970.

Abstract only. Aggregation of human blood platelets was initiated by the addition of epinephrine to platelet-rich plasma. The addition of 45 ng/ml of PGE_1 partially reduced the degree of aggregation and after a 10 min incubation increased the level of cAMP by 76%. Similar effects were produced by the addition of theophylline. The addition of PGE_1 and theophylline together completely abolished aggregation and increased the level of cAMP by

225%. These data support the concept that PGE_1 and theophylline inhibit platelet aggregation by increasing the level of cAMP, and that the level of cAMP is an important determinant of the tendency of platelets to aggregate. (ART) 2118

1263

COLLET, A. and J. JACQUES.

Analogues simplifies des prostaglandines. [Simplified analogs of prostaglandins].
Chimica Therapeutica. 5: 163-168. 1970.

The authors describe the syntheses (from 1-nitroso 2 naphtol) of 3-[o-(3'-hydroxy-1'-alkenyl) phenyl] propionic acid and 5-[...] pentanoic acid, simplified analogs of prostaglandins, in which there is a benzene ring instead of a pentagonal cycle in natural products. (Authors) 2419

1264

COLLIER, H.O.J.

Endogenous broncho-active substances and their antagonism.
Advances in Drug Research. 5: 95-107. 1970.

The author reviews the literature exploring the role of endogenous substances, adrenaline and PGE_2 as bronchodilators, and histamine, kinins, $PGF_{2\alpha}$, and a chemically unidentified slow reacting substance (SRS-A) as bronchoconstrictors in allergy. The results of experiments on both human and guinea pig tissues in vivo and in vitro are cited and compared. Specific citations of prostaglandin literature include: Piper and Vane (1968), release of PGE_2 and $PGF_{2\alpha}$ from sensitized guinea pig lung on perfusion with antigen, Sweatman and Collier (1968), James (1969), Berry and Collier (1964), and Collier and Sweatman (1968) on the identity of SRS-A as a non-prostaglandin and probably not a prostaglandin releaser. James is further cited with reference to the effect of $PGE_2\alpha$ on the guinea pig lung as a constrictor of the larger airways, an effect which is not blocked by meclofenamate and is slightly enhanced by propanolol. The effect on isolated human bronchial muscle is similar (Sweatman and Collier, 1968). Rosenthale et al. (1968), and Collier and James (1970), are cited on the effect of PGE_2 on the lung which is the opposite of $PGF_{2\alpha}$, preventing responses to histamine, bradykinins, acetylcholine and 5-hydroxy tryptamine. Observations from the works previously cited are mentioned again by the author as indicating further PGE_2 activity as an aerosol against bronchoconstrictor aerosols and the relaxation of isolated human bronchial muscle. The author notes that there is no logical reason why antagonists do not give symptomatic relief from human asthma and discusses some of the reasons which have been advanced for the failure of antihistamines and fenamates. The ability of aspirin, an inhibitor of $PGF_{2\alpha}$, to produce occasional relief from asthma is briefly commented upon. (RMS) 2066

1265

COLLIER, H.O.J. and G.W.L. JAMES

Humoral factors in airway function, with particular reference to anaphylaxis in the guinea pig.
In: Bouhuys, A., ed., "Airway Dynamics, Physiology and Pharmacology," p. 239-252.
Springfield, C.C. Thomas, 1970.

The authors review 50 papers concerned with the effects of kinins, prostaglandins, SRS-A, adrenaline, and histamine on airways resistance in biological preparations of the guinea pig. Also discussed are the effects of selective antagonists, propranolol, mepyramine, meclofenamate and tachyphylaxis to kinins and to SRS-A. It is concluded that fenamate adrenaline, histamine, kinins, and an unidentified active constituent of SRS-A appear to be humoral factors involved in anaphylactic airways response in the heavily anesthetized or decentralized guinea pig, and that PGE_2 and $PGF_{2\alpha}$, which are liberated from the lungs during anaphylaxis, may well participate in anaphylaxis. Figures presented show the effect of PGE_2 on the response to histamine of the guinea pig lung in vivo and the effect of $PGF_{2\alpha}$ on the separate responses of the trachea and of the pulmonary airways and the failure of meclofenamate to antagonize $PGF_{2\alpha}$ in artificially ventilated guinea pig. (RMS) 2499

1266
COREY, E.J., Z. ARNOLD, and J. HUTTON.
Total synthesis of prostaglandins E_2 and $F_{2\alpha}$ (dl) via a tricarbocyclic intermediate.
Tetrahedron Letters. 307-310. 1970.

The authors describe a total synthesis of PGE_2 and $PGF_{2\alpha}$. The starting compound was 6-methoxy-bicyclo [3.1.0] hexene-2. Step by step structural formulas are given. The authors conclude that this approach is inferior to an earlier described synthesis because of the low selectivity of one of the steps. (JRH) 2409

1267
COREY, E.J., R. NOYORI, and T.K. SCHAAF
Total synthesis of prostaglandins $F_{1\alpha}$, E_1, $F_{2\alpha}$, and E_2 (natural forms) from a common synthetic intermediate.
Journal of the American Chemical Society. 92: 2586-2587. 1970.

The authors describe the synthesis of the prostaglandins $F_{1\alpha}$ and E_1, using the intermediate 11,15-bis-tetrahydropyranyl ether of prostaglandin $F_{2\alpha}$. The transformations reported also mark the realization of the primary prostaglandins $F_{1\alpha}$, E_1, $F_{2\alpha}$, and E_2 from the same intermediate. (GW) 2279

1268
COREY, E.J., T.K. SCHIR, W. HUBER, U. KOELLIKER, and N.M. WEINSHENKE
Total synthesis of prostaglandins $F_{2\alpha}$ and E_2 as the naturally occurring forms.
Journal of the American Chemical Society. 92: 397-398. 1970.

Outline of the first synthesis of prostaglandin $F_{2\alpha}$ and E_2 in the naturally optically active forms is presented. The efficiency of this process of the conversion of the optically active hydroxyl acid to the immediate precursor of prostaglandin $F_{2\alpha}$ and E_2 can be effected under present conditions in ca. 50% yield. (GW) 2278

1269
COREY, E.J. and R. NOYORI.
A total synthesis of prostaglandin $F_{2\alpha}$ (dl) from 2-oxabicyclo [3.3.0] oct-6-en-3-one.
Tetrahedron Letters. 311-313. 1970.

The authors describe the synthesis of racemic $PGF_{2\alpha}$. The starting compound is 2-oxabicyclo [3.3.0] oct-6-en-3-one. A step by step synthesis is described with structures of the intermediates given. It is concluded that the main problem with this synthesis is the occurrence of ring opening in one of the steps. (JRH) 2408

1270
COX, J.S.G., J.E. BEACH, A.M.J.N. BLAIR, A.J. CLARKE, J. KING, T.B. LEE, D.E.E. LOVEDAY, G.F. MOSS, T.S.C. ORR, J.T. RITCHIE, and P. SHEARD
Disodium cromoglycate (Intal®).
Advances in Drug Research. 5: 115-196. 1970.

This review of disodium cromoglycate (registered trademark "Intal"), which represents an important stage in the exploration of the biological properties of chromone 2-carboxylic acids, a research program that had its origins with a known drug, khellin, and had as its target a more specific treatment for bronchial asthma, cites 114 references. The only mention of prostaglandins occurs on page 159 in the discussion of smooth muscle activity and antagonism studies in vitro. It is mentioned that studies were carried out on human bronchial strip contracted by histamine, SRS-A, acetylcholine, and $PGF_{2\alpha}$, and relaxed by PGE_1. (ART) 2175

1271

CRAWFORD , M.A., M.M. GALE , M.H. WOODFORD , and N.M. CASPED
Comparative studies on fatty acid composition of wild and domestic meats.
International Journal of Biochemistry. 1: 295-305. 1970.

In discussing the implications of changing the ratio of different classes of fats in the human diet, the authors briefly mention that the prostaglandins among other compounds are synthesized from polyenoic acids and their unsaturated elongation products. The changing of the ratio of these fats in the diet could cause changes in tissues in which such compounds are important. (JRH) 2434

1272

CROWSHAW, K. and J.Z. SZLYK
Distribution of prostaglandins in rabbit kidney.
Biochemical Journal. 116: 421-424. 1970.

Three prostaglandins (PGE_2, $PGF_{2\alpha}$, and PGA_2) are present in rabbit kidney medulla. An acidic lipid extract (0.165g) obtained from 2kg of frozen rabbit kidney cortex was separated by silicic acid chromatography to yield eluates containing fatty acids, possible non-polar prostaglandin metabolites. PGA, PGE and PGF compounds. Ultraviolet spectra of the eluates before and after treatment with sodium hydroxide did not yield chromophores typical of any known prostaglandins or related metabolites. By using more sensitive bioassay procedures (contraction of rabbit duodenum) weak activity equivalent to 60 µg of PGE_2 and 10 µg of $PGF_{2\alpha}$ was detected in the PGE and PGF eluates respectively. Extraction and bioassay of fresh kidney cortex revealed no prostaglandin-like activity. Attempts to biosynthesize prostaglandins in fresh homogenates of rabbit kidney cortex from endogenous precursors and from added arachidonic acid were unsuccessful. When freshly prepared homogenates of rabbit kidney cortex were incubated with added PGE_1 no evidence of enzymic breakdown was obtained. It is concluded that rabbit kidney prostaglandins are present predominantly in the medulla and there are no cortical mechanisms for their biosynthesis or inactivation under normal conditions. (Author) 2202

1273

CROWSHAW, K., J.C. McGIFF, J.C. STRAND, A.J. LONIGRO, and N.A. TERRAGNO
Prostaglandins in dog renal medulla.
Journal of Pharmacy and Pharmacology. 22: 302-304. 1970.

This letter to the editor concerns the isolation of three biologically active lipids from dog renal medulla which exhibited the typical chromatographic and pharmacological properties of PGE_2, $PGF_{2\alpha}$, and PGA_2. The medulla from 20 dogs was extracted and taken for preliminary chromatographic and biological testing. Smooth-muscle stimulating activity was determined using rat stomach strip, rat colon and chick rectum superfused by Krebs solution. Extract of 453 g. of dog renal medulla by column chromatography resulted in the recovery of 21 µg PGA_2, 40.5 µg PGE_2 and 25 µg $PGF_{2\alpha}$. The demonstration that prostaglandins are present in the kidney medulla suggests that these biologically active compounds may be important renal hormones. (GW) 2302

1274
CROWSHAW, L., J.C. McGIFF, J.C. STRAND, A.J. LONIGRO, and N.A. TERRAGNO
A sensitive vasodepressor assay for the detection of prostaglandin-like substance in dog renal medulla.
Federation Proceedings. 29: 745 abs. 1970.

Abstract only. PGA_2 is usually detected by its vasodepressor response when injected intravenously into an anesthetized, vagotomized, pentolinium-treated rat. PGE_2 is also depressor in the rat whereas $PGF_{2\alpha}$ is pressor. In contrast to PGE_2 and $PGF_{2\alpha}$, PGA_2 is not removed by the lung. The vasodepressor responses to intravenous (i v) and intra-arterial (i a) injections of these prostaglandins have been studied: PGA_2 (79 to 100 ng) was vasodepressor on i v and i a injection. PGE_2 (30 to 60 ng) was vasodepressor and $PGF_{2\alpha}$ (40 to 60 ng) was pressor only on i a injections, much larger doses ($\times$ 10) being required, when given i v to produce comparable responses. An acidic lipid extract prepared from 453 g. of dog renal medulla and containing smooth-muscle stimulating activity equivalent to 66.7 μg of PGE_2 was separated by chromatographic methods and yielded eluates of PGE_2 $PGF_{2\alpha}$ and PGA_2 in concentrations estimated to be 90 ng, 55 ng, and 33 ng, respectively, per gram of medulla. (ART) 2126

1275
CULLITON, B.J.
Something for everyone.
Science News. 98: 306-307. 1970.

It is proposed that the prostaglandins have a whole spectrum of potential uses from abortions to heart disease. These ubiquitous substances exert their activity intracellularly, functioning as regulators of cellular behavior. Prostaglandins that leak out of the cell are quickly metabolized or broken down, a feature that makes them ideal drugs. They present a paradox, for in some tissues these compounds mimic the action of neurohormones and hormones, while in others they inhibit hormonal response. (GW) 2342

1276
CUNLIFFE, W., P. HUDGSON, J.J. FULTHORPE, M.M. BLACK, R. HALL, I.D.A. JOHNSTON and S. SCHUSTE.
A calcitonin-secreting medullary thyroid carcinoma associated with mucosal neuromas, marfanoid features, myopathy and pigmentation.
American Journal of Medicine. 48: 120-126. 1970.

In this case history report of a patient with medullary carcinoma of the thyroid it is mentioned that prostaglandins and other humoral agents have been shown to be secreted by these tumors. It is suggested that the prostaglandins could be responsible for the gastrointestinal symptoms seen in this patient. (JRH) 2468

1277
DAVIS, L.J., P.A. SHEFFIELD, and R.T. JACKSON
Drug-induced patency changes in the eustachian tube. A comparison of routes of administration.
Archives of Otolaryngology. 92: 325-328. 1970.

Dose related responses to a number of important pharmacological agents were observed in a study on patency of the dog eustachian tube. Responses to levarterenol and PGE_1 did not follow this pattern; progressive dose response curves were obtained upon intraarterial administration up to the point where large blood pressure changes interfered with response. With topical administration, the direction of response was dose dependent. Low doses (less than

0.05 mg) of PGE_1 decreased patency; high doses increased it. Increased patency was seen with epinephrine, levarterenol, phenylephrine, dopamine, angiotensin amide, methoxamine, and glucagon. Decreased patency was associated with PGE_1, isoproterenol, histamine, acetylcholine, paverine and chlorobutanol. One or more of the agents in either case is probably responsible for the control of blood flow in the eustachian mucosa. (RMS) 2271

1278
DAWSON, W., S.J. JESSUP, W. McDONALD-GIBSON, P.W. RAMWELL, and J.E. SHAW
Prostaglandin uptake and metabolism by the perfused rat liver.
British Journal of Pharmacology. 39: 585-598. 1970.

An investigation was made of the speed of removal of PGE_1 and $PGF_{1\alpha}$ from the circulating blood and their subsequent metabolism by the isolated perfused rat liver. Injection of radiolabeled $1^{-14}C$ PGE_1 into the hepatic artery or portal vein indicated that the liver could efficiently remove 89-95% of circulating PGE_1 on a single passage. It is suggested that biliary excretion may be a major route for elimination of compounds smaller than C20 prostaglandins. The results indicate rapid uptake of circulating prostaglandins by the rat liver. Decarboxylation of prostaglandins results in pharmacological inactivation. The products are excreted into the bile and venous effluent. (GW) 2033

1279
DEISSEROTH, A., S.M. WOLFE, and N.R. SHULMAN
Platelet phosphorylase activity in the presence of activators and inhibitors of aggregation.
Biochemical and Biophysical Research Communications. 39: 551-557. 1970.

Phosphorylase activity in suspensions of human platelets has been assayed before and after addition of thrombin and epinephrine which aggregate platelets, prostaglandins E_1 and E_2 and dibutyryl-3′,5′-AMP which inhibit aggregation, or sodium fluoride and isoproterenol which alter adenine nucleotide metabolism. These various reagents did not change phosphorylase a activity in intact platelets, nor did cyclic-3′,5′-AMP affect platelet phosphorylase activity in subcellular fractions. In contrast to many other tissues, platelet phosphorylase activity does not appear to be modulated by a direct effect of cyclic AMP. (Authors) 2024

1280
DEJONGH, D.C.
Mass Spectrometry.
Analytical Chemistry. 42: 169R-205R. 1970.

In this extensive review of the uses of mass spectrometry it is briefly mentioned that prostaglandin metabolism has been followed with this technique. (JRH) 2469

1281
DEKKER, A. and J.B. FIELD
Correlation of effects of thyrotropin, prostaglandins and ions on glucose oxidation, cyclic-AMP, and colloid droplet formation in dog thyroid slices.
Metabolism. 19: 453-464. 1970.

The in vitro effects of thyroid stimulating hormone (TSH), PGE_1, PGA_1, PGB_1, and $PGF_{1\alpha}$, and varying ionic content of the incubating medium were studied on glucose oxidation cyclic-AMP (cAMP) concentration, and colloid droplet formation in thyroid slices from dogs treated with thyroxine. Small doses (1 μg/ml) of PGE_1 and TSH (5 mU/ml) increase glucose oxidation, cAMP, and colloid droplet formation. $PGE_{1\alpha}$ (10 μg/ml) increased glucose oxidation but had little or no effect on cAMP and colloid droplet formation, while higher doses (100 μg/ml) did. Similarly, high doses of PGA$_1$ and PGB$_1$ increased glucose oxidation and colloid droplet formation but only PGA$_1$ may have increased cAMP. These results and the effects of TSH are discussed in the relation to the hypothesis that TSH and prostaglandins may regulate thyroid metabolism via the generation of cAMP. (Authors modified) 2238

1282
DEPPERMAN, W.H.
Up-to-date scalp tonic.
New England Journal of Medicine. 283: 1115. 1970.

Tha author in his letter to the editor proposes the use of prostaglandin E$_1$ to halt male-pattern baldness. Hair falls out when the highly vascular subcutaneous adipose layer under the influence of androgens decreases drastically in thickness, thereby decreasing the number of blood vessels. Prostaglandin E$_1$ combined with estrogens may act as an inhibitor of androgens which in turn may act synergistically whether by summation or potentiation. (GW) 2296

1283
DeVISSCHER, M.
Les Prostaglandines. [Prostaglandins.]
Louvain Medica. 89: 297-303. 1970.

The structure and history of prostaglandins is briefly reviewed. This is followed by a short summation of their role in the areas of reproduction, the nervous system, cyclic AMP and smooth muscle. (NES) 2421

1284
DIASSI, P.A. and Z.P. HOROVITZ
Endocrine Hormones.
Annual Review of Pharmacology. 10: 219-236. 1970.

The tissue prostaglandins are a group of unsaturated lipid-soluble acids derived from dihomo-γ-linolenic and arachidonic acids. Although originally found in human seminal fluid, they now have been identified in most, if not all, mammalian tissue. They are formed by cyclization and oxidation of these fatty acids, and enzyme systems that are particularly active in effecting these transformations are found in human and sheep seminal fluid. The prostaglandins have a wide spectrum of biological activities varying from lowering blood pressure, stimulating smooth muscle, affecting sperm transport, inhibiting lipolysis, aggregating platelets, to causing uterine movements and the secretion of gastric juice. Excellent reviews on the physiological role of the prostaglandins have been published recently. Although their precise physiological function is not yet known, their variety of effects, both qualitative and quantitative, may serve as biological regulators. Prostaglandins liberated by nerve stimulation, which then have actions opposite to the nerve stimulation, suggest a role as feed-back inhibitors. The total synthesis of both natural and enantromeric forms of prostaglandin E$_1$ has been accomplished. The main urinary metabolite of prostaglandins F$_{2\alpha}$ in the guinea pig is 5α,7α-dihydroxy-11-ketotetra-nor-prostanoic acid. In man, prostaglandin E$_2$ is metabolized to the dicarboxylic acid IV and prostaglandin F$_{2\alpha}$ to the dicarboxylic acid V. Recent work has described the synthesis of oxygen-containing analogues of prostaglandins that inhibit the activity of natural prostaglandins on smooth muscle. At certain concentrations these compounds can manifest agonistic activity on the gerbil colon. (Authors) 2065

1285
DJERASSI, C.
Birth control after 1984.
Science. 169: 941-951. 1970.

The author selected three topics for new approaches to contraception that can be developed to control the world's burgeoning population. 1) A new female contraceptive; 2) a male contraceptive pill; 3) a draconian agent. Our knowledge of the female reproductive cycle provides more limits about rational approaches to contraception than our knowledge of the male process does. European clinical studies have appeared on the use of $PGF_{2\alpha}$ and PGE_2 as abortifacients but time consuming work must be overcome before prostaglandins can be considered practical candidates. The following are a few of the more obvious problems. 1) The prostaglandins act on almost all body systems. 2) Research has not yet found a way to synthesize prostaglandins on a large scale. 3) The requirements of intravenous infusion limit use of the prostaglandin to hospitalized patients. 4) Intramuscular administration leads to a definite possibility of incomplete follow-up and raises the spector of potential teratogenesis. (GW) 2027

1286
van DORP, D.A.
Recent research in prostaglandins.
Journal of the American Oil Chemists' Society. 47: 326A-327A. 1970.

Abstract only. Much attention is still being paid in this laboratory to investigations concerning the relationship between essential fatty acids and prostaglandins. Previously we had obtained evidence that biologically active prostaglandins can be formed only from polyunsaturated fatty acids which are effective in curing the symptoms of essential fatty acid deficiency in rats. We concluded that $\omega 5$, $\omega 6$ or $\omega 7$ acids with double bonds in the positions 5,8,11,14 are the structures of the ultimate essential fatty acids in the rat. As both the $\omega 4$ and $\omega 8$ acids with three double bonds in the positions 8,11,14 did not show EFA-activity and were not converted into prostaglandins, we thought it possible that the rat in these cases was not able to dehydrogenate $18:3\omega 4$ and $22:3\omega 8$ fast enough into the desired 5,8,11,14 structures. Therefore the possiblity remained that the tetraenoic acids could display EFA-activity in contrast to their trienoic precursors. Another question which remained to be settled was whether acids containing double bonds in a nonskipped position to the 8,11,14 system might have EFA-activity and might be precursors of prostaglandins. Because there are so many possible isomers we restricted ourselves to a number of eicosatetraenoic acids: all-cis 4,8,11,14 all-cis 8,11,14, 18, 2-trans, 8-cis, 11-cis, 14-cis, and 3-trans, 8-cis, 11-cis, 14-cis. In addition, two stereoisomers of arachidonic acid each containing one trans double bond were synthesized: 5-trans, 8-cis, 11-cis, 14-cis and 5-cis, 8-cis, 11-cis, 14-trans. This work led to the discovery of yet another highly biologically active prostaglandin. A number of unsaturated fatty acids have been tested for inhibitory activity in the enzymic conversion of all-cis 8,11,14-eicosatrienoic acid into prostaglandin E_1. In 8-cis, 12-trans, 14-cis-eicosatrienoic acid Nugteren discovered a very powerful inhibitor. In vitro and in vivo studies with this inhibitor will be reported. As a broad objective we see the total metabolic steady state of unsaturated fatty acids such as EFA in the animal body. These studies should naturally also include the metabolic fate of the prostaglandins, and in connection with this we developed a highly sensitive method for determining ng quantities of prostaglandins. Gas liquid chromatography with electron capture proved to be very suitable. The method will be discussed in detail and also some of the results already obtained with it, with special reference to the skin and the kidney. Another aspect of the objective cited above is the study of the mechanism dealing with keeping up a constant level of free fatty acid in the animal organism. To this end we studied the influence of prostaglandins on the cyclic AMP system in adipose tissue. It could be shown that adipose tissue does contain enough prostaglandin precursors and also enough prostaglandin synthetase, to make the physiological role of prostaglandins in adipose tissue, as deduced by Ramwell and Shaw from the finding of Steinbert, likely. (Author) 2355

1287
DOUGLAS, W.W.
Prostaglandins.
In: Goodman, L.S. and A. Gilman, eds., "Pharmacological Basis of the Therapeutics," p. 672-674.
London. Collier MacMillian, 1970.

The historical, chemical and pharmacological properties on reproductive, gastrointestinal, respiratory, cardiovascular, and nervous systems are discussed in a review citing more than 80 references on the vasoactive polypeptides and prostaglandins. (RMS) 2068

1288
DOWNING, D.T., D.G. AHERN, and M. BACHTA
Enzyme inhibition by acetylenic compounds.
Biochemical and Biophysical Research Communications. 40: 218:223. 1970.

5,8,11,14-eicosatetraynoic acid was found to be an efficient and apparently irreversible inhibitor of prostaglandin synthetase in sheep vesicles. Incubation with mixtures containing 4×10^{-5} M acetylenic acid totally inhibited the oxydation of polyethylenic acids. Although the polyethylenic acids are inhibitors of prostaglandin synthesis, their effect does not proceed to total inhibition as is the case with the acetylenic acid. It is postulated that inhibition may result from conversion of the acetylenic compound to a reactive allene by the enzymes studied. (GW) 2245

1289
DUNCAN, G.W. and B.B. PHARRISS
Effect of nonsteroidal compounds on fertility.
Federation Proceedings. 29: 1232-1239. 1970.

The work of many investigators, represented by 68 references, is reviewed. It is stated in summary that "we now have at our disposal agents which can control reproduction in the female by acting at the level of the hypothalamus, the pituitary, the ovary, or the uterus, and we have substances that, in the male, control spermatogenesis, posttesticular sperm fertility, and in utero sperm fertility . . ." Among the agents discussed in 13 of the cited references is $PGF_{2\alpha}$, with regard to which, findings of fundamental significance to the understanding of the regulation of the reproductive cycle can be reduced to a practical application for fertility control. $PGF_{2\alpha}$ initiates a chain of events resulting in regression of the corpus luteum and the shedding of the uterine edometrium, in effect initiating menstruation and ensuring continuation of the menstrual cycle. The $PGF_{2\alpha}$ then offers a totally new approach to contraception, one that should allow a simplification in dosing regimen and one that might be quite "physiological" in action. (ART) 2136

1290
DUNHAM, E.W. and B.G. ZIMMERMAN
Release of prostaglandin-like material from dog kidney during nerve stimulation.
American Journal of Physiology. 219: 1279-1285. 1970.

Dog kidneys, autoperfused in situ at a constant flow produced a low basal efflux of prostaglandin-like material tentatively identified as PGE based on thin layer chromatographic behavior and continuous bioassay using rat fundus, chick rectum, and gerbil colon in a blood-bathed organ technique. Results of studies on renal vascular constriction elicited by renal nerve stimulation at 2-10 cycles/second or infusion of norephinephrine show markedly enhanced prostaglandin production. Indications are that renal production of PG occurs duing an increase of renal vascular resistance not necessarily dependent upon changes in total renal blood flow, but upon alternations of the distribution of intrarenal flow. (GW) 2061

1291
DUNHAM, E.W. and B.G. ZIMMERMAN
Release of prostaglandin-like material from the dog kidney during renal nerve stimulation.
Federation Proceedings. 29: 745 abs. 1970.

Abstract only. Samples of renal venous effluent from dog kidneys perfused in situ at constant flow were collected both before and during renal nerve stimulation. Bioassay for prostaglandin activity (on rat fundus and gerbil colon) tentatively identified a prostaglandin of the E series. The results indicate that the release of prostaglandin-like material is enhanced during renal vascular constriction elicited by renal nerve stimulation and that the release is not necessarily dependent upon a decrease in total renal blood flow. (ART) 2124

1292
DUVAL, D.L. P. DIDISHEIM, J.A. SPITTELL, and C.A. OWEN
Effects of monoamine oxidase inhibitors, glyceryl guaiacolate, and ethanol on experimental arterial thrombosis.
Mayo Clinic Proceedings. 45: 579-585. 1970.

It is briefly mentioned in the discussion section of this paper that PGE_1 among other agents have been found to reduce platelet adhesiveness. (JRH) 2503

1293
DYER, D.C.
Comparison of constricting actions produced by serotonin and prostaglandins on isolated sheep umbilical arteries and veins.
Gynecologic Investigation. 1: 204-209. 1970.

Sheep umbilical arteries and veins contracted to serotonin at concentrations less than those of prostaglandins. Also, serotonin produced a greater maximal contraction relative to the prostaglandins. $PGF_{2\alpha}$ and PGE_2 were essentially equiactive in inducing contractions while PGE_1 and $PGF_{1\alpha}$ were clearly less active than $PGF_{2\alpha}$ or PGE_2. (Author) 2453

1294
EAKINS, K.E., S.M.M. KARIM, and J.D. MILLER
Antagonism of some smooth muscle actions of prostaglandins by polyphloretin phosphate.
British Journal of Pharmacology. 39: 556-563. 1970.

The antagonism actions of $PGF_{2\alpha}$ and PGE_2 by polyphloretin phosphate (PPP) was studied on smooth muscle preparations and on blood pressure of the rabbit. PPP markedly antagonized contraction responses produced by PGE_2 and $PGF_{2\alpha}$. Intravenous injections of PPP antagonized the fall in blood pressure produced by intravenous injections of $PGF_{2\alpha}$ but did not antagonize vasodepressor responses produced by PGE_2 and acetylcholine. (GW) 2032

1295
EAKINS, K.E.
Increased intraocular pressure produced by prostaglandins E_1 and E_2 in the cat eye.
Experimental Eye Research. 10: 87-92. 1970.

Doses of 1 mg or more of either PGE_1 or PGE_2 in cats were found to produce an extremely variable increase in intraocular pressure, vasodilation and miosis, and an increase in protein content of the aqueous humor. Smaller doses produced only miosis. Marked tachyphylaxis was noted in the pressure response to PGE. Discussion notes that responses were similar to those of the rabbit though the rabbit eye was much more sensitive. The rise in intraocular pressure and increased protein content are taken to indicate vasodilation and increased blood-aqueous permeability as important factors in the response. (RMS) 2058

1296
EAKINS, K.E. and S.M.M. KARIM
Polyphloretin phosphate—a selective antagonist for prostaglandins $F_{1\alpha}$ and $F_{2\alpha}$.
Life Sciences. 9: 1-5. 1970.

The compound polyphloretin phosphate (PPP) was tested as a possible inhibitor of PGE_1, PGE_2, $PGF_{1\alpha}$, $PGF_{2\alpha}$, acetylcholine and serotonin stimulation of the jird colon. It was found that PPP 5-10 μg/ml) completely or partially blocked the stimulation by F prostaglandins but had no effect on E prostaglandin stimulation. PPP did not block smooth muscle stimulation by acetylcholine or serotonin. (JRH) 2138

1297
EAKINS, K.E., J.D. MILLER, and S.M.M. KARIM
Polyphloretin phosphate, a selective prostaglandin antagonist.
Federation Proceedings. 29: 745 abs. 1970.

Abstract only. The selective antagonism of the smooth muscle stimulating actions of prosta-glandins of the F series by polyphloretin phosphate (PPP) was observed in vitro on the gerbil colon, rabbit jejunum, and rabbit uterus preparations. Concentrations of PPP required to antagonize prostaglandins were without effect on responses to acetylcholine, serotonin, and histamine. On the gerbil colon, responses to $PGF_{2\alpha}$ were more readily antagonized than those produced by PGE_2 and shifts to the right of the prostaglandin dose-response curves in the presence of PPP suggest a competitive relationship. PPP (2.5 to 10 μg/ml) antagonized $PGF_{2\alpha}$ and PGE_2 equally on the rabbit jejunum and uterus. In vivo, the vasodepressor effects of $PGF_{2\alpha}$ but not PGE_2 on the anesthetized rabbit blood pressure was antagonized by PPP (25 to 100 mg/kg., i.v.) (Author modified) 2125

1298
ECKSTEIN, P.
Mechanisms of action of intra-uterine contraceptive devices in women and other mammals.
British Medical Bulletin. 26: 52-59. 1970.

Prostaglandins are briefly mentioned as being one of the substances which may be involved in the mechanism of action of intrauterine devices. (JRH) 2467

1299

EFFENDIC, S. and J. OSTMAN

Effects of glucose and prostaglandin E_1 on catecholamine induced lipolysis in human adipose tissue in vitro.

Diabetologia. 6: 43-44. 1970.

Abstract only. The authors discuss research designed to establish whether glucose as a stimulant and PGE_1 as an inhibitor of lipolysis act via the α or β adrenergic receptors in catecholamine-induced glycerol release from human omental tissue in an in vitro preparation containing noradrenaline, isopropylnoradrenaline, theophylline or N^6-$2'0$-dibutyryl-AMP. Data indicate 2 sites of glucose action on stimulated lipolysis: direct on the triglyceride splitting enzyme system, and stimulation of the responsivity of the β receptors. Results show PGE_1 inhibition of α and β receptors; the net effect depending upon the activity of the receptors. (RMS) 2098

1300

EFENDIC, S. and J. OSTMAN

Effetti del glucosio e della prostaglandina E_1 sulla lipolisi indotta da catecolamine nel tessuto adiposo umano in vitro. [Effects of glucose and prostaglandin E_1 on lipolysis induced by catecholamine in human adipose tissue in vitro.] (in Italian)

Acta Diabetologica Latina. 7: 174-175. 1970.

Abstract only. The results presented by the authors indicate, in the first place, that glucose has two points of attack in the stimulation of lipolysis: 1) direct stimulation of the enzymatic system of the triglycerides; 2) stimulation of beta-adrenergic receptors. Secondarily, the results demonstrate that PGE_1 inhibits the alpha- and beta-adrenergic receptors and acts on the lipolysis itself by means of the activity of such receptors in vitro. (ART—translation) 2377

1301

EFENDIC, S.

Influence of prostaglandin E_1 on lipolysis induced by noradrenaline, isopropylnoradrenaline, theophylline, and dibutyryl cAMP in human omental adipose tissue in vitro.

Acta Medica Scandinavica. 187: 503-507. 1970.

Human adipose tissue was incubated in vitro with noradrenaline (NA), isopropylnoradrenaline(ISPNA), a beta-adrenergic stimulator, phentolamine, PGE_1, dibutyryl cAMP, and theophylline. After a period of incubation with the compound or compounds being tested, the amount of glycerol formed was taken as a measure of the amount of lipolysis. In the first series of experiments it was found that PGE_1 (1 μg/ml) almost completely blocked the lipolytic activity of both ISPNA and theophylline. It was found that PGE_1 was much more effective in blocking ISPNA than NA. When PGE_1 was tested against a mixture of NA plus phentolamine it greatly reduced the amount of lipolysis. PGE_1 had no effect on the lipolysis induced by dibutyryl cAMP. The authors feel these results indicate that the antilipolytic effect of PGE_1 in human adipose tissue is localized at the level of adenyl cyclase. They also feel that PGE_1 inhibits both alpha and beta-adrenergic receptors. (JRH) 2143

1302

EGGENA, P., I.L. SCHWARTZ and R. WALTER

Action of aldosterone and hypertonicity on toad bladder permeability to water.

In: Cort, J.H., and B. Lichardus eds., "Regulation of Body Fluid Volumes by the Kidney," p. 182-192. Basel, S. Karger. 1970.

It has been discovered that exposure of the serosal side of an isolated toad bladder to hypertonic solutions causes a change in permeability of mucosal epithelium similar to that caused by vasopressin. To determine if these two stimuli have a different mode of action, the serosal side of a toad bladder was exposed to PGE_1 $(2.7 \times 10^{-7}M)$ which is known to inhibit the response to vasopressin. The PGE_1 had no effect on the response to hypertonicity. On the basis of these results and those of other parts of the experiment, the authors conclude that hypertonicity activates some part of the same train of events that vasopressin activates, probably at some point after the formation of cAMP. (JRH) 2513

1303
EGGENA, P., I.L. SCHWARTZ, and R. WALTER
  Threshold and receptor reserve in the action of neurohypophyseal peptides. A study of synergists
    and antagonists of the hydroosmotic response of the toad urinary bladder.
  Journal of General Physiology. 56: 250-271. 1970.

Prostaglandin E_1 was one of four substances used to study the interrelationship of several physiological receptors which influence the hydroosmotic response of the toad urinary bladder. The catalytic activity of a series of neurohypophyseal peptides declined when the hydroxy radical of tryrosine residue in oxytocin is replaced by a methoxy and then an ethoxy radical. Prostaglandin E_1 was found to be a non-competitive inhibitor of neurohypophyseal peptides and theophylline; whereas the maximal hydroosmotic response of the bladder to [2-Omethyl-tryosine]-oxytocin and theophylline was greatly depressed by PGE_1; the response to saturating concentrations of oxytocin was only slightly diminished—a finding which reveals a "receptor nerve" for oxytocin. It is concluded that neurohypophyseal peptides are capable of producing graded effects on adenyl cyclase both below and above the range of enzyme activity which evokes graded changes in membrane permeability. (Authors modified) 2316

1304
ELIASSON, R. and Z. BRZDEKIEWICZ
  Effects of various agonists on the tachyphylactic response of the isolated rat uterus to
    prostaglandin E_1.
  Life Sciences. 9: 925-930. 1970.

The tachyphylactic response pattern of the rat uterus to prostaglandins E_1 (PGE_1) has been further studied. When the myometrium was desensitized by repeated applications of PGE_1 the reactivity could be restored by intervening contractions of oxytocin or various other prostaglandins, but not by vasopressin, acetylcholine, histamine or 5-HT. Atropine, adenosine diphosphate or adenosine tripohosphate did not modify that tachyphylactic response pattern. The myometrium responded in a regular fashion when exposed to a test solution containing both oxytocin and PGE_1, but a marked tachyphylaxis was developed for test solutions containing PGE_1, PGE_2, $PGE_{1\alpha}$ $PGF_{2\alpha}$, or oxytocin vasopressin PGE_1. Magnesium was not essential for the desensitization phenomenon. (Authors) 2018

1305
ELIASSON, R.
  Inhibitory effect of p-hydroxyphenylisopropylarterenol on the isolated human myometrium.
  Journal of Pharmacy and Pharmacology. 22: 619-620. 1970.

This letter to the editor presents two characteristics of p-hydroxyphenylisopropylarterenol not previously described. The drug at 1 to $2.5 \times 10^{-7}g/ml$ inhibited the amplitude and frequency of contractions of myometrial strips from 16 non-pregnant patients. Restoration of the spontaneous activity after washing always resulted in complete refraction to a second dose, even if this was 10 times larger than the first. A subsequent dose of PGE_1 always inhibited the motility indicating a normal reactivity to other inhibitors. (GW) 2287

1306

ELSBACH, P. and M.A. RIZACK

The effect of collagenase preparations contaminated with phospholipase C activity on adipose tissue lecithin.

Biochemica et Biophysica Acta. 198: 82-87. 1970.

Collagenase prepared from bacteria is standardly used in the preparation of adipose cell cultures. The purity of the collegenase varies considerably depending on the commercial source. Experiments were performed to determine if crude and purified collegenase might contain significant phospholipase C activity. It was found that crude collegenase preparations contained varying amounts of phospholipase C activity which greatly altered the phospholipid content of plasma membranes of adipocytes. It is suggested that the presence of this enzyme and its alteration of the plasma membrane might account for the failure of PGE_1 to cause a rise in cAMP levels of adipocytes, even though the level does increase when PGE_1 is added to adipose tissue. (JRH) 2456

1307

EMBREY, M.P.

Effect of prostaglandins on human uterus in pregnancy.

Journal of Reproduction and Fertility. 23: 372-373. 1970.

Abstract only. The author reports the use of the PGE series for induction of labor and abortion. He finds PGE's to be more effective that the PGF_α series. Twenty-five near term women were infused with PGE (0.5-2.0 μg/min) intravenously until induction of labor ($\frac{1}{2}$ to $10\frac{1}{2}$ hours). In two cases labor was not induced; one of them was attibuted to a shortage of PGE. The only side effect was vomiting in two cases. For induction of abortion, PGE was given i.v. for $2\frac{1}{2}$-26 hr (2.0-5.0 μg/min). The gestational ages ranged from 9 to 28 weeks. Induction was successful in 9 of 11 cases. Again failure in one case was attributed to inadequate dosage of PGE. (JRH) 2015

1308

EMBREY, M.P.

Effect of prostaglandins on the human uterus in pregnancy.

International Journal of Gynaecology and Obstetrics. 8: 159-160. 1970.

Abstract only. Studies of the physiological effects of individual prostaglandins have been concentrated on nonpregnant myometrium. These have indicated that the prostaglandins of the "E" series generally inhibit, while the "F" prostaglandins usually stimulate myometrial activity. However, recent studies showed that the actions of the prostaglandins on the human pregnant uterus are quite different. These effects were first observed in vitro. Lower segment myometrium was relatively inactive, but on uppersegment myometrium well marked spasmogenic effects were produced by $PGF_{2\alpha}$ and (contrary to expectation) PGE_2 in the dose range 50-100 ng/ml. The stimulatory effects of the prostaglandins in pregnancy were later confirmed in the intact uterus. Using both "E" prostaglandins (PGE_1, PGE_2) and those of the "F" series ($PGF_{1\alpha}$, $PGF_{2\alpha}$) intravenously, in the dose range 2-8 μg/min. stimulation of myometrial contractility was observed, the oxytocic properties of "E" prostaglandins were particularly striking. Their stimulatory properties in pregnancy and the successful induction of labour by "E" and "F" prostaglandins suggests they may have a physiological role in parturition. (Author) 2381

1309
EMBREY, M.P.
Induction of abortion by prostaglandins E_1 and E_2.
British Medical Journal. 2: 258-260. 1970.

PGE_1 and PGE_2 were used to induce abortion in 11 patients whose gestational age was 9 to 28 weeks. The prostaglandin was administered as an intravenous infusion at a rate of 2 and 5 μg/min. The infusion time varied from 2¼ hours to 26 hours. Abortion was successfully induced in 9 of the 11 women. One of the failures was attributed to the small amount of prostaglandin given. No side effects were noticed. (JRH) 2252

1310
EMBREY. M.P.
Induction of labour with prostaglandins E_1 and E_2.
British Medical Journal. 2: 256-258. 1970.

The potential value of PGE_1 and PGE_2 was explored as to their usefulness in the induction of labor. Studies were conducted with 25 women at or near term (36 to 42 weeks) who were recommended for interruption of pregnancy. Intravenous infusions of PGE_1 (4 patients) or PGE_2 (21 patients) resulted in the successful induction of labor in 23 of the 25 subjects. As the infusion continued, there was a progressive augmentation of uterine activity, the tocographic pattern of contractility resembling that seen in normal labor. No undersirable side effects occurred. (ART) 2251

1311
EMBREY, M.P.
Prostaglandins.
Lancet. 2: 874-875. 1970.

In a letter to the editor the author questions the opinion that early stages of pregnancy may be more susceptible to the abortifacient action of prostaglandins than later weeks (Wiqvist and Bydgeman, Lancet, October 3, 1970, p. 716) on the grounds that $PGF_{2\alpha}$ was used. He notes his own successful use of PGE_2 in mid-trimester. Work is also cited on a small series using intrauterine PGE_2 at doses of one-tenth those required intravenously; early observations showed a 40% failure rate, but are called promising. Some of the problems involved before marketing of prostaglandins for mass use are reviewed. (RMS) 2062

1312
EMBREY, M.
Prostaglandine zur wehenstimuleirung. [Prostaglandins in the induction of labor.]
Gynakologische Rundschau. 10: 268-274. 1970.

Article in German. Abstract not available at present. 2523

1313

FELDBERG, W., R.D. MYERS, and W.L. VEALE

Perfusion from cerebral ventricle to cisterna magna in the unanaesthetized cat. Effect of calcium on body temperature.

Journal of Physiology. 207: 403-416. 1970.

A method is described for the perfusion of the liquor space from a lateral cerebral ventricle to the cisterna magna in the unanaesthetized cat. Perfusions were carried out for 30-40 minutes using various physiological salt solutions while rectal temperature was recorded. The effluent collected from the cisterna contracted the fundus strip of the rat stomach, regardless of what fluid had been used for the perfusion. The contractions produced by the effluent were not due to 5-HT since they were little affected by BOL. They are thus attributed mainly to an action of PGE_1. (GW) 2235

1314

FICHMAN, M.P.

Natriuretic and vasodepressor effect of prostaglandin (PGA_1) in man.

Clinical Research. 18: 149. 1970.

Abstract only. To determine its effect on blood pressure (BP), H_2O and Na^+ excretion (U_{Na}), PGA_1 (.25-5 μg/kg/min) was infused in 35 patients. PGA_1 produced no consistent change in urine volume (Uv), urine osmolality (Uosm), or Tc_{H_2O} during H_2O restriction, and no inhibition of the antidiuretic effect of a 200 mug/hr vasopressin (ADH) infusion. During H_2O loading, PGA_1 increased creatinine clearance (Ccr), Uv, and C_{H_2O}, and Cosm. PGA_1 ($< .5$ μg/kg/min) failed to lower BP in 20 normotensive and 5 hypertensive subjects, but lowered systolic by 20 - 40 and diastolic BP by 10-30 mm Hg in 4 anephric dialysis patients. PGA_1 (2.5-5 μg/kg/min) lowered systolic by 20-90 and diastolic by 20-40 mm Hg in 3 1° hyperaldosteronism and 1 essential hypertensive patients. In 20/28 subjects, PGA_1 increased U_{Na} from 1-100 μeq/min by 1-36 fold to 3-200 μeg/min, and urine K^+ 1-5 fold. In 12/18 subjects, greater augmentation of U_{Na} by 2-100 fold occurred when PGA_1 was given during ADH infusion, the greatest response shown in a cirrhotic with ascites in whom U_{Na} increased 100 fold from 2.4 to 240 μeg/min. In 2 patients with the "hepatorenal" syndrome, Ccr (10 ml/min) doubled, Uv increased 2-3 fold, and U_{Na} (1-6 μeg/min) increased 10-20 fold to 20-60 μeg/min with PGA_1 and ADH. These data suggest that while PGA_1 failed to alter the effect of ADH on H_2O excretion, its natriuretic effect in man is potentiated by ADH infusion, and its vasodepressor effect enhanced in anephric states. (Author) 2163

1315

FILSHIE, G.M.

Therapeutic abortions using prostaglandin E_2.

Journal of Reproduction and Fertility. 23: 371-372. 1970.

Abstract only. PGE_2 was administered i v to 52 patients at a rate of 5 μg/min. The gestational ages ranged from 9 to 22 weeks. The intrauterine pressure of 22 of the 35 patients whose gestational ages were 15 weeks or more were recorded during the PG infusion. There was a 1-5 min latent period followed by an increase in uterine tone. The hypertonus then decreased and a rhythmic contraction of the uterus began and continued until abortion occurred. 50 of the 52 patients aborted within 48 hours and the remaining 2 were aborted by other means. In 7 of the patients there was a 30-min to 10-hour delay in explusion of the placenta and in an additional 7 patients the products of conception had to be evacuated. Side effects: 13 patients had nausea, 4 had diarrhea, 6 had venous erythema along the infused vein, 1 had tachycardia and one had transient blurred vision. Bleeding of more than 50 ml occurred in only 2 patients and 2 patients had a temperature of more than 38°C. (JRH) 2014

1316

FINE, L.G.
Acquired prostaglandin E_2 (medullin) deficiency as the cause of oliguria in acute tubular necrosis. A hypothesis.
Irish Journal of Medical Science. 6: 346-350. 1970.

Based on documented observation, a mechanism of oliguria in acute tubular necrosis is proposed. It is suggested that renal ischemia damages the tubules of the outer medulla which is the site of prostaglandin production in the kidney. Intrarenal production of prostaglandin falls, leading to selective cortical ischemia, diminished glomerular filtration and oliguria which persist despite reversal of the precipitating factors. The oliguric phase is only terminated once prostaglandin production is restored in the damaged tubules, causing a return of cortical perfusion and glomerular filtration to normal. (Author) 210l

1317

FINKEL, A.L. and D.R. SMITH
Parameters of renal functional capacity in reversible hydroureteronephrosis in dogs.
Investigative Urology. 8: 299-310. 1970.

The work of others who found that infusion of PGE_1 into the renal artery of dogs caused an increase in urine volume, sodium excretion, free water clearance and renal plasma flow is briefly mentioned in the discussion section of this paper. (JRH) 2464

1318

FISHBEIN, M.
Medicine 1969.
Postgraduate Medicine. 47: 172-175. 1970.

A review of some of the important medical advancements of 1969 is presented. The prostaglandins are mentioned as being one of the new hormone-like substances under investigation. Their effects on various tissues of the body are being experimentally studied in the treatment of such widely different conditions as nasal congestion and peptic ulcer. They are also being studied in relation to the induction of labor, contraception, and male sterility. (GW) 2284

1319

FISHER, J.W.
Kidney, especially its hormones.
In: Eigenmann, R. ed., "Proceedings of the 4th International Congress on Pharmacology," July, 1969. p. 111-114. Basle, Schwabe, 1970.

In this chairman's report on the discussion at a conference on pharmacological effects of various hormonal agents produced by the kidney it is mentioned that PGE_2 was found to be the most abundant prostaglandin in the kidney of rabbits. PGA_2 and $PGF_{2\alpha}$ were also found. Prostaglandin secretion was doubled by infusion of norepinephrine. Neither α-adrenergic, β-adrengeric blockade nor atropine blocked the effects of PGE on renal function. It was suggested that the natriuretic and diuretic effects of norepinephrine are due to its stimulation of prostaglandin release. Unilateral renal ischemia in dogs caused a release of PGE_2, $PGF_{2\alpha}$ and PGA_2 from the contralateral kidney in anesthetized dogs and it was suggested that angiotensin II was the mediator of this release. (JRH) 2483

1320

FOX, C.A.

Reduction in the rise of systolic blood pressure during human coitus by the β-adrenergic blocking agent, propranolol.

Journal of Reproduction and Fertility. 22: 587-590. 1970.

A continuous record of systolic blood pressure during coitus was maintained in both male and female partners in a test of the effect of β-adrenergic blocking drugs on the rise of systolic blood pressure and tachycardia of orgasm. The author considers the cause of the systolic rise during coitus and subsequent sharp fall to below resting level and mentions that no rise of prostaglandins E_1 and E_2 among other agents have been detected in the blood. His speculation is that the increase is due largely to muscular tension and exertion. (RMS) 2072

1321

FREDHOLM, B.B. and S. ROSELL

Effects of prostaglandin E_1 in canine subcutaneous adipose tissue in situ.

Acta Physiologica Scandinavica. 80: 450-458. 1970.

The effect of PGE_1 on the uptake of glucose and the release of FFA and glycerol before and after sympathetic nerve stimulation (4 cps) was investigated in perfused canine subcutaneous adipose tissue in situ. Glucose uptake was significantly increased by PGE_1 at all concentrations used (5×10^{-10} to 7×10^{-7} M in blood). The effect of PGE_1 on the release of FFA and glycerol in unstimulated adipose tissue was inconsistent. Increases as well as decreases were observed. Lipolysis, as measured by glycerol release, induced by nerve stimulation was inhibited dose-dependently. A 50% inhibition was produced by approximately 1.2×10^{-7} M PGE_1. Stimulated FFA release was also inhibited but there was no clear dose-response relationship. It is concluded that PGE_1 has similar effects in canine subcutaneous adipose tissue with an intact blood supply as are known to be produced in vitro. (Authors) 2039

1322

FREDHOLM, B.B., B. OBERG, and S. ROSELL

Effects of vasoactive drugs on circulation in canine subcutaneous adipose tissue.

Acta Physiologica Scandinavica. 79: 564-574. 1970.

Effects of acetylcholine, bradykinin, isoprenaline, histamine, 5-hyrdoxytryptamine, noradrenaline, and prostaglandin E_1 were studied in adipose tissue preparations done in 24 female mongrel dogs. PGE_1 was the most potent, producing clear vasodilation at blood concentrations of 10^{-9} M making it 1000 times more potent than acetylcholine and 10 times more potent than histamine. A moderate increase in flow with a venoconstriction was observed with 5-HT. Noradrenaline induced vasoconstriction with increased capillary filtration coefficient (CFC). CFC increases were also noted with bradykinin, acetylcholine, isoprenaline, 5-HT, PGE_1, and sympathetic nerve stimulation. The latter paired with bradykinin or histamine produced a decrease or no change in CFC. Paired with the other agents, increases were seen on sympathetic stimulation which are held to be due to increased capillary membrane permeability. (RMS) 2170

1323

FREDHOLM, B.B., S. ROSELL and K. STRANDBERG

Release of prostaglandin-like material from canine subcutaneous adipose tissue by nerve stimulation.

Acta Physiologica Scandinavica. 79: 18A-19A. 1970.

The amount of prostaglandin-like substance in the venous outflow from dog adipose tissue was determined before and after nerve stimulation. No PG could be detected by bioassay before nerve stimulation, but significant amounts (0.9-8.2 ng) were found after nerve stimulation in 7 out of 11 dogs. This PG was characterized as being of the E type. When PGE_1 was infused into the adipose tissue only 5 to 17% could be recovered in the venous blood. When adipose tissue was extracted for PG, the amount was similar in stimulated and unstimulated tissue (1.3 to 5.5 ng/g tissue). It was concluded that a prostaglandin-like substance is released in response to nerve stimulation from adipose tissue and that PG's are rapidly inactivated in adipose tissue. (JRH) 2171

1324

FREDHOLM, B.B.

Studies on the sympathetic regulation of circulation and metabolism in isolated canine subcutaneous adipose tissue.

Acta Physiologica Scandinavica. Supp. 354: 1-47. 1970.

In addition to a bibliography of 162 citations, it is stated that the present survey is based on studies which, except for some hitherto unpublished material, have been presented in 6 papers which are referenced (4 in press, 2 published). The survey consists of an introduction; a discussion of methodology, including experimental procedure and analytical methods; a chapter on results and discussion, including subheadings as follows: composition of adipose tissue, metabolism during basal conditions, blood circulation, adrenergic transmission, metabolic effects of nerve stimulation, modification by receptor antagonists, role of cyclic AMP, role of prostaglandins, changes in re-esterification, and modification of the metabolic responses by vasoconstriction; and a summary. The presentation concerning the role of prostaglandins is headed "Effects of prostaglandins on lipolysis during nerve stimulation" and is a summary of one of the papers in press referenced at the beginning of the survey, as follows: "Effects of prostaglandin E_1 in canine subcutaneous adipose tissue *in situ*. Acta Physiol. Scand. 1970. In press. (Together with S. Rosell)." This paper is summarized as follows: "We have been able to confirm the in vitro observation that PGE_1 inhibits lipolysis stimulated by nerve activity (Berti and Usardi 1961) and extended it by demonstrating dose-response characteristics in a blood perfused organ stimulated with frequencies more similar to those occurring physiologically. We have also confirmed the results of Shaw and Ramwell (1968) that prostaglandin-like material is released from adipose tissue upon nerve stimulation . . . . " (ART) 2109

1325

FRIBOURG, J. and S. CARRIERE

Correlation entre les effets vasodilateur et natriuretique de l'acetylcholine et de la prostaglandine. [Correlation between vasodilator and natriuresis effects of acetylcholine and of prostaglandin.]

L'Union Medicale du Canada. 99: 325. 1970.

Natriuresis and intrarenal blood flow changes induced by PGE_1 and acetylcholine were compared by diminishing curves of Kr^{85} and autoradiography. The effects of both substances were comparable. Natriuresis, urinary flow and renal plasma circulation were augmented significantly while glomerular filtration remained unchanged. Cortical blood flux was altered upward, while medullary flux was unchanged; however, the total medullary and cortical flow were increased because of relative increase of the total outer medullary volume. The increase in cortical flux can cause relative glomerular filtration increases in the short cortical nephrons with possibly limited sodium resorption ability, thus accounting for the observed natriuresis. An alternate explanation is hemodynamic modifications at the head of the peritubular plexi inhibit sodium resorption. (RMS) 2073

1326
FRIED, J., M.M. MEHRA, W.-L. KAO, and C.H. LIN
Synthesis and biological activity of 7-oxaprostaglandins and related substances.
In: Abstracts, 5th Mid Atlantic Regional Meeting of the American Chemical Society. 1-3 April, Newark. p. 60. 1970.

Following up on our earlier synthesis of 7-oxa-prostaglandin $F_{1\alpha}$ and related substances (Chem. Comm. 634 (1968)), all of which caused contraction of smooth muscle (Life Sciences, *8*, 983 (1969)), and of the corresponding 9,11 and 15-deoxy compounds, which proved to be competitive antagonists of both PGE_1 and $PGF_{1\alpha}$ (Nature, *223*, 208 (1969)), we have completed the synthesis of 7-oxa-PGE_1 by several routes. These will be described and the biological activities of the resulting compounds will be discussed. There appears to be a consistent correlation between the degree of hydroxylation and the type of activity observed, antagonism being maximal when no hydroxyl groups are present. (Authors) 2370

1327
FRIED, J., M.M. MEHRA, W.L. KAO and C.H. LIN.
Synthesis of (±)-7-oxaprostaglandin E_1.
Tetrahedron Letters. 2695-2698. 1970.

The authors describe the total synthesis of racemic 7-oxaprostaglandin E_1. Starting with all-cis-1,2-epoxycyclopentane-3,5-diol a step by step synthesis is described with the structures of intermediates given. (JRH) 2411

1328
FUJIMOTO, S. and M.F. LOCKETT
The diuretic actions of prostaglandin E_1 and of noradrenaline, and the occurrence of a prostaglandin E_1-like substance in the renal lymph of cats.
Journal of Physiology. 208: 1-19. 1970.

Prostaglandin E_1, noradrenaline and angiotensin II-amide produced diuresis and natriuresis in cats under chloralose anesthesia. PGE_1, 0.13-0.19 mg/kg/min increased glomerular filtration rate (GFR) but did not alter the filtration fraction (FF). Noradrenaline and angiotensin produced rises in both GFR and FF. Perfusion pressure was lowered 10mm Hg by PGE_1 and raised by noradrenaline and angiotensin. All 3 raised renal lymph flow; noradrenaline and PGE_1 raised the equivalent content of PGE_1-like activity of the lymph. Block of α and β adrenergic receptors prevented vasopressor and FF effects without influencing the diuresis, natriuresis, or lymph effects of noradrenaline. PGE_1 and angiotensin were unaffected by α and β blockage as were noradrenaline-caused diuresis and release of PGE_1-like material. It is suggested that part of the actions attributed to noradrenaline are due to intrarenal release of a PGE_1-like material. (RMS) 2004

1329
FULGHUM, D.D.
The skin and the prostaglandins.
Archives of Dermatology. 102: 225-226. 1970.

When PGE_1 and PGE_2 are intradermally injected into normal skin, prolonged edema and erythema are produced. Similar but lesser effects are seen with PGF_1 and PGF_2. In patients with atopic dermatitis, an erythematous reaction like that seen in normal subjects is produced. Other reactions in chick embryos, rabbits and cats are described. (GW) 2091

1330
FUNDER, J.W., J.R. BLAIR-WEST, M.C. CAIN, K.J. CATT, J.P. COGHLAN, D.A. DENTON, J.F. NELSON, B.A. SCOGGINS and R.D. WRIGHT
Circulatory and humoral changes in the reversal of renovascular hypertension in sheep by unclipping the renal artery.
Circulation Research. 27: 249-258. 1970.

It is mentioned in the discussion section of this paper that others have shown prostaglandins to be released from kidneys after various experimental manipulations. However, the exact role of these compounds cannot be determined at present. (JRH) 2461

1331
GEWIRTZ, G.P. and I.J. KOPIN
Effect of intermittent nerve stimulation on norepinephrine synthesis and mobilization in the perfused cat spleen.
The Journal of Pharmacology and Experimental Therapeutics. 175: 514-520. 1970.

Norepinephrine-C^{14} is synthesized in the cat spleen perfused with tyrosine-C^{14} and is released during nerve stimulation. There is preferential release of newly synthesized norepinephrine after nerve stimulation and this release is dependent upon the interval between trains of stimuli. In the cat spleen perfused with blood containing prostaglandin E_1 to prevent platelet aggregation the mean specific activity of the released norepinephrine was the same as that in the spleen. Phenoxybenzamine enhances norepinephrine release and increases its synthesis. (CWS) 2277

1332
GILDER, S.S.B.
Prostaglandins and uterine activity (London letter).
Canadian Medical Association Journal. 102: 656. 1970.

In this brief editorial review the author discusses the use of prostaglandins to induce labor and abortion. (JRH) 2460

1333
GILES, T.D. and G.E. BURCH
Anesthesia, dogs, and cardiovascular data.
American Heart Journal. 79: 141-142. 1970.

The authors discuss the problems involved in interpreting the results of cardiovascular experiments in anesthetized dogs due to the cardiovascular effects of the anesthetic. As an example, the authors describe the results others have obtained with PGE_1 in dogs under different kinds of anesthetic. (JRH) 2410

1334
GILLESPIE, A., and J.M. Beazley
Ideal means of fertility control?
Lancet. 1: 717. 1970.

In this letter to the editor the authors state that prostaglandins may not prove to be the ideal fertility control substance that others have proposed. Because of its side effects, it may prove to be too toxic for general or unsupervised use. Because of the amount of prostaglandin

346

required to induce abortion, it would be difficult to find a solvent that would hold an effective dose in solution in a small enough volume to impregnate a tampon. Finally it would be impossible to rapidly remove the prostaglandin when administered intravaginally if undesirable side effects developed. (JRH) 2049

1335
GIMENO, A.L., R. SANTILLAN DE TORRES and M.F. GIMENO
Efecto de prostaglandina E_1 sobre la contraccion y los trigliceridos del miocardio auricular de ratas alimentadas y ayunadas. [The effect of prostaglandin E_1 on the contraction and the triglycerides of the auricular myocardium of fed and unfed rats.]
Medicina (Buenos Aires). 30: 626-627. 1970.

The amount of triglycerides (TG) and the effect of PGE_1 on the amplitude of isometric contractile tension (AICT) was measured in isolated auricles from fed and unfed rats both immediately after sacrifice and after periods of incubation in vitro. In the absence of exogenous substrate there was a greater drop in AICT in auricles from fed animals than in those from unfed rats. Additon of PGE_1 (5 μg/ml) under similar conditons had no effect on the drop in AICT in auricles from fed animals but significantly increased the drop in those from unfed animals. The initial TG levels were higher in the tissue from unfed rats. When incubated for 60 min in the absence of substrate, the TG levels in auricles from fed rats declined while those in unfed rats were unchanged. Addition of PGE_1 to the medium blocked the drop in TG in fed rats. (JRH)2452

1336
GIROUD, J.P. and D.A. WILLOUGHBY
The interrelations of complement and a prostaglandin-like substance in acute inflammation.
Journal of Pathology. 101: 241-249. 1970.

It has been shown that a causal relation exists between C' activation and the formation of a possible mediator of inflammation. The mediator is active in increasing vascular permeability and causes a slow contraction of guinea-pig ileum. It can be formed in plasma by the addition of a variety of irritants, and is not histamine, 5-HT or acetylcholine. There are resemblances to bradykinin, but also some important differences. It would seem to resemble the prostaglandins most clearly, hence the description "prostaglandin-like" activity. It has been shown that C'-deficient serum is incapable of forming this material in vitro. It is suggested that the interaction of serum and damaged tissue proteins leads both to activation of the complement system and to the formation of this "prostaglandin-like" activity; this provides one explanation for the mediation of the delayed phase of vascular permeability in acute inflammation in the rat. (Authors) 2285

1337
GOTO, Y.
The effects of prostaglandins on lipid transport.
Shinryo. 23: 1364-1369. 1970.

Article in Japanese. Abstract not available at present.

1338
GOTTENBOS, J.J. and G. HORNSTRA
The influence of prostaglandin E_1 on experimental platelet aggregation in rats.
In: Jones, R.L., ed. "Atherosclerosis," Proceedings of the 2nd International Symposium. p. 130-133. New York, Springer-Verlag. 1970.

The influence of PGE_1 on the intraarterial thrombosis in rats in vivo was investigated. A method was developed for continuously measuring the degree of platelet aggregation in the circulating blood. A filter was connected to the arterial circulation so that the flow of aggregates in the blood is obstructed. The degree of aggregation can then be measured as the ratio of the blood pressures distal and proximal to the filter after ADP administration. Microscopic examination of the occlusion on the filter showed it almost exclusively was caused by thrombocytes. After mixing PGE_1 with ADP, the aggregation-inhibiting action of PGE_1 can therefore be calculated simply by subtracting the aggregation index caused by the PGE_1/ADP mixture from that caused by ADP alone. To prevent intravascular thrombosis, PGE_1 must be administered continuously in doses as low as possible. (GW) 2331

1339

GREAVES, M.W. and J. SONDERGAARD

A new pharmacological finding in human allergic contact eczema.
Archives of Dermatology. 101: 659-661. 1970.

Contact eczema in 22 patients was studied by a perfusion technique which resulted in the isolation of a smooth muscle contracting agent extractable by ethyl acetate at pH 3. The activity of this agent is compatible with prostaglandins which are noted to be vasoactive in human skin. (RMS) 2182

1340

GREAVES, M.W. and J. SONDERGAARD

Pharmacological studies in cutaneous inflammation in man using an in vivo perfusion method.
British Journal of Dermatology. 82 (supp. 6): 82-85. 1970.

A continuous skin perfusion method for use in human subjects which enables direct recovery of pharmacological substances from inflamed skin was developed. Sixteen control subjects were studied. These were patients with localized non-inflammatory skin conditions such as virus warts, naevi and neoplasms. Fourteen patients with contact dermatitis received a patch test $6cm^2$ on the forearm and perfusion of duration 90 minutes was carried out on the inflamed skin 48 hours later. In 10 subjects an agent was found which contracted the guinea-pig ileum. The possibility that prostaglandin-like fatty acids are responsible for the smooth muscle contracting activity in these perfusates was raised when the perfusates from 4 patients were extracted by partition chromatography between ethyl-acetate and water at pH 3. Residue from the ethyl acetate fraction contracted the guinea-pig colon and the rat uterus. (GW) 2248

1341

GREAVES, M.W.

Recent immunopharmacological developments in immediate hypersensitivity reactions.
Acta Dermato-Venereologica. 50(supp. 64): 5-14. 1970

The supplement makes mention of the prostaglandins under the heading of "the pharmacology of immediate hypersensitivity." The importance of prostaglandins in human hypersensitivity reactions in entirely unknown but their pharmacological properties justify full investigation of their role in this context. A pharmacologically active fatty acid with prostaglandin-like properties in allergic contact eczema using a skin perfusion technique has been recently reported. Prostaglandins are possible mediators of immediate hypersensitivity in the skin and gut causing sustained erythema and hypermotility respectively. (GW) 2293

1342

GREAVES, M.W. and J. SONDERGAARD
Urticaria pigmentosa and factitious urticaria: direct evidence for release of histamine and other smooth muscle-contracting agents in dermographic skin.
Archives of Dermatology. 101: 418-425. 1970.

The analysis of pharmacological agents released in inflamed human skin is described. A substance was found in the perfusate from dermographic skin of urticaria pigmentosa that could contract the isolated uterus. Winkelmann and his co-workers concluded that the agent is bradykinin, but their finding is inconclusive since several other naturally occurring smooth muscle contracting substances cause contraction of the rat uterus. These include acetylcholine, adrenaline, tyramine and prostaglandins. (GW) 2358

1343

GREENGARD, P. and E. COSTA, eds.
Role of cyclic AMP in cell function. Advances in Biochemical Psychopharmacology 3. 386p. New York, Raven Press. 1970.

This book on the role of cyclic AMP in cell function contains 20 articles and an index. Included are 6 articles on prostaglandins. Abstracts of these articles are filed under Appleman, M.; Birnbaumer, L.; Butcher, R.W.; Hoffer, B.J.; Robison, G.A. and Rosen, O.M. 2363

1344

GRIECO, M.H.
Current concepts of the pathogenesis and management of asthma.
Bulletin of the New York Academy of Medicine. 46: 597-610. 1970.

There is only brief mention that $PGF_{2\alpha}$ has been identified in the human lung in this review on asthma. (JRH) 2459

1345

GROSSI, F. and M. GRASSI
Fattori umorali e fisiopatologia del processo digestivo. [Humoral and physiological factors of the digestive process.]
Clinica Terapeutica. 53: 547-562. 1970.

The concept of humoral regulation of the motor and secretory functions of digestion based on recent investigations into some of the active chemical groups, actions, synergisms, and antagonisms of digestive hormones, histamine, bradykinin, and of the prostaglandins is discussed. The possible therapeutic employment of these substances is also considered. One hundred twelve papers are included in an international bibliography including 12 which mention prostaglandins by title. (RMS) 2089

1346

GUTKNECHT, G.D., G.W. DUNCAN, and L.J. WYNGARDEN
Effect of prostaglandin $F_{2\alpha}$ on ovarian blood flow in the rabbit as measured by hydrogen desaturation.
Physiologist. 13: 214. 1970.

Abstract only. Luteolysis, as determined by peripheral plasma progestin levels, corpora lutea progestin content, and corpora lutea weights occurred when prostaglandin $F_{2\alpha}$ (200-400 μg/kg) was injected intravenously in the rabbit twice a day on days 9 through 12 of pseudopregnancy. This initial dose reduced ovarian blood flow by 50% in 9 day pseudopregnant rabbits. Subcutaneous administration of $PGF_{2\alpha}$ required a higher dose to induce a similar decline in blood flow. The simultaneous induction of $PGF_{2\alpha}$ of luteolysis and reduced ovarian blood flow supports the concept that a relationship exists between the ovarian vascular system and luteal regulation. (GW) 2274

1347
HADDY, F.J.
Pathophysiology and therapy of the shock of myocardial infarction.
Annals of Internal Medicine. 73: 809-827. 1970.

Clinical methods for increasing myocardial oxygen tension are reviewed. The use of PGE_1 as a thrombolytic agent is briefly mentioned, and two references are cited. (RMS) 2178

1348
HALDAR, J., H. MAIWEG and C.E. GROSVENOR
Inhibition by prostaglandin of oxytocin-induced contraction of breast myoepithelium.
Abstracts of the 52th Meeting of the Endocrine Society. 10-12. June. p. 130. 1970.

Abstract only. Injection of 1 μg of PGE_1 intravenously or intraarterially to urethane anaesthetized rats totally inhibited the normal rise of intramammary pressure (IMP) induced by intraarterially injected oxytocin. Less than total, though significant depression of IMP followed injection of 0.2 μg PGE_1. PGE_1 also inhibited the oxytocin-induced contraction of mammary gland strips in vitro when added to the fluid bathing the gland. The inhibitory effect of PGE_1 lasted 5-15 min and could be repeated several times in the same in vitro or in vivo preparation. The degree of inhibition of IMP and PGE_1 in vivo was not influenced by prior spinal cord section but was completely prevented by the adrenergic blocking agent, phenoxybenzamine (dibenzyline). Prior addition of the adrenergic blocking agent, propranolol, to the bathing fluid also totally prevented the in vitro inhibitory effect of PGE_1. These results suggest that PGE_1 inhibits the contractile response of the rat mammary gland myoepithelium to oxytocin by a direct action on the gland and that the inhibition involves activation of β receptors. (Authors modified) 2525

1349
HAMBERG, M. and U. ISRAELSSON
Metabolism of prostaglandin E_2 in guinea pig liver.
Journal of Biological Chemistry. 245: 5107-5114. 1970.

Seven metabolites were isolated by reversed phase partition chromatography and thin layer chromatography after incubation of tritium-labeled prostaglandin E_2 with the soluble fraction of homogenates of guinea pig liver. The two major compounds (forming about 61% of the recovered radioactivity) were identified as 11α, 15-dihydroxy-9-keto-prost-5-enoic acid and 11-hydroxy-9,15-diketoprost-5-enoic acid. Three compounds (together about 26% of the recovered radioactivity) belonged to the F series of prostaglandins, and were identified as prostaglandin $F_{2\alpha}$, 9α, 11α, 15-trihydroxyprost-5-enoic acid, and 9α, 11α-dihydroxy-15-keto-prost-5-enoic acid. The isolation of these compounds after incubation of prostaglandin E_2 for the first time showed that prostaglandin F_α compounds can be formed from prostaglandin E compounds in animal tissue. Two minor compounds (together about 5% of the recovered radioactivity) were identified as 8-isoprostaglandin E_2 and, tentatively, 8-isoprostaglandin $F_{2\alpha}$. (Authors) 2259

1350
HAYASHI, Y.
  [Standard for the practice of planned delivery from the standpoint of cervical canal and
    sensitivity factor.]
  Sanfujinka no Jissai. 19: 679-691. 1970.

  Article in Japanese. Abstract not available at present. 2077

1351
HEDQVIST, P.
  Antagonism by calcium of the inhibitory action of prostaglandin E_2 on sympathetic neurotrans-
    mission in the cat spleen.
  Acta Physiologica Scandinavica. 80: 269-275. 1970.

  Experiments were carried out with the isolated, perfused cat spleen and bovine splenic nerve
  trunk to study the target for prostaglandin E_2 induced inhibition of noradrenaline release from
  sympathetic nerves. Increasing the calcium concentration in the medium perfusing the spleen
  was found to increase the outflow of noradrenaline to nerve stimulation and to antagonize the
  inhibitory action of prostaglandin E_2 on this parameter. Prostaglandin E_2 did not affect the
  noradrenaline releasing effect of tyramine. Atropine did not alter the inhibitory effect of
  prostaglandin E_2 on noradrenaline release. Prostaglandin E_2 did not change the amplitude of
  action potentials in bovine splenic nerves. There is thus no support for a blocking action on
  nerve conduction. It is concluded that prostaglandin E_2 may prevent calcium from gaining
  access to those sites in sympathetic nerve terminals where it promotes secretion of noradrenaline.
  (Author) 2030

1352
HEDQVIST, P.
  Control by prostaglandin E_2 of sympathetic neurotransmission in the spleen.
  Life Sciences. 9: 269-278. 1970.

  Concentrations of PGE_2 ranging from 6×10^{-8} to 6×10^{-6} M were infused intraarterially into
  isolated cat spleen, to study the effect of PGE_2 on the release of noradrenaline (NA) and on
  the smooth muscle contraction in response to nerve stimulation. The isolated spleen was
  perfused at a constant rate of about 10ml/min at $37°C$ with a slightly modified Krebs-
  Heuseleit's medium, and the effluent divided into 10 ml fractions. The NA stores were labelled
  by intra-arterial infusion of 30 μC of ^{3}H-dl-NA. The NA content of the perfusate was measured
  fluorimetrically and the radioactivity in the perfusate fractions was determined by liquid
  scintillation counting. The splenic nerves were electrically stimulated by platinum electrodes at
  2-3 m sec, 10-15 V, 10 pulses/sec. Smooth muscle and splenic contraction was measured by
  transducer. PGE_2 did not materially change the outflow of NA, but markedly depressed the
  outflow in response to nerve stimulation, the inhibition being dose related. The pressor as well
  as splenic contraction responses to nerve stimulation were inhibited by low doses of PGE_2 but
  progressively diminished with higher concentrations of PGE_2 and returned almost to control
  levels at 6×10^{-6}. On the basis of these data, it seems likely that endogenous PGE_2 may play
  a regulatory role in the splenic sympathetic neuro-effector system. (JWJ) 2137

1353
HEDQVIST, P. and A. WENNMALM
  Inhibition by prostaglandin E_1 of the sympathetic neuromuscular transmission in the rabbit
    heart.
  Acta Physiologica Scandinavica. 79: 19A-20A. 1970.

Abstract only. An isolated rabbit heart with intact sympathetic nerves was suspended in Tyrode solution. The effect of PGE_1 on the nerve induced increase in contractile force and rate of beat was measured. The amount of noradrenaline (NA) in the perfusate collected for 90 sec after stimulation of the nerve was determined. It was found the PGE_1 greatly reduced the inotropic and chronotropic effects of sympathetic nerve stimulation. It also greatly reduced the increase in NA found in the outflow after nerve stimulation. This was taken to indicate that the PGE_1 in low doses inhibits the release of transmitter substance in sympathetic nerves. (JRH) 2052

1354

HEDQVIST, P., L. STJARNE, and A. WENNMALM

Inhibition of prostaglandin E_2 of sympathetic neurotransmission in the rabbit heart.

Acta Physiologica Scandinavica. 79: 139-141. 1970.

PGE_2 was perfused into the isolated rabbit heart preparation with intact sympathetic nerve supply. Concentrations of 7×10^{-8} to $5.5 \times 10^{-7}M$ did not appreciably alter flow through the organ; the heart rate remained unchanged in nine experiments, increased moderately in 4 and decreased in 1. Contractile force was unchanged in 10 experiments, decreased in 3 and increased in 1. Noradrenaline (NA) release was very low or undetected. A frequency-dependent increase in heart rate, contractile force and release of considerable amounts of NA accompanied sympathetic stimulation for 30-second periods at frequencies of 1,5, and 10/sec. PGE_2 counteracted the increases in NA production and heart rate, while contractile force was less affected. Atropine did not materially change this PGE_2 effect on sympathetic stimulation of the heart. The chronotropic and inotropic responses of the heart to NA were found to be slightly and inconsistently diminished by PGE_2. This work confirms results cited from work on other sympathetically innervated tissues and leads to the conclusion that endogenous PGE_2 acts as a modulator of sympathetic nerve stimulation. (RMS) 2057

1355

HEDQVIST, P.

Inhibition by prostaglandin E_1 of vascular response to sympathetic nerve stimulation in vivo.

Acta Physiologica Scandinavica. 80: 6A. 1970.

The author investigated the effect of PGE_1 on the vascular response to sympathetic nerve stimulation in the anesthetized cat. Infusion of PGE_1 $6 \times 10^{-9}M$, into the left femoral artery produced a rapid reduction of the perfusion pressure in the hindleg. The inhibition progressively increased with the dose of PGE_1, $47\pm 3\%$ at $6 \times 10^{-8}M$. The pressor response to noradrenaline was largely unaffected in the presence of PGE_1. The absence of affect of PGE_1 on the vascular response to exogenous NA strongly suggests that PGE_1 inhibits the sympathetic neuro-vascular system prejunctionally and acts on the process of NA release. (GW) 2173

1356

HEDQVIST, P.

Studies on the effect of prostaglandins E_1 and E_2 on the sympathetic neuromuscular transmission in some animals.

Acta Physiologica Scandinavica. Supp. 345: 1-40. 1970.

The aim of the study was to analyze the inhibitory effects of PGE_1 and PGE_2 on sympathetic neuromuscular system in cat spleen, guinea pig vas deferens and rabbit heart. It was found that PGE_1 and PGE_2 depress the function of the sympathetic neuromuscular system both pre- and post-junctionally. It was tentatively suggested that the two compounds might act pre-junctionally by interfering with the transport of noradrenaline into the extracellular space. (GW) 2036

1357
HEDWALL, P.R., W.A. ABDEL-SAYED, P.G. SCHMID, and F.M. ABBOUD
  Inhibition of venoconstrictor responses by prostaglandin E_1.
  Proceedings of the Society for Experimental Biology and Medicine. 135: 757-759. 1970.

The literature pertaining to the interaction of the vasodilator effect and the inhibition of venoconstrictor responses by PGE_1 is discussed. A dissociation of these two effects previously seen is confirmed by the present work. PGE_1 and 7×10^{-9} g/ml caused venodilation in the isolated, perfused gracilis muscle of chloralose anesthetized dogs but venoconstrictor responses to nerve stimulation, norepinephrine, and angiotensin were not significantly inhibited. PGE_1 had no venodilator effects on the perfused hindpaw but inhibited venoconstrictor responses significantly and uniformly. (RMS) 2256

1358
HEDWALL, P.R., F.M. ABBOUD, and P.G. SCHMID
  Potentiation of the vasoconstrictor response to nerve stimulation by a prostaglandin (PGE_1).
  Physiologist. 13: 219. 1970.

In the vas deferens, low concentrations of PGE_1 inhibit the responsiveness to nerve stimulation (NS) whereas high concentrations of PGE_1 enhance this responsiveness. The interaction between PGE_1 and the effect of NS was tested in two vascular beds of the dog. Changes in perfusion pressure in the isolated perfused gracilis muscle and in the isolated perfused hind paw (PP paw) were measured in response to NS and to intra-arterial injection of norepinephrine (NE). PGE_1 altered responsiveness in the paw, but not in the muscle. After intra-arterial injection of a high concentration of PGE_1 (2-5 μ), the average increase in PP paw in response to NS (6Hz) was augmented from 56±14 (SE) to 95±22mmHg, wheras the response to NE (0.3 μg) decreased from 65±15 to 48±13 mmHg. The ratio of responses to NS/NE increased from 0.86 before to 1.98 after PGE_1 (p<0.05). Subsequent infusion of a low concentration of PGE_1 (0.37 μg/min) reduced responsiveness to both NS and NE to 48±10 and 26±12 mmHg respectively, but the ratio of responses to NS/NE remained elevated. The results indicate the PGE_1 depresses vascular responsiveness to NE, but potentiates responsiveness to NS in the paw. This potentiation cannot be attributed to interference with a cholinergic component of adrenergic transmission, to inhibition to re-uptake of the neurotransmitter nor to antagonism of adrenergic beta receptors. Potentiation may be related to facilitation of release of neurotransmitter. (GW) 2275

1359
HEDWALL, P.R., W. ABDEL-SAYED, P.G. SCHMID, and F.M. ABBOUD
  Selective interaction between prostaglandin E_1 and adrenergic stimuli.
  Circulation. 42 (supp. 3): III 89. 1970.

The hypothesis that the vasodilator effect of PGE_1 is mediated through specific interference with the constrictor effects of catecholamines or peptides was tested in a muscular and a cutaneous bed of the dog. Infusion of PGE_1 (7×10^{-9}/ml) caused equivalent dilation in muscle and paw. In the paw, PGE_1 caused a reduction in responsiveness to norepinephrine by a factor of 10, to nerve stimulation by a factor of 3 and to angiotensin by a factor of 2. In the muscle however, corresponding reductions in responsiveness were only 1.9, 1.5 and 1.6 respectively, Glyceryltrinitrate did not alter the responsiveness to adrenergic stimuli in the paw or muscle. The results indicate that the inhibitory effect of PGE_1 on the constrictor responses to adrenergic stimuli is selective in that it is most pronounced with norepinephrine in the paw and is negligible in the muscle. (GW) 2359

1360
HEDWALL, P.R., F.M. ABBOUD, and W. ABDEL-SAYED
Vasodilator effect of a prostaglandin (PGE$_1$): the role of inhibition of adrenergic vasoconstriction.
Federation Proceedings. 29: 387 abs. 1970.

Abstract only. Studies were carried out on the hind limb of the cat, perfused with blood at a constant rate, in order to obtain information on the interaction between PGE$_1$ and adrenergic vasoconstrictor stimuli. PGE$_1$, perfused intra-arterially at a rate of 0.035 μg/min reduced hind limb perfusion pressure by 20 percent; a 5-fold higher infusion rate produced no further decrease in perfusion pressure. Vasoconstrictor responses induced by intra-arterial injection of norepinephrine (0.03 to 0.3 μ), angiotensin-II-amide (0.01 to 0.1 μg), or stimulation of carotid baroreceptors during infusion of PGE$_1$ were unaltered as compared to the corresponding responses during infusion of saline. The results indicate that inhibition of adrenergic vasoconstriction does not contribute to the vasodilator action of PGE$_1$. (ART) 2123

1361
HERD, J.A.
Overall regulation of the circulation.
Annual Review of Physiology. 32: 389-412. 1970.

This chapter reviews the literature concerning the many control systems affecting cardiovascular function. The prostaglandins are included under the heading of "Humoral Substances." When large amounts of prostaglandins were infused intraarterially, some of these substances caused arteriolar dilation, some arteriolar constriction, and some venous constriction. Prostaglandins had no effect on cardiac contractility. Any changes that occurred were a reflex response to peripheral vasodilation and hypotension. (GW) 2180

1362
HIGGINS, C., S. VATNER, D. FRANKLIN, and E. BRAUNWALD
Augmentation of coronary blood flow and cardiac output in conscious dogs by intravenous prostaglandin A$_1$.
Circulation. 42 (supp. 3): III 123. 1970.

Abstract only. One μg/kg of PGA$_1$ injected intravenously into 14 conscious dogs elevated mean cardiac output (3033-4488 ml/min), heart rate (80-130/min), coronary flow (40-71 ml/min), mesenteric flow (498-762 ml/min), renal flow (298-420 ml/min.), iliac flow (263-415 ml/min) while mean aortic pressure fell (98-66mm Hg). Total peripheral resistance fell from .033-.015mm Hg/ml/min as did resistance in the coronary (2.45 to .94 mm Hg/ml/min), mesenteric, renal and iliac beds. PGA$_1$ produced similar reductions after beta blockade and after combined beta and cholinergic blockade. (GW) 2264

1363
HIGGINS, C.B., S.V. VATNER, D. FRANKLIN, and E. BRAUNWALD
Effects of intravenous prostaglandin A$_1$ (PGA) on circulatory dynamics of conscious dogs.
Pharmacologist. 12: 247. 1970.

In order to characterize the circulatory action of prostaglandin A$_1$, the effects of i v PGA$_1$ (.01-1.0 μg/kg) were determined on flow and resistance in the aorta, left circumflex coronary, mesenteric, renal and external iliac arteries in 14 conscious dogs. 1.0 μg/kg elevated cardiac output, heart rate coronary, mesenteric, renal and iliac flows, while mean aortic pressure fell.

Total peripheral resistance fell as did the resistances in the coronary, mesenteric, renal and iliac beds. PGA_1 also decreased coronary rate after beta adrenergic blockage with propranolol and combined beta and cholinergic blockade. PGA_1 in doses well-tolerated by conscious animals is a potent vasodilator which raises cardiac output and peripheral blood flow in all major vascular beds. (GW) 2190

1364

HILLIER, K., and S.M.M. KARIM

The human isolated cervix: a study of its spontaneous motility and responsiveness to drugs.
British Journal of Pharmacology. 40: 576P-577P. 1970.

Experiments were performed to determine if the human cervix could show contractility. Isolated strips of cervix were tested in vitro with various drugs including PGE_2 (0.025-0.5 $\mu g/ml$) and $PGF_{2\alpha}$ (0.5-2.0 $\mu g/ml$) in vitro. PGE_2 inhibited activity in the strips. $PGF_{2\alpha}$ produced a variable response which did not depend on dose. (JRH) 2199

1365

HIMMS-HAGEN, J.

Adrenergic receptors for metabolic responses in adipose tissue.
Federation Proceedings. 29: 1388-1401. 1970.

Prostaglandins are only briefly mentioned in this review article. (JRH) 2462

1366

HINMAN, J.W.

Prostaglandins: a report on early clinical studies.
Postgraduate Medical Journal. 46: 562-575. 1970.

In this review article the author discusses the results of studies involving the clinical uses of prostaglandins. The bibliography has 77 references. (JRH) 2491

1367

HODGMAN, R.E., G.W. JELKS, S. SWINDALL, and R.M. DAUGHERTY, Jr.

Effects of local infusion of $PGF_{2\alpha}$ on the skin and muscle vasculature of the dog forelimb.
Clinical Research. 18: 312. 1970.

Abstract only. $PGF_{2\alpha}$ was infused intraarterially into an isolated innervated dog forelimb. Blood, arterial and venous pressures and volume from the skin and skeletal muscles were recorded as well as aortic blood pressure. $PGF_{2\alpha}$ (0.776 to 38.9 $\mu g/min$) caused a progressive fall in skin outflow but muscle outflow was unaffected. Total limb resistance rose, due mainly to the rise in skin artery and vein resistance (8 fold). Aortic pressure remained unchanged. Limb weight increased. Similar measurements were made in a limb where infusion was held constant with a blood pump. The results were similar to those above. Thus local infusion of $PGF_{2\alpha}$ caused a rise in total forelimb resistance due mainly to an increase in skin resistance. (JRH) 2162

1368

HOFFER, B.J., G.R. SIGGINS and F.E. BLOOM

Possible cyclic AMP mediated adrenergic synapses to rat cerebellar Purkinje cells: Combined structural, physiological, and pharmacological analyses.

In: Greengard, P. and E. Costa, eds., "Role of Cyclic AMP in Cell Function: Advances in Biochemical Psychopharmacology 3," p. 349-370.
New York, Raven Press. 1970.

This review article on the role of cAMP in adrenergic synapses in rat Purkinje cells contains a section on the effects of prostaglandins. PGE_1 and PGE_2 caused the opposite effect of norepinephrine and cAMP. PGE's also inhibited the response to norepinephrine by Purkinje cells. (JRH) 2543

1369
HOFFER, B.J., G.R. SIGGINS and F.E. BLOOM.
Prostaglandins E_1 and E_2 antagonize norepinephrine effects on cerebellar Purkinje cells: microelectrophoretic study.
Journal of The American Oil Chemists Society. 47: 130A. 1970.

Abstract only. In microelectrophoretic experiments, prostaglandins E_1 and E_2 antagonize the reduction in discharge rate of cerebellar Purkinje cells produced by norepinephrine. Slowing of discharge evoked by $3'5'$-adenosine monophosphate or gamma aminobutryric acid is not antagonized. These data provide the first indication that endogenous prostaglandins may physiologically function to modulate central noradrenergic junctions. (Authors) 2440

1370
HOGG, J.A.
Drugs from natural products—animal sources.
In: Abstracts, 160th Meeting of the American Chemical Society, 14-18 September, Chicago. Medi 21. 1970.

Abstract only. Prostaglandins are depicted as natural products that are derived from animal sources and are not yet commercially available for therapeutic use. An attempt will be made to identify distinct phases in their development. (GW) 2371

1371
HOLDSTOCK, D.J. and J.J. MISIEWICZ
Factors controlling colonic motility: colonic pressures and transit after meals in patients with total gastrectomy, pernicious anaemia or duodenal ulcer.
Gut. 11: 100-110. 1970.

It is briefly mentioned in the discussion section of this paper that prostaglandins are not likely to be the mediators of the increased colonic pressure activity that occurs after eating since others have found that the increased rate of intestinal transit caused by PGE_1 was not associated with an increase in colonic pressure. (JRH) 2482

1372
HOLMES, S.W.
The spontaneous release of prostaglandins into the cerebral ventricles of the dog and the effect of external factors on this release.
British Journal of Pharmacology. 38: 653-658. 1970.

Prostaglandins E_1, E_2, $F_{1\alpha}$ and $F_{2\alpha}$ have been identified in perfusates of the cerebral ventricles of anaesthetized dogs. Infusions of serotonin into the lateral ventricle caused a four-fold increase in the release of prostaglandins E into the ventricles and this increase was dissociated from the hyperthermic action. Intraventricular infusions of adrenaline and noradrenaline had no effect on the level of prostaglandin release. Neither electrical stimulation of a hind foot pad nor the intraperitoneal administration of chlorpromazine, amphetamine, tranylcypromine or imipramine had any consistent effect on the amounts of prostaglandins released into the cerebrospinal fluid. When prostaglandin E_1 was added to the fluid perfusing the ventricular system, respiratory changes were observed but almost all the added prostaglandin was recovered from the perfusate leaving the cisterna. (Author) 2233

1373
HORNSTRA, G.
Einfluss von prostaglandin E_1 auf experimentell induzierte thrombo-zytemaggregation in ratten.
[Influence of prostaglandin E_1 on experimentally induced thrombocyte-aggregation in rats.]
Fette-Seifen-Anstrichmittel. 72: 960-963. 1970.

Two methods for experimentally inducing thrombocyte-aggregation in rats is described. Using these methods it is shown that prostaglandin E_1 (PGE_1), on intravasal, intramuscular, subcutaneous or intraperitoneal application, strongly inhibits the thrombocyte-aggregation in flowing blood and consequently the process of haemostasis resulting from the aggregation is considerably slowed down. These results suggest the possible application of PGE_1 for prevention of intravasal thrombosis. (Author) 2542

1374
HORNSTRA, G.
A method for the determination of thrombocyte aggregation in circulating rat blood.
Experientia. 26: 111-112. 1970.

A method is described for measuring platelet aggregation and "des-aggregation" in circulating rat blood continuously during a prolonged period without any loss of blood. Thrombocyte aggregation was induced by infusion of a solution of ADP in the mechanical apparatus involved and a solution of PGE_1 was used as an aggregation inhibitor. A rectilinear relationship is shown between PGE_1-dose and aggregation inhibiting action of the doses tested. (ART) 2129

1375
HORTON, E.W.
Physiological roles of the prostaglandins.
Scientific Basis of Medicine: Annual Reviews, p. 57-60 London, Arthlone Press 1970.

This review on the physiological role of prostaglandins contains sections on prostaglandins and the reproductive tract, prostaglandins as central nervous transmitters, prostaglandins and the adenyl cyclase system, and prostaglandins and the corpus luteum. The bibliography lists 53 references. (JRH) 2425

1376
HUGHES, J. and J.R. VANE
Relaxations of the isolated portal vein of the rabbit induced by nicotine and electrical stimulation.
British Journal of Pharmacology. 39: 476-489. 1970.

A pharmacological analysis of the inhibitory innervation of the isolated portal vein of the rabbit has been made. The vein was stimulated with rectilinear pulses of 1 ms duration and supramaximal voltage (15-18 V across the tissue). The prostaglandins were one of the agonists used. PGE_1 (10^{-7} to 10^{-6} g/ml) and PGE_2 (10^{-8} to 10^{-7} g/ml) reduced half-maximal contractions elicited by noradrenaline and electrical stimulation by 30-90%. PGE_1, E_2, or $F_{2\alpha}$ did not produce relaxations but higher concentrations (5×10^{-7} to 5×10^{-6} g/ml) induced contractions which increased with the concentration of the drug. (GW) 2354

1377
HUGUES, J. and P. MAHIEU
Platelet aggregation induced by basement membranes.
Thrombosis et Diathesis Haemorrhagica. 24: 395-408. 1970.

A study was made of platelet adhesion to the basement membrane using suspensions of membranes of renal origin. As with collagen, the addition of human platelet-rich plasma (PRP) preincubated for five minutes with promethazine to the basement membrane suspension markedly inhibited its aggregating activity as shown by turbidimetry. PGE_1 (1.2 ml PRP incubated for 5 minutes with 0.1 ml PGE_1) produced a more irregular inhibition. (GW) 2290

1378
IINO, K.
The effects of prostaglandins on the cerebral circulation.
Shinryo. 23: 1347-51. 1970.

Article in Japanese. Abstract not available at present. 2415

1379
ILINOV, P.P.
Prostaglandins.
Vutreshni Bolesti. 9: 1-8. 1970.

Article in Russian. Abstract not available at present. 2514

1380
ISHIZAWA, M., K. SATO, H. KINOSHITA and T. WADA
Effects of prostaglandin on gastric secretion and on the gastrointestinal smooth muscle.
Gastroenterologia Japonica. 5: 320. 1970.

Abstract only. The effects of prostaglandin E_1 (PGE_1) and E_2 (PGE_2) on gastric secretion in man and rats, and on the motilities of circular and longitudinal muscle strips of guinea-pig stomach were investigated. The results obtained were as follows; 1. PGE_1 and PGE_2 had an inhibiting action on basal gastric acid secretion. However, both of these showed an increasing effect on gastric pepsin secretion when 0.04γ/kg/min of each of these was infused for 30 min intravenously in man. The same results were also observed in rats. 2. Continuous infusion of PGE_1 or PGE_2 from 2 to 4γ/kg/hr in man and no anti-gastrin effect when the latter of doses 4γ/kg b.w. was injected percutaneously for the stimulation of gastric secretion. 3. PGE_1 and PGE_2 had an effect of showing a tonic contraction on the longitudinal muscle strip and phasic contraction on the circular muscle strip of guinea-pig stomach, and the effectiveness of PGE_2 was more potent than that of PGE_1. (Authors) 2529

1381
ITO, H.
[Seminal vesicle contractile response and prostaglandin E_1.]
Igaku No Ayumi. 75: 486-487. 1970.

Abstract only. We re-examined the report by Eliasson (1966) that a small dose of PGE_1 enhanced the contractile response of the isolated seminal vesicle by catecolamine and acetylcholine. Upon confirming the above observation, we studied the effect of PGE_1 on the response of the human seminal vesicle. Longitudinally sectioned pieces (about 2 mm in width) of the apical portion of the seminal vesicles were obtained from the guinea pig (b.w. 300-400 g) and from humans who had operations for cancer of the bladder. These pieces of seminal vesicles were placed in an "organ bath" containing 15 ml of Tyrode's solution and gased with 5% CO_2, 95% O_2. The contractile curve was recorded by an ink recorder using a frontal lever (X 11 magnification). PGE_1 (pure, Ono Pharmaceutical Co.) was injected one minute prior to the addition of smooth muscle stimulants. PGE_1 was dissolved in 0.2 M phosphate buffer pH 7.2. The contractile response of the guinea pig seminal vesicle to acetylcholine and adrenaline was enhanced by PGE_1 (25 ng/ml or less) added prior to the catecolamines. This enhancement was not increased by increased PGE_1. Human seminal vesicle did not respond to acetylcholine. The final concentration of PGE_1 25 ng/ml did not alter the contractile response of the human seminal vesicle to adrenaline. However, when the concentration reached 1.3 μg/ml, PGE_1 inhibited the response to adrenaline. This inhibitory effect was stronger when the concentration was increased to 6.7 μg/ml. (Author) 2495

1382
JACKSON, R.T.
Prostaglandin E_1 as a nasal constrictor in normal human volunteers.
Current Therapeutic Research. 12: 711-717. 1970.

Four doses (37,50,75, and 100 μg) of prostaglandin E_1 were tested on 15 human subjects as a nasal vasoconstrictor. Three subjects gave no response. Seven subjects gave both no response and good response. Five subjects gave good responses. When a response occurred, it lasted from 3 to 12 hours. The 100-μg. dose was uniformly irritating. (Author) 2380

1383
JACOBSEN, S.
De allergiske reaksjoners mediatorsubstanser. [The mediator substances of allergic reactions.]
Tidsskrift for den Norske Laegeforening. 90: 547-549. 1970.

This Norwegian review briefly mentions prostaglandins in regard to their formation and functions. They are considered as possible mediators in allergic reactions. (RAP) 2431

1384
JACOBSON, E.D.
Comparison of prostaglandin E_1 and norepinephrine on the gastric mucosal circulation.
Proceedings of the Society for Experimental Biology and medicine. 133: 516-519. 1970.

Secretion of gastric mucosa was stimulated with pentagastrin or histamine to about one third maximal in conscious dogs provided with gastric fistulae. PGE_1 or norepinephrine, 0.1 or 1.0 μg/kg^{-1}/min^{-1} was administered. The high dosage of either agent decreased secretion regardless of pentgastrin or histamine stimulation, but the ratio of blood flow to secretory rate was decreased only by norepinephrine indicating that its action is through limitation of mucosal blood flow. PGE_1 apparently limits secretion by a direct action on secreting cells. Thus PGE_1 is of possible therapeutic value in the control of hypersecretion by non-ischemic, thus non-ulcerogenic, means. (RMS) 2257

1385
JACOBSON, E.D., G.F. BROBMANN, and G.A. BRECHER
Intestinal motor activity and blood flow.
Gastroenterology. 58: 575-579. 1970.

This editorial reviews briefly the 3 modes (neurohumoral, local metabolite, and mechanical) by which alteration in the function of an organ could evoke a change in blood flow. Studies are cited in which an in situ preparation in the dog was reported; chloralose was the anesthetic, blood flow was freely variable, vascular integrity was preserved, and the entire small gut and its blood supply were studied. It is merely mentioned that both motor activity and blood flow were diminished by norepinephrine, whereas blood flow was increased and motor activity decreased by PGE_1. No further mention is made of prostaglandins. (ART) 2133

1386
JAGER, F.C.
Effect of fatty acids on the contraction of guinea-pig ileum in vitro.
Experientia. 26: 731-732. 1970.

Investigation of the activity of essential fatty acids on the contraction on isolated guinea pig ileum demonstrated that the direct precursor of prostaglandin E_1 and E_2, namely dihomo-gamma-linolenic acid and arachidonic acid are active. (GW) 2128

1387
JESSUP, S.J., W.J. McDONALD-GIBSON, P.W. RAMWELL, and J.E. SHAW
Biosynthesis and release of prostaglandins on hormonal treatment of frog skin, and their effect on ion transport.
Federation Proceedings. 29: 387abs. 1970.

Abstract only. PGE_1 and PGE_2 have been isolated from ventral frog skin and a basal release of these compounds has been detected across both the outer and inner surfaces on perfusion in vitro. PGE_1 when applied to the inside surface of ventral frog skin increases short circuit current (SCC) after a latent period of one to 5 minutes; this effect is associated with mobilization of tissue calcium and an inward movement of sodium across the outer surface of the skin. Increased biosynthesis and release of prostaglandins occur on stimulation of SCC with isoprenaline but not with dibutyryl cyclic AMP in the presence of theophylline. Prostaglandins are known to modify the adenyl cyclase system, which is itself susceptible to the ionic environment, and cAMP is believed to mediate hormonal effects on ion transport in epithelial membranes. Results of the study here reported suggest that endogenous prostaglandins, which are mobilized under hormonal stimulation, may modify ion movements and thereby contribute to the increased ion response and cAMP accumulation seen with isoprenaline. (Authors modified) 2120

1388
JOHNSON, M.C.R. and L. SAUNDERS
Physical studies of aqueous solutions of two prostaglandins.
Biochemica et Biophysica Acta. 218: 543-545. 1970.

The degree of association of $PGF_{2\alpha}$ and PGE_1 with water was investigated in a preliminary series of experiments. Surface tension as a function of concentration, light scattering, and vapor pressure osmometry are reported as are results of potentiometric titration of $PGF_{2\alpha}$. Results are graphed and discussed. Neither exhibits a critical micelle concentration in surface tension results; light scattering studies of $F_{2\alpha}$ are in agreement. The data indicate that $F_{2\alpha}$ exists as monomer units in aqueous solution and it is assumed that PGE_1 is similar. Surface tension curves are similar; a calculated area per molecule at the surface was 73Å^2 taken as an averaged value for ionized and undissociated forms. A 3-dimensional model was arranged with the terminal carboxyl and 3 hydroxyl groups in the same plane giving an estimated area 80Å^2 in reasonable agreement with the experimental calculation. Potentiometric titration of $F_{2\alpha}$ with 10 mM NaOH gave a typical weak acid-strong base curve with a pK_A of 5.44 at $20°$. (RMS) 2010

1389
JOHNSTON, C.I., T.J. MARTIN, and J. RIDDELL
Medullary thyroid carcinoma: a functional peptide secreting tumour.
Australasian Annals of Medicine. 19: 50-53. 1970.

Brief mention is made of Williams et al. 1968 who reported that medullary carcinomas of the thyroid secrete prostaglandin. (JRH) 2494

1390
JOHNSTON, J.O. and K.K. HUNTER
Prostaglandin $F_{2\alpha}$ mode of action in pregnant hamsters.
Physiologist. 13: 235. 1970.

Abstract only. A single subcutaneous injection of 300 μg of $PGF_{2\alpha}$ in the hamster on day 8 of pregnancy induced fetal resorption. This same dose lowered progestin values (2.6±0.5 ng/ml). 300 μg of prostaglandin can induce uterine contractions, which associated with declining progestin levels can cause fetal disruption. The failure of 100 μg $PGF_{2\alpha}$ to interrupt gestation in 8-day pregnant hamsters may be due to increased placental luteotropic support. (GW) 2276

1391
JONES, J.
The vogue in renal physiology: a review article.
The Central African Journal of Medicine. 16: 167-173. 1970.

The author reviews 123 articles on renal physiology in a discussion of the salt and water balance mechanisms of the kidney. He concludes: that vasopressin regulates the conservation of water; the renin-angiotensin-aldosterone system regulates sodium excretion; and that the sympathetic vasomotor nerves can prevent the loss of both sodium and water. The possiblitity of a natriuretic hormone in addition to renin, angiotensin, and aldosterone is considered. Medullin is mentioned, and the action of prostaglandin A_2 as well as PGA_1 and PGE_1 in renal vasodilation are noted without details citing Lee (1967), Carr (1968), Grantham and Orloff (1968), Orloff and Handler (1965), Weeks (1969), and Wooster (1969). (RMS) 2110

1392
JONES, R.L.
A prostaglandin isomerase in cat plasma.
Biochemical Journal. 119: 64P-65P. 1970.

The metabolism of prostaglandin A_1 by cat blood plasma was investigated. Prostaglandin A_1 was incubated with cat plasma at $37°C$ for 1 hour and the plasma extracted for polar fatty acids. U.V. spectrophotometry and silicic acid chromatography revealed after separation from any residual PGA_1 by g.l.c. prostaglandin B_1, a structural isomer of PGA_1. The isomerization of prostaglandin A_1 to B_1 results in the considerable loss of biological activity. For example, prostaglandin B_1 has only 2% of the potency of prostaglandin A_1 on the peripheral blood vessels of the cat. (GW) 2289

1393
JOUVENAZ, G.H., D.H. NUGTEREN, R.K. BEERTHUIS, and D.A. van DORP
  A sensitive method for the determination of prostaglandins by gas chromatography with
    electron-capture detection.
  Biochimica et Biophysica Acta. 202: 231-234. 1970.

  A method of identification of prostaglandins from biological materials sensitive to nanogram
  quantities is described. The method uses a gas chromotograph equipped with an electron
  capture detector. The prostaglandins were purified by column chromatography. (JRH) 2195

1394
JOY, M.D. and R.D. LOWE
  Abolition of the central cardiovascular effects on angiotensin by ablation of the area postrema.
  Clinical Science 39: 3P. 1970.

  Abstract only. The caudal half of the medulla of the greyhound mediates an increase in blood
  pressure, heart rate and cardiac output when small doses of angiotensin are infused into the
  vertebral artery. This area of postrema was ablated in 9 greyhounds and the response to
  vertebral artery infusion of prostaglandin $F_{2\alpha}$ was studied. This procedure abolished the
  response to vertebral artery infusion of angiotensin while the response to $PGF_{2\alpha}$ remained.
  These results show that the central cardiovascular effects of blood borne angiotensin are
  dependent on the integrity of the area postrema. (GW) 2353

1395
JOY, M.D. and R.D. LOWE
  Evidence that the area postrema mediated the central cardiovascular response to angiotensin II.
  Nature. 228: 1303-1304. 1970.

  Greyhound dogs were pretreated with morphine then anesthetized with chloralose. The area
  postrema was bilateraly ablated by thermal coagulation. This treatment had no effect on base
  line heart rate or blood pressure but it abolishes the effect normally produced by small doses of
  angiotensin II infused into the vertebral artery. The ablation had no effect on the hemodynamic
  response to intraarterial infusions of $PGF_{2\alpha}$ (8ng/kg/min), acetylcholine, or bilateral occlusion
  of the common carotid arteries. (JRH) 2307

1396
JUHLIN, L. and G. MICHAELSSON
  Vascular reactions in urticaria.
  British Journal of Dermatology 82(supp.5): 66-73. 1970.

  Prostaglandins are only briefly mentioned in this review of mediators of urticaria. It has been
  found that patients with chronic urticaria have a prolonged period of erythyma following
  intradermal injection of PGE and sometimes an edema develops. (JRH) 2458

1397

KADOWITZ, P.J., C.S. SWEET, and M.J. BRODY

Facilitation of adrenergic venoconstrictor responses by angiotensin, cocaine, and prostaglandin $F_{2\alpha}$.

Pharmacologist. 12: 249. 1970.

The effect of angiotensin (ANG), cocaine and prostaglandin $F_{2\alpha}$, on venoconstrictor responses to norepinephrine, tyramine and sympathetic nerve stimulation was studied in the constant flow perfused saphenous vein of the dog. Cocaine 2 μg/min and $PGF_{2\alpha}$ 1 μg/min markedly enhanced constrictor response to both sympathetic nerve stimulation and norepinephrine in a parallel manner. Prostaglandin $F_{2\alpha}$ also enhanced the venoconstrictor response to tyramine. The potentiating action of cocaine and prostaglandin declined rapidly when the infusions were terminated. These data suggest that both naturally occurring substances possess the ability to modulate adrenergic neurotransmission in venous smooth muscle. (GW) 2192

1398

KANNEGIESSER, H. and J.B. LEE

Hemodynamic differences between prostaglandins A and E.

Clinical Research. 18: 314. 1970.

Abstract only. PGA_1, PGA_2, PGE_1 or PGE_2 (80 mμg) was injected into the thoracic aorta of pentobarbital anesthetized cats. Blood pressure with respect to time of injection was recorded. Some cats were pretreated with pentolinium (10 mg i p) It was found that prostaglandins of the E type produced a rapid fall in blood pressure (20 sec). Prostaglandins of the A type required significantly longer (60 sec) to cause a decrease in blood pressure. The PGE effect was not affected by pretreatment with pentolinium but the response to PGA was greatly reduced. These results are taken to indicate that PGE's produce their effects by direct vasodilation while PGA's produce their effect by an indirect mechanism. (JRH) 2161

1399

KADOWITZ, P.J., C.S. SWEET, and M.J. BRODY

Modulation of adrenergic vasoconstriction in the dog hindpaw by prostaglandins E_1 and $F_{2\alpha}$.

Clinical Research. 18: 593. 1970.

Abstract only. The effect of intra-arterial (i.a.) infusion of prostaglandin E_1 and $F_{2\alpha}$ (PGE_1 and $PGF_{2\alpha}$) on adrenergic vasoconstrictor responses was evaluated in the vascularly isolated dog hindpaw perfused at constant flow. Infusion of PGE_1, 1 μg/min decreased vascular resistance in the paw and markedly reduced vasoconstrictor responses to sympathetic nerve stimulation and i.a. norepinephrine, epinephrine, tyramine, angiotensin, serotonin and vasodilator responses to nitroglycerin. Interruption of sympathetic nerves to the paw decreased the vasodilator action of PGE_1 but did not interfere with the antagonism of constrictor responses to nerve stimulation and norepinephrine. Infusion of $PGF_{2\alpha}$, 1 μg/min, greatly enhanced constrictor responses to nerve stimulation without altering vascular resistance in the paw. Responses to i.a. norepinephrine and tyramine were unchanged. When $PGF_{2\alpha}$ was infused into a paw in which vasoconstrictor responses were blocked by the previous administration of PGE_1, responses to nerve stimulation returned to control values within 5 min., whereas responses to norepinephrine remained depressed. Results of these studies suggest the PGE_1 antagonizes vasoconstriction in the paw by a non-specific effect on vascular smooth muscle whereas $PGF_{2\alpha}$ enhances responses to nerve stimulation by facilitation of transmitter release from sympathetic nerve terminals. It is postulated that these two naturally occurring substances possess the ability to modulate vasoconstrictor tone in the cutaneous vessels of the paw. (Author) 2154

1400
KAPLAN, E.L., N. SAXENA, and G.W. PESKIN
Prostaglandins: a possible mediator of diarrhea in endocrine syndromes.
Surgical Forum. 21: 94-95. 1970.

Everted ileal sacs from male hamsters were incubated for one hour at 37°C in Krebs-Ringer medium with 3-10 mg/ml of PGE_2. The activity of the PGE_2 was a net reduction in the absorption of fluid and electrolytes which corresponds qualitatively to the results obtained with extracts of Zollinger-Ellison tumor extracts and indicates that PGE_2 may be a mediator of the diarrheal phase of endocrine syndromes. (RMS) 2080

1401
KARIM, S.M.M., K. HILLIER, R.R. TRUSSELL, R.C. PATEL, and S. TAMUSANGE
Induction of labour with prostaglandin E_2.
Journal of Obstetrics and Gynaecology of the British Commonwealth. 77: 200-210. 1970.

The effect of prostaglandin E_2 on the activity of the pregnant human uterus in vivo has been studied in 50 women at or near term and in need of induction. Labour was successfully induced in all cases with a continuous infusion of 0.5 μg/min. The uterine activity produced by prostaglandin infusion resembled that of normal spontaneous labour. No unphysiological increase in uterine tonus was observed. The average infusion time was 5.5 hours and the average infusion-delivery interval was 10 hours. (Authors) 2319

1402
KARIM, S.M.M.
Prostaglandins.
Lancet 1: 425. 1970.

I should like to point out two inaccuracies in your leading article on prostaglandins (Jan. 31, p. 223). 1. You credit me with the observation that amniotic fluid collected during labour is particularly rich in prostaglandin E_1. The concentration of this prostaglandin in amniotic fluid during labour is, however, not significantly more than before labour. Amniotic fluid during labour is rich in prostaglandins E_2, $F_{1\alpha}$, and $F_{2\alpha}$. All these prostaglandins have oxytocic activity and have been successfully used for the induction of labour at or near term. 2. Human umbilical blood has not been shown to contain prostaglandins. Only the umbilical blood-vessels contain prostaglandins. (Author modified) 2283

1403
KARIM, S.M.M. and K. HILLIER
Prostaglandins and spontaneous abortion.
Journal of Obstetrics and Gynaecology of the British Commonwealth. 77: 837-839. 1970.

High concentrations of prostaglandins E_2 and $F_{2\alpha}$ have been found in human amniotic fluid following spontaneous abortion at 13-23 weeks gestation. Prostaglandin-like activity is also present in maternal venous blood obtained during spontaneous abortion. Detectable amounts of prostaglandins E_2 and $F_{2\alpha}$ were not found in samples of amniotic fluid obtained from women during abdominal termination of pregnancy on therapeutic grounds. The possible role of prostaglandins in premature labour leading to spontaneous abortion is discussed. (Authors) 2441

1404
KARIM, S.M.M.
Prostaglandins in fertility control.
Lancet. 1: 1115. 1970.

The occurrence of side effects depends on the particular prostaglandin used, route of administration and concentration. Nausea and vomiting occurred when $PGF_{2\alpha}$ was used for abortion but rarely with PGE_2. Vasodilation, migraine, and chest pain occurred with PGE_1 infusion but were absent even with 10 times the concentration when PGE_2 and $PGF_{2\alpha}$ were used. (GW) 2047

1405
KARIM, S.M.M.
Prostaglandins in fertility control.
Lancet. 2: 610. 1970.

A letter to the editor reports that the oral administration of prostaglandin E_2 and $F_{2\alpha}$ can induce labor in women at or near term without producing side effects. Much higher doses are required to stimulate the non-pregnant uterus or uterus in early pregnancy when the intravenous route is used. The vaginal route of administration is effective in stimulating the uterus throughout pregnancy without producing any side effects. (GW) 2298

1406
KARIM, S.M.M. and G.M. FILSHIE
Therapeutic abortion using prostaglandin $F_{2\alpha}$.
Lancet. 1: 157-159. 1970.

Fifteen consecutive cases of termination of pregnancy with intravenous infusions of prostaglandin $F_{2\alpha}$ (50 μg. per minute) are described. Abortion was successful in fourteen cases and complete in thirteen. Diarrhea and vomiting were the only side-effects noted. (Author) 2086

1407
KARIM, S.M.M. and G.M. FILSHIE
Use of prostaglandin E_2 for therapeutic abortion.
British Medical Journal. 3: 198-200. 1970.

Results of the use of 5 mg/min of intravenous PGE_2 in 52 therapeutic abortions is reported. The women were 13-40 years of age with 0-10 previous pregnancies. The gestation periods were 9-22 weeks. Abortion was successful in 50 and complete in 43 cases. Of the 2 failures, one was nulliparous, 10 weeks pregnant, the other a primagravida, 12 weeks pregnant. No serious side effects are reported. Ten cases showed nausea and vomiting. Diarrhea appeared in 4 patients; 2 slight cases of pyrexia and one genital tract infection with a pyrexia of 38°C are reported. One incidence of blurring of vision subsided with temporary discontinuation of the infusion. The hazards of therapeutic abortion are discussed with the observation that the termination of pregnancy with prostaglandin is simple, safe, and effective. (RMS) 2108

1408
KARIM, S.M.M.
Use of prostaglandin E_2 in the management of missed abortion, missed labour, and hydatidiform mole.
British Medical Journal. 3: 196-197. 1970.

Six patients with fetal death in the first and second trimesters of pregnancy, 1 with hydatidiform mole, and 15 patients with death of the fetus in third trimester of pregnancy were treated with 5mg. of PGE_2/min intravenously. Results were a complete abortion in all cases. Two out of the 6 with missed abortion showed the only side effects with vomiting during the PG infusion. The method is noted to be safe, reliable, and rapid in managing missed abortion, missed labor, and hydatidiform mole. (RMS) 2107

1409
KATORI, M., K. TAKEDA, and S. IMAI
Effect of prostaglandins E_1 and $F_{1\alpha}$ on the heart-lung preparation of the dog.
Tohoku Journal of Experimental Medicine. 101: 67-75. 1970.

In the heart-lung preparation of the dog, injections of PGE_1 into the left atrium resulted in a positive inotropic effect. The heart rate remained unchanged. Concomitant with this change, there was a marked increase in coronary sinus outflow with only a slight increase in myocardial oxygen consumption, indicating that the substance has a direct dilator effect upon the coronary vasculature. By comparing the increase in coronary flow produced by injections of PGE_1 into the left atrium with that produced by injections into the venous-supply tubing, it was concluded that 98-99% of PGE_1 could be removed from the circulating blood during a single passage through the lung. Cardiac actions of $PGF_{1\alpha}$ were found to be quite feeble, although the effect produced by large doses of this compound was not qualitatively different from that produced by PGE_1. (Authors) 2081

1410
KAY, J.E. and S.D. HANDMAKER
Uridine incorporation and RNA synthesis during stimulation of lymphocytes by PHA.
Experimental Cell Research. 63: 411-421. 1970.

PGE_1 (1 µg/ml) was used in uridine incorporation and RNA synthesis experiments during stimulation of lymphocytes by incubation for one to two days with phytohemagglutinin (PHA) of *Phaseolus vulgaris* because it (PGE_1) is known to inhibit uridine incorporation without causing cell death under the conditions employed, and because it was considered most probable that the action of PGE_1 is primarily on RNA metabolism rather than on uridine kinase activity. Inhibition of tritiated uridine incorporation was observed in resting lymphocytes but the activity of uridine kinase was unaffected by the PGE_1. (ART) 2186

1411
KESSINGER, J.M.
Comparative effects of a prostaglandin E_1, 8-iso-prostaglandins and prostaglandin $F_{2\alpha}$ on the systemic and pulmonary circulations.
Texas Reports on Biology and Medicine. 28: 395-396. 1970.

It has been well established that prostaglandin E_1 (PGE_1) is a vasodilator and prostaglandin $F_{2\alpha}$ ($PGF_{2\alpha}$) is a vasoconstrictor in dogs. The present study was undertaken to compare the cardiovascular effects of 8-iso-prostaglandin (8-iso-PGE_1) with those of PGE_1 and $PGF_{2\alpha}$ in dogs. In open-chest dogs anesthetized with sodium pentobarbital (30 mg/kg), systemic and pulmonary arterial pressures were continuously measured with Statham pressure transducers. Myocardial contractile force was measured with a Walton-Brodie strain gauge arch sutured to the right ventricle. It was found that the i.v. injection of 2-16 µg/kg of 8-iso-PGE_1 increased heart rate, myocardial contractile force and pulmonary arterial pressure and decreased systemic arterial pressure. The systemic hypotensive action of 8-iso-PGE_1 is 1/125 to 1/250 of that of

PGE$_1$ in dogs. In dogs in which pulmonary arterial blood flow was kept constant by means of a Sigma-motor pump, the i.a. (pulmonary artery) injection of 2 μg/kg of PGE$_1$ and PGF$_{2\alpha}$ decreased and increased significantly the pulmonary arterial pressure, respectively. The i.a. injection of the same dose of 8-iso-PGE$_1$ also increased pulmonary arterial pressure. The pulmonary hypertensive effect of 8-iso-PGE$_1$ was approximately ½ of that of PGF$_{2\alpha}$. The present study indicates that 8-iso-PGE$_1$ is a unique prostaglandin which causes a mild hypotensive action in the systemic circulation and a marked hypertensive action in the pulmonary circulation of dogs. (Author) 2329

1412
KIMBERG, D.V., and M. FIELD
Adenyl cyclase in gut mucosa: effects of cholera toxin and prostaglandins.
Clinical Research. 18: 680. 1970.

Abstract only. Both prostaglandins and cholera toxin (CT) were tested for their effect on adenyl cyclase activity and this was correlated with the effects of these substances on chlorine secretion in rabbit and guinea pig small intestinal mucosa. PGE$_1$ and PGE$_2$ increased adenyl cyclase 2 fold. Other prostaglandins had less effect. The prostaglandins also increased the short-circuit current (SCC) of guinea-pig mucosa with PGE$_1$ and PGE$_2$ being most effective. Giving theophylline which also increases SCC, inhibited the response to prostaglandins, and vice-versa. This was taken to indicate that both PGE$_1$ and theophylline increase SCC by stimulating active chlorine secretion. CT also stimulated adenyl cyclase on a time dependent basis. Phosphodiesterase activity in the crude membrane fraction was unaffected by either PGE$_1$ and CT. It was concluded that both prostaglandins and CT stimulate adenyl cyclase and thus activate the chlorine secretory pump. (JRH) 2147

1413
KING, C.G.
Biological and medical aspects of fats.
Journal of the American Oil Chemists' Society. 47: 419A-420A, 442A. 1970.

A brief history of lipid biochemistry and some points on the health relationships of the fatty acids are related in remarks from a paper presented at the March 1969 Oil Chemists' Society meeting. The formation of prostaglandins from fatty acids and some of the functions of the prostaglandins are reviewed very briefly. (RMS) 2082

1414
KINLOUGH-RATHBONE, R.L., M.A. PACKHAM, and J.F. MUSTARD
The effect of prostaglandin E$_1$ on platelet function in vitro and in vivo.
British Journal of Haematology. 19: 559-571. 1970.

Low concentrations of prostaglandin E$_1$ (PGE$_1$) inhibit ADP-induced aggregation in pig and rabbit citrated platelet-rich plasma and in suspensions of washed platelets. Higher concentrations also inhibit the initial change in shape induced by ADP and the release of platelet ATP, ADP and serotonin caused by stimuli such as collagen, thrombin, antigen-antibody complexes and gamma-globulin-coated polystyrene particles. PGE$_1$ is not taken up by platelets and its effects can be removed by resuspending platelets in fresh medium. Immediately following an intra-arterial injection, ADP-induced platelet aggregation is suppressed, but after 5 min the response returns to normal. PGE$_1$ inhibits haemostasis in rabbits when given as a continuous infusion. It is concluded that the effect of PGE$_1$ on haemostasis involves inhibition of the release of ADP from platelets exposed to collagen and thrombin, and inhibition of ADP-induced aggregation. (Authors) 2360

1415

KIRTON, K.T., B.B. PHARRISS, and A.D. FORBES
Luteolytic effects of prostaglandin $F_{2\alpha}$ in primates.
Proceedings of the Society for Experimental Biology and Medicine. 133: 314-316. 1970.

Prostaglandin $F_{2\alpha}$ was luteolytic in rhesus monkeys when injected 30 mg/day subcutaneously b.i.d. for 5 days, if the injections were initiated on day 11, 12, or 13 postovulation of fertile cycles. At this time circulating progestin levels are elevated, presumably due to a trophic stimulus from the blastocyst. Such an effect was not obtained by injection earlier in the luteal phase of the reproductive cycle. This indicates that the corpus luteum of this species is more vulnerable to luteolysis in the later stages of the reproductive cycle. (Authors) 2244

1416

KIRTON, K.T., B.B. PHARRISS, and A.D. FORBES
Some effects of prostaglandins E_2 and $F_{2\alpha}$ on the pregnant rehesus monkey.
Biology of Reproduction. 3: 163-168. 1970.

Prostaglandin E_2 (PGE_2) or prostaglandin $F_{2\alpha}$ ($PGF_{2\alpha}$) was injected subcutaneously or infused intravenously into pregnant rhesus monkeys. Animals were between day 30 and 127 of pregnancy; pregnancy termination, serum progestin levels, and uterine contractility were parameters measured to determine efficacy. Subcuatenous injections of $PGF_{2\alpha}$ or PGE_2 terminated pregnancy in 8 of 13 animals when given between day 30 and 41 and beyond day 100 of pregnancy. Intravenous infusion terminated 3 of 7 animals between day 30 and 40 of pregnancy. Progestin levels were usually depressed within 24-48 hr after the initial administration of prostaglandin, whereas uterine contractility was increased within 20-30 min. Prostaglandin E_2 was about 10 times as potent as $PGF_{2\alpha}$ in causing uterine contractions; intravenous infusion of 0.8 and 8.0 μg/min, respectively, caused maximal uterine contractions during the latter half of pregnancy. The relationship of these compounds to normal pregnant uterine physiology in the human and rhesus monkey is discussed. (Authors) 2361

1417

KISCHER, C.W. and J.S. KEETER
Prostaglandin B_1, embryonic skin, and the dermo-epidermal junction.
Journal of Cell Biology. 47: 303-310. 1970.

The down feather organ derived from the dorsal skin of the chick embryo fails to develop from presumptive sites when treated in vitro with prostaglandin B_1 (PGB_1). Examination of the cultured tissues by electron microscopy reveals many breaks or gaps in the dermo-epidermal junction. In some cases, these areas of disjunction are continuous with complete perforations of the epidermis. Mesenchymal cells fill the perforated area and form plaques on the dorsal surface of the skin. It has yet to be determined whether or not this phenomenon is directly related to the morphogenetic block of the feather organ produced by PGB_1. (Authors) 2311

1418

KLEIN, D.C. and L.G. RAISZ
Prostaglandins: Stimulation of bone resorption in tissue culture.
Endocrinology. 86: 1436-1440. 1970.

In vivo and in vitro studies on the effect of PGE_1, PGE_2, PGA_1, and $PGF_{1\alpha}$ on bone reabsorption were performed. The prostaglandins were compared to parathroid hormone (PTH) and epinephrine. The in vitro studies used rat embryo radius or ulna that had been labeled with

^{45}Ca. Low doses of PGE$_1$ (0.004 to 0.04 µg) were more effective on a weight basis that PTH, but larger doses of PTH produced a greater maximal effect. Epinephrine produced no measurable effect. The prostaglandins in order of effectiveness in increasing bone reabsorption were PGE$_1$>PGE$_2$>PGA$_1$. PGF$_{1\alpha}$ did not produce a significant increase. The time course response to PGE$_1$ and E$_2$ and PTH was not signifcantly different. Thyrocalcitonin and cortisol blocked PGE$_1$ to about the same degree as it did PTH. There was no synergistic effect of PGE$_1$ and PTH. In the in vitro studies parathyroidectomized rats were injected with 4 or 40 µg of PGE$_1$. Thee was no change in blood calcium levels after 5 hours. When the rats were given 400 µg PGE$_1$ there was a slight decrease in calcium perhaps due to hypertensive effects since the treated animals were quite ill. The authors conclude the in vitro experiments indicated that prostaglandins and PTH apparently affect the same or a closely related mechanism which leads to an increase in cAMP. Failure of PGE$_1$ to work in vivo may have been due to rapid degradation of the prostaglandin in the animal. (JRH) 2227

1419
KLOEZE, J.
Influence of prostaglandins E$_1$ and E$_2$ on coagulation of rat blood.
Experientia. 26: 307-308. 1970.

The effects of PGE$_1$ and PGE$_2$ on thromboelastograph values in vitro were studied. Results are shown in a table. Citrated platelet-poor or platelet-rich plasma was prepared for testing. Prostaglandin was introduced instead of or in addition to thromboplastin. The 'one-stage' prothrombin-time test showed that PGE$_1$ or PGE$_2$ when introduced in addition to thromboplastin, did not affect the prothrombin time of the diluted plasma. Omission of thromboplastin increased the coagulation time 4.5 to 5.5 times both for normal and diluted plasma. Addition of PGE$_1$ or PGE$_2$ instead of thromboplastin had no effect. It is concluded, therefore, that PGE$_1$ and PGE$_2$ did not interfere with thromboplastic activity nor do they possess such an activity. The calcium clotting times of rat plasma were not affected by PGE$_1$ at any of the CaCl$_2$ concentrations used for recalcification of the plasma. (ART) 2127

1420
KLOEZE, J.
Prostaglandins and platelet aggregation.
In: Schletter, G., ed., "Platelets and the Vessel Wall-Fibrin Deposition: Symposium of the European Atherosclerosis Group, Heidelberg, 1969," p. 54-59. Sutttgart, Georg Thieme, 1970.

In this review article the author discusses the role of prostaglandins in platelet aggregation. The bibliography has 10 entries. (JRH) 2501

1421
KLOEZE, J.
Prostaglandins and platelet aggregation in vivo. I. Influence of PGE$_1$ and ω-homo-PGE$_1$ on transient thrombocytopenia and of PGE$_1$ on the LD$_{50}$ of ADP.
Thrombosis et Diathesis Haemorrhagica. 24: 286-292. 1970.

Prostaglandins E$_1$ and ω-homo-E$_1$ which were shown previously to inhibit adenosine diphosphate (ADP)-induced platelet aggregation in vitro have now been found to inhibit this process also in vivo. Both prostaglandins inhibit transient thrombocytopenia induced by intravenous infection of ADP; PGE$_1$ increases also the LD$_{50}$ of ADP when injected in high amounts into young rats. In both cases platelet aggregation in vivo is the primary cause of the phenomena observed. The symptoms observed on overdosing ADP by intravenous injection are unconsciousness within 10-20 sec after completion of the injection immediately followed by respiratory arrest and eventual death of the animal, generally within 10 min. It seems that blocking of the supply of blood to the brain by platelet thrombi, which on histological examination were found to occlude blood vessels in different organs, is the most important causative factor of the symptoms observed. (Author) 2292

1422
KLOEZE, J.
Prostaglandins and platelet aggregation in vivo. II. Influence of PGE_1 and $PGF_{1\alpha}$ on platelet thrombus formation induced by an electric stimulus in veins on the rat brain surface.
Thrombosis et Diathesis Haemorrhagica. 23: 293-298. 1970.

It was shown previously that prostaglandin E_1 (PGE_1) inhibits the aggregation of blood platelets in vitro whereas $PGF_{1\alpha}$ does not; in vivo, PGE_1 inhibits in rats the action of intravenously injected adenosine diphosphate. Locally administered PGE_1 is found in the present experiment to inhibit in rats the formation and growth of platelet thrombi, induced by an electric stimulus in cortical veins. $PGF_{1\alpha}$ appeared to be inactive. The morphology of platelet thrombi induced by an electric stimulus is shown. The probability of a primary role of erythrocytes in the formation of this type of thrombi is discussed. (Author) 2291

1423
KLOK, R., H.J.J. PABON, and D.A. van DORP
Synthesis of 2-(6'-carboxyhexyl)-4-hydroxy-3-(3'-hydroxyoctyl)-1-cyclo-pentanone (racemic dihydroprostaglandin E_1).
Recueil des Trauvaux Chimiques des Pays-Bas et Belgique. 89: 1043-1053. 1970.

By the reaction of 3-(3'-acetoxyoctyl-2-(6'carboxyhexyl)-2-cyclopenten-1-one, the acetoxy derivative of Prostaglandin E_1 - 237 (PGE_1 - 237), with N-bromosuccinimide in carbon tetrachloride, followed by treatment of the reaction product with silver acetate in acetic acid, an acetoxy group was introduced at the allyl position of the ring. After hydrolysis of both acetoxy groups, 2-(6'carboxyhexyl)-4-hydroxy-3-(3'-hydroxyoctyl)-2-cyclo-penten-1-one was obtained. In an attempt to convert the latter into 2-(6'-carboxyhexyl)-4-hydroxy-3-(3'-hydroxyoctyl)-1-cyclopentanone (racemic dihydro-PGE_1), hydrogenation with a 5% rhodium on carbon catalyst was carried out. Both reduction of the double bond in the ring and hydrogenolysis of the hydroxy group attached to the ring occurred, leading to 2-(6'-carboxyhexyl)-3-(3'-hydroxyoctyl)-1-cyclopentanone. However, racemic dihydro-PGE_1 was also obtained in low yield. It showed some biological activity on blood-platelet aggregation. (Authors) 2382

1424
KOCH, G.K. and J.W. DALENBERG.
Preparation of ^3H labeled prostaglandin E_1 by hydrogenation of prostaglandin E_2 over RhCl $(PPH_3)_3$. An example of selectivity in homogeneous hydrogenation.
Journal of Labelled Compounds. (Brux) 6: 395-398. 1970.

Labelled prostaglandin E_1 (PGE_1) can be prepared biosynthetically from labelled dihomo-γ-linolenic acid or, particularly when high specific activities are required, by catalytic hydro-genation of prostaglandin E_2 (PGE_2) over a Pd-C-catalyst. The latter method, however, gives a complex mixture from which we could obtain PGE_1 in only 10% yield. We have now found that homogeneous hydrogenation of PGE_2 with RhCl $(PPh_3)_3$-catalyt (sic) leads to much better results (yields up to 50%). (Authors modified) 2337

1425

KOROLKIEWICZ, Z. and L. SZAJDUKIS-SZADURSKI

Effect of prostaglandin E_1 on active sodium transport induced by noradrenaline.

Dissertationes Pharmaceuticae et Pharmacologicae. 22: 530-531, 1970.

The role of lipolysis in the action of CA on the active sodium transport was investigated. The experiments were carried out on frog skin with the method of Brown. PGE_1, which blocks release of FFA and stimulates glycolysis was employed in the experiments. The action of NA in this experiment was studied in preparations with glycolytic phosphorylation inhibited by iodoacetate and with lipolysis inhibited with PGE_1. PGE_1 alone did not block the stimulating action of NA. The action of NA is neither blocked by iodacetate. Simultaneous inhibition of glycolytic phosphorylation and lipolysis abolished, however, the action of NA. The obtained results indicate the role of FFA as the substrate in the process of oxidative phosphorylation. It should be also noted that simple block of glycolytic substrate phosphorylation with iodoacetate as well as simple block of lipolysis by PGE_1 did not abolish the action of NA on the active sodium transport which may indicate that in the process of stimulation of sodium transport by NA both FFA and carbohydrates may be used as substrates. (Authors) 2451

1426

KOROLKIEWICZ, Z.A., M. MATUSZEK, and E. POCWIARDOWSKA

Influence of pyrogen (LPS), Dinitrophenol (DNP), some antipyretic drugs and prostaglandins E_1 (PGE_1) on plasma free fatty acids and blood glucose in rabbits.

Dissertationes Pharmaceuticae et Pharmacologicae. 22: 257-261. 1970.

Rabbits which had constant temperatures and stable serum-free fatty acid (FFA) levels were used to determine the effect of antilipolytic agents on the alterations of serum FFA and blood glucose induced by the hyperthermizing agents, progen (from *E. coli*), dinitrophenol (DNP), and theophylline. Some rabbits were pretreated with sodium salicylate 3-5 days before getting the hyperthermizing agents. Others received PGE_1 (5 gamma/Kg/10 min.) intravenously at the same time as the hyperthermizing agents. It was found that the three hyperthermizing agents caused a rise in serum FFA above control values and both sodium salicylate and PGE_1 inhibited the effect of these agents. Pyrogen and DNP caused a decrease in blood glucose but theophylline caused an increase in glucose levels. PGE_1 inhibited the effects of all three agents on glucose levels (i.e. the decrease caused by pyrogen and DNP and the increase caused by theophylline) but sodium salicylate enhanced the effects of all three. (JRH) 2450

1427

KOROLKIEWICZ, Z., S. GORA, E. RUCINSKA, and L. SZAJDUKIS-SZADURSKI

Zastosowanie prostaglandyny (PGE_1) do analizy mechanizmu dzialania amin katecholowych. [The use of prostaglandin E_1 (PGE_1) for the analysis of the mechanism of action of catecholamines.]

Polski Tygodnik Lekarski. 25: 1415-1417. 1970.

For the analysis of the mechanism of action of catecholamines, PGE_1 was used trying to show the correlation between the effect of these amines on the isolated striated muscle and the bioelectric properties of the cell membrane. PGE_1 owing to its property of selective blockade of lipolysis can be applied for an analysis of the metabolic and myotropic activity of catecholamines. The experiments carried out as yet suggest that during physical effort as well as a result of the action of exogenous catecholamines a mobilization of the metabolic substrates of glycolysis occurs but, mainly that of as well as a direct action on the cell membrane. Catecholamines activate the transport of sodium using mainly the process of oxidative-phosphorylation as a source of energy, although it seems to be a secondary process in relation to the observed increase in passive permeability of the cell membrane activating the Na^+ and K^+-dependent ATP-ase. (Author) 2078

1428
KRANTZ, J.C., Jr.
  Drug-induced abortion.
  Current Medical Digest. June: 589, 592. 1970.

In this editorial the author briefly reviews the history of abortifacient drugs. He discusses the use of prostaglandins as abortifacients. (JRH) 2563

1429
KUEHL, F.A., J.L. HUMES, J. TARNOFF, V.J. CIRILLO, and E.A. HAM
  Prostaglandin receptor site: Evidence for an essential role in the action of luteinizing hormone.
  Science. 169: 883-886. 1970.

Isolated mouse ovaries were cultured in vitro with $(8\text{-}^{14}C)$ adenine to produce intracellular (^{14}C) ATP. Theophylline was then added to inhibit the action of cAMP phosphodiesterase which degrades cAMP to 5'-AMP. The ovaries were then incubated for an additional period of time with the prostaglandin and/or gonadotropin to be tested. The reactions were terminated and the amount of cyclic (^{14}C)-AMP was measured chromatographically. It was assumed that the labeled cAMP was produced by conversion of the labeled ATP by adenyl cyclase. It was found that PGE_1 and PGE_2 increased the amount of cAMP with PGE_2 being more effective at lower doses (both achieved the same maximum level of stimulation). Luteinizing hormone LH was found to produce only half as much increase in cAMP as either prostaglandin. It appears that both prostaglandins activate the same receptor since combined maximal doses produced no additive effect. The prostaglandin inhibitor 7 oxa-13-prostynoic acid antagonized the action of these prostaglandins on cAMP inrease in a competitive manner. This inhibitor also antagonized the action of LH stimulation of cAMP formation to the same degree. It was also found that incubation of the tissue with arachidonic acid led to a four-fold stimulation of cAMP formation. On the basis of this data the authors feel the prostaglandin receptor functions as a necessary intermediate in the action of LH to raise cAMP formation. (JRH) 2144

1430
KUEHL, F.A., D.J. PATANELLI, J. TARNOFF, and J.L. HUMES
  Testicular adenyl cyclase: stimulation by the pituitary gonadotrophins.
  Biology of Reproduction. 2: 154-163. 1970.

The effect of FSH and LH on cAMP synthesis in rat seminiferous tubules was measured. It was found that neither PGE_1 nor $PGF_{2\alpha}$ were effective in inhibiting the cAMP increase induced by gonadotropins in these experiments. (JRH) 2214

1431
KUNZE, H.
  Formation of $[I\text{-}^{14}C]$ prostaglandin E_2 and two prostaglandin metabolites for $[I\text{-}^{14}C]$
    archidonic acid during vascular perfusion of the frog intestine.
  Biochimica et Biophysica Acta. 202: 180-183. 1970.

The conversion of $[I\text{-}^{14}C]$ arachidonic acid during vascular perfusion was studied in frog intestine preparations. $[I\text{-}^{14}C]$ Prostaglandin E_2 was rapidly formed from $[I\text{-}^{14}C]$ arachidonic acid and partly metabolized into two compounds which have not yet been further identified. The amount of prostaglandin E_2 formed from the labeled arachidonic acid during a 10 min perfusion was about 3%; 25% of the prostaglandin E_2 appeared as a less polar metabolite and 10% as a more polar one. (Author) 2196

1432
KUPIECKI, F.P.
    Stimulation of lipolysis by inhibitors of lipid mobilization.
    Federation Proceedings. 29: 897abs. 1970.

    Abstract only. PGE_1 reduces the plasma free fatty acid level without influencing lipolysis in incubated adipose tissue of rats. (ART) 2169

1433
LABHSETWAR, A.P.
    Effects of prostaglandin $F_{2\alpha}$ on pituitary luteinizing hormone content of pregnant rats: a possible explanation for the luteolytic effect.
    Journal of Reproduction and Fertility. 23: 155-159. 1970.

    Prostaglandin $F_{2\alpha}$ induced degeneration of embryos in intact pregnant rats but not in ovariectomized rats maintained on progesterone. The pituitary LH stores in the intact, treated rats were significantly higher than those in the pregnant, pseudopregnant or cyclic controls. The prostaglandin failed to inhibit the ovulatory release of LH in the cyclic rat. It is suggested that the luteolytic effect of $PGF_{2\alpha}$ may be mediated in part by the increased synthesis and secretion of LH. (Author) 2084

1434
LAITY, J.L.H. and D. MOORE
    The effect of histamine releasers on the output of prostaglandins from rat diaphragms.
    Journal of Pharmacy and Pharmacology. 22: 384-385. 1970.

    Antigen reportedly causes the release of prostaglandins E_2 and $F_{2\alpha}$ from sensitized perfused guinea-pig lungs. It has been demonstrated that (+)-tubocurarine causes the release of polar acidic lipids from rat diaphragm. In three experiments with tolazoline hydrochloride (100 mg/ml) and in two experiments with pethidine hydrochloride (100 mg/ml), using the same technology, there was a release of polar acidic lipids from rat diaphragms. The polar acidic lipids, released by these three compounds, consist of prostaglandins of the E and F series with the amount of PGF being larger than that of PGE. It is concluded that in rat diaphragms histamine releasers cause a release of prostaglandins similar to that of the antigen-antibody reaction in perfused guinea-pig lungs. (FDB) 2303

1435
LARSSON, C. and E. ANGGARD
    Distribution of prostaglandin metabolizing enzymes in tissues of the swine.
    Acta Pharmacologica et Toxicologica 28 (supp. 1): 61. 1970.

    Prostaglandin Δ 13 reductase (PGR) and 15-hydroxy prostaglandin dehydrogenase (PGDH) were assayed in 15 different tissues of the swine. The assay of PGDH was based on the development of alkali of the 500 nm chromophore of 15-keto-PGE_1 after incubation of the tissue with PGE_1 and NAD+. PGR was determined by following the disappearance of 15-keto-PGE_1 using NADH as cofactor. The values for the enzymatic activities were calculated from the initial velocities of the oxidation of PGE_1 and the reduction of 15-keto-PGE_1 respectively. Highest PGDH activities were found in the lung, spleen, and kidney with lower activities observed in the stomach, testicle, small intestine, heart, and adipose tissue. The renal cortex contained about three times more PGDH than the medulla. Highest PGR activities were observed in the liver, spleen, small intestine, and kidney, with fairly high activities in the adrenal, pancreas, adipose tissue, brain, lung, and stomach. The highest specific activity was found in adipose tissue. The wide distribution of PGDH and PGR support the concept that these enzymes catalyze biologically important initial steps in the catabolism of the prostaglandins. (JWJ) 2141

1436

LAVERY, H.A., R.D. LOWE, and G.C. SCROOP

Cardiovascular effects of prostaglandins mediated by the central nervous system of the dog.

British Journal of Pharmacology. 39: 511-519. 1970.

Infusion of prostaglandin A_1, E_1, $F_{1\alpha}$, and $F_{2\alpha}$ into the vertebral artery of a greyhound and the resulting cardiovascular responses were compared with those obtained on intravenous and intracarotid infusion in the same dose range. Infusion of $PGF_{2\alpha}$ intravertebrally caused an increased of blood pressure and a fall of central venous pressure. $PGF_{1\alpha}$ had similar effects but was less potent. PGE_1 infused into the vertebral artery had no significant effect on blood pressure. A tachycardia was greater than that obtained with intracarotid or intravenous infusion. PGA_1 caused a small fall in blood pressure accompanied by an increase of heart rate. The dose responses were similar for all three routes of administration. (GW) 2031

1437

LEE, J.B.

Hypertension and the lupus syndrome.

In: Aach, R. and J. Kissane, eds., "Clinicopathologic Conference."

American Journal of Medicine. 49: 519-528. 1970.

Abstract only. A summary of the role of prostaglandins in relation to the vascular system and hypertension is presented as a part of the discussion following the case history of a patient who died with chronic hypertension and nephrosclerosis. Experiments on rats are described briefly which indicated the role of the kidney medulla in sustained lowering of the blood pressure. Clinical studies are mentioned which demonstrated the hemodynamic properties of PGA_1 and PGA_2. A hypothetical role for PGA_2 in the regulation of blood pressure is outlined in which systemic hypertension is seen as a result of prostaglandin deficiency. An alternate hypothesis is also mentioned in which PGA_2 and PGE_2 function in regulating the intrarenal distribution of blood flow. An intrarenal prostaglandin deficiency would lead to systemic hypertension due to the release of renal pressor substances. (RMS) 2183

1438

LEE, J.B.

Prostaglandins.

Physiologist. 13: 379-397. 1970.

In this review, the author discusses the role of prostaglandins in renal and hemodynamic physiology. (JRH) 2506

1439

LEFEBVRE, P.J., J.C. DODYEZ, A.S. LUYCKX, and P.P. FOA

Adipose tissue lipolysis in golden hamsters with chronic hypoglycemia and hyperinsulinemia due to a transplantable islet cell tumor.

Hormone and Metabolic Research. 2: 64-67. 1970.

The epididymal fat pads of hamsters which were hypoglycemic due to a transplanted islet tumor were incubated in Krebs-Ringer bicarbonate buffer. The incubation mixture in some cases contained PGE_1 (1 or 5 μg/ml). The amount of glucose uptake, glycerol release and free fatty acid release were measured. In normal control animals, PGE_1 produced a decrease in glycerol release from adipose tissue from both fasted and non-fasted animals. However, in islet tumor bearing animals, PGE_1 only produced a decrease in glycerol release of the non-fasted animals. This was taken to indicate that PGE_1 was ineffective in fasted tumor bearing animals because of the high levels of lipolytic substances in these animals. (JRH) 2229

1440
LEVINE, L. and VUNAKIS, H. van
Antigenic activity of prostaglandins.
Biochemical and Biophysical Research Communications. 41: 1171-1177. 1970.

Prostaglandins E_1 and $F_{2\alpha}$ were coupled to poly-L-lysine using carbodiimide. The conjugates, complexed to succinylated hemocyanin, were used to immunize rabbits. The antibodies elicited with polylysine-prostaglandin $F_{2\alpha}$ showed specificity for $F_{2\alpha}$ but reacted, although less effectively, with heterologous prostaglandins. Antibodies elicited with polylysine-prostaglandin E_1 were most effective serologically with prostaglandin A_1. They cross reacted with prostaglandin E_1 and $F_{2\alpha}$ but less effectively. A radioimmunoassay for measurement of prostaglandins in nanogram amounts has been developed. (Authors) 2187

1441
LEVINE, R.A.
The role of cyclic AMP in hepatic and gastrointestinal function.
Gastroenterology. 59: 280-300. 1970.

This review of the role of cyclic AMP in hepatic and gastrointestinal function contains a section on the interrelationships between prostaglandins and cyclic AMP. The bibliography lists 18 prostaglandin articles. (JRH) 2230

1442
LEWIS, G.P. and N.J. P. WINSEY
The action of pharmacologically active substances on the flow and composition of cat hind limb lymph.
British Journal of Pharmacology. 40: 446-460. 1970.

Experiments are discussed which were designed to measure lymph flow, blood flow and lymph protein concentration in cat hind limb with the particular object of examining the inter-relationship between these three parameters. Prostaglandins, among others, were used as vasodilator agents only, and were found not be very active in increasing lymph flow. PGE_1 and $PGF_{1\alpha}$ were used at high venous pressures and it was found that infusion of PGE_1 at the rate of 1 µg/0.4 ml) per minute caused a 500 percent increase in blood flow (ml/min), a 33 percent increase in lymph flow (µl/min), and a 24 percent increase of protein concentration (mg/ml). Infusion of $PGF_{1\alpha}$ at the rate of 3 µg/0.4 ml per minute caused a 100 percent increase in blood flow, a 20 percent increase in lymph flow, and a 10 percent decrease in protein concentration. At normal venous pressure, infusion of PGE_1 caused a 59 percent increase in blood flow, a 5.5 percent decrease in lymph flow, and a 2.4 percent increase in protein concentration. Since it was considered possible that histamine, bradykinin or even a prostaglandin plays a role in the vascular reactions during tissue injury, the possibility that they increased not only the permeability of the blood vessels but perhaps the permeability of the cell membranes as well was investigated. There was no significant increase in the concentration of any of the intracellular enzymes measured, namely, lactate dehydrogenase, glutamic oxaloacetic transaminase, glutamic pyruvic transaminase, and acid phosphatase. The concentrations of most of the enzymes in the lymph remained constant during the infusion in spite of an increased lymph flow. Thus it is apparent that infusions of histamine, bradykinin, or PGR_1 did not increase the leakage of intracellular enzymes into the lymph. (ART) 2210

1443

LIETZ, H. and K. DONATH

Zur Ultrastruktur und Entstehung des Amyloides im medullaren Schilddrusen-carcinom. [Ultrastructure and origin of amyloid in medullary thyroid carcinoma] Virchows Archiv; Abteilung A: Pathologische Anatomie. 350: 261-274. 1970.

Article in German with an English summary. Only brief mention is made of prostaglandins in the discussion section of this case history report on a patient with medullary carcinoma of the thyroid. (JRH) 2480

1444

LIMA, F., M.F. GIMENO, A. GOLDRAIJ, A.S. BEDNERS, and A.L. GIMENO

Motilidad y prostaglandina $F_{2\alpha}$ en el miometrio aislado de ratas castradas. [Motility and prostaglandin $F_{2\alpha}$ in the isolated myometrium of ovariectomized rats.] Medicina (Buenous Aires). 30: 629-630. 1970.

Abstract only. A study was made of the presence of $PGF_{2\alpha}$ and of motility in the uterine horns of ovariectomized and intact rats. The $PGF_{2\alpha}$ was studied by thin layer chromatography and it was found to be present in all the uterine horn extracts from ovariectomized rats but not in any from intact animals. It was concluded 1) that the myometrium of ovariectomized rats possessed a clearly seen motility different (greater magnitude and stability) from that in intact rats; 2) that only in the myometrium of ovariectomized rats could the presence of $PGF_{2\alpha}$ be verified; and 3) that it is reasonable to suppose that ovarian hormones or gonadotropin exercise some control in the synthesis, degradation or storage of the myometrial $PGF_{2\alpha}$ and that its presence there is related to the peculiarities of automatism and the contraction of the isolated uterine horns of ovariectomized rats. (ART) 2378

1445

LINDHEIMER, M.D.

Further characterization of the influence of supine posture on renal function in late pregnancy. Gynecologic Investigation. 1: 69-81. 1970.

In this discussion, the author reviews factors which affect sodium excretion, mentioning prostaglandins without details. (MT) 2389

1446

LINDSEY, H.E. and J.H. WYLLIE

Release of prostaglandins from embolized lungs. British Journal of Surgery. 57: 738-741. 1970.

Prostaglandin E_2 was released from isolated, perfused lungs of guinea-pigs and rats as a result of embolizing the lungs with several types of particle in the size range $1-120\mu$ diameter. This observation may partly explain the hemodynamic effects of pulmonary embolism in man in which the release of powerful vasoactive substances other than histamine has long been suspected. (Authors) 2273

1447

LIPPMANN, W.

Inhibition of gastric acid secretion by a potent synthetic prostaglandin. Journal of Pharmacy and Pharmacology. 22: 65-67. 1970.

The synthetic prostaglandin 9-oxo-15ξ-hydroxy-prostanoic acid (AY-22,093) has been examined for its effects on gastric acid secretion in rats. After ligation under anaesthesia at the pyloric end of the stomach and also at the esophagus, five injections of AY-22,093 at 20 minute intervals beginning 20 minutes after ligation were given. The synthetic PGE prevented the increase in gastric acid secretion at 6.4 μg/kg, s.c. but not at 3.2 μg/kg s.c. The synthetic PGE as well as PGE_1 inhibit the increase in gastric acid secretion caused by pentagastrin. Unlike PGE_1, the synthetic PGE is a racemate with 4 possible isomers and one of these might prove even more active. (GW) 2301

1448

MacLEOD, R.M. and J.E. LEHMEYER
Prostaglandin-mediated release of growth hormone.
Clinical Research. 18: 366. 1970.

Abstract only. Pituitary glands from female rats were incubated in vitro with radio-labeled leucine. Growth hormone (GH) and prolactin were separated electrophoretically from tissue homogenates after incubation with the test prostaglandin. It was found that PGE_1 and PGE_2 (10^{-6}M) increased GH release 100-300% while 10^{-7} to 10^{-8}M PGE_1 increased synthesis but not release. PGA_1 (10^{-6}M) increased synthesis but not release of GH. $PGF_{2\alpha}$ was found to be inactive. No reproducible results were found for prolactin production. (JRH) 2159

1449

MacLEOD, R.M. and J.E. LEHMEYER
Release of pituitary growth hormone by prostaglandins and dibutyryl adenosine cyclic 3'5'-monophosphate in the absence of protein synthesis.
Proceedings of the National Academy of Sciences. 67: 1172-1179. 1970.

Incubation of rat adenohypophyses with 10^{-6}M PGE_1 in medium with 4,5-^{3}H leucine for 7 hours caused a 40-300% increase in the release of labelled growth hormone. PGE_1 at 10^{-8}M increased synthesis but not release. PGE_2 had similar effects to PGE_1; PGA_1 stimulated only synthesis, and $PGF_{2\alpha}$ had no effect. In the pituitary, prostaglandins at 10^{-4}M increased adenyl cyclase activity without change in phosphodiesterase activity; no change was seen in prolactin synthesis and release. Dibutyryl cAMP with or without caffein increased labeled growth hormone release with no consistent effect on prolactin. Hypertonic potassium caused up to 215% increase in growth hormone release; in combination with 10^{-6}M PGE_1 the release was increased by 325% suggesting independent action. Theophylline increased both synthesis and release. Fluoride stimulated release while inhibiting leucine incorporation; puromycin inhibited synthesis without blocking prostaglandin, dibutyryl cAMP, theophylline, or fluoride stimulated release of hormone. It is concluded that prostaglandins increase pituitary adenyl cyclase with increase in growth hormone release independently of protein synthesis. (RMS) 2011

1450

MANNAIONI, P.F.
Influence of bradykinin and prostaglandin E_1 on the uptake and release of histamine by murine neoplastic mast cells in vitro.
Biochemical Pharmacology. 19: 1159-1163. 1970.

Experiments were carried out on a clone of murine neoplastic mast cells maintained as an ascitic tumor in mice. The histamine levels of the cells continuously grown in mice were fairly constant over a period of one year. Cells from one of the mice bearing the ascitic tumor were removed by aspiration of peritoneal fluid, appropriately prepared for study, and histamine was estimated biologically on guinea-pig ileum. Bradykinin was found capable of blocking histamine uptake without causing the release of histamine. PGE_1 did not interfere either with the release or with the uptake of histamine. PGE_1, which has the structure of unsaturated hydroxy-carboxylic acid, does not possess any of the molecular requirements for occupying the histamine binding sites, or for displacing the already bound histamine from the granules in the mastocytoma cells. Possible explanations are discussed for the reported finding by others that PGE_1 is capable of producing degranulation and histamine release from rat peritoneal mast cells. (ART) 2212

1451
MARK, A.L., P.G. SCHMID, J.W. ECKSTEIN, and M.G. WENDLING
Saphenous and colic venomotor responses to prostaglandin $F_{2\alpha}$.
Journal of Clinical Investigation. 49: 61a. 1970.

Abstract only. The effectiveness of $PGF_{2\alpha}$ as a venoconstrictor was compared to norepinephrine (NE). The importance of the sympathetic nervous system in the response was also measured. It was found that NE was 19.2 to 13.6 times more effective than $PGF_{2\alpha}$ in venoconstriction. The venomotor responses to $PGF_{2\alpha}$ were not dependent on integrity of the sympathetic nervous system. (JRH) 2219

1452
MARKOV, K.M.
Prostaglandins.
Uspekhi Fizicheskikh Nauk. 1: 98-125. 1970.

This review article in Russian discusses all aspects of prostaglandin research from chemical structure and synthesis to clinical applications. (JRH) 2526

1453
MARMO, E., F. DIMEZZA, A. IMPERATORE, and S. DIGIACOMO
Metoclopramid und die muskulatur von osophagus, magen, darm, milz, trachea, gallen—und harnblase: Untersuchungen in vitro.
[Metoclopramide and the muscular systems of esophagus, stomach, duodenum, spleen, trachea, gall and urinary bladders: studies in vitro.]
Arzneimittel-Forschung. 20: 18-27. 1970.

The authors report in vitro investigations on the activity of N-(di-ethylaminoethyl)-2-methoxy-4-amino-5-chlorobenzamide toward the smooth muscles of a number of human organs and those of frogs, rats, guinea pigs, cats, rabbits, and dogs. Low concentrations had a stimulating effect; high concentrations had a depressant effect which was aspecfic, antagonistic, and papaverine-like in comparison to acetylcholine, methacholine, histamine, 5-hydroxy-tryptamine, noradrenalin, bradykinin and prostaglandin E_1. (RMS) 2083

1454
MARQUIS, N.R., R.L. VIGDAHL and P.A. TAVORMINA
Biochemical role of prostaglandin E_1 in the inhibition of platelet aggregation.
In: Jones, R.L., ed., "Atherosclerosis," Proceedings of the 2nd International Symposium. p. 134-142, 633-700. New York, Springer-Verlag, 1970.

In this paper presented at a symposium, the authors discuss the mechanism by which several compounds influence human platelet aggregation in vitro. PGE_1, cAMP and dibutyrl cAMP were found to inhibit the stimulation of platelet aggregation caused by ADP, collagen and epinephrine. Using platelet membrane preparations, PGE_1 in concentrations as low as 10^{-8} M stimulated cAMP synthesis. PGA_1 and PGF_1 also increase cAMP synthesis and inhibit platelet aggregation but not as effectively as PGE_1. These effects were not due to inhibition of phosphodiesterase. Epinephrine only slightly decreases basal cAMP synthesis but causes considerable inhibition of PGE_1 stimulated cAMP synthesis. Pentolamine causes a slight but consistent stimulation of basal cAMP synthesis but prevents epinephrine from inhibiting PGE_1 stimulating cAMP synthesis. Propranolol had no effect. These results suggest that the adenyl cyclase system is involved in the alpha-adrenergic response. (JRH) 2332

1455

MARQUIS, N.R., J.A. BECKER, and R.L. VIGDAHL

Platelet aggregation. III. An epinephrine induced decrease in cyclic AMP synthesis.
Biochemical and Biophysical Research Communications. 39: 783-789. 1970.

Epinephrine decreases prostaglandin E_1-stimulated and basal cyclic AMP synthesis of intact platelets and of platelet membrane fractions. This effect of epinephrine is in turn inhibited by the α-adrenergic antagonist, phentolamine, but not by the β-adrenergic antagonist, propranolol. The epinephrine-mediated decrease in cyclic AMP in relation to platelet function is discussed. (Authors) 2246

1456

MARSH, J.M.

Effect of prostaglandin E_2 on adenyl cyclase of the corpus luteum.
Federation Proceedings. 29: 387abs. 1970.

Abstract only. Prostaglandins stimulate steroidogenesis in rat and rabbit ovaries and in slices of bovine corpora lutea. The role of cyclic AMP in this stimulation was assessed by study of the effect of PGE_2 on the adenyl cyclase activity of homogenates of bovine corpora lutea. Addition of PGE_2 (10 μg/ml) to the assay solution significantly increased the accumulation of α-^{32}P-cyclic AMP formed from an α-^{32}P-ATP. Since no phosphodieterase could be demonstrated, this increase must have been due to a stimulation of adenyl cyclase activity. PGE_2 was also incubated with slices of some of the same corpora lutea and its effect on progestin synthesis was measured. PGE_2 increased steroidogenesis and the extent of this increase correlated well with the extent of adenyl cyclase activation. (Author Modified) 2121

1457

MARSH, J.M.

The stimulatory effect of prostaglandin E_2 on adenyl cyclase in the bovine corpus luteum.
FEBS Letters. 7: 283-286. 1970.

The effect of PGE_2 was assessed on the adenyl cyclase of homogenates of bovine corpora lutea. Capsular tissue was removed from the corpus luteum of each of six cows slaughtered in the first six months of pregnancy and 1 gram of luteal tissue was minced and homogenized. The formation of cAMP was assessed by measuring the conversion of α-^{32}P-ATP into ^{32}P-cAMP. Aliquots of an homogenate equivalent to 20 mg. of luteal tissue were incubated under control conditons with PGE_2 (10 mg/ml) or with LH (10 mg/ml). Both PGE_2 and LH significantly increased the amount of cAMP formed. These results suggest that PGE_2 stimulates steroidogenesis via the mediation of cAMP. They also demonstrate that PGE_2 increases cAMP by activating the adenyl cyclase rather than by inhibiting the phosphodiesterase system. (JWJ) 2135

1458
MATSUDA, S.
Practical use of prostaglandins in gynecological and obstetrical fields.
Shinryo. 23: 1388-1392. 1970.

Article in Japanese. Abstract not available at present. 2427

1459
MATSUKURA, S.
Action mechanism of ACTH with special reference to the significance of adrenergic agents.
Folia Endocrinologica Japonica. 46: 618-624. 1970.

Article in Japanese. Abstract not available at present. 2430

1460
MAYER, H.E., F.M. ABBOUD, P.G. SCHMID, and A.L. MARK
Release of norepinephrine by prostaglandin E_1.
Clinical Research. 18: 594. 1970.

Abstract only. These experiments were performed to determine if PGE_1 causes the release of endogenous norepinephrine (NE). Dog hind paws were perfused with 1-norepinephrine-7-T (200 uCi) to label the endogenous NE. Then 5-μg of PGE_1 was injected intraarterially. There was a rapid increase in the radioactivity in effluent from the paw. Chromatographic analysis indicated that the radioactivity was due to NE not its metabolites. Results of this and previous experiments lead the authors to conclude that the dialator action of PGE_1 represents the net result of several mechanisms of action. (JRH) 2153

1461
McCRAKEN, J.A., M.E. GLEW, and R.J. SCARAMUZZI
Corpus luteum regression induced by prostaglandin $F_{2\alpha}$.
Journal of Clinical Endocrinology and Metabolism. 30: 544-546. 1970.

This paper reports the effect on ovarian blood flow and progesterone secretion rate of infusing $PGF_{2\alpha}$ into the sheep ovary autotransplanted with vascular anastomoses to the vessels of the neck. Low infusion rates for one hour caused a slow rise in blood flow with some stimulation of progesterone secretion followed by a slight fall. An infusion rate of 25 μg/hour caused a rapid rise and fall of the blood flow; the progesterone secretion already diminished to 50% of the controls, rose slightly. A $PGF_{2\alpha}$ infusion for a period of six hours at 50 μg/hour caused an initial rise in blood flow and an immediate decline in the rate of progesterone secretion and continued to fall slowly during the six-hour course. Infusion during the mid-luteal phase with 100 μg/hour $PGF_{2\alpha}$ for six hours gave similar results with the additional effect of a pronounced vasoconstriction. The results indicate that $PGF_{2\alpha}$ is capable of inducing complete luteal regression, and causing behavioral estrus in sheep bearing left ovarian transplants. (GW) 2308

1462
McGIFF, J.C., N.A. TERRAGNO, K. CROWSHAW and A.J. LONIGRO
Inhibition of the renal actions of pressor systems by prostaglandins E_2 and A_2.
Clinical Research. 18: 510. 1970.

Abstract only. It was found that PGE_2 and PGA_2 infused in the renal artery immediately reversed an experimental antidiuresis that had been produced either by renal nerve stimulation, angiotensin II, or norepinephrine in dogs. PGE_2 was 3 to 5 times more effective that PGA_2. $PGF_{2\alpha}$ had no effect. The effective concentration of PGE_2 varied between 0.7 and 1.6 ng/ml blood, which is about the same as its concentration in renal venous blood during intraarterial infusion of angiotensin II. There was no effect of PG infusion on the opposite kidney. The authors feel that these data support the hypothesis that PG's function as an intrarenal hormone system for the control of sodium balance and arterial pressure. (JRH) 2156

1463

McGIFF, J.C., A.J. LONIGRO, and N.A. TERRAGNO
Interactions of renal prostaglandins with pressor systems.
Circulation. 42 (supp. 3): III 63. 1970.

Abstract only. Prostaglandins (PGs) E_2, A_2 and $F_{2\alpha}$ have been isolated from canine renal medulla. Buffering of vasoconstrictor systems by PGs has been proposed. In 14 morphine-chloralose anesthetized dogs, renal blood flows (RBFs) (electromagnetic flow meters), urine flows (photoelectric drop counters) and mean aortic blood pressure were recorded simultaneously. Infusion of PGE_2 (100 ng/min) or PGA_2 (500 ng/min) increased RBF by $35\pm 3\%$ and $36 \pm 3\%$ from mean control values of 230 and 218 ml/min, respectively. PGE_2 given during renal nerve stimulation (RNS) or during intravenous infusions of either angiotensin II(AT-II) or norepinephrine (NE) resulted in reversal of the renal vasoconstriction and antidiuresis produced by AT-II and RNS, whereas similar changes produced by NE were undiminished by PGE_2. In contrast, PGA_2 opposed the renal vasoconstriciton and antidiuresis produced by RNS, but not these actions of NE and AT-II. $PGF_{2\alpha}$ did not alter the renal effects of these vasoconstrictor stimuli. Concentrations of infused PGE_2 in renal blood (0.75 ± 0.05 ng/ml) were similar to those reported in renal venous blood during renal ischemia or AT-II infusions. Thus renal PGs may regulate the renal effects of these pressor systems. Since PGE_2 and PGA_2 opposed the renal effects of RNS but not those of NE, they presumably inhibit release of NE. (Authors) 2265

1464

McGIFF, J.C., K. CROWSHAW, N.A. TERRAGNO, and A.J. LONIGRO
An intrarenal hormonal system regulating the renal actions of angiotensin II and norepinephrine.
Journal of Clinical Investigation. 49: 63a. 1970.

Abstract only. Arterial blood pressure, renal blood flow, and urine flow in chloralose-anesthetized dogs was measured to determine the effect of vasoconstrictor hormones and the release of prostaglandin-like substances (PLS) in renal venous blood. It was found that angiotensin II and norepinephrine caused a release of PLS in renal venous blood which coincided with the loss of their vasoconstrictor and antidiruetic actions. Chromatographic analysis of the PLS showed it to be PGE. The log of the threshold dose of angiotensin II which released PLS was plotted against control values of plasma renin activity (PRA). It was found that when the sodium balance was negative (high PRA) the threshold dose of angiotensin was high. It was concluded that renal prostaglandins are oriented to salt and water metabolism and that they modulate vasoconstrictor hormones. (JRH) 2218

1465

McGIFF, J.C., K. CROSHAW, N.A. TERRAGNO, A.J. LONIGRO, J.C. STRAND, M.A. WILLIAMSON, J.B. LEE, and K.K.F. NG
Prostaglandin-like substances appearing in canine renal venous blood during renal ischemia.
Circulation Research. 27: 765-782. 1970.

Renal prostaglandins might mediate an antihypertensive function of the kidney. The blood-superfused organ technique possesses the sensitivity (threshold 0.4 ng/ml blood) and specificity required for identification of PG's in blood. Induction of unilateral renal ishemia in 14 chloralose-anesthetized dogs reduced renal blood flows from a mean value of 257 to 109 ml/min on the ischemic side and from 250 to 209 ml/min on the contra-lateral side. Concomitantly, PG-like substances were detected by assay organs in the venous blood of ischemic (13 experiments) and contralateral (11 experiments) kidneys. In one experiment, in a spontaneously hypertensive dog PG's were not detected during renal ischemia. Renal venous blood and renal medullary tissue were extracted for acidic lipids and assayed for PG-like substances. Extracts of venous blood collected during renal ischemia and extracts of renal medulla yielded substances with biological activity indistinguishable from PG-like substances or PG standards. Chromatographic characterization of PG-like substances suggest that they are predominantly a mixture of PGE_2 and $PGF_{2\alpha}$. (Authors) 2034

1466
McGIFF, J.C., K. CROWSHAW, N.A. TERRAGNO, and A.J. LONIGRO
Prostaglandin-like substances released from the kidney in response to angiotensin II.
Federation Proceedings. 29: 841abs. 1970.

Abstract only. Prostaglandin-like substances (PLS) were suggested to be released from the kidney during renal ischemia in response to increased generation of antiotensin II. Angiotensin II was infused into the renal artery (IRA) in chloralose-anesthetized dogs and release of PLS (detected by the superfused blood-bathed organ technique of Vane) was related to changes in renal blood flow (RBF) (electromagnetic flowmeter) and urine formation (photoelectric drop counters). With the appearance of PLS in renal effluent, samples of renal venous blood were removed and subjected to acidic lipid extraction and TLC which characterized PLS as a mixture of PGE_2 and $PGF_{2\alpha}$. In all experiments, angiotensin II released PLS although the threshold dose varied greatly (0.08 to 4.4 ng of angiotensin II per ml of arterial blood). In 5 or 6 experiments these concentrations (<1.0 ng/ml blood) fell within the range reported in renal vascular hypertension. A mean reduction in RBF of 35% of control and antidiuresis were observed initially in response to angiotensin II (IRA). Despite continued angiotensin II infusion, RBF and urine flow recovered coincident with the appearence of PLS in renal venous blood. The effects of angiotensin II are nonspecific since norepinephrine also released PLS. These results suggest that renal prostaglandins function in a renal feedback control mechanism regulating depressor and antinatriuretic systems. (Authors) 2114

1467
McGIFF, J.C., K. CROWSHAW, N.A. TERRAGNO, and A.J. LONIGRO
Release of a prostaglandin-like substance into renal venous blood in response to angiotensin II.
Circulation Research. 27 (supp. 1): I 121-130. 1970.

A prostaglandin-like substance (PLS) was released into renal venous blood by angiotensin II infused into the renal artery in all nine chloralose-anesthetized dogs. A bioassay procedure, the superfused blood-bathed organ technique, was used to detect PLS. This method possesses the required sensitivity and specificity for the detection of nanogram quantities of prostaglandins in blood. The threshold doses of angiotensin II required to release PLS were variable: 0.5 to 20 ng/kg/min (concomitant mean reduction of renal blood flow SEM: 35 10%). Release of PLS coincided with recovery of renal blood flow in 26 of 30 observations. The amount of PLS appearing in renal effluent was not related to the dose of angiotensin II. Tentative identification of PLS released by angiotensin II was obtained by 1) matching the activity of assay organs produced by angiotensin II wth PGE_2 standards, and 2) collection of renal venous blood simultaneously with the appearance of PLS for biochemical characterization. Acidic lipid extracts of the blood and purified samples, prepared by thin-layer chromatography, produced effects on the assay organs indistinguishable from PGE$_2$ standards. These results suggest that a renal prostaglandin, PGE$_2$, which was been identified in canine renal medulla, might function as a regulator of renal blood flow and urine formation when circulating levels of angiotensin II are increased. (Authors) 2267

1468

McGIFF, J.C., K. CROWSHAW, N.A. TERRAGNO, and A.J. LONIGRO
Renal prostaglandins: possible regulators of the renal actions of pressor hormones.
Nature. 227: 1255-1257. 1970.

The prescence of prostaglandins in renal venous blood in response to infusions of angiotensin II or noradrenaline into the renal artery was investigated using the superfused blood-bathed organ technique. The acidic lipids were purified by thin-layer chromatography and bioassayed on rat stomach strip, rat colon and chick rectum. Concentrations of prostaglandin-like substances in renal venous blood, when assayed as prostaglandin E$_2$ equivalents, ranged from 0.1 to 3.3 ng/ml. Extracts of renal venous blood obtained during infusion of the pressor hormones and samples obtained from prostaglandin E zones of the chromatographed extracts showed activity similar to prostaglandin E$_2$ standards. The release of PGA$_2$ could not be detected. (GW) 2205

1469

McKEEL, D.W. and L. JARETT
Distribution and characterization af adenyl cyclase in subcellular fractions of pig adenohyphysis.
Journal of Cell Biology. 47: 135a. 1970.

Abstract only. Subcellular fractions of pig anterior pituitary gland were prepared and assayed for the presence of various enzymes believed to be involved with secretion. Adenyl cyclase was found to be associated with the plasma membrane, Golgi apparatus and secretory granule membrane fractions. The adenyl cyclase was stimulated by PGE$_1$, NaF, and EGTA. It was inhibited by Ca^{++}. Relative levels of adenyl cyclase in these fractions and sensitivity to various substances is discussed. (JRH) 2447

1470

MELVIN, K.E.W.
Histaminase and medullary thyroid carcinoma.
New England Journal of Medicine. 283: 1286-1287. 1970.

Williams et al (1968) are cited briefly with reference to the secretion of prostaglandins by medullary tumors of the thyroid and its possible relationship to the diarrhea occurring in association with this tumor. (RMS) 2476

1471

MERTZ, D.P.
Nucleotidstoffwechsel und magensaureskretion. [Nucleotide material exchange and stomach acid secretion.]
Klinische Wochenschrift. 48: 831-838. 1970.

The author cites 15 papers regarding the role of PGE$_1$ in a review of how far hormones and drugs, which are known to change the intracellular level of adenosine-3',5'-cyclic monophosphate (cAMP) within various tissues, may interfere with the formation of gastric acid in humans during maximal stimulation of gastric mucosa. Substances capable of activating or inactivating the enzymes adenyl cyclase and 3',5'-AMP-phosphodiesterase in the responsive tissues may involve interference with the formation of gastric acid only in certain conditions. It is suggested that there may exist a second pathway for the metabolism of cyclic AMP except the degradation by 3',5'-AMP-phosphodiesterase. Experimental findings are discussed with respect to the possibility that cAMP participates on a molecular basis in the secretory mechanism of gastric acid. In defined conditions direct measurements of the content of cAMP within the cells of gastric mucosa are lacking at present excepting stimulation by methylxanthines. The problem remains unsolved whether cAMP or a metabolite of cAMP is a mediator of gastric acid formation. It is not known whether cAMP is the final effector of some hormone or drug-induced alterations of acid secretion by the stimulated gastric mucosa. (Author modified) 2105

1472
MICHAELSSON, G.
  Effects of antihistamines, acetylsalicylic acid and prednisone on cutaneous reactions to
    kallikrein and prostaglandin E_1.
  Acta Dermato-Venereologica. 50: 31-36. 1970.

The effects of antihistamines, acetylsalicylic acid and prednisone on the reactions to intradermally injected kallikrein, prostaglandin E_1, bradykinin and histamine were studied in patients with various minor skin disorders and in patients with chronic urticaria. The erythema produced at one hour by 0.5 μg of PGE_1 was reduced after cypropeptadine, but not after mepyramine. The erythema produced by 0.1 μg of PGE_1 was, however, reduced by mepyramine. (GW) 2294

1473
MICHELI, H.
  Some characteristics of lipolysis in rabbit adipose tissue. Effects of noradrenaline, ACTH,
    theophylline and prostaglandin E_1.
  Acta Physiologica Scandinavica. 79: 289-298. 1970.

Perirenal adipose tissue from rabbits of different strains was incubated in vitro and the glycerol release determined to evaluate the lipolysis. Noradrenaline was found to stimulate lipolysis, but to a lower degree than ACTH. From one strain of rabbits it was apparent that lighter (younger) rabbits were more responsive to noradrenaline than heavier (older) animals, some of which were completely unresponsive to the catecholamine. Theophylline as well as PGE_1 were ineffective when added alone to the incubation medium. However, theophylline increased the glycerol release induced by noradrenaline and low concentrations of ACTH and PGE_1 inhibited lipolysis under these conditions. The results suggest that there is a low rate of formation of cyclic AMP in rabbit adipose tissue under basal conditions of incubation while the hormonal stimulations might well operate by means of increasing the accumulation of cyclic AMP. (Author) 2056

1474
MILCU, ST.-M.
  Prostaglandinele. [Prostaglandins.]
  Studii si Cercetari de Endocrinologie. 21: 455-456. 1970.

This is a very brief review article in Romanian. (JRH) 2093

1475

MILTON, A.S. and S. WENDLANDT

A possible role for prostaglandin E_1 as a modulator for temperature regulation in the central nervous system of the cat.

Journal of Physiology. 207: 76P-77P. 1970.

Various prostaglandins (PGA_1, PGE_1, $PGF_{1\alpha}$, $PGF_{2\alpha}$) were injected in volumes of 10 or 100 ml into the third ventricle of conscious cats to see if any change occurred in rectal temperature. $PGF_{2\alpha}$ was without effect in doses of 0.28 and 2.8 μ-moles. In contrast, $PGF_{1\alpha}$ in similar doses produced slight elevations of temperature in some but not all of 6 cats. PGA_1 produced a rise in rectal temperature in all 4 cats tested with a dose of 28 μ-moles and in 2 out of 4 cats tested with a dose of 2.8 μ-moles. An antipyretic drug, 4-acetamidophenol (4-Ac), administered intraperitoneally in a dose of 0.3 m-moles/kg 30 minutes before the prostaglandins, was found to prevent the temperature rise due to PGA_1 and $PGF_{1\alpha}$, and it reduced the established fever when given after administration of these prostaglandins. These effects of 4-Ac on centrally induced fever were similar to those reported when the fever was induced by serotonin or by pyrogen. PGE_1 was administered at 4 different dose levels from 28 n-moles to 6 cats. The increase in temperature was dose-dependent; 28 μ-moles produced increases as large as $3°C$. 4-Ac was both unable to prevent the onset of fever or to reduce a fever already induced by PGE_1. It is speculated that PGE_1 may act as a modulator in temperature regulation and that the action of antipyretics may be to interfere with the release of PGE_1 by serotonin or pyrogen. This would explain why serotonin fever but not that induced by PGE_1 is affected by 4-Ac. (ART) 2237

1476

MINEMURA, T., W.W. LACY and O.B. CROFFORD

Regulation of the transport and metabolism of amino acids in isolated fat cells. Effect of insulin and a possible role for adenosine 3',5'-monophosphate.

Journal of Biological Chemistry. 245: 3872-3881. 1970.

Prostaglandin E_1 is only briefly mentioned in the discussion section of this paper. The authors feel that the inhibition of protein synthesis by insulin may be due to its effect on cAMP. If this is true, then its action should be mimicked by antilipolytic agents including PGE_1. (JRH)2474

1477

MISIEWICZ, J.J., S.L. WALLER and E.W. HORTON

Prostaglandins and colonic motility.

Proceedings of the Royal Society of Medicine. 63: 6. 1970.

Abstract only. The effect of ingestion of 2 mg of prostaglandin E_1 on intestinal transit and motor activity was studied in four normal volunteers. The transit rate through the small intestine and the colon was markedly increased after oral prostaglandin, with the passage of fluid and faeces per rectum and the production of abdominal colic(Misiewicz et al. 1969). (Authors Modified) 2510

1478

MITCHELL, J.R.A.

Restraint of thrombogenesis.

Thrombosis et Diathesis Haemorrhagica. Suppl. 40: 285-288. 1970.

A search for an arterial antithrombotic agent should turn to agents which modify platelet behavior. If an agent prevents white thrombi from forming, then in this sense it is indeed antithrombotic. Several agents have been shown to inhibit white body formation; prostaglandin E_1 probably acts on the circulating cells and, at present, is the most powerful inhibitor of aggregation available. The quest for an effective antithrombotic agent can be resolved in two ways: first by identifying the cause of thrombosis and then searching for an agent, and second by measuring the effectiveness of agents in order to see whether the modifications which they induce are beneficial. (GW) 2320

1479
MIYANI, M.
  Prostaglandins. III. Synthesis of methyl esters of 15-dehydro-PGB_1, 15-dehydro-PGE 237, and
    dl-PGE 237.
  Journal of Organic Chemistry. 35: 2314-2318. 1970.

  The stereochemistry of bicyclo (2.2.1) hept-5-ene-3-n-hexanoyl-2-carboxylic acid was elucidated and the chloro ketone was prepared. The methyl ester of 15-dehydro-PGB_1 was synthesized in 10% overall yield from the chloro ketone. Also presented are the unequivocal proof of the methyl 9,15,dioxo-prosta-8(12),13-dienoate structure and evidence for the erroneous structural assignment to the free acid in the literature. The dienoate was converted into the methyl esters of 15-dehydro-PGE 237 and racemic PGE 237. (Author modified) 2374

1480
MOGHISSI, K.S. and C.P. MURRAY
  The function of prostaglandins in reproduction.
  Obstetrical and Gynecological Survey. 26: 281-296. 1970.

  This survey of the function of prostaglandins in reproduction includes a discussion on the source and structure of prostaglandins, their general effects on such things as the gastrointestinal tract and the vascular system and their effect on the female reproductive tract. (GW) 2323

1481
MONTES, J. and J. CUEVA
  Prostaglandinas consideraciones generales. [Prostaglandins. General considerations.]
  Algeria. 18: 91-95. 1970.

  A review of the classes of prostaglandins, their sturcture and function, is given on a very elementary level. (RMS) 2094

1482
MOREAU, D.
  New gold rush for the drug industry?
  New Scientist. Sept. 3: 468-469. 1970.

  In this editorial review the author briefly outlines the history of the development of present knowledge about prostaglandins and briefly discusses their pharmacological applications in abortion, birth control, and asthma therapy. (JRH) 2383

1483

MORIWAKI, T.

[The effects of prostaglandins on glucose metabolism.]
Shinryo. 23: 1323-1327. 1970.

Article in Japanese. Abstract not available at present. 2399

1484

MOSKOWITZ, J., J.P. HARWOOD, W.D. REID, and G. KRISHNA

Interaction between norepinephrine (NE) and prostaglandin E_1 (PGE_1) on adenyl cyclase (AC) in blood platelets.
Federation Proceedings. 29: 602abs. 1970.

Abstract only. Rabbit and human platelets were incubated with tritiated adenine in order to label the intracellular ATP. They were then resuspended in their own plasma containing theophylline and incubated with PGE_1. In the presence of PGE_1 a marked dose-dependent increase in the rate of cAMP formation was observed. Noradrenaline alone had little effect on the basal activity but inhibited the effect of PGE_1 both in human and rabbit platelets (more in the rabbit). The interaction of PGE_1 and norepinephrine on the adenyl cyclase in platelets provides a unique model to study the actions of neurohormones on adenyl cyclase systems in sympathetic nerve endings. (ART) 2116

1485

MUEHRCKE, R.C., A.K. MANDAL, and F.I. VIOLINI

Renal interstitial cells: prostaglandins and hypertension. A pathophysiological review of the renal medullary interstitial cells and their relationship to hypertension.
Circulation Research. 27 (supp. 1): I 109-119. 1970.

One of the functions of the medullary interstitial cells may be the production of the antihypertensive factor. One of three lipid factors extracted from the renal medulla which lowers the blood pressure of hypertensive animals has been identified as a prostaglandin E_2-217, and is now termed prostaglandin A_2. (GW) 2313

1486

MUGGIA, A.

Le prostaglandine. [The prostaglandins.]
Minerva Medica. 61: 29. 1970.

This non-technical survey cites the contributions of some of the major research in the field of prostaglandins. The article is in Italian. (GW) 2388

1487

MUIRHEAD, E.E., G.B. BROWN, G.S. GERMAIN and B.E. LEACH

The renal medulla as an antihypertensive organ.
Journal of Laboratory and Clinical Medicine. 76: 641-651. 1970.

Experiments were performed on experimentally hypertensive inbred rats to determine if the renal medulla or the renal cortex can function as an antihypertensive organ. It was found that genetically compatible grafts (isografts) of kidney medulla tissue caused a reduction in blood pressure in the hypertensive recipient rats. Nongenetically compatible grafts (allografts) also caused a reduction in blood pressure, but the hypertension returned when the grafts were rejected immunologically. Surgical removal of the isografts also caused a return of the hypertension. Renal cortex or killed medullary cells did not lower blood pressure. It is mentioned that prostaglandins and other vasoactive compounds have been isolated from the renal medulla. On the basis of these experiments it is suggested that the renal medulla may be an anithypertensive organ. (JRH) 2479

1488
MURAD, F., H.B. BREWER, and M. VAUGHAN
  Effect of thyrocalcitonin on adenosine 3',5'-cyclic phosphate formation by rat kidney and
    bone.
  Proceedings of the National Academy of Sciences. 65: 446-453. 1970.

The authors report that 5 μg/ml PGE$_1$ incubated with washed, particulate preparation of whole rat kidneys produced 0.17 mμ moles cAMP/mg kidney protein/10 min. The significance of this was not investigated. Thyrocalcitonin caused 0.22 mμ moles cAMP/mg protein/10 min to be accumulated. (RMS) 2253

1489
MURAD, F., V. MANGANIELLO, and M. VAUGHAN.
  Effects of guanosine 3',5'-monophosphate on glycerol production and accumulation of
    adenosine 3',5'-monophosphate by fat cells.
  Journal of Biological Chemistry. 245: 3352-3360. 1970.

The fact that PGE$_1$ has been shown to decrease cAMP in adipocyte cell cultures but increase cAMP concentration in fat pad fragments is given as an example of the pitfalls that are encountered when biochemical events which occur in a tissue are compared with those of a single cell type. (JRH) 2475

1490
MURAI, S., N. MIZUSHIMA, A. NUKAGA, S. KOYAMA, and T. SANO
  Effect of prostaglandin E$_1$ on the acidity of gastric juices.
  Gastroenterologica Japonica 5: 274-275. 1970.

The effect of PGE$_1$ on gastric secretion is briefly reported. Subcutaneous PGE$_1$ inhibited gastric secretion in rats starting 10 minutes after the injection of 100 μg and lasting 3 to 5 hours with peak inhibition at 30 minutes. Histamine, histalog, and gastrin administered during the period did not cause any acid secretion. Hypersecretion induced by prior administration of these agents was rapidly inhibited by PGE$_1$. That 100 μg of PGE$_1$ injected subcutaneously caused an inhibition of gastric acid secretion was also shown by an analysis of collected gastric juices 4 hours after pyloric ligation. Subcutaneous PGE$_1$ at 50 μg every 4 hours held the production of Shay ulcers in a 17-hour period to 1 to 3 in contrast to multiple ulcers seen without PGE$_1$. In guinea pigs, PGE$_1$ did not inhibit gastric secretion. Human gastric secretion accelerated by intravenous gastrin was not inhibited by subcutaneous injection of 200 μg of PGE$_1$. (RMS) 2496

1491
MURAI, S.
  [The effects of prostaglandins on gastric functions.]
  Shinryo. 23: 1379-1387. 1970.

Article in Japanese. Abstract not available at present. 2397

1492

MURPHY, G.P., V.E. HESSE, J.L. EVERS, G. HOBIKA, J.W. MOSTERT, A. SZOLNOKY, R. SCHOONEES, J. ABRAMCZYK, and J.T. GRACE
The renal and cardiodynamic effects of prostaglandins (PGE_1, PGA_1) in renal ischemia.
Journal of Surgical Research. 10: 533-541. 1970.

Dogs with chronically implanted electromagnetic aortic, superior mesenteric, and renal arterial transducers were studied for renal and cardiodynamic responses to two prostaglandins PGA_1 and PGE_1. Animals were normotensive or hypertensive as a result of recent stenosis of the right renal artery. Extraction ratios of PAH, creatinine, and insulin were also determined before and after drug infusion in hypertensive or normotensive dogs. The response to PGE_1 and PGA_1 differ in dogs with renal ischemia and hypertension. PGE_1-treated hypertensive animals had a marked rise in renal arterial flow associated with decreased renal extraction and intrarenal shunting of blood from noncortical tissue. PGA_1-treated hypertensive dogs had improved renal extraction ratios, increased urine flow rates, and elevated clearances of PAH, insulin, and creatinine without elevation of renal arterial flow. The improved cortical perfusion was evidently a result of decreased afferent glomerular arteriolar resistance. (Authors) 2295

1493

MUSTARD, J.F. and M.A. PACKHAM
Factors influencing platelet function: Adhesion, release and aggregation.
Pharmacological Reviews. 22: 97-187. 1970.

In this extensive review article on platelet function, it is mentioned that prostaglandins (PGE_1) are the most potent inhibitors of platelet aggregation ever developed. (JRH) 2471

1494

MUSTARD, J.F., and M.A. PACKHAM
Thromboembolism. A mainfestation of the response of blood to injury.
Circulation. 42: 1-21. 1970.

In this review of thromboembolism, it is briefly mentioned that PGE_1 is the most potent inhibitor of platelet aggregation discovered so far. It is suggested that this is due to its stimulation of platelet cAMP. (JRH) 2488

1495

NAJAK, Z., K. HILLIER, and S.M.M. KARIM
The action of prostaglandins on the human isolated non-pregnant cervix.
Journal of Obstetrics and Gynaecology of the British Commonwealth. 77: 701-709. 1970.

The isolated non-pregnant cervix has been shown to exhibit spontaneous isotonic and isometric contractility and to respond to drugs. PGE_2 causes a marked relaxation whilst the effect of $PGF_{2\alpha}$ is more variable. Oxytocin and adrenaline cause contraction of the muscle. The implications of the finding that the cervix can show isotonic and isometric contractility and the effects of prostaglandins are discussed. (Authors) 2318

1496

NAKANO, J. and J.M. KESSINGER
Cardiovascular effects of the metabolites of prostaglandin E_1 (PGE_1) in dogs.
Clinical Research. 18: 595. 1970.

Abstract only. The cardiovascular effects of three metabolites of PGE_1: 15-keto-PGE_1; dihydro-PGE_1; and 15-keto-dihydro-PGE_1 were compared in dogs. It was found that graded doses of these metabolites (0.25-4 μg/kg) given i v decreased systemic arterial pressure and increased heart rate and myocardial contractile force. When injected i v PGE_1 (0.6-10 ng/kg) and the three metabolites (10-2500 ng/kg) caused a decrease in perfusion pressure of a hind limb perfused with a sigma-motor pump, in proportion to dose. The magnitude of the vasodilator action of dihydro-PGE_1, 15-keto-PGE_1 and 15-keto-dihydro-PGE_1 was approximately 1/5, 1/100, and 1/1000 of that of PGE_1 in dogs. This suggests that saturation of the 13 double bond of PGE_1 leaves considerable activity but oxidation of the secondary alcohol group at 15-C almost abolishes its vasodilator activity. (JRH) 2152

1497

NAKANO, J. and J.M. KESSINGER
Effect of 8-isoprostaglandin E_1 on the systemic and pulmonary circulations.
Clinical Research. 18: 322. 1970.

Abstract only. In open heart dogs anesthetized with sodium pentobarbital, systemic and pulmonary arterial presure was recorded. Myocardial contractile force was also measured. The relative effects of PGE_1, $PGF_{2\alpha}$, and 8-iso-PGE_1 were compared. It was found the i v injection (2-16 μg/kg) of 8-iso-PGE_1 increased heart rate, myocardial contractile force, pulmonary arterial pressure and decreased systemic arterial pressure. The systemic effectiveness was 1/125 to 1/250 of that of PGE_1. When the pulmonary arterial pressure was kept constant by means of a pump, injection of 2 μg/kg of PGE_1 and $PGF_{2\alpha}$ decreased and increased significantly the pulmonary arterial pressure respectively. The same amount of 8-iso-PGE_1 also increased the pulmonary arterial pressure (½ that of $PGE_{2\alpha}$). Thus it appears that 8-iso-PGE_1 is unique in that it is hypotensive in the systemic vessels and hypertensive in the pulmonary circulation. (JRH) 2160

1498

NAKANO, J. and J.M. KESSINGER
Effect of 15-R-PGA_2 on the cardiovascular responses to PGA_2 in dogs.
Clinical Research. 18: 117. 1970.

Abstract only. This experiment was made to determine if 15-R-PGA_2 would block the cardiovascular effects of PGA_2 (15-S-PGA_2) in dogs anesthetized with sodium pentobarbital. PGA_2 (0.025-0.8 μg/kg) given i v decreased blood pressure and increased heart rate and myocardial contractile force on a dose dependent basis. The same dose of 15-R-PGA_2 did not cause any significant changes. When the two PGA_2s were given together it was found that the 15-R-PGA_2 had no effect on the response caused by PGA_2. (JRH) 2165

1499

NAKANO, J. and J.M. KESSINGER
Effects of 8-isoprostaglandin E_1 on the systemic and pulmonary circulations in dogs.
Proceedings of the Society for Experimental Biology and Medicine. 133: 1314-1317. 1970.

The effects of 8-iso-prostaglandin E_1 on the systemic and pulmonary circulations were studied and compared with those of PGE_1 and $PGF_{2\alpha}$ in anesthetized dogs. It was found that the i.v. administration of 8-iso-PGE_1 decreased systemic arterial pressure slightly and increased heart rate and myocardial contractile force slightly. The magnitude of the systemic hypotensive effect of 8-iso-PGE_1 was equivalent to approximately 1/125 to 1/250 of that of PGE_1 in dogs. On the other hand, the pulmonary hypertensive action of 8-iso-PGE_1 was 5 times greater than that of PGE_1. The present study indicates that 8-iso-PGE_1 increases pulmonary arterial pressure through its vasoconstrictor action on the pulmonary vascular bed. (Authors) 2242

1500

NAKANO, J. and L.J. GREENFIELD
Metabolic degradation of prostaglandin E_1 in human lungs and plasma.
Journal of Laboratory and Clinical Medicine. 76: 1018. 1970.

Abstract only. Experiments were performed to determine the rate of degradation of PGE_1 in human lung tissue and plasma in vitro. Tritiated PGE_1 was incubated with a homogenate of human lung tissue and with human plasma. The amount of labeled PGE_1 left in the culture medium was measured to determine the rate of metabolism. No significant breakdown occurred in the plasma after one hour. The lung tissue homogenate metabolized 95% of the prostaglandin in 20 min. (JRH) 2053

1501

NAKANO, J.
Metabolism of prostaglandin E_1 in dog kidneys.
British Journal of Pharmacology. 40: 317-325. 1970.

An average 43% of PGE_1 passing through isolated, perfused dog kidneys was converted to the less polar metabolite I in a single passage. With continued recirculation, a lesser polar metabolite II appeared in addition to metabolite I, but at a significantly slower rate. Maximum degradation usually required 6 passages. Further separation by chromatography showed 2 individual metabolites, IIa and IIb comprising metabolite II. (Author modified) 2020

1502

NAKANO, J.
Metabolism of prostaglandin E_1 (PGE_1) in kidney and lung.
Federation Proceedings. 29: 746abs. 1970.

Abstract only. Guinea pig lungs and dog kidneys were perfused with Tyrode solution containing tritiated PGE_1. Both organs metabolize about 30 to 40 percent of the tritiated PGE_1 to *less* polar metabolites in only one circulation through these organs. PGE_1 is not metabolized to *more* polar metabolites by these organs, however. (ART) 2115

1503

NAKANO, J., B. MONTAGUE, and B. DARROW
Metabolism of prostaglandin in human uterus and placenta and swine ovary.
Clinical Research. 18: 625. 1970.

Abstract only. Homogenates of human uterus, placenta, and swine ovary were incubated for one hour with radiolabeled PGE_1. The remaining PGE_1 and its metabolites were purified by silic acid chromatography. It was found that human uterus and placenta metabolized 20% and 75% of the PGE_1 respectively. Less that 5% of the PGE_1 was metabolized to less polar metabolites by human uterus and plasma and swine ovary. The authors feel this study suggests that released or injected prostaglandins may be inactivated by human placenta as well as lung but the uterus does not significantly degrade prostaglandins. (JRH) 2150

1504

NAKANO, J., N.H. MORSY, and M. DISTLER
Metabolism of prostaglandin in rat plasma, brain, heart, lung, kidney and testicle.
Clinical Research. 18: 675. 1970.

Abstract only. The amount of metabolic degradation of PGE_1 by homogenates of various rat tissues was measured. It was found that rat plasma degraded very little PGE_1 (less that 5%) in one hour of incubation. Other tissues which exhibit very little degradation are brain and heart. Rat kidney and lung were very active in degrading PGE_1 (95% in 20 min.). There was also rapid degradation in rat testicle. (JRH) 2149

1505
NAKANO, J., N.H. MORSY, and L.D. HENDERSON
Tissue distribution and subcellular localization of H^3-prostaglandin E_1 (H^3-PGE_1).
Clinical Research. 18: 676. 1970.

Abstract only. Rats and dogs were injected i v with H^3-PGE_1. After one hour, the animals were sacrificed and the amount of H^3-PGE_1 in homogenates of various organs and tissues measured. The highest total radioactivity was in the liver and kidneys; the least was in the spleen, stomach, brain, pancreas, thyroid, adrenals, uterus, ovary and testicles. The percentage of H^3-PGE_1 metabolites was much higher than H^3-PGE_1 in the lungs, kidneys and spleens indicating the high activity of prostaglandin degrading enzymes in these organs. The amount of radioactivity was highest in the soluble subcellular fraction and lowest in the microsomal fraction. This is taken to indicate that PGE_1 is effectively metabolized into 15-keto-PGE_1 by microsomal enzyme. (JRH) 2148

1506
NAKANO, J.
Tissue distribution and subcellular localization of H^3-prostaglandin E_1 in rats and dogs.
Archives Internationales de Pharmadocynamie et de Therapie. 187: 106-119. 1970.

The tissue and plasma distribution and subcellular localization of the intravenously injected H^3-PGE_1 were studied in rats and in dogs. It was found that the highest total radioactivity was in the liver and kidneys, and the least in the spleen, stomach, brain, pancreas, thyroid, adrenals, uterus, ovary and the testicles. The percentile concentrations of H^3-PGE_1 were markedly smaller than those of H^3-PGE_1 metabolites in lungs, kidneys, and livers, indicating the marked activity of prostaglandin-degrading enzymes in these tissues as compared with that in the brain, spleen and testicle. The subcellular total radioactivity was highest in the soluble fraction of the cells in brain, heart, lung, liver and kidney, and the least in the lysosomal and microsomal fractions. This observation is in agreement with those by the previous workers that PGE_1 is effectively metabolized into 15-keto-PGE_1 by microsomal enzyme. (Author) 2025

1507
NATARAJAN, S., V.V.S. MURTI, T.R. SESHADRI, and A.S. RAMASWAMY
Some new pharmacological properties of flavonoids and biflavonoids.
Current Science. 39: 533-535. 1970.

Experiments were performed to determine some of the pharmacological properties of some flavonoids and biflavonoids prepared from *Gingko biloba* and *Cupressus torulosa* leaves. One of the tests was to determine the ability of these compounds to inhibit the spasmotic activity of guinea pig ileum in vitro caused by PGE_1. It was found that quercetin, naringenin, myricetin, morin and quercetin-7-methyl ether produced 100% inhibition; homo eriodicytol produced 50% inhibition and cyanidia hydrochloride caused no inhibition. No speculations about the significance of these results are given. (JRH) 2429

1508

NEKRASOVA, A.A., F.M. PALEEVA and S. Sh. KHUNDADZE

[The content of depressor prostaglandinolike (sic) substances and the activity of renin in the kidneys in patients suffering from renovascular hypertension.]

Kardiologiia. 10: 88-92. 1970.

The author investigated the content of prostaglandin-like substances (PPS) and renin activity in 16 kidneys removed during the operation in patients with renovascular hypertension and 3 kidneys of healthy persons taken immediately after their death. The content of PPS in 12 ischemic kidneys was reduced more than twofold in comparison with healthy ones, and only in 4 kidneys it was augmented. The activity of renin in the majority of kidneys was increased and only in 2 kidneys it was normal. In ischemic kidneys with a high content of PPS the activity of renin was low, in kidneys with a high activity of renin PPS was present in the form of traces. It is possible that PPS by increasing the circulation in the renal cortex may inhibit the synthesis of renin. The level of the arterial pressure in patients with renovascular hypertension before the operation did not depend upon the renin activity and PPS content in the ischemic kidneys. (Authors) 2404

1509

NEKRASOVA, A.A., A. SEREBROVSKAYA, L.A. LANTSBERG, and A. UCHITEL

[Depressor prostaglandinolike (sic) substances and renin in the kidneys in experimental renal hypertension.

Kardiologiia. 10: 31-36. 1970.

Levels of "prostaglandinolike substances" and renin activity were studied in 20 rabbits with induced ischemia of one kidney and in 20 control animals. On the fourth day, prostaglandin levels dropped in intact and ischemic kidneys while renin activity was unchanged. On the 14th day, renin activity rose in the ischemic kidneys and dropped in the intact ones, while prostaglandin activity of both kidneys approached control levels; simultaneously, maximal rise in arterial pressure was observed. On the 30th day, prostaglandin content of ischemic kidneys rose signficantly with a concurrent reduction of renin activity. The authors assume that increasing prostaglandin activity produces a drop in renin activity causing a mild course in this form of hypertension. (RMS) 2085

1510

NELSON, N.A.

The synthesis of prostaglandins.

In: Abstracts, 5th Mid Atlantic Regional Meeting of the American Chemical Society, 1-3 April. p. 59. Newark. 1970.

Abstract only. This introduction to a lecture discusses in general terms the merits of total chemical synthesis of prostaglandins over biosynthetic methods. (JRH) 2394

1511

NEVE, P. and J.E.DUMONT

Effects in vitro of thyrotropin, cyclic 3'5'-AMP, dibutyryl cyclic 3',5'-AMP, and prostaglandin E_1 on the ultrastructure of dog thyroid slices.

Experimental Cell Research. 63: 285-292. 1970.

Phagocytosis of colloid droplets by apical pseudopods in dog thyroid slices in vitro was studied by electron microscopy. Phagocytosis was induced by incubation with TSH, cAMP and caffein, dibutyryl cAMP, and PGE$_1$ suggesting some common mode of action among these substances. (RMS) 2006

1512

NEZELOF, C., D. GUY-GRAND and E. THOMINE

Les megacolons avec hyperplasie des plexus myenteriques. Une entite anatomo-clinique, a propos de 3 cas. [Megacolon with hyperplasia of the myenteric plexus. A pathological entity, a report of three cases.]

Presse Medicale. 78: 1501-1506. 1970.

Brief mention is made of Williams' et al paper on secretion of prostaglandins from medullary thyroid carcinomas in this case history report on 3 patients. The article is in French with an English summary. (JRH) 2485

1513

NG, K.K.F., K.H. SIT, and W.C. WONG

Relaxant effect of prostaglandin E$_1$ on the isolated intestine of the toad (Bufo melanostictus). Agents and Actions. 1: 227-230. 1970.

The pharmacological actions of prostaglandins E$_1$, F$_{1\alpha}$ E$_2$ and A$_1$ were studied on the longitudinal and circular muscles of the isolated small intestine of the toad. Prostaglandin E$_1$ caused relaxation, whereas prostaglandin A$_1$ was inactive and prostaglandins F$_{1\alpha}$ and E$_2$ caused contractions. Other substances which relaxed the toad intestine were adrenaline, noradrenaline and isoprenaline. On a molar basis, prostaglandin E$_1$ was seven times less potent than isoprenaline, but it was three times more potent than adrenaline and seventeen times more potent than noradrenaline. Phentolamine and propranolol blocked the relaxant effects of catecholamines in concentrations which did not alter the relaxant effect of prostaglandin E$_1$. The results suggested that prostaglandin E$_1$ acted directly on the intestinal smooth muscle of the toad rather than by the local release of catecholamines. (Authors) 2407

1514

NICOLAESCU, V.

Mediatia farmacologica a conflictului antigen-anticorp. [Pharmacological mediation of the antigen-antibody conflict.]

Studii si Cercetari de Medicina Interna. 11: 379-392. 1970.

This review on the mediators of anaphylactic shock contains a brief section on prostaglandins. (JRH) 2422

1515

NOVOGRODSKY, A. and E. KATCHALSKI

Effect of phytohemagglutinin and prostaglandins on cyclic AMP synthesis in rat lymph node lymphocytes.

Biochimica et Biophysica Acta. 215: 291-296. 1970.

Phytohemagglutin does not increase the level of cyclic adenosine 3',5'-monophosphate (cAMP) and does not stimulate adenyl cyclase activity in rat lymph node lymphocytes. N^6-2'-O-Dibutyryl cyclic AMP stimulates RNA synthesis but does not cause cell transformation.

394

Prostaglandin E_1 markedly increases the accumulation of cAMP and stimulates adenyl cyclase activity of rat lymph node lymphocytes. It does not, however, cause cell transformation. It is tentatively concluded that cyclic AMP does not mediate the effect of phytohemagglutinin in the transformation of rat lymph node lymphocytes. (Authors) 2324

1516
NUGTEREN, D.H.
Inhibition of prostaglandin biosynthesis by 8*cis*, 12*trans*, 14*cis*-eicosatrienoic acid and 5*cis*, 8*cis*, 12*trans*, 14*cis*-eicosatetraenoic acid.
Biochimica et Biophysica Acta. 210: 171-176. 1970.

A number of unsaturated fatty acids have been tested for inhibitory activity in the enzymic conversion of all-*cis* 8, 11, 14-eicosatrienoic acid into prostaglandin E_1. The highest inhibitory action was displayed by 8*cis*, 12*trans*, 14*cis*-eicosatrienoic acid and 5*cis*, 8*cis*, 12*trans*, 14*cis*-eicosatetraenoic acid. Enzyme-kinetic studies suggest that the inhibition is competitive. (Author) 2041

1517
NUKAGA, A., S. MURAI, N. MIZUSHIMA, S. KOYAMA, T. SANO and M. ABE
Effect of prostaglandin E_1 on gastric secretion and experimental gastric ulcer.
Gastroenterologia Japonica. 5: 321. 1970.

Abstract only. Synthesized prostaglandin E_1 (PGE_1) was investigated regarding its inhibitory effect on gastric juice secretion and experimental ulcer formation. Continuous increase of gastric juice secretion was produced by continuous injection of histalog to the dogs with pylorus ligation. The gastric juice secretion was remarkably inhibited by intravenous injection of PGE_1 2 µg/kg/min, for 20 minutes. The secretion started again 20 minutes after discontinuation of the injection. Gastric ulcer was produced experimentally on rats to observe the inhibitory effect of PGE_1. PGE_1 remarkably inhibited gastric juice secretion to produce Shay's ulcer and Reserpin ulcer, but only mildly to produce steroid ulcer. (Authors) 2520

1518
NUTTER, D.O. and H. CRUMLY
Coronary-myocardial responses to prostaglandins E_1 and A_1.
Circulation. 42(supp. 3): III 124. 1970.

Abstract only. The effects of prostaglandin (PG) types E_1 and A_1 on coronary resistance and LV function were studied in 18 anesthetized open chest dogs (8 beta blocked). PGE_1 and A_1 (0.1,0.5,2.5,12.5µg) was injected to the left circumflex coronary during autoperfusion at constant flow in the beating and electrically arrested heart. Mean coronary perfusion pressure (CPP), mean aortic pressure (MAP), LV contractile force (LVCF, Walton-Brodie gauge), LV internal diameter (LVD, sonomicrometer), aortic and inferior caval blood flow was recorded. In beating hearts all PGE_1 and A_1 doses produced immediate, but transient (<10 min) coronary vasodilation PGE_1 and A_1 (0.1µ) decreased CPP 109 to 88 mmHg (-19%) and 128 to 112 mmHg (-12%) respectively, without affecting MAP. PGE_1 and A_1 (12.5µ) decreased CPP 94 to 49 mmHg (-48%) and 130 to 108 mmHg (-17%) while MAP fell 56 to 44 mmHg (-12%) and 87 to 55 mmHg (-37%). In the arrested heart PGE_1 (12.5µ) produced an equivalent fall in CPP (99 to 61 mmHg,-38%) whereas PGA_1 did not significantly alter CPP. PGE_1 and A_1 augmented LVCF at all doses but did not alter heart rate. PG in doses 0.5 to 12.5µ decreased LVD while small transient increases in cardiac output and caval blood flow occurred. Beta blockage did not affect the responses to PGE_1 or A_1. PGE_1 is a potent, direct coronary vasodilator (CVD) and inotropic agent. The less potent CVD effects of PGA_1 may be secondary to positive inotropism. (Authors) 2262

1519

O'CONNELL, J.M.B. and R.J. SHALHOUB

Studies on the mechanism of inhibition of P-amino hippuric acid (PAH) by renal vasodilator drugs.

Clinical Research. 18: 64. 1970.

Abstract only. The mechanism of inhibition of para-amino hippuric acid (PAH) transport in rat renal tissue by PGE_1, aminophylline, dopamine, and papaverine were investigated by 1) correlating the degree of inhibition with medium concentration of the drug 2) measuring oxygen consumption during incubation (qO_2) and 3) determining the rate of PAH run out at drug concentrations which produce approximately 50% inhibition. Results indicated no significant change in qO_2 or PAH run out. This would indicate that the drugs inhibit active transport. Since the inhibitory concentrations of drugs were in the same range as the PAH it might indicate a competitive inhibition. Cellular poisoning was ruled out since there was no change in qO_2. (JRH) 2167

1520

OKADA, F.

The hypotensive effects of prostaglandins.

Shinryo. 23: 1370-1378. 1970.

Article in Japanese. Abstract not available at present. 2428

1521

OKADA, F., Y. YAMAUCHI, K. SUGIYASU, Y. SUGITANI, K. TAKAYASU, M. KARITA and T. MORIWAKI

The hypotensive effects of prostagrandin (sic) E_1.

Japanese Circulation Journal. 34: 857. 1970.

Abstract only. The effects of prostaglandin E_1 (PGE_1) on human blood pressure, the level of serum lipids (NEFA, total cholesterol, triglyceride) and blood glucose have been studied in seven healthy adults, twenty-one essential hypertensives, ten renal hypertensives and five hypertensives with diabetes mellitus. After the intravenous administration of $50\mu g$ PGE_1 in 4 ml of saline for 3 minutes, the blood pressure was measured until 60 minutes after the injection. The level of serum lipids and blood glucose also were measured before and 5,10,20,30,40, and 60 minutes after the administration. Both the systolic and diastolic blood pressures of hypertensives were depressed significantly, although the blood pressures of healthy subjects remained unchanged by the administration of PGE_1. In essential hypertensives the depressor effects of patients with PSP results above thirty per cent were less than those of the ones with PSP results below twenty-nine per cent. The depressor effects of the patients with renal hypertension were alike of essential hypertensives with PSP results below twenty-nine per cent. The levels of serum lipids and blood glucose were not altered significantly by the administration of PGE_1 in all subjects. (Authors) 2516

1522

OLDHAM, S.

The prostaglandins.

Manufacturing Chemist and Aerosol News. 41: 43-47. 1970.

This comprehensive review covers all areas of prostaglandin research starting with a historical section and ending with speculations about the future value of prostaglandins. (JRH) 2432

1523

ONAYA, T. and D.H. SOLOMON

Stimulation by prostaglandin E_1 of endocytosis and glucose oxidation in canine thyroid slices.
Endocrinology. 86: 423-426. 1970.

Prostaglandin E_1 (PGE_1) stimulated canine thyroid tissue in vitro as exemplified by a 2- to 4-fold increase in flucose oxidation and a 40- to 100-fold increase in the number of intracellular colloid droplets. The effect on colloid droplets was detectable with a concentration of PGE_1 as low as 10^{-7} M and that on glucose oxidation at 10^{-6} M. Both actions of PGE_1 were inhibited by the lysosome stabilizer, chlorpromazine, and were additive with submaximal doses of TSH. The similarity of PGE_1 action to that of TSH was also manifest in vivo; PGE_1 stimulated both intracellular colloid droplet formation and ^{131}I release. (Authors) 2224

1524

O'REILLY, R.A. and P.M. AGGELER

Determinants of the response to oral anticoagulant drugs in man.
Pharmacological Reviews. 22: 35-96. 1970.

In this review of oral anticoagulant drugs, it is briefly mentioned that PGE_1 has been shown to reduce platelet adhesiveness. (JRH) 2470

1525

OSTERGARD, D.R.

The physiology and clinical importance of amniotic fluid. A review.
Obstetrical and Gynecological Survey. 25: 297-319. 1970.

The purpose of this review is to correlate the known chemical and biological components of the amniotic fluid with various clinical situations that may be encountered during the course of pregnancy. Prostaglandins are present in lipid extracts of amniotic fluid and are known to have smooth muscle stimulating activity. Recently these substances have been subdivided into 4 groups labeled E_1, E_2, $F_{1\alpha}$, and $F_{2\alpha}$. During early gestation only the E series are found in small quantities in the amniotic fluid. With the advent of labor all four substances are present, but the F group is present in the greatest concentration. In vitro the E group inhibits isolated myometrical activity, whereas the F group stimulates activity of uterine muscle. The exact role of these substances as initiators of labor is currently conjectural and is the subject of intensive investigation. (GW) 2322

1526

PACE-ASCIAK, C. and L.S. WOLFE

Biosynthesis of prostaglandins E_2 and $F_{2\alpha}$ from tritium-labelled arachidonic acid by rat stomach homogenates.
Biochimica et Biophysica Acta. 218: 539-542. 1970.

Mass spectral evidence is presented showing that PGE_2 and $PGF_{2\alpha}$ are synthesized by the homogenized rat stomach from tritiated arachidonic acid thus demonstrating the presence of synthetase activity. Additionally the formation of quantities of a prostanoic derivative migrating in the E region was noted. Specific activity of the PGE_2 formed shows a 10-fold dilution in specific activity of added arachidonic acid by endogenous arachidonic acid and that PGE_2 is formed from both. The relationship of the present experiment to previous results on the regulation of fatty acid precursors and prostaglandin synthesis in the stomach is discussed. (RMS) 2008

1527

PACE-ASCIAK, C., K. MORAWSKA, and L.S. WOLFE
Metabolism of prostaglandin $F_{1\alpha}$ by the rat stomach.
Biochimica et Biophysica Acta. 218: 288-295. 1970.

The metabolism of $PGF_{1\alpha}$ was shown to be by the prostaglandin 15-hydroxy-dehydrogenase pathway rather than by β-oxidation in the rat stomach. Using the supernatant fraction of homogenized rat stomach, the authors found $PGF_{2\alpha}$ to be converted to 15-keto-$PGF_{1\alpha}$ and dihydro-15-keto-$PGF_{1\alpha}$ as established by chromatographic analysis, preparation of sodium borodenteride derivatives, and mass spectrophotometry of trimethyl silyl derivatives. No metabolism was found in the particulate (10,000 g or 100,000 g) fractions. (RMS) 2007

1528

PACE-ASCIAK, C. K. MORAWSKA and L.S. WOLFE
Metabolism of $PGF_{1\alpha}$ by rat stomach.
Proceedings of the Canadian Federation of Biological Societies 13: 144. 1970.

Abstract only. The supernatant fraction (30,000 X G) of a homogenate of rat stomach was shown to convert tritium labelled prostaglandin $F_{1\alpha}$ (5,6-$^3H_2$-$PGF_{1\alpha}$, Rf 0.27) into two metabolites: 15-keto-$PGF_{1\alpha}$ (I, Rf 0.45) and dihydro-15-keto-$PGF_{1\alpha}$ (II, Rf 0.56). Sodium borohydride reduction of I gave two isomers (Rf 0.27 and 0.34) whereas reduction of II gave only one product (Rf 0.35). The mass spectra of the reduced products of I were identical to that of $PGF_{1\alpha}$, whereas that of the reduced product of II indicated that the Δ^{13} double bond was saturated. Only one deuterium atom was incorporated in I and II upon reduction with sodium borodeuteride. The location of the deuterium atom and therefore the ketone in I and II was established by mass spectrometry to be at position 15. The formation of these compounds from $PGF_{1\alpha}$ indicates that the prostaglandin 15-hydroxy dehydrogenase previously found in guinea pig lung (Anggard, E., and Samuelsson, B., J. Biol. Chem. 239: 4097, 1964) also occurs in rat stomach. (Authors) 2418

1529

PACE-ASCIAK, C. and L.S. WOLFE
Polyhydroxylated by-products of the enzymatic conversion of tritiated arachidonic acid into prostaglandins by sheep seminal vesicles.
Chemical Communications. 1235-1236. 1970.

In the biosynthesis of prostaglandins by sheep seminal vesicles, the prostaglandin E fraction after treatment with alkali to convert PGE_2 into PGB_2 contained four major compounds still migrating on t.l.c. in the PGE region: three of the compounds were derived from triatiated arachidonic acid and one was derived from endogenous eicosatrienoic acid. (Authors— modified) 2268

1530
PACE-ASCIAK, C. and L.S. WOLFE
  A prostanoic acid derivative formed in the enzymatic conversion of tritiated arachidonic acid
     into prostaglandins by rat stomach homogenates.
  Chemical Communications. 1234-1235. 1970.

  The structure is reported of a new compound which behaves chromatographically like
  prostaglandin E_2 but is not dehydrated by alkali to PGB_2, formed during the enzymatic
  conversion of arachidonic acid into PGE_2 by rat stomach homogenates. (Authors) 2269

1531
PALMER, M.A., P.J. PIPER, and J.R. VANE
  The release of rabbit aorta contracting substance (RCS) from chopped lung and its antagonism
     by anti-inflammatory drugs.
  British Journal of Pharmacology. 40: 581P-582P. 1970.

  The authors report detecting the release of histamine, slow reacting substance, prostaglandins,
  and rabbit aorta contracting substance from chopped lung tissue from sensitized guinea pig
  when challenged. These same substances were also released when the chopped lung from normal
  or sensitized animals was mechanically stirred with a blunt rod. Since equal amounts of these
  substances were released upon three stirrings, it is suggested that this might provide a sensitive
  test for blocking agents to the release of these substances. It is suggested that release of the
  same substances caused by anaphylatic shock and mechanical stirring might be due to damage
  to cell membranes in both cases. (JRH) 2198

1532
PALMER, M.A., P.J. PIPER, and J.R. VANE
  Release of vaso-active substances from lungs by injection of particles.
  British Journal of Pharmacology. 40: 547P-548P. 1970.

  Lungs were removed from guinea pigs which had been sensitized to ovalbumen and were
  arranged for perfusion. Effluent from the lungs was bioassayed by a series of assay tissues,
  which would detect histamine, rabbit aorta contracting substance and prostaglandins. The lungs
  were then perfused with various emulsions with known particle size. It was found that
  prosparol, sephadex G 10, and Bacto-latex caused the release of prostaglandins and rabbit aorta
  contracting substance and in some cases histamine. Thus, it appears that injection of particles
  into sensitized lung can cause the release of some, perhaps all, the mediators of anaphylaxis in
  the guinea pig. This type of response may be involved in the changes in lung function which
  follows pulmonary embolism in dogs or man. (JRH) 2197

1533
PAOLETTI, R. and C. SIRTORI
  Recent investigations on drug interaction with lipid transport.
  Journal of the American Oil Chemists' Society. 47: 84A. 1970.

  Abstract only. The authors report a line of investigation on the inhibitors of triglyceride
  hydrolysis and adipose tissue free fatty acid release with the adenyl cyclase-phosphodiesterase
  system. The activity of lipase was compared under experimental conditions involving nicotinic
  acid, antiadrenergic agents, and prostaglandins. The increased sensitivity to adrenergic
  stimulation in fasting and in essential fatty acid deficiency was investigated; in both cases, the
  role of prostaglandin is diminished or eliminated thus establishing the significance of the

antilypolytic role of prostaglandins. The significance of dietary essential fatty acids in modulating lipase response to adrenergic stimuli is discussed. (RMS) 2095

1534
PAQUOT, C.
  Acides gras et derives a activite biologique. [Fatty acids and biologically active derivatives.]
  Rivista Italiana delle Sostanze Grasse. 47: 296-307. 1970.

  Included in this extensive review of fatty acids is a short section on prostaglandins. The author is mainly concerned with their structure, but also summarizes some of their physiological aspects. (NES) 2522

1535
PATEL, N.
  Prostaglandins: Newest of the hormones?
  Annals of Internal Medicine. 73: 483-485. 1970.

  Twenty four papers are cited in a review of the literature of PGE_1, PGE_2, $PGF_{2\alpha}$, and PGA_1 covering history, cardiovascular and renal effects, induction of labor, therapeutic abortion, male fertility, treatment of asthma, blood pressure regulation, and gastric secretion. (RMS) 2177

1536
PENG, T.-C., K.M. SIX, and P.L. MUNSON
  Effects of prostaglandin E_1 on the hypothalamo-hypophyseal-adrenocortical axis in rats.
  Endocrinology. 86: 202-206. 1970.

  These experiments were undertaken to determine if prostaglandins might be the corticotropin-releasing factor of the hypothalmus which mediates ACTH release from the anterior pituitary during stress. It was found that i v injection of 0.5 μg of PGE_1 into intact rats caused an increase in adrenal cortex function. PGA_1 and $PGF_{2\alpha}$ did not have this effect even at 10 times the PGE_1 dose. PGE_1 had no ACTH stimulating effect in hypophysectomized rats. Pretreatment of intact rats with cortisol blocked the effect of PGE_1. Morphine had an inhibitory effect on PGE_1s effect on the adrenal cortex. These experiments seem to indicate that prostaglandins are not the corticotropin-releasing factor of the hypothalmus but that at least PGE_1 stimulates the release of this substance in the hypothalmus which in turn causes the release of ACTH. It was thought unlikely the effectiveness of PGE_1 was due to its vasodilator action since PGA_1, which is also a vasodilator, had no effect. (JRH) 2225

1537
PENTO, J.T., R.J. CENEDELLA and E.K. INSKEEP
  Effects of prostaglandins E_1 and $F_{1\alpha}$ upon carbohydrate metabolism of ejaculated and epididymal ram spermatozoa in vitro.
  Journal of Animal Science. 30: 409-411. 1970.

  The effects of PGE_1 and $PGF_{1\alpha}$ on oxygen consumption and glucose metabolism of ejaculated and epididymal sperm cell were studied in vitro. Samples were taken from four rams for each experiment. Prostaglandins had no effect on ejaculated sperm. Both PGE_1 and $PGF_{1\alpha}$ increased ($P < .0$ and $P > .05$, respectively) oxidation of glucose to CO_2 by epididymal sperm. $PGF_{1\alpha}$ reduced oxygen consumption by epididymal sperm ($P < .05$) while PGE_1 had no effect. (Authors) 2099

1538
PERRIER, C.V. and L. LASTER
Adenyl cyclase activity of guinea pig gastric mucosa.
Journal of Clinical Investigation. 49: 73a. 1970.

Abstract only. The effect of various biologically active substances on adenyl cyclase in mammalian stomach preparations was measured in vitro. Adenyl cyclase activity was detected in guinea pig, hamster, rat, rabbit, cat, monkey and human stomach preps. Histamine stimulated gastric cyclase activity (guinea pig) but did not stimulate cyclase activity in other organs. PGE_1 was nearly as active as histamine. It was concluded histamine or a closely related substance might be the final stimulus for gastric acid secretion. The role of prostaglandin is not clear. (JRH) 2217

1539
PHARRISS, B.B.
The effect of prostaglandins on luteal steriodogenesis.
Abstracts of the 3rd International Congress on Hormonal Steroids. 7-12 September 1970. p. 51-52.

Abstract only. The author reviews current theories on the effects of prostaglandins on steroidogenesis. (JRH) 2505

1540
PHARRISS, B.B.
The possible vascular regulation of luteal function.
Perspectives in Biology and Medicine. 13: 434-444. 1970.

It has been proposed that if the uterus could reduce the blood flow through the anterior uterine vein and subsequently the utero-ovarian vein by secreting a venoconstrictor, then conceivably the arterial perfusion of the ovary on that side could be controlled, which would then terminate luteal activity. Prostaglandin $F_{2\alpha}$, a venoconstrictor substance, infused at a rate of 1 mg/kg/day on days 5 and 6 of pseudopregnancy caused a dramatic shift in the steriod pattern of the ovaries. This shift of depressed progesterone levels and elevated 20 hydroxypregn-4-ene-3-one levels has been described as an early indication of luteal degeneration. (GW) 2280

1541
PHARRISS, B.B. and K.T. KIRTON
Prostaglandin $F_{2\alpha}$: A new contraceptive approach.
Advances in Planned Parenthood. 5: 168-169. 1970.

In this brief editorial review, the author describes the history of animal research that has led to the proposed use of $PGF_{2\alpha}$ as a method of terminating pregnancy in humans. (JRH) 2532

1542
PHARRISS, B.B.
The role of prostaglandins in obstetrics and gynecology.
Year Book of Obstetrics and Gynecology. 139-148. 1970.

In this review, the author discusses the role of prostaglandins in the menstrual cycle, abortion and induction of labor. There are 40 references cited. (JRH) 2413

1543
PHARRISS, B.B., J.C. CORNETTE, and G.D. GUTKNECHT
Vascular control of luteal steroidogenesis.
Journal of Reproduction and Fertility. Supp. 10: 97-103. 1970.

A mechanism whereby a uterine horn could affect the adjacent ovary preferentially without benefit of a blood portal system, lymphatic connection, or any other known means of local communication, has been suggested. The hypothesis makes use of a venoconstrictor which is released from the uterus during the luteal stage of the cycle. The venoconstrictor causes venospasms in the uterine and utero-ovarian veins; the latter being shared by the ovary and uterus and constituting the only route of ovarian venous drainage. A reduction in blood flow at such a critical time (increased luteal activity) could induce luteolysis by various means. This concept was tested by infusion of $PGF_{2\alpha}$ into pseudopregnant rats. $PGF_{2\alpha}$ is a specific, potent venoconstrictor and is found in uterine tissue; it caused a rapid demise of luteal metabolism and a return to estrus. It was later shown that $PGF_{2\alpha}$ is not directly toxic to luteal steroidogenesis, and that neither the pituitary nor the uterus is involved in this response. However, no evidence existed that $PGF_{2\alpha}$ could alter utero-ovarian vein blood flow. This report presents preliminary attempts to measure this parameter in rats and rabbits. Results indicate that $PGF_{2\alpha}$ can significantly reduce blood flow in the utero-ovarian vein. Further studies, however, are needed for reasons that are discussed. (ART) 2207

1544
PICKLES, V.R.
Intrauterine administration of prostaglandins.
Lancet. 2: 779. 1970.

This letter to the editor comments upon the use of intrauterine dosages of $PGF_{2\alpha}$ at one-tenth the quantity required for intravenous administration (Wiqvist and Bygdeman, Lancet Oct. 3, 1970, p. 716) noting that Embrey in 1965 showed human myometrial contractions induced by $PGF_{2\alpha}$ in quantities of 0.5mg, less by a factor of a thousand than those of Wiqvist and Bygdeman. (RMS) 2063

1545
PICKLES, V.R.
Prostaglandins.
Lancet. 1: 358. 1970.

The author, in this letter to the editor, points out that an editorial which quoted his article stated that he discovered that some prostaglandins act via adenyl cyclase. This is incorrect; his article is just a general account, intended primarily for non-biologists. (GW) 2281

1546
PIKE, J.E.
Total synthesis of prostaglandins.
Fortschritte der Chemie Organischer Naturstoffe. 28: 313-342. 1970.

This is a very complete review article on the chemical synthesis of prostaglandins. It contains sections on structure and chemical transformations of the prostaglandins, general approaches to prostaglandin synthesis, synthetic routes to structurally simplified prostaglandins, synthetic routes to prostaglandin analogs, miscellaneous synthetic approaches, and resolution of racemic prostaglandins. The bibliography has 68 entries. (JRH) 2395

1547

PIPER, P.J., S.I. SAID, and J.R. VANE

Effects on smooth muscle preparations of unidentified vasoactive peptides from intestine and
lung.

Nature. 225: 1144-1146. 1970.

Extracts of lung and small intestine from the pig contain previously unidentified vasoactive
peptides which, when given intravenously, increase peripheral blood flow and cause a systemic
hypotension, both effects being long lasting. The peptides (10^{-7} g/ml) were active on six
tissues, the rat stomach strip, chick rectum, chick rectal caecum, guinea pig trachea and
longitudinal strips of guinea pig gall bladder. The spectrum of activity distinguished them from
the known substances tested, including the prostaglandins. (GW) 2305

1548

PIPER, P.J., J.R. VANE, and J.H. WYLLIE

Inactivation of prostaglandins by the lungs.

Nature. 225: 600-604. 1970.

Prostaglandins E_1, E_2, $F_{2\alpha}$ and A_2 are enzymatically inactivated to different extents during
passage through the pulmonary circulation. The enzyme responsible is probably 15-hydroxy-
prostaglandin dehydrogenase. (Authors) 2026

1549

POLGAR, N.

Fatty-acids and related compounds.

Annual Reports on the Progress of Chemistry. Section B. Organic Chemistry. 67:
523-533. 1970.

In this review of the chemistry of fatty acids and related compounds, the syntheses of several
prostaglandins are briefly mentioned. (JRH) 2390

1550

POLIS, B.D., R.P. MILLER, A.M. PAKOSKEY, H.P. SCHWARTZ, E. POLIS, L. DREISBACH and
D.P. MORRIS

Prostaglandin induced, stress related, phospholipid changes in the rat.

Johnsville, Pa. Naval Development Center, Aerospace Medical Research Department, June 10,
1970. 23p. (NADC-MR-7006) (Bureau of Medicine and Surgery Work Unit no.
MR005.06.-0011B Report no.3)

The injections of the prostaglandin isomers (PGE_1, $PGF_{1\alpha}$, PGB_1, and PGB_x) into rats caused
changes in plasma and brain phosphatidyl glycerol and related phospholipids that mimic the
changes found in accelerated rats and in the plasma of physically or psychically stressed
humans. The prostaglandin effects on normal rat plasma phospholipids were abolished in the
hypophysectomized rat. A similar block in phospholipid change was observed in
hypophysectomized rats subjected to acceleration stress. All four prostaglandin isomers caused
significant increases in plasma and brain phosphatidyl glycerol. Differences were observed in the
effects on other phospholipids. Thus PGE_1 decreased the total plasma phospholipid and
phosphatidyl choline levels while $PGF_{1\alpha}$ increased both levels. PGE_1 caused severe symptoms
of lassitude and diarrhea in both normal and hypophysectomized rats. These effects were absent
with the other prostaglandin iosmers. In contrast, PGB_x appeared to enhance the state of well
being and lively behavior of the rats. These results in conjunction with previous work on
phospholipids in stress, implicate the prostaglandins in adaptive response to stress which
involves the mobilization of energy yielding molecular components and a gearing of metabolic
events for survival. (Authors) 2509

1551
POSNER, J.
The release of prostaglandin E_2 from the bovine iris.
British Journal of Pharmacology. 40: 163P-164P. 1970.

Prostaglandins have been extracted from the iris of sheep, cat and rabbit. The author collected bath fluid from a Krebs solution containing bovine iris and tested it on isolated preparations. Rat fundus, chick rectum and rabbit duodenum responded with contractions. Thin-layer chromatography indicated the presence of PGE_2. Release of the active material was inhibited by incubation with amyl nitrate, 2, 4-dinitrophenol, Ca^{2+} free Krebs solution, anoxia or stimulation with high pulse widths. These results suggest that the release of the prostaglandin-like substance is not a passive process. (GW) 2022

1552
POTTS, M.
The prostaglandins—a new factor in fertility control.
IPPF Medical Bulletin. 4: 1-4. 1970.

This conference traces the history of the prostaglandins. The first hurdle to be overcome in the future development of prostaglandins is to synthesize it in large quantities. It has been determined that prostaglandin E_2 is the most effective in inducing labor. The most signficant potential of prostaglandins may lie in the production of a once-a-month contraceptive since there is evidence from several animal species that prostaglandins destroy the corpus luteum. There is a need for a satisfactory technique to assay physiological quantities oof prostaglandins in experimental animals and human beings, and for intensive investigation of any possible teratoligal effect. (GW) 2044

1553
POYSER, N.L., E.W. HORTON, C.J. THOMPSON, and M. LOS
Identification of prostaglandin $F_{2\alpha}$ released by distension of guinea-pig uterus in vitro.
Journal of Endocrinology. 48: 43. 1970.

Abstract only. The uterine horns of 35 guinea pigs were removed after the animals were killed on day 3 of the estrous cycle. One horn from each animal was distended by the insertion of a polyethylene tube while the other horn was left untouched and acted as the control. Each horn was incubated in Tyrode's solution. The test sample contained the equivalent of 1 μg $F_{2\alpha}$ the contol sample the equivalent of 0.1 μg $PGF_{2\alpha}$. It is concluded that distension of the guinea pig uterus in vitro releases $PGF_{2\alpha}$. (GW) 2347

1554
PRESL, J.
Prostaglandiny. [Prostaglandins.]
Ceskoslovenska Gynekologie. 35: 383-384. 1970.

Article in Czech. Abstract not available at present. 2088

1555
PRESL, J.
Prostaglandiny. II. [Prostaglandins. II.]
Ceskoslovenska Gynekologie. 35: 436-437. 1970.

This brief article in Czech concentrates mainly on the possible roles of prostaglandins in reproductive physiology. (JRH) 2096

1556

RAMWELL, P.W. and J.E. SHAW
Biological significance of the prostaglandins.
Recent Progress in Hormone Research. 26: 139-187. 1970.

The authors develop a comprehensive working hypothesis for prostaglandin action within a number of systems to enable workers in various fields to determine the possible significance and interactions of prostaglandins in their own areas of biology and medicine. Topics discussed are: structure and metabolism; tissues in which prostaglandin release has been detected; pharmacology and interactions with the adrenal system, corpus luteum, thyroid, pancreas, platelets, and central nervous system; and effects on epithelial membranes in general and frog skin in particular. A detailed account of the mechanism of action concludes the presentation which was based on 114 references. A discussion of the paper follows; comments by B.W. O'Malley, J.D. Flack, J. Fried, J.R. Gill, J. Kowal, D.H. Solomon, R. Gaunt, K. Sterling and R. Horton and the responses of the authors are presented. The synthesis and activity of analogs and derivatives are discussed in some detail. (RMS) 2255

1557

RAMWELL, P.W.
Release of prostaglandins.
In: Eigenmann, R., ed., "Proceedings of the Fourth International Congress on Pharmacology," July, 1969. p. 32-41. Basle, Schwabe, 1970.

This review article deals with factors affecting the release of prostaglandins. (JRH) 2437

1558

ROBERTS, G., A. ANDERSON, J. McGARRY, and A.C. TURNBULL
Absence of antidiuresis during administration of prostaglandin $F_{2\alpha}$.
British Medical Journal. 2: 152-154. 1970.

The authors compare the antidiuretic effects of $PGF_{2\alpha}$ and oxytocin in a series of 6 patients admitted for therapeutic abortion. Results show that oxytocin in dosages varying from 8 to 120m U. (B.P.)/minute induced pronounced antidiuresis while $PGF_{2\alpha}$ at 0.05mg/kg/min produced none. One patient receiving $PGF_{2\alpha}$ vomited 3 times during 4 3/4 hours of infusion. Discussion notes that advantages of $PGF_{2\alpha}$ over oxytocin must be shown before it becomes more widely used. In this case an absence of risk of water intoxication is demonstrated. (RMS) 2104

1559

ROBERTS, G.
Induction of labour using prostaglandins.
Journal of Reproduction and Fertility. 23: 370-371. 1970.

Abstract only. PGE_1 and $PGF_{2\alpha}$ were used to induce labor in term pregnancies. PGE_1 was given at a rate of 1.5 to 3 μg/min and $PGF_{2\alpha}$ at 3 to 6 μg/min for 3 to 18 hours. Labor was successfully induced in all cases. In six patients the effects of $PGF_{2\alpha}$ were compared with the use of oxytocin. It was found that $PGF_{2\alpha}$ did not have the pronounced antidiuretic effect that oxytocin does and therefore might be the method of choice especially in cases of pre-eclampsia or chronic renal disease where the antidiuretic effect might be harmful. (JRH) 2013

1560
ROBISON, G.A.
  Cyclic AMP as a second messenger.
  Journal of Reproduction and Fertility. Supp. 10: 55-74. 1970.

  The catecholamines and the prostaglandins are at present the only hormones known which
  produce some of their effects by way of an increase in the level of cyclic adenosine
  monophosphate (cAMP) and others by way of a decrease. Certain aspects of the role of cAMP
  as a second messenger are discussed. Prostaglandins and insulin are capable of producing a
  dramatic fall in the intracellular level of cAMP when added to intact fat cells. It is presumed
  that the effect must involve either inhibition of adenyl cyclase or stimulation of the
  phosphodiesterase, although neither effect could be demonstrated when the hormones were
  added to broken cell preparations. It would appear, therefore, that these hormones exert a
  primary action which secondarily influences the accumulation of cAMP. The nature of this
  primary action is unknown, however. PGE_1 has been used in the presence of theophylline to
  increase the level of cAMP above its normal baseline in a study of the cAMP-lowering effect of
  alpha-adrenergic agonists in human blood platelets, where such agonists promote aggregation.
  PGE_1 stimulates platelet adenyl cyclase while PGE_2 has a lower potency as an inhibitor of
  platelet aggregation. The effects of epinephrine and PGE_1 on the level of cAMP in platelets are
  just the opposite of their effects in rat adipocytes. The increase in the level of cAMP in
  adipocytes in response to epinephrine is mediated by beta-receptors, while the fall in the level
  of cAMP in platelets is mediated by alpha-receptors. Whether the receptors mediating the
  divergent effects of the prostaglandins can be associated with different surface characteristics
  seems much less clear. This has not been studied in relation to the divergent effects of these
  hormones on the level of cAMP. (ART) 2206

1561
ROBISON, G.A., M.J. SCHMIDT, and E.W. SUTHERLAND
  On the development and properties of the brain adenyl cyclase system.
  In: Greengard, P. and E. Costa, eds., "Role of Cyclic AMP in Cell Function: Advances in
    Biochemical Psychopharmacology 3," p. 11-30.
  New York, Raven Press. 1970.

  The author summarizes currently available information about adenyl cyclase and mentions that
  the observed effects of the prostaglandins on rat brain cyclic AMP levels have to date been
  small and variable. (GW) 2365

1562
ROBISON, G.A. and E.W. SUTHERLAND
  Sympathin E, sympathin I, and the intracellular level of cyclic AMP.
  Circulation Research. 27(supp. 1): I 147-161. 1970.

  Cyclic AMP is a versatile regulatory agent that acts to control the rate of a number of cellular
  processes. The level of cyclic AMP is determined by the activity of at lease two enzymes. One,
  adenyl cyclase, is the catylist for the formation of cAMP from ATP. The most striking feature
  of adenyl cyclase obtained from the cells of multicellular organisms is its sensitivity to
  stimulation by hormones. One of the stimulators listed is PGE_1 when used with thyroid slices.
  Studies on the formation and metabolism of cAMP, on the role of cAMP in hormonal
  responses, and on the mechanism of action of cAMP are reviewed. (CWS) 2260

1563
ROSEN, O.M.
Preparation and properties of a cyclic 3',5'-nucleotide phosphodiesterase isolated from frog erythrocytes.
Archives of Biochemistry and Biophysics. 137: 435-441. 1970.

This communication describes the purification and properties of a cyclic 3',5'-nucleotide phosphodiesterase (cPDE) isolated from the frog erythrocyte. cPDE activity was measured by the rate of formation of ^{3}H-5'-AMP from ^{3}H-cAMP. Prostaglandin E_1 at concentrations ranging from 10^{-8} to 10^{-6}M had no effect on the activity of cPDE. At higher concentrations slight inhibition was noted. (GW) 2351

1564
ROSEN, O.M., E.N. GOREN, J. ERLICHMAN, and S.M. ROSEN.
Synthesis and degradation of cyclic 3',5'-adenosine monophosphate in frog erythrocytes.
In: Greengard, P. and E. Costa, eds., "Role of Cyclic AMP in Cell Function. Advance in Biochemical Psychopharmacology 3." pp. 31-50. New York, Raven Press. 1970.

The studies summarized here describe the development and basic properties of the enzyme systems which synthesize and degrade cyclic AMP in the frog erythrocyte. In the absence of fluoride ions the addition of catecholamines activated the adenyl cyclase which was present in hemolysates of adult frog erythrocytes. The addition of prostaglandin (E_1, E_2, A_1, B_1, $F_{1\alpha}$) (0.1-25 μg/ml) did not influence adenyl cyclase activity. Prostaglandin E_1 (1.4×10^{-6}M) inhibited the hydrolysis of cyclic AMP. (GW) 2366

1565
ROSENTHALE, M.E., A. DERVINIS, A.J. BEGANY, M. LAPIDUS, and M.I. GLUCKMAN
Bronchodilator activity of prostaglandin E_2 when administered by aerosol to three species.
Experientia. 26: 1119-1121. 1970.

This study evaluates the response of the guinea pig, dog and monkey to the bronchodilator properties of prostaglandin E_2. It was shown that PGE_2 in the form of an aerosol acts as a bronchodilator in these species. Its effectiveness, natural origin and fast metabolism speak for its use as a bronchodilatory aerosol. (GW) 2352

1566
ROSENTHALE, M.E., A. DERVINIS, and J. KASSARICH
Further studies on the bronchodilator properties of the prostaglandins PGE_1 and PGE_2.
Pharmacologist. 12: 264. 1970.

Abstract only. The bronchodilating properties of PGE_1, PGE_2, and isoproterenol are reported when administered i v or by aerosol in the cat. An aerosol dose of 0.01 mcg of PGE_1 or PGE_2 is sufficient to significantly reverse the increased pulmonary resistance and decreased compliance brought about by i v administration of succinylcholine and neostigmine. PGE_1 and PGE_2 were 10 to 100 times as potent as isoproterenol by the aerosol route, whereas isoproterenol was 10 to 50 times as potent as the prostaglandins when administered i v PGE_1 and PGE_2 are potent bronchodilator substances when administered by aerosol to cats with cholinergic-induced bronchoconstriction. (GW) 2189

1567

ROTH-BRANDEL, U., M. BYGDEMAN, and N. WIQVIST

A comparative study on the influence of prostaglandin E_1, oxytocin and ergometrin on the pregnant human uterus.

Acta Obstetricia et Gynecologica Scandinavica. 49: 1-7. 1970.

The influence of prostaglandin E_1 (PGE_1), oxytocin and methyl ergometrin maleate (Methergin®) on the contractility of the pregnant human uterus was compared in in vivo experiments on 8 early-pregnant and 14 midpregnant women. The substances were administered intravenously in separate injections. With the doses used PGE_1 and oxytocin caused a marked elevation of tone whereas Methergin® had a less significant effect. The duration of the elevation of tone after PGE_1 was approximately three times longer than that after oxytocin but considerably shorter than after Methergin®. The motility pattern following Methergin® was characterized by small irregular contractions which differed completely from that after PGE_1 and oxytocin. The pharmacological effect of PGE_1 on the midpregnant uterus suggests that this substance might reduce blood loss and shorten the period of time for placental separation during the third stage of labour. However, this assumption could not be confirmed as judged from a small clinical pilot study carried out on 188 women. (Authors) 2339

1568

ROTH-BRANDEL, U., M. BYGDEMAN, and N. WIQVIST

Effect of intravenous administration of prostaglandin E_1 and $F_{2\alpha}$ on the contractility of the non-pregnant human uterus in vivo.

Acta Obstetricia et Gynecologica Scandinavica. 49: 19-25. 1970.

The effect on uterine contractility of prostaglandin E_1 (PGE_1) and $F_{2\alpha}$ ($PGF_{2\alpha}$) was studied in 19 non-pregnant women. Uterine motility was recorded by the micro-balloon method and the prostaglandins administered by single intravenous injections in increasing doses. Doses above threshold levels of both PGE_1 and $PGF_{2\alpha}$ always resulted in stimulation of the uterus primarily in terms of tone elevation. A short drop of tone during 40-60 seconds immediately following the injection was obtained in a few experiments after high doses of PGE_1. This response preceeded the normal stimulatory effect. The threshold dose for PGE_1 was approximately 20 μg and that for $PGF_{2\alpha}$ around 50 μg or lower. The non-pregnant uterus is probably more sensitive to $PGF_{2\alpha}$ than the midpregnant uterus. The number of experiments were too few to allow conclusions as to differences in uterine sensitivity during the various phases of the menstrual cycle. Administration of oxytocin and vasopressin did not enhance the uterine sensitivity to stimulation by PGE_1. (Authors) 2341

1569

ROTH-BRANDEL, U. and M. ADAMS

An evaluation of the possible use of prostaglandin E_1, E_2 and $F_{2\alpha}$ for induction of labour.

Acta Obstetricia et Gynecologica Scandinavica. 49: 9-17. 1970.

The stimulatory effect on uterine contractility of prostaglandin E_1, E_2, and $F_{2\alpha}$ was studied in 29 term-pregnant women and 4 midpregnant volunteers admitted for therapeutic abortion. The substances were administered intravenously and uterine contractility recorded by measuring the amniotic pressure. In a careful analysis of the quantitative tracings it was found that suitable doses of the prostaglandins induce efficient uterine contractions. However, these contractions are characterized by a comparatively low intensity in relation to frequency and also occurrence of incoordinated contraction complexes. The results do not support the opinion that the prostaglandins are superior to oxytocin for induction of labour at term pregnancy. (Authors) 2340

..

408

1570

ROTH-BRANDEL, U., M. BYGDEMAN, N. WIQVIST, and S. BERGSTROM
Prostaglandins for induction of therapeutic abortion.
Lancet. 1: 190-191. 1970.

This letter to the editor suggests a method for inducing therapeutic abortion after the 12th week of pregnancy. It would be preferable to induce labor-like uterine contractions by intravenous or subcutaneous injection of a substance with strong oxytocic activity. Intravenous infusions of prostaglandin E_1 or prostaglandin $F_{2\alpha}$ induced considerable uterine hypertonicity and strong, labor-like contractions appeared. Subcutaneous injections caused a similar contractility pattern. Pregnancy was terminated in 3 out of 11 women. Administration of sufficient prostaglandin to produce long-term uterine hyperactivity will probably result in explusion of the conceptus in most cases. Subjective side effects (slight nausea with PGE_1 and diarrhea with $PGF_{2\alpha}$) were noted in some of the patients on high doses. (GW) 2349

1571

RUBIN, R.P.
The role of calcium in the release of neurotransmitter substances and hormones.
Pharmacological Reviews. 22: 389-428. 1970.

The author briefly mentions that it is unlikely that prostaglandins have any direct role in the calcium-dependent secretory process. (JRH) 2433

1572

RUDICK, J., M. GONDA, and H.D. JANOWITZ
Prostaglandin E_1: an inhibitor of electrolyte and stimulant of enzyme secretion in the pancreas.
Federation Proceedings. 29: 445abs. 1970.

Abstract only. Five dogs with duodenal and gastric fistulae were studied (a) in the resting state; (c) during secretin and pancreozymin infusion; (b) during intravenous infusion of secretin; and (d) during single injections of secretin or secretin and pancreozymin. PGE_1 was administered by single intravenous injection and by continuous i v infusion. In the secretin and secretin plus pancreozymin-stimulated pancreas, PGE_1 in single injections decreased volume of pancreatic secretion and, at higher dosage, decreased HCO_3 concentration. Continuous PGE_1 infusion inhibited volume and HCO_3, almost completely suppressing secretion at dosage of 5 μ/kg/min. Enzyme output was stimulated by PGE_1. Thus PGE_1 has a dual effect on the pancreas—inhibition of electrolytes and stimulation of enzymes, of which it is a partial agonist. (ART) 2117

1573

SABATINI-SMITH, S.
Action of prostaglandins E_1 and $F_{2\alpha}$ on calcium (CA) flux in the isolated guinea pig atria and fragmented cardiac sarcoplasmic reticulum.
Pharmacologist. 12: 239. 1970.

Abstract only. The objective of the present study was to examine the effects of PGE_1 and $PGF_{2\alpha}$ (1 μgm/ml) on Ca transport in the rate controlled guinea pig left atria and isolated cardiac sarcoplasmic reticulum. PGE_1 increased contractile tension 62.3±8.0% (± S.E.); this was associated with an uptake of 8.9158±0.5387 μmol Ca (mgm tissue/min ($\times 10^{-6}$). Control uptake was 3.9265±0.5605 μmol Ca (mgm tissue/min ($\times 10^{-6}$) (P<0.01). $PGF_{2\alpha}$ increased contractile tension 35.9±19.2% with a concomitant uptake of 6.7903±1.3720 μmol Ca/mgm tissue/min ($\times 10^{-6}$) ($P<0.05$). Total tissue Ca did not differ significantly from control. Incubation of PGE_1 and $PGF_{2\alpha}$ for 10 min significantly enhanced the uptake of Ca by fragments of cardiac sarcoplasmic reticulum. These results would indicate that the positive inotropic effect produced by these two prostaglandins is mediated by an increase in the Ca available to the contractile apparatus. (Author) 2193

1574
SAID, S.I. and V. MUTT
Long-acting vasoactive peptides from lung and small intestine.
Clinical Research. 18: 72. 1970.

Abstract only. Authors briefly mention that vasoactive peptides they isolated from the hog gut produced as prolonged vasodilation as PGE_1 and PGE_2 produce. (JRH) 2166

1575
SAID, S.I. and V. MUTT
Potent peripheral and splanchnic vasodilator peptide from normal gut.
Nature. 225: 863-864. 1970.

The extraction from the small intestine of the hog of a highly active vasodepressor peptide substance is reported. Purification of the gut extract included the removal of digestive hormones and prostaglandins. Intraarterial injection of the extract into a dog caused a marked increase in femoral arterial blood flow. This effect was qualitatively similar to that of prostaglandin E_1 and E_2. An intravenous infusion of the extract caused a fall in arterial blood pressure and an increase in peripheral blood flow. Infusion of the extract into the inferior vena cava increased the blood flow in the splanchnic vessels along with lowering the arterial blood pressure. The strong vasodilator action of the extract on the peripheral and splanchnic circulations raises the possibility that it might have a physiological role in the regulation of blood flow in the digestive organs. (GW) 2304

1576
SAID, S.I.
Pulmonary responses to inhaled particles: role of vasoactive substances.
Archives of Internal Medicine. 126: 475-476. 1970.

The prostaglandins are reported as being among the biologically active compounds which are concentrated in the lung. In anaphylaxis there is evidence for the release of prostaglandins. (GW) 2346

1577
SAITO, K.
[The effects of prostaglandins on lipolytic metabolism.]
Shinryo. 23: 1328-1333. 1970.

Article in Japanese. Abstract not available at present. 2398

1578
SALM, R. and M.A. VOYCE
Renal papillary necrosis in a neonate (with an hypothesis as to its aetiology).
British Journal of Urology. 42: 277-283. 1970.

In the discussion of the etiology of renal papillary necrosis observed at necropsy in this infant who died at the age of 15 days, it is postulated that, since such conditions associated with renal papillary necrosis as acidosis, metabolic diseases (diabetes), collagen diseases, mechanical obstruction of urinary tract, toxicity (bacterial agents, drugs), and diminished blood flow, cannot be held to be directly responsible for the production of renal papillary necrosis, it is possible they may do so indirectly by acting as a release mechanism for substances in the outer renal medulla which reduce the supply of arterial blood to the pyramids. Such substances have of late been identified; they are the prostaglandins. The structure, properties and distribution of the prostaglandins are discussed briefly. In view of the selective distribution of necrosis in renal papillary necrosis, the known pattern of the arterial blood supply, and the newly discovered properties and intrarenal localization of vasodepressor lipids, a working hypothesis is developed to the effect that conditions associated with renal papillary necrosis stimulate production of prostaglandins in the outer medulla. As a consequence, the straight arteries become grossly dilated, the local blood pressure, which is normally below that of the systemic circulation, drops even further and leads to hemostasis, causing medullary ischemia and necrosis. (ART) 2250

1579
SAMUELSSON, B.
Biosynthesis and metabolism of prostaglandins.
In: Eigenmann, R., ed., "Proceedings of the Fourth International Congress on Pharmacology," July, 1969. p. 12-31. Basle, Schwabe, 1970.

This article is a comprehensive review of the literature on the biosynthesis and metabolism of prostaglandins. There are 45 references cited. (JRH) 2438

1580
SAMUELSSON, B., M. HAMBERG, and C.C. SWEELEY
Quantitative gas chromatography of prostaglandin E_1 at the nanogram level: use of deuterated carrier and multiple-ion analyzer.
Analytical Biochemistry. 38: 301-304. 1970.

Prostaglandin E_1 was used as a sample in the demonstration of a novel procedure for quantitative gas chromatography at the nanogram level involving the addition of a deuterated carrier and determination of the protium-deuterium ratio by a combination of gas chromatography-mass spectrometry. (RMS) 2184

1581
SANDER, S.
Prostaglandiner. En ny gruppe hormoner? [Prostaglandins. A new group of hormones?]
Tidsskrift for den Norske Laegeforening. 90: 1068-1073. 1970.

The comprehensive review article in Norwegian covers all areas of prostaglandin research. The bibliography lists 31 articles from 1949 to 1968. (JRH) 2097

1582
SANO, T.
[The effects of prostaglandins on fatty tissue.]
Shinryo. 23: 1352-1363. 1970.

Article in Japanese. Abstract not available at present. 2400

1583

SALZMAN, E.W., E.B. RUBINO and R.V. SIMS

Cyclic 3',5'-adenosine monophosphate in human blood platelets. III.

The role of cyclic AMP in platelet aggregation.

Series Hemotologica. 3: 100-113. 1970.

The relation of cyclic 3',5'-AMP to platelet function has been studied by investigating the influence of this compound and of its N^6-2'-0-dibutyryl derivative on platelet aggregation and other aspects of platelet behavior and by studying the adenyl cyclase and phosphodiesterase activity of homogenized platelets and the cyclic AMP content of intact platelets. Dibutyryl cyclic AMP and cyclic AMP inhibited platelet aggregation induced by ADP, epinephrine, collagen, and thrombin. The platelet release reaction was also inhibited: specifically, there was inhibition of the induction of platelet factor 3 activity and of the release of labelled 5-hydroxytryptamine. Platelet swelling produced by ADP was not inhibited. The activity of dibutyryl cyclic AMP was associated with uptake of the compound by platelets. Platelet cyclic AMP was increased by prostaglandin PGE_1 by stimulation of adenyl cyclase and by caffeine by inhibition of phosphodiesterase. Epinephrine reduced platelet cyclic AMP by inhibition of adenyl cyclase, and α-adrenergic effect. NaF increased adenyl cyclase of platelet homogenates but did not alter the cyclic AMP content of platelets. Adenyl cyclase activity was inhibited by collagen, 5'-hydroxytryptamine, and thrombin. ADP did not alter adenyl cyclase activity. (Authors) 2493

1584

SAMBHI, M.P. and C.E. WIEDEMAN

In vitro influence of prostaglandins on angiotensin generation by human renin.

Clinical Research 18: 124. 1970.

PGE_1, PGE_2, and PGA_1 (5 to 100 ng/ml) of human plasma were incubated with a constant amount of human renin for 20-180 min. The amount of angiotensin produced was measured. It was found that under these conditions the prostaglandins tested did not influence the amount of angiotensin produced. (JRH) 2164

1585

SCHOFIELD, J.G.

Prostaglandin E_1 and the release of growth hormone in vitro.

Nature. 288: 179-180. 1970.

The effect of PGE_1 on the release of growth hormone was tested by incubating pituitary slices obtained from heifers with PGE_1 and theophylline. At the end of the incubation period (45 min), the slices were dried, weighed and assayed for growth hormone. In the absence of theophylline, PGE_1 increased the release of growth hormone by about 30% in the largest concentration tested, but had no effect in lower concentrations. In the presence of theophylline, the release of growth hormone was increased at a PGE_1 concentration of 5×10^{-8} M. This stimulation of growth hormone in vitro by PGE_1 demonstrates that the stimulation by PGE_1 is not specific to one cell type or to one hormone. (GW) 2357

1586

SCHOR, J.M.

Agents affecting thrombosis.

In: Cain, C.K., ed., "Annual Reports in Medicinal Chemistry," 237-245. New York, Academic Press, 1970.

Prostaglandins are mentioned among the most potent inhibitors of platelet aggregation known. (JRH) 2455

1587
SCHUTTERLE, G. and P. DICKER
Renale hypertonie. [Renal hypertension.]
Munchener Medizinische Wochenschrift. 112: 2129-2135. 1970.

In this review article on renal hypertension it is briefly mentioned that the role of prostaglandins in the kidney is not known. (JRH) 2106

1588
SCHWEPPE, J.S. and R.A. JUNGMANN
Prostaglandins: The inhibition of hepatic cholesterol ester synthesase in the rat.
Proceedings of the Society of Experimental Biology and Medicine. 133: 1307-1309. 1970.

The effect of the prostaglandins E_1 and $F_{1\alpha}$ on the in vitro biosynthesis of cholesteryl palmitate, oleate, and linoleate by rat liver microsomes from cholesterol and free fatty acids was studied. Both prostaglandins exerted an inhibitory action on the enzymatic esterification of cholesterol. The degree of inhibition increased with increasing concentrations of PGE_1 and $PGF_{1\alpha}$ in the incubation medium. At the lowest concentration used ($4.6 \times 10^{-8} M$) cholesteryl linoleate synthesis showed the least inhibition (about 15%) while cholesteryl palmitate formation was inhibited by about 25%. At higher concentrations of PGE_1 and $PGF_{1\alpha}$ used, no difference in the degree of inhibition of the individual esters was observed. (Authors) 2243

1589
SCOTT, R.E.
Effects of prostaglandins, epinephrine and NaF on human leukocyte, platelet, and liver adenyl cyclase.
Blood. 35: 514-516. 1970.

The effect of adenyl cyclase stimulators on the cAMP levels in homogenized human leucocytes, platelets, and liver cells was measured. The compounds tested were PGE_1, $PGF_{1\beta}$, epinephrine and NAF. No cAMP was found in untreated leucocyte and platelet homogenates. PGE_1 (200 μg/ml) produced the greatest increase in cAMP in leucocytes and platelets, (150 nM/10^6 leucocytes and 6.5 nM/10^6 platelets.). $PGF_{1\beta}$ was much less effective at the same concentration 3.5 nM/10^6 leucocytes and 1.0 nM/10^6 platelets. NaF ($1 \times 10^{-2} M$) produced an increase (8.5 nM/10^6 leucocytes and 3.5 nM/10^6 platelets). Epinephrine was interesting in that it increased the cAMP in leucocytes (7.0 nM/10^6 leucocytes) but produced little effect in platelets (0.5 nM/10^6 platelets). Prostaglandins were not tested against liver cells. (JRH) 2249

1590
SCOTT, R.S. and P.I.C. RENNIE
Factors controlling the life-span of the corpora lutea in the pseudopregnant rabbit.
Journal of Reproduction and Fertility. 23: 415-422. 1970.

Female rabbits were hysterectomized on day 12 of an experimentally induced pseudopregnancy. On days 15 to 18 they were injected i p twice daily with 0.25 mg of $PGF_{2\alpha}$. These rabbits were killed and autopsied on day 21. When compared to controls which were hysterectomized at the same time, it was found that the $PGF_{2\alpha}$ caused rapid leuteal regression. It was also found the hysterectomy prolonged the life of corpus leutea both ovarian and transplanted indicating a leutolytic factor was produced by the uterus. However, this leutolytic substance was not proven to be prostaglandin. (JRH) 2092

1591
SEKHAR, N.C.
Effect of eight prostaglandins on platelet aggregation.
Journal of Medicinal Chemistry. 13: 39-44. 1970.

Eight prostaglandins, PGE_1, PGE_2, PGA_1, PGA_2, $PGF_{1\alpha}$, $PGF_{2\alpha}$, $PGF_{2\beta}$, were tested for their effect on platelet aggregation-adhesion induced by ADP, thrombin, and collagen. All compounds inhibited aggregation in platelet-rich rat plasma and human plasma to varying degrees. PGE_1 was the most active compound in the group. In addition, PGE_1 showed very potent thrombolytic effect against ADP-induced platelet thrombi in vitro. A single intravenous injection of 3 mg of PGE_1/kg in rats inhibited platelet aggregation in blood samples withdrawn from animals 30 min following the injection. Platelet aggregation was also inhibited significantly in rats given infusion of PGE_1 at 1.8 mg/kg per day for 30 days. (Authors) 2314

1592
SEKHAR, N.C.
Prostaglandins and platelet function.
Thrombosis and Diathesis Haemorrhagica. Supp. 42: 305-314. 1970.-

Eight prostaglandins were studied for their effect on platelet aggregation. PGE_1 was found to be the most active compound in the group. A comparison of the activities on platelets, blood pressure, and smooth muscle showed that there is a split between the biological activities of the prostaglandins. PGE_2 revealed a biphasic response, inhibiting platelet aggregation at high concentrations and potentiating ADP-induced rat platelet aggregation at low concentrations. PGE_1 produced a significant prolongation of whole blood clotting time of blood samples from man, dog, rabbit and rat. PGE_1 also showed a significant inhibition of clot retraction. This compound did not produce any significant effect on fibrinolysis in rats given a single injection or a continuous infusion. PGE_1 is not inactivated by any of the blood components for at least 72 hours at room temperature. On incubation with blood, PGE_1 binds to albumin and cannot be dialysed out. Albumin-bound PGE_1 is stable and is as potent as the free PGE_1 itself, against platelet aggregation. (Author) 2321

1593
SELLNER, R.G. and E.W. WICKERSHAM
Effects of prostaglandins on steroidogenesis.
Journal of Animal Science. 31: 230. 1970.

Abstract only. Nine bovine corpora lutea of pregnancy were used to determine the effects of prostaglandins (PG) E_1, E_2, and $F_{2\alpha}$ on steroidogenesis during 2-hr in vitro incubation. De novo progesterone production by each of three glands was determined for each prostaglandin at concentrations of .001, .100, and 10.0 μg PG per ml of KRB plus glucose medium; alone and in combination with luteinizing hormone (NIH-LH-B5, 0.05 μg/ml medium). Analyses of variance showed the overall mean de novo progesterone production by luteal tissue incubated with PG's alone to be significantly greater ($P<.05$) than the mean for incubated control tissue. Differences in mean responses between the three PG's were not statistically significant. There was a highly significant (P<.01) dose-response relationship between PG concentrations and de novo progesterone production. Mean response to combinations of PG and LH was greater (P<.05) than mean response to PG alone. (Authors) 2087

1594
SHEPPARD, H., and C.R. BURGHARDT
The stimulation of adenyl cyclase of rat erythrocyte ghosts.
Molecular Pharmacology. 6: 425-429. 1970.

In the work reported here, it was shown that catecholamines and PGE_2 were the only hormones that stimulated the production of cyclic adenosine monophosphate by stimulating rat erythrocyte adenyl cyclase activity. (ART) 2208

1595
SHIMIZU, H., C.R. CREVELING, and J.W. DALY
The effect of histamines and other compounds on the formation of adenosine 3',5'-monophosphate in slices from cerebral cortex.
Journal of Neurochemistry. 17: 441-444. 1970.

Incubation of labelled cerebral slices of rabbit or guinea pig in the presence of prostaglandin E_1 (2.0 mM) for 10 or 15 minutes failed to change levels of cyclic $[^{14}C]$ AMP by more than 30 percent. (GW) 2315

1596
SHIO, H., A.M. PLASSE and P.W. RAMWELL
Platelet swelling and prostaglandins.
Microvascular Research. 2: 294-301. 1970.

The swelling and subsequent aggregation of rat platelets induced by ADP was systematically studied. Platelet stickiness occurred during the swelling phase only and was transient. Prostaglandin E_1 (PGE_1), which is one of the strongest inhibitors of platelet aggregation, (1) inhibited ADP-induced swelling of platelets and (2) stimulated recovery process from the swollen state without affecting metabolism of added ADP. PGE_2 had no significant effect on cell swelling in contrast to its effect in promoting aggregation. The significance of PGE_1 on platelet behavior is discussed with reference to platelet adenyl cyclase and the platelet contractile system. (Authors) 2527

1597
SHIO, H., P.W. RAMWELL, N.H. ANDERSEN, and E.J. COREY
Stereospecificity of the prostaglandin 15-dehydrogenase from swine lung.
Experientia. 26: 355-357. 1970.

The stereochemical requirements of swine lung prostaglandin-15-dehydrogenase using synthetic substrates is explained with the object of determining whether structural modifications would reduce the rate of biological degradation without reducing the pharmacological activity. The racemic 15-epimer of PGE was not dehydrogenated by the enzyme. A 15 (S)-hydroxyl group appears to be a major stereo structural requirement for both smooth muscle stimulating activity and dehydrogenase activity. (GW) 2130

1598

SIGGINS, G.R., D.J. WOODWARD, B.J. HOFFER, and F.E. BLOOM
    Responsiveness of cerebellar Purkinje cells (P-cells) to norepinephrine (NE), cyclic AMP (cAMP),
        and prostaglandin E_1 (PGE_1) during synaptic morphogenesis in neonatal rat.
    Pharmacologist. 12: 198. 1970.

Abstract only. The development of cerebellar Purkinje cells in newborn rats was followed and
the effect of norepinephrine, cAMP, serotonin, γ-amino butyrate and glutamate on their activity
during development was followed physiologically and morphologically (electron microscope). It
was found that norepinephrine (NE) was highly active in these cells before and after synapse
formation. PGE_1 selectively blocked the effects of NE in these cells as it does in the adult.
(JRH) 2194

1599

SIH, C.J., C. TAKEGUCHI, P. FOSS
    Mechanism of prostaglandin biosynthesis. III. Catecholamines and serotonin as coenzymes.
    Journal of the American Chemical Society. 92: 6670. 1970.

This communication presents experimental evidence to show that catecholamines and serotonin
can act as coenzymes by providing the reducing equivalents for prostaglandin synthesis. Both
serotonin and 5-hydroxyindole-3-acetic acid (5-OH-IAA) were capable of supporting
prostaglandin synthesis. This was confirmed when approximately 2 mole of 5-OH-IAA was
consumed per mole of prostaglandin formed thus showing that 5-OH-IAA is the source of
reducing equivalents. Since catecholamines, serotonin and prostaglandins are widely distributed
throughout various tissues, one could speculate the catecholamines and serotonin may be the
natural coenzymes in regulating the in vivo biosynthesis of prostaglandins. (GW) 2288

1600

SIOUFI, A., F. PERCHERON, and J.E. COURTOIS
    L'AMP-cyclique. [Cyclic AMP.]
    Pathologie et Biologie. 18: 79-94. 1970.

This review article in French concerning cyclic AMP mentions that the prostaglandins have a
double action: 1) light stimulation of lipolysis in the absence of lipolytic hormones; and 2)
antagonism in the presence of these hormones. According to Bergstrom, the prostaglandins seem
to act as a general modulator of some reactions with adenyl cyclase in the different tissues.
(GW) 2379

1601

SJOSTRAND, N.O. and G. SWEDIN
    Potentiation by various smooth muscle stimulants of an isolated sympathetic nerve-seminal
        vesicle preparation from the guinea-pig.
    Acta Physiologica Scandinavica. 80: 172-177. 1970.

The contractions of the seminal vesicle in response to hypogastric nerve stimulation or coaxial
stimulation were recorded with a balloon inserted in the lumen of the seminal vesicle.
Adrenaline (A), noradrenaline (NA), acetylcholine (Ach), histamine (Hi), serotonin (5-HT),
barium chloride and high concentrations of angiotensin were found to potentiate the motor
response of the seminal vesicle to sympathetic nerve stimulation. This potentiation occurred
with concentrations 10-100 times smaller that those giving a direct contraction of the organ.
Prostaglandins (PGE_1, PGE_2 and $PGF_{2\alpha}$) in low concentrations (1-10 ng/ml) exerted no effect

or caused a slight decrease in the response to sympathetic nerve stimulation, while higher concentrations (100 ng-1 mg/ml) enhanced the response to nerve stimulation. This effect was pronounced with PGE_1, but rather weak with the other prostaglandins. Oxytocin, vasopressin and felypressin were without effect on the seminal vesicle preparation. It is concluded that the seminal vesicle responds to smooth muscle stimulants in a similar way as the vas deferens. (Authors) 2038

1602
SKOREPA, J.
Metabolismus Mastnych kyselin. III. Prostaglandiny.
Prakticky lekar. 50: 765-766. 1970.

Article in Czech. Abstract not available at present. 2385

1603
SMEJKAL, V., O. ROUGIER, and D. GARNIER
The influence of adrenergic and some other cardioactive drugs on electric activity of the isolated frog atrium.
Physiologica Bohemoslovaca. 19: 23-25. 1970.

The effect of some adrenergic and other cardioactive drugs in the isolated atrium trabeculae of the frog was studied in vitro. The catecholamine action is opposite to that of acetylcholine. The catecholamine threshold decrease seems to be of a mixed alpha-beta-type, because there is not a great difference between the alpha- and beta-adrenomimetics in this effect; the beta-adrenomimetic active drugs (isoprenaline, adrenaline) increase the height and duration of action potential. Prostaglandin E_1 mimics the catecholamine action, but its effect on the electric threshold is masked by the presence of the pharmacological solvent (ethanol). Procainamide antagonizes the isoprenaline decrease in electric threshold. (Authors) 2424

1604
SMEJKAL, V., J. LENFANT, and Y.M. GARGOUIL
Search for a cardioresuscitating drug (microelectrode in isolated rabbit atrium).
Physiologia Bohemoslovaca. 19: 350-351. 1970.

Abstract only. The classical microelectrode method in the isolated rabbit atrium (1) can be used for studying the effect of cardioactive drugs, especially those which are able to restore arrested mechanical and electrical cellular activities after various physical and chemical agents. Anoxia, unchanged physiological solution with fatigue tissue metabolites, Mn^{++}, changes in pH and different drugs can produce a halt in activity. On the other hand, adrenaline is a relatively effective medium for restitution of electrical atrium action potentials (sinoatrial potentials) and of mechanical phenomena. Other adrenergic drugs such as methoxamine, noradrenaline, isoprenaline, orciprenaline, other substances such as angiotensamid, prostaglandin E_1 have a weaker or no effect. Adrenaline acts after Mn^{++} or in anoxia with an effective concentration of the order of $10^{-8}M$ per 15 ml bath. The pacemaker activity once started persists with no further adrenaline or electrical stimulation. On the other hand, in the case of Mn^{++} intoxication, the stimulation of the pacemaker zone with rectangular pulses of 5 volts, frequency 1.5 c/sec, reintroduces the electrical and mechanical heart activities in a more potent way than the adrenaline. But this activity disappears as soon as the stimulation is interrupted. In the presence of propranolol 10^{-6} to $10^{-5}M$ pro bath, the effect of adrenaline fails and no effect is enregistered after electro-stimulation. General effects of cardioactive drugs on transmembrane heart activity are discussed, then the experimental model is compared to the clinical situation (aarhythmia precedes the heart stoppage). (Authors) 2375

1605

SMITH, J.B. and A.L. WILLIS
Function and release of prostaglandins by platelets in response to thrombin.
British Journal of Pharmacology. 40: 545P-546P. 1970.

This study shows that human platelets under the influence of thrombin form and release two prostaglandins, PGE_2 and $PGF_{2\alpha}$. Platelets in buffered saline solution were shaken for 5 minutes with bovine thrombin, sedimented by centrifugation and the supernatent decanted. Samples were cooled to $4°C$, adjusted to pH 3, extracted twice with ethyl acetate and evaporated to dryness. The dry extracts were dissolved in Krebs solution and assayed. Reconstituted extracts were re-extracted, dried, dissolved in ethanol and chromatographed in the AI, AII, and CII solvent systems. Extracts in the AI system showed activity consistent with E and F prostaglandin types. Activity in the AII and CII systems was located in zones corresponding to PGE_2 and $PGF_{2\alpha}$. Control suspensions contained little or no PG activity while thrombin treated suspensions all contained activity. (ICM) 2021

1606

SMITH, J.B. and D.C.B. MILLS
Inhibition of adenosine 3':5'-cyclic monophosphate phosphodiesterase.
Biochemical Journal. 120: 20P. 1970.

Abstract only. We have compared the activities of the following drugs as inhibitors of cyclic AMP (adenosine 3':5:-cyclic monophosphate) phosphodiesterase from three sources: theophylline, caffeine, papaverine, dipyridamole and RA233[2,6-bis(diethanolamino)-4-piperidinopyrimido[5,4d] pyrimidine]. Papaverine and dipyridamole were more active than caffeine and theophylline as inhibitors of phosphodiesterase activity in all three preparations, and theophylline was more active than caffeine. RA233 and papaverine were particularly effective against the human platelet phosphodiesterase. Prostaglandin E_1 inhibits the aggregation of platelets induced by ADP (Kloeze, 1967) and increases the incorporation of radioactivity into cyclic AMP in human platelets containing labelled nucleotides (Marquis et al. 1969; Mills, Smith & Born, 1970.) Both of these effects are increased by phosphodiesterase inhibitors, and the activities of the five drugs tested as potentiators of prostaglandin E_1 were closely related to their abilities to inhibit the phosphodiesterase from platelets when a concentration of 3μM-cyclic AMP was used for the assay. (Authors modified) 2392

1607

SMITH, J.W. and C.W. PARKER
The responsiveness of leukocyte cyclic adenosine monophosphate to adrenergic agents in patients with asthma.
Journal of Laboratory and Clinical Medicine. 76: 993-994. 1970.

Abstract only. One theory of the mechanism of allergic asthma is that there is adrenergic imbalance, with increased α-adrenergic activity, decreased β-adrenergic activity or both (Szentivanyi, J. Allergy 42: 203, 1968). We have searched for a convenient in vitro model by which adrenergic responsiveness and adenyl cyclase activity could be evaluated in patients with asthma. We have found that human peripheral blood leukocytes exhibit marked increases in cyclic adenosine monophosphate (CAMP) concentration in response to norepinephrine and isoproterenol. We have compared the leukocyte CAMP responses to these agents, prostaglandin E_1 (PGE_1) and hydrocortisone in 3 groups of patients: Group 1, 6 patients with recent severe asthma; Group 2, 7 patients with severe seasonal pollinosis, with or without a history of asthma; and Group 3, 6 normal controls. Leukocytes were isolated from heparinized blood by dextran sedimentation, resuspended in Krebs-Ringer bicarbonate (5×10^6 cells per tube), incubated at 37°C for one hour as indicated below and assayed for CAMP by radio-immunoassay (Steiner and associates, Proc. Nat. Acad. Sci. 64: 367, 1969). Group I patients exhibited markedly decreased responsiveness to high concentrations of isoproterenol, norepinephrine, and hydrocortisone, whereas PGE_1 responsiveness was essentially normal. Group I patients had recently received (within 1 to 2 months) relatively aggressive antiallergic therapy which may have contributed to the response pattern; however, cells from a patient in Group II whose last sympathomimetic treatment was 18 months before also failed to respond. This in vitro system should be of value for further elucidation of the metabolic abnormality in patients with epinephrine refractory asthma. Whether or not the leukocyte response pattern would be a useful indicator of in vivo responsiveness to treatment with these agents remains to be established. (Author) 2356

1608
SMITH, S.D.
   Inhibitors of PAH uptake: uremic sera compared with hippurate and prostaglandin.
   Texas Reports on Biology and Medicine. 28: 414. 1970.

Abstract only. We studied active PAH uptake (slice PAH minus medium PAH) by slices of rat kidney at varying medium [PAH] in the absence or presence of human uremic sera diluted 1:4 (n=5), 2.2×10^{-4} M hippurate (n=6) or 1.0×10^{-4} M prostaglandin (PGA_1)(n=10). We obtained the apparent medium [PAH] at which uptake was half maximal (Km) and the apparent maximal uptake capacity (Cmax) by two methods: plotting 1/PAH and plotting [PAH]/PAH uptake vs. [PAH]. The mean Km ($\mu g/ml$) and Cmax ($\mu g/gm$) values derived from the plots for the inhibitor-exposed and paired control slices are shown in the table.

	Km	Cmax
Control	31	245
Uremic	81	319
Control	11.4	124
PGA_1	13.5	81
Control	17.4	161
Hippurate	14.7	66

The data are consistent with competitive inhibition of Cmax PAH uptake by uremic sera, with non-competitive inhibition by PGA_1 and surprisingly, with non-competitive inhibition by hippurate. Because of the different type of inhibition, we conclude that the inhibitor of PAH uptake in human uremic sera is neither hippurate nor PGA_1. (Author) 2330

1609
SMYTHIES, J.R.
   The chemical nature of the receptor site. A study in the stereochemistry of synaptic mechanisms.
   International Review of Neurobiology. 13: 181-222. 1970.

Certain portions of the neuronal membrane are specialized so that they can bind transmitters with the consequences that ionic channels are opened. If these channels conduct sodium, the membrane becomes depolarized, and if they conduct chloride or potassium, the membrane becomes hyperpolarized, resulting in excitation or inhibition of the neuron, respectively. A good deal is now known about the molecular specificity of transmitters, but almost nothing is known about the molecular specification of receptor sites. One theme developed in this essay is to explore the consequences of the basic speculative hypothesis that the primary receptor site for transmitters is membrane RNA. A molecular model of helical RNA suggests two potential binding sites for transmitters capable of disrupting the Watson-Crick hydrogen bond. Factors that could differentiate one segment of helical RNA from another, and thus specify the site, include, among others, the degree of torsion of the helix and the degree of involvement of other molecules bound to the RNA. The most difficult of these variables to specify is the degree of torsion of the helix; this may, however, be specified within resonable limits if the further postulate is made that the receptor consists not of naked RNA or even ribonucleoprotein, but of a prostaglandin-ribonucleoprotein complex. This complex is discussed. Comparison of the molecular model of prostaglandins with the RNA model shows that with a particular degree of torsion of the helix the two molecules bear a specific stereochemical relationship to each other such that two prostaglandins can bind to a four-base-pair segment of helical RNA by four hydrogen bonds as illustrated in figures in the text. This concept and its implications are developed at some length. Discussion of the prostaglandin-ribonucleoprotein complex includes consideration of the specification of the cholinergic receptor, including the muscarinic and nicotinic sites, and cholinergic antagonists; specification of the catecholamine and serotonin receptors; amino acid transmitters; and the disulfide bond. (ART) 2213

1610
SMYTHIES, J.R., F. BENINGTON and R.D. MORIN
    The mechanism of action of hallucinogenic drugs on a possible serotonin receptor in the brain.
    International Review of Neurobiology. 12: 207-233. 1970.

    The possible role of prostaglandins as part of a receptor site due to its structural ability to bind to RNA is discussed. (JRH) 2453

1611
SMYTHIES, J.R.
    A prostaglandin-RNA complex as a potential receptor site.
    Neurosciences Research Progress Bulletin. 8: 123-147. 1970.

    RNA may be involved in the serotonin receptor site. If this is true it may also be involved in other receptors. Working with molecular models, the author describes a RNA-prostaglandin complex that could serve as a receptor site for several adrenergic agents and explain how several psychotomimetic drugs produce their effect. (JRH) 2524

1612
SOMLYO, A.P. and A.V. SOMLYO
    Biophysics of smooth muscle excitation and contraction.
    In: Bouhuys, A., ed., "Airway Dynamics, Physiology and Pharmacology". p. 209-228.
        Springfield, C.C. Thomas. 1970.

    Reference to prostaglandins is unspecific and without discussion in this review. Sweatman and Collier, 1968 are cited. (RMS) 2500

1613
SOMLYO, A.P. and A.V. SOMLYO
    Vascular smooth muscle. II. Pharmacology of normal and hypertensive vessels.
    Pharmacological Reviews. 22: 249-353. 1970.

This extensive review on vascular smooth muscle contains a section on prostaglandins. It lists the effects of various prostaglandins on vascular smooth muscle from various species and factors which affect the response. (JRH) 2472

1614

SOMOVA, L., and D. DOCHEV

Changes in the renin activity in rats with experimental hypertension treated with PGE_1 and PGE_2.

Comptes Rendus de l'Academie Bulgare des Sciences. 23: 1581-1584. 1970.

Experiments performed to measure the direct effect of long term treatment with PGE_1 and PGE_2 on renin secretion in rats with experimental hypertension (produced by clamping the renal arteries for two weeks). Experimental rats were injected intraperitoneally daily with PGE_1 (15 μg/kg) or PGE_2 (30 μg/kg) for 30 days. Normal and hypertensive controls were injected with an equal volume of vehicle. Blood pressure was measured and blood samples from the renal and jugular veins were collected. The experimental animals were sacrificed 24 hours after the last prostaglandin injection. There was a large increase in renin activity before the rise in blood pressure, but after the blood pressure rose, renin activity returned to normal values. Both PGE_1 and PGE_2 completely corrected the hypertension, with PGE_1 being twice as potent as PGE_2. (JRH) 2512

1615

SOMOVA, L. and D. DOCHEV

Tracing the metabolic effect of different PGE_1 and PGE_2 doses in rats with experimental hypertension.

Comptes Rendus de l'Academie Bulgare des Sciences. 23: 1461-1464. 1970.

The effect of PGE_1 and PGE_2 on blood pressure, blood levels of free fatty acids (FFA), esterified fatty acids (EFA) and glucose in normal and experimentally hypertensive rats was measured. PGE_1 (15 and 10 μg/kg) or PGE_2 (30 and 20 μg/kg) was injected intraperitoneally, daily for 30 days into normal and hypertensive rats (one month after the onset of moderate hypertension). The experimental animals were sacrificed 24 hours after the last dose of prostaglandin. The lower doses of both prostaglandins produced no changes in blood pressure in normal or hypertensive rats. The larger doses prevented the experimental hypertension but had no effect on the blood pressure in normotensive rats. Low doses of PGE_1 (10 μg/kg) caused an increase in FFA and EFA but produced no change in glucose levels. However, the larger dose caused a decrease in FFA and EFA and a decrease in blood sugar. These effects were the same in normal and hypertensive rats and were apparent in the hypertensive animals even before the blood pressure began to decline. PGE_2 did not cause any changes in FFA, EFA or glucose. All of the animals receiving either prostaglandin showed a considerable loss of weight (86% mean). The authors feel that the lipolytic actions of low doses of PGE_1 may be due to stimulation of the sympathetic nervous system leading to the release of lipolytic substances, while the antilipolytic effect of high doses is due to the direct effect on adipose tissue. The effect on glucose levels could be due to its stimulation of glycogenesis or its effect on factors regulating glucose metabolism. (JRH) 2511

1616

SONDERGAARD, J. and M.W. GREAVES

Recovery of a pharmacologically active fatty acid during the inflammatory reaction, invoked by patch testing in allergic contact dermatitis.

International Archives of Allergy and Applied Immunology. 39: 56-61. 1970.

Using allergic contact eczema as a model delayed hypersensitivity response, an agent hitherto unrecognized in delayed hypersensitivity reactions has been detected using a skin perfusion technique. Pharmacological studies of perfusate from eczematous skin revealed that this agent, recovered in 21 of 30 patients, is an ethyl acetate extractable fatty acid of the prostaglandin type. Histamine was found in 7 and kinin activity in 12 of the 30 patients. Esterase activity was detected in one of 5 patients. (Authors) 2240

1617

SONDERGAARD, J. and M.W. GREAVES

Recovery of a pharmacologically active fatty acid from human allergic contact eczema using a skin perfusion method.

Federation Proceedings. 29: 419abs. 1970.

Abstract only. A method of continuous in vivo cutaneous infusion using warm Tyrode solution has been developed which permits direct analysis of pharmacological agents released in inflamed human skin. Aliquots of perfusate were subjected to assay. Smooth contracting activity (SMCA) has been recovered. Its solubility in ethyl acetate at pH3 indicates that it is a fatty acid of the prostaglandin type. (GW) 2350

1618

SOTES, M.R.

Prostaglandinas. Revision de conjunto. [Prostaglandins. General Survey.]

Hospital Generale. 10: 45-58. 1970.

The authoress makes a general revision of prostaglandin, some biological substances which at present are the object of interesting studies. The nature and properties of the various elements making up this group as well as their action on human tissues and organs are analyzed. The most probable hypotheses about their way of acting are stated and finally the possible relations existing between these fatty acids and some pathological situations such as the medullary cancer of thyroid and other new formations of endocrine character, producers of biologically active substances (pheochromocytomas, carcinoid, tumor, etc) are stressed. (Author) 2384

1619

SPEIDEL, J.J., and R.T. RAVENHOLT

Ideal means of fertility control?

Lancet. 1: 565. 1970.

This letter to the editor defines the ideal means of fertility control as "A non-toxic and completely effective substance which when self-administered on a single occasion would ensure the non-pregnant state." Recent evidence suggests that prostaglandins may meet this definition. (GW) 2282

1620

SPEROFF, L. and P.W. RAMWELL

Prostaglandins in reproductive physiology.

American Journal of Obstetrics and Gynecology. 107: 1111-1130. 1970.

The purpose of this comprehensive review is to evaluate the status of prostaglandins as related to reproductive physiology. It contains sections on nomenclature and structures, biosynthesis, separation, analysis, studies with seminal prostaglandins, effects on the uterus, studies related to pregnancy, prostaglandins and the fallopian tube, the ovary and prostaglandins, and recent publications. (JRH) 2179

1621

SPEROFF, L. and P.W. RAMWELL

Prostaglandin stimulation of in vitro progesterone synthesis.

Journal of Clinical Endocrinology and Metabolism. 30: 345-350. 1970.

Bovine corpora lutea slices were incubated with prostaglandins and gonadotropins. All the prostaglandins tested (PGE_2, PGE_1, $PGF_{2\alpha}$, and PGA_1) were found to be steroidogenic, stimulating both the production of progesterone as measured in micrograms and the incorporation of radioactivity from acetate-1-^{14}C. Prostaglandin E_2 (PGE_2) gave the greatest effect, being approximately half as effective as luteinizing hormone (LH) on a molar basis. There were similarities between gonadotropin and prostaglandin. Cycloheximide (lmM) equally blocked the steroidogenic response to both PGE_2 and LH, the specific activities of the progesterone formed in the presence of prostaglandins and that formed in the presence of gonadotropins were approximately of the same order of magnitude, the time-response curves of PGE_2 and LH were similar, and there was no additive effect when prostaglandins were added to luteal slices incubated with saturating doses of either LH or human chorionic gonadotropin (HCG). These results are in contrast to the in vitro luteolytic effect of $PGF_{2\alpha}$, and it is unlikely, therefore, that the luteolytic effect of $PGF_{2\alpha}$ is due to a direct inhibition of luteal steroidogenesis. Thus, the in vitro and in vivo effects of prostaglandins may represent 2 different mechanisms of action. (Authors) 2309

1622

SPEROFF, L.

Relationship of prostaglandins to reproductive function.

Bulletin of the Sloane Hospital for Women. 16: 117-122. 1970.

This lecture reviews all aspects of prostaglandin research, but mainly concentrates on aspects that are clinical in nature. (JRH) 2402

1623

STEWART, J.M

Substance P: a hormone in search of a function.

Pharmacologist. 12: 265. 1970.

Samples of the present substance P available from two different sources were found to be completely stable in the pulmonary circulation of the rat. "Local" hormones are destroyed upon passage through the lungs. Application of the "pulmonary inactivation rule" to the prostaglandins would indicate that prostaglandin A is a circulating hormone, while prostaglandin E and F are local hormones. (GW) 2188

1624

STOSSEL, T.P., F. MURAD, R.J. MASON, and M. VAUGHAN

Regulation of glycogen metabolism in polymorphonuclear leukocytes.

Journal of Biological Chemistry. 245: 6228-6234. 1970.

Data on the effects of glucose, theophylline, NaF glucose-6-P, MG^{++}, ATP, orthophosphate and prostaglandin E_1 on glycogen metabolism in guinea pig polymorphonuclear leucocytes are presented. Prostaglandin E_1 in the presence of theophylline increased phosphorylase activity and cAMP levels and decreased the percentage of glucose-6-P-independent synthetase. PGE_1 effect on phosphorylase activity was observed whether or not either glucose or theophylline or both were present. New glycogen breakdown was not influenced by either NaF or PGE_1. (RMS) 2003

1625
STRIKE, D.P. and H. SMITH
  A novel approach to the total synthesis of prostaglandins. Preparation of a stereoisomeric mixture containing (±)-13,14-dihydro prostaglandin E_1.
  Tetrahedron Letters. 4393-4396. 1970.

The authors describe a total synthesis of prostaglandins. Starting with an appropriately substituted levulic aldehyde (structure given) a step by step synthesis is described which leads to a stereoisomeric mixture containing racemic 13,14-dihydroprostaglandin E_1 which is an enantiomorph of a natural PGE_1 metabolite. (JRH) 2412

1626
SUZUKI, T.
  [The effects of prostaglandins on thrombogenesis.]
  Shinryo. 23: 1334-1338. 1970.

Article in Japanese. Abstract not available at present. 2386

1627
SWEET, C.S., P.J. KADOWITZ, and M.J. BRODY
  Facilitation of adrenergic transmission by a new humoral factor released during acute renal ischemia.
  Pharmacologist. 12: 247. 1970.

Abstract only. The possibility that renal venous blood (RVB) contains a factor which enhances the ability of angiotensin (ANG) to facilitate adrenergic vasoconstriction was examined in the perfused dog hind paw (HP) in which the maximally facilitating effect of ANG on responses to sympathetic nerve stimulation (SNS) was present. Vasoconstrictor responses to SNS and intraarterial norepinephrine (NE) were examined in two HP's, one perfused with RVB and one with arterial blood before and during reduction of renal pressure (RRP). Responses to SNS were significantly enhanced during RRP in both HP's while responses to NE were enhanced only in the HP perfused with RVB. Prostaglandin E_2 (PGE_2) (reported to be released during renal ischemia) could be the renal potentiating factor elaborated since (1) responses to SNS were facilitated by PGE_2 in the presence of ANG and (2) no facilitated responses to SNS were seen when RVB was first passed through an isolated lung. These data suggest that RRP releases factors which facilitate adrenergic transmission beyond the level achievable by ANG alone, and might explain why renal hypertension persists at a time when ANG levels are not measurably elevated. (Authors) 2191

1628
SWIDERSKA-KULIKOWA, B.
  Rola prostaglandyn w patogenezie nadcisnienia tetniczego. [Role of prostaglandins in the pathogenesis of arterial hypertension.]
  Polskie Archiwum Medycyny Wewnetrznej. 45: 823-826. 1970.

Article in Polish. Abstract not available at present. 2401

1629
SYKES, J.A.
  Pharmacologically active substances in malignant ascites fluid.
  British Journal of Pharmacology. 40: 595P-596P. 1970.

Abstract only. Mice were inoculated intraperitoneally with ascitic fluid. Seven days later the peritoneal fluid was aspirated and analyzed for prostaglandins. The fluid was presumed to contain PGE_1 and PGE_2 with traces of $PGF_{2\alpha}$. (JRH) 2201

1630
SZUPERSKA-OCETKIEWICZ, A., K. BIERON and R. GRYGLEWSKI
The release of an unknown substance contracting smooth muscle by noradrenaline.
Dissertationes Pharmiceuticae et Pharmcologicae. 22: 542. 1970.

Abstract only. The experiments were carried out on cats. NA (0.025-10 μg/kg/min) was infused i.v. or intraarterially with simultaneous recording of arterial blood pressure and the blood content of endogenous NA using the rat stomach strip preparation (RSS), endogenous A using RSS and chick rectum preparation (CR), and endogenous unknown substance (CS) using CR. RSS and CR were superfused with mixed venous blood or arterial blood (Vane's technique). NA, similarly to IP caused the appearance of CS in blood. CS produced contraction of CR and RSS. The CS content was always lower in arterial than in venous blood. During i.v. infusion of NA the liberation of CA was constant but not always proportional to the dose of NA. During the period of i.v. infusion of NA the liberation of CS exhausted. When NA was infused intraarterially the CS liberation usually started after completion of the infusion. The blockade of NA reuptake by tissues potentiated circulatory effects of NA but did not alter the liberation of CS. The attempts to establish the site of production of CS were unsuccessful. The most probable source of CS production is spleen, as intraarterial microinfusions of NA in spleenic artery, straining of the vessels of spleen pedicle, and the pressing on the spleen caused the appearance in blood of an agent contracting CR and RSS. In a normotensive cat the cessation of an infusion lasting a few minutes caused hypotension and simultaneous appearance of CS in blood. The hypotension vanished parallel to disappearance of CS from blood. The presented observations seem to suggest that CS may be identical with prostaglandines, particularly with PGE_1 or PGE_2. (Authors modified) 2445

1631
TAKAO, M.
[Iris extract in relation to ocular inflammation.]
Acta Societatis Ophthalmologicae Japonicae. 74: 1094-1099. 1970.

Amount of irin in the rabbit iris and the aqueous humor was assayed, using rat colon. An experimental uveitis was produced in rabbit eyes by an intravitreous injection of crystalline egg albumin. Iris extract from inflamed eyes showed an increased smooth-muscle stimulating activity than those from normal iris. Aqueous humor of inflamed eyes showed higher irin content than normal aqueous which hardly stimulated rat colon. (Author) 2403

1632
TAUB, D., R.D. HOFFSOMMER, C.H. KUO, H.L. SLATES, Z.S. ZELAWSKI, and N.L. WENDLER
A steroselective total synthesis of prostaglandin E_1.
Chemical Communications. 1258-1259. 1970.

A total synthesis of prostaglandin E_1 incorporating stereochemical control at the nuclear chiral centres is presented. (Authors) 2270

1633
TERRAGNO, N.A., A.J. LONIGRO, J.C. STRAND, and J.C. McGIFF
Renal prostaglandins: Regulators of the renal actions of pressor hormones?
Journal of Laboratory and Clinical Medicine. 76: 1046-1047. 1970.

Abstract only. Morphine-chloralose-anesthetized dogs were infused with either PGE_2, PGA_2 or $PGF_{2\alpha}$ in conjunction with renal nerve stimulation(RNS) or infusion of angiotensin II or norepinephrine. It was found that infusion with PGE_2 or PGA_2 (100 and 500 ng/min) respectively in the renal artery produced increases in renal blood flow (RBF) of 35 and 36%. RNS, angiotensin II, and norepinephrine produced a reduction in RBF. PGE_2 produced a reversal of the vasoconstrictor responses to RNS and angiotensin II but only slightly inhibited norepinephrine. PGA_2 had a greater effect on RNS effects and a lesser effect on angiotensin II and norepinephrine. $PGF_{2\alpha}$ had no significant effect on any of these vasoconstrictors. The concentrations of PGE_2 used to produce these responses was in the same range as found in the venous blood of kidneys during renal ischemia or during angiotensin II infusions. It is postulated that prostaglandins may regulate the renal effects of pressor systems. (JRH) 2051

1634
THIBAULT, P.H.
Complexite des prostaglandines. [Prostaglandin complexity.]
Presse Medicale. 78: 1019-1024. 1970.

This review cites an editorial and four reports of original research in a short discussion of the role of prostaglandins in hypertension, pulmonary control of angiotensin-prostaglandin relationships, and the use of prostaglandins in therapeutic abortion. (RMS) 2258

1635
THOMPSON, C.J., M. LOS, and E.W. HORTON
The separation, identification and estimation of prostaglandins in nanogram quantities by combined gas chromatography-mass spectrometry.
Life Sciences. 9: 983-988. 1970.

A method is described for identifying prostaglandins and their major metabolites from biological sources in nanogram quantities by combined gas-chromatography and mass spectrometry. Extraction of prostaglandins from biological material is accomplished by solvent partition followed by silicic acid column chromatography. This separates prostaglandins into 3 groups. 1) PGA and PGB, 2) PGE 3) PGF and 19-hydroxy PGA and PGB. The prostaglandins are then converted to their respective methyl esters and then trifluoroacetates. They are then subjected to gas chromatography. The material from each peak of the chromatograph is confirmed by mass spectrometry. The identification can be made in spite of other acidic substances which are simultaneously extracted from the biological material. (JRH) 2017

1636
TOBIAN, L. and J. VIETS
Potentiation of in vitro norepinephrine vasoconstriction with prostaglandin E_1.
Federation Proceedings. 29: 387 abs. 1970.

Abstract only. PGE$_1$ inhibition of the vasoconstrictor action of norepinephrine in vivo has been noted by many investigators. This paper reports an opposite effect during perfusion of a preparation of the mesenteric arterial bed in vitro. By itself, PGE$_1$ neither constricted nor dilated the arterial preparation. In 29 experiments, however, the combination of PGE$_1$ and norepinephrine produced a vasoconstrictive response that averaged 2.05 times greater than the response with norepinephrine alone. This potentiation occurred with all doses of PGE$_1$ used, varying from 10-40 ng/ml in the muscle bath. (ART) 2122

1637
TREISTER, G., and E.H. BARANY
Degeneration mydriasis and hyperemia of the iris after superior cervical ganglionectomy in the rabbit. Evidence for release of more than norepinephrine during degeneration of adrenergic terminals.
Investigative Ophthalmology. 9: 873-887. 1970.

The authors found that ocular sympathetic denervation in rabbits caused a hyperemia and mydriasis. Other workers have attributed the hyperemia found in the "intraocular irritation syndrome" to release of "irin" which seems to be composed in part of PGE$_2$ and PGF$_{2\alpha}$. The authors conclude that the results of sympathetic denervation of the eye are caused by release of a vasoactive substance which could be a prostaglandin, adenosine derivative, histamine, kinin or other polypeptide. (JRH) 2481

1638
TSUKATANI, H., T. ITAMI, and K. MATSUDA
Prostaglandin E$_2$ like activity in monkey lung.
Chemical and Pharmaceutical Bulletin. 18: 206-209. 1970.

The occurrence of a compound which has a PGE$_2$-like activity in the lung of monkeys *Macaca irus* is reported. The acidic lipid substance was separated from the lung by the use of organic solvent distribution, column chromatography and thin-layer chromatography. The active fraction had strong smooth muscle stimulating activity and a marked depressor activity on rats and rabbits. Judging from the results, this fraction was assumed to be PGE$_2$. The occurrence of PGF$_{2\alpha}$ in the monkey lung has been reported previously by Anggard (1965).(GW) 2328

1639
TUCKER, A.
Abortion drug could help the heart patients.
Guardian (Manchester). Oct. 5: 5. 1970.

This newspaper article reviews the potential uses of prostaglandins as birth control substances and describes some other future uses. (JRH) 2536

1640
TURKER, R.K. and A. OZER
The effect of prostaglandin E$_1$ and bradykinin on normal and depolarized isolated duodenum of the rat.
Agents and Actions. 1: 124-127. 1970.

The action of prostaglandin E$_1$ and bradykinin on the rat isolated normal and K Cl-depolarized duodenum was studied and compared with that of isoprenaline. These three drugs cause a relaxation of the normal duodenum. Prostaglandin E_1 and bradykinin produce a contraction when the duodenum is depolarized by isotonic K Cl-Tyrode's solution. The relaxing effect of isoprenaline is not changed by depolarization. The contractile effect of bradykinin but not that of prostaglandin E_1 in the depolarized duodenum is dependent on the calcium content of the external medium. The relaxing effect of isorprenaline is reduced but not completely abolished when the duodenum is depolarized in calcium-free isotonic K Cl-Tyrode's solution. (Authors) 2376

1641

VANASIN, B., W. GREENOUGH, and M.M. SCHUSTER
Effects of prostaglandins (PG) on electrical and motor activity of colon muscle.
Clinical Research. 18: 682. 1970.

The effect of prostaglandins on electrical and motor activity in isolated human and dog colon muscle was studied. PGE_1, PGE_2, $PGF_{2\alpha}$ and PGA (10^{-8} to 10^{-8} M) significantly increased the frequency and amplitude of electrical slow waves, which was associated with increased tone and also increased spike potentials associated with strong phasic contractions, in longitudinal colon muscle. Hyoscine and atropine given before the PG reduced its effect on phasic contraction. Hexamethonium had no effect. In circular muscle $PGF_{1\alpha}$, $PGF_{2\alpha}$ and PGA produced contraction while PGE_1 and PGE_2 produced relaxation. PG antagonist SC 19220 inhibited the action of PGE_1 and PGE_2. All the prostaglandins studied produce contraction of colon longitudinal muscle. PGE's relax while PGF's and PGA's contract circular colon muscle. PG appears to act on postganglionic cholinergic receptors and also directly on smooth muscle. (JRH) 2146

1642

VANASIN, B., W. GREENOUGH, and M.M. SHUSTER
Effect of prostaglandin (PG) on electrical and motor activity of isolated colonic muscle.
Gastroenterology. 58: 1004. 1970.

Abstract only. Isolated muscle strips resected from human and dog distal colon were studied. Motility was recorded from both electrical and motor activity simultaneously. Two types of spontaneous electrical activity were recorded from longitudinal muscle; slow sinusoidal waves which correlated with muscle tone were present 15 to 60 percent of recording time and rapid spike potentials preceded phasic contractions. PGE_1, PGE_2, $PGF_{1\alpha}$, $PGF_{2\alpha}$ and PGA (0.05 to 10 μg/ml) produced an increase in frequency and amplitude of the slow waves and of spike potentials. PGA produced an extremely regular rhythmical pattern. In circular muscle (in which electrical activity is not present), $PGF_{1\alpha}$, $PGF_{2\alpha}$, and PGA produced contraction while PGE_1 and PGE_2 produced relaxation or had no effect. The prostaglandins appeared to act on postganglionic cholinergic receptors and also to act directly on smooth muscle. It is concluded that prostaglandins in physiological amounts may play a significant role in colonic activity. (ART) 2131

1643

VANE,J.R., and K.I. WILLIAMS
A sensitive method for the assay of oxytocin in blood.
British Journal of Pharmacology. 38: 444P-445P. 1970.

A search has been made for an assay tissue selectively sensitive to oxytocin and suitable for use as a blood-bathed organ. The duck pulmonary vein was sensitive to oxytocin 2×10^{-6} μ/ml) but insensitive to PGA_1, PGE_1, PGE_2 and $PGF_{2\alpha}$ (all at 10 ng/ml). To detect the release of endogenous oxytoxin, strips of duck pulmonary vein were bathed in internal maxillary vein blood and in femoral arterial blood. (GW) 2234

1644

VAUGHAN, M., N.F. PIERCE, and W.B. GREENOUGH, III
Stimulation of glycerol production in fat cells by cholera toxin.
Nature. 226: 658-659. 1970.

An investigation on the use of glycerol production by adipocytes from rabbit ileal mucosa as a sensitive bioassay for the fluid transport active factor (choleragen) from lyophilized culutres of *V. cholerae* is reported. It is mentioned without elaboration that PGE_1 added to the culture at 0.1 μg/ml inhibited the effect of toxin during the subsequent hour. (RMS) 2306

1645

VELASQUEZ, M.T., A.V. NOTARGIACOMO and J.N. COHN
Mechanism for the fall in PAH extraction (^{E}PAH) induced by renal vasodilator drugs.
Clinical Research. 18: 519. 1970.

Abstract only. This study was performed to determine what causes the reduction in para amino hippuric acid (PAH) extraction caused by vasodilators. The intrarenal distribution of blood flow and the cortical transit time (CTT) were measured simultaneously with extraction of PAH, before, and during intrarenal infusion of acetylcholine (ACH) and PGE (the specific PGE was not given nor the dose levels) in anesthetized dogs. The cortical fraction (CF) of renal blood flow (RBF) was calculated from dye curves after injection of indocyanine green into the renal artery. During ACH infusion RBF increased and the CF rose from 73 to 81%. CTT was reduced (9.0 sec to 2.8 sec) and extraction of PAH decreased (0.68 to 0.21). PGE did not produce a change in CF, but did cause a sharp decrease in PAH extraction (0.70 to 0.21) and CTT (10.3 sec to 5.3 sec). Extraction of PAH was significantly lower with PGE than with ACH. The authors conclude that the reduction in extraction of PAH is directly related to reduced CTT and that PGE may have an additional direct inhibitory effect on PAH extraction. (JRH) 2155

1646

VERGROESEN, A.J.
De rol van de lipiden in de voeding. [The role of lipids in nutrition.]
Chemische Weekblad. 66: 24-25. 1970.

Special consideration of prostaglandins is given in this Dutch editorial on the role of lipids in nutrition since the essential fatty acids are precursors of prostaglandins. (JRH) 2545

1647

VIGDAHL, R.L. and N.R. MARQUIS
Cyclic AMP mediated inhibition of platelet aggregation by vasodilators.
Circulation. 42(supp.3): II 150. 1970.

Abstract only. We have postulated that prostaglandin E_1 (PGE_1), a vasodilator, inhibits platelet aggregation (PA) by stimulating adenyl cyclase (AC) and increasing platelet cyclic AMP. To clarify the mode of action of vasodilators which inhibit PA, studies were conducted to determine if their effects were also mediated by an increase in cyclic AMP. PA was measured turbidimetrically; AC and phosphodiesterase (PDE) activities were radiochemically assayed. The vasodilators, papaverine, ethaverine, dioxyline and dipyridamole, are potent inhibitors of platelet PGE (ki's 2×10^{-5}- 7×10^{-6} M) as well as platelet adenosine uptake. In contrast, caffeine and theophylline stimulate adenosine uptake at concentrations inhibitory to PDE. All of these drugs potentiate PGE_1-stimulated cyclic AMP synthesis by platelets as well as PGE_1-induced inhibition of PA. Lidoflazine, verapamil, nylidrin, isoetharine and nitroglycerine have lesser or no effects on platelet PDE activity and adenosine uptake. These observations suggest that the inhibition of PDE may mediate the inhibition of PA by certain vasodilators. (Authors) 2338

1648

VILHARDT, H. and P. HEDQVIST

A possible role of prostaglandin E_2 in the regulation of vasopressin secretion in rats.
Life Sciences. 9: 825-830. 1970.

An attempt was made to determine if the effect of PGE_2 on vasopressin (ADH) secretion is caused by direct action on the neural lobes or by stimulation of higher levels in the brain. Rat neural lobes were incubated with PGE_2 then with potassium both with and without PGE_2. The bathing solution was then bioassayed for ADH content. In another series of experiments PGE_1 was injected into either the carotid artery or the systemic aorta proximal to the renal arteries in varying doses (1-20 μg). Urine flow and osmolarity were measured to determine is ADH was produced. It was found that PGE_2 did not stimulate ADH secretion in isolated neural lobes nor did it effect the ADH secretion stimulated by potassium. When injected into the carotid artery there was a marked increase in ADH secretion. However, injection into the systemic aorta did not produce any change. From these results it was concluded that PGE_1 action on the secretion of ADH occurs in the hypothalmic nuclei or in ganglia at higher levels of the brain. (JRH) 2016

1649

WADA, T. and M. ISHIZAWA

[Effects of prostaglandin on the function of the gastric secretion.]
Japanese Journal of Clinical Medicine. 28: 2465-2468. 1970.

Article in Japanese. Abstract not available at present. 2396

1650

WADE, D.R., T.M. CHALMERS, and C.N. HALES.

The effects of Mg^{++}, and Ca^{++}, ATP and cyclic 3',5'-AMP on the hormone-sensitive lipase of adipose tissue.
Biochimica et Biophysica Acta. 218: 496-507. 1970.

Prostaglandin E_1 is mentioned as having antilipolytic properties on adipocytes in vitro due to its inhibition of cAMP, even though it has been shown to stimulate cAMP in adipose tissue. (JRH) 2457

1651

WAITZMAN, M.B.

Possible new concepts relating prostaglandins to various ocular functions.
Survey of Ophthalmology. 14: 301-326. 1970.

A survey was made of the very ubiquitous nature of prostaglandins in biological systems and the close physiological and biochemical involvement of these lipids in many vital body functions. There are innumerable questions yet to be answered relative to some effects of prostaglandins on ocular function. Perhaps the most provocative area of interest concerning prostaglandins and the eye is the area concerning possible involvement of prostaglandins in the etiology of diabetic retinopathy. (GW) 2387

1652

WAITZMAN, M.B.

Pupil size and ocular pressure after sympathectomy and prostaglandin-catecholamine treatment.

Experimental Eye Research. 10: 219-222. 1970.

Studies that have demonstrated a dose-related physiological antagonism between PGE_1 and catecholamines relative to both intraocular pressure (IOP) and pupil size suggest the hypothesis that an (experimentally produced) animal supersensitive to catecholamines might require a smaller dose than usual of one of such substances to antagonize PGE_1-induced miosis and IOP elevation. This hypothesis was tested in male albino rabbits in which unilateral superior cervical sympathetic ganglionectomy had been performed. The animals were pretreated with phenoxybenzamine prior to each study of the prostaglandin-norepinephrine interaction. Results are shown in a figure. That dose of norepinephrine which was ineffective in altering ocular effects of PGE_1 in unoperated animals was found to be effective in antagonizing the ocular effects of PGE_1 in ganglionectomized animals. The eyes on both the operated and unoperated sides became supersensitive to norepinephrine but to a significantly greater degree in the eyes ipsilateral to the side on which the ganglionectomy was performed. (ART) 2134

1653

WAITZMAN, M.B.

Pupil and ocular pressure after sympathectomy and prostaglandin-catecholamine treatment.

Investigative Ophthalmology. 9: 158. 1970.

Abstract only. Cervical sympathetic ganglionectomized and phenoxybenzamine-pretreated rabbits show physiological antagonism between PGE_1 and norepinephrine, with responsiveness (supersensitivity) to a dose of norepinephrine which was ineffective in nonganglionectomized animals. This supersensitivity to norepinephrine was shown relative to pupil response and intraocular pressure. (Author) 2334

1654

WALKER, R.J., G.N. WOODRUFF, and G.A. KERKUT

The action of four prostaglandins, E_1, E_2, $F_{1\alpha}$, and $F_{2\alpha}$, on the spontaneous activity and on the response to acetylcholine of neurons from the snail *helix aspersa*.

Comparative General Pharmacology. 1: 69-72. 1970.

The effect has been investigated of four prostaglandins, E_1, E_2, $F_{1\alpha}$, and $F_{2\alpha}$, on the activity of neurons from the suboesophageal ganglionic mass of *helix aspersa*. Twenty-five per cent of the cells tested responded to the application of one or more prostaglandins either by excitation or inhibition of their activity but the doses required were 1-100 μg. (Authors) 2142

1655

WALLACE, S., C.S. HILL, D.D. PAULUS, M.L. IBANEZ, and R.L. CLARKE

The radiologic aspects of medullary (solid) thyroid carcinoma.

Radiologic Clinics of North America. 8: 463-474. 1970.

Brief mention is made of Williams (1968) work which indicated abnormally high levels of prostaglandins in patients with medullary carcinoma of the thyroid. It is suggested that the prostaglandins might be responsible for the diarrhea often seen in these patients. (JRH) 2414

1656

WEEKS, J.R.

Pharmacology of prostaglandin $F_{2\alpha}$ on the uterus and corpus luteum.

In: Eigenmann, R., ed., "Proceedings of the Fourth International Congress on Pharmacology," July, 1969. p. 49-53.Basle, Schwabe, 1970.

This article reviews the literature that indicates that $PGF_{2\alpha}$ may have luteolytic properties. (JRH) 2436

1657

WEINHEIMER, A.J. and R.L. SPRAGGINS

Two new prostaglandins isolated from the gorgonian plexaura homomalia (esper): Chemistry of coelenterates XVI.

In: Youngken, H.W., ed. "Food-Drugs from the Sea." pp. 311-314. Proceedings of Marine Technology Society 1969. 1970.

The authors report the occurrence of two new prostaglandin derivatives in the gorgonian, *Plexaura homomalia*, in high concentration. The air-dried cortex of the animal contains 0.2% 15-epi-PGA_2 and 1.3% of its diester. (GW) 2343

1658

WENNMALM, A., and P. HEDQVIST

Prostaglandin E_1 as inhibitor of the sympathetic neuroeffector system in the rabbit heart. Life Sciences. 9: 931-937. 1970.

The effect of PGE_1 on the sympathetically innervated isolated rabbit heart was measured. PGE_1 8×10^{-8} to 3×10^{-7}M slightly increased the myocardial contractile force of the isolated hearts and in a few cases caused a slight increase in heart rate. Electrical stimulation of the sympathetic nerve to the heart caused a pronounced increase in contractile force, heart rate and the out flow of noradrenaline (NA). When PGE_1 was given just prior to stimulation of the nerve there was a marked decrease in the inotropic effects and in the out flow of NA from the heart. Adding NA directly to the physiological bath solution produced an effect similar to stimulation of the nerve. PGE_1 was not effective in blocking the effects of NA. It is theorized that the effect of PGE_1 is prejunctional, perhaps inhibiton of NA release from the neuron. (JRH) 2019

1659

WENZL, M.R., and E.J. SEGRE

Physiology of human menstrual cycle and early pregnancy. A review of recent investigations. Contraception. 1: 315-338. 1970.

This review cites 76 references but lists only 73. It is noted that one of the prostaglandins ($PGF_{2\alpha}$ actually) has luteolytic effect when administered during the luteal phase to rhesus monkeys and that prostaglandins have been detected in the human menstrual endometrium. It is also noted that in the monkey it appears that $PGF_{2\alpha}$ interferes with corpus luteum function sufficiently to interrupt embryonic development and induce bleeding. (ART) 2145

1660

WERNING, C., W. VETTER, P. WEIDMANN, H.U. SCHWEIKER, D. STIEL and W. SIEGENTHALER

Die wirkung von prostaglandin E_1 auf die plasmareninaktivitat, die ^{51}Cr-EDTA-und ^{125}J-o-Jodippursaureclearance und die elektrolytenkretion bein narkotisierten hund. [The

influence of PGE_1 on plasma-renin activity, the ^{51}Cr-EDTA and ^{125}I-o-iodiohippuric acid clearance, and the excretion of electrolytes in the anesthetized dog.]
Verhandlungen der Deutsch Gesellschaft fur Innere Medizin. 76: 672-674. 1970.

Twelve male dogs weighing about 20 kg. each were anesthetized with pentobarbital and received intraarterial infusions of 25,000 ng of PGE_1 in saline solution. Plasma renin activity (PRA), hematocrit, and the serum electrolyte sodium, potassium, calcium and magesium were determined before injection and 2,15,30,60,75,90,120,135,150,180 and 210 minutes afterwards. Urine sodium, potassium and clearances were determined 15,30,60,75,90,120,135, 150,180 and 210 minutes after injection. PRA, sodium excretion, and potassium excretion and urine volumes showed significant increases in the first hour after injection, while ^{51}Cr-EDTA- and ^{125}I-o-iodichippuric acid clearances showed slight increases by 75 minutes, which were not considered statistically significant. Levels of serum potassium, sodium, calcium, magnesium, and creatinine remained normal as did the hematocrit. The authors attribute the observed phenomena to decreased tubular absorption of sodium, together with increased water elimination. Potassium excretion may also be important. (MEMH) 2405

1661
WERNING, C. and W. SIEGENTHALER
Prostaglandine und niere. [Prostaglandins and kidney.]
Deutsche Medizinische Wochenschrift. 95: 2345-2349. 1970.

A review of the effects of prostaglandins on various metabolic processes is presented in this article in German. The role of prostaglandin in various kidney functions is considered to be of prime importance. One table summarizes the effects of various prostaglandins on the central nervous system, the circulatory system, the intestinal tract, the sex organs, the endocrine glands and metabolism while another table depicts the chemical structure of prostaglandins E_1, E_2, E_3, $F_{1\alpha}$, $F_{2\alpha}$, and $F_{3\alpha}$. The effects of the prostaglandins on natriuresis, vasodilation and plasma-renin activity are also summarized. (GW) 2204

1662
WESTURA, E.E., H. KANNEGIESSER, J.D. O'TOOLE, and J.B. LEE
Antihypertensive effects of prostaglandin A_1 in essential hypertension.
Circulation Research. 27 (supp. 1): I 131-140. 1970.

Prostaglandin A_1 was infused intravenously into six patients with essential hypertension and its antihypertensive properties evaluated. At administration rates of 1.0 $\mu g/kg/min$ and 2.0 $\mu g/kg/min$, no direct effect on cardiac performance could be detected, although there was a significant fall in blood pressure. The decrease in blood pressure was associated with reflex changes in cardiac performance consistent with those normally associated with decreases in peripheral vascular resistance. (Authors) 2266

1663
WILLEMS, C., P.A. ROCMANS, and J.E. DUMONT
Stimulation in vitro by thyrotropin, cyclic 3',5'-AMP, dibutyryl cyclic 3',5'-AMP and prostaglandin E_1 of secretion by dog thyroid slices.
Biochimica et Biophysica Acta. 222: 474-481. 1970.

Proceeding on the hypothesis that the in vitro release of ^{131}I radioactivity in dog thyroid slices corresponds to secretion, it was found that secretion in vitro as well as intracellular colloid droplet formation was enhanced by thyroid stimulating hormone (TSH) and TSH-mimicking agents such as cyclic AMP (cAMP), dibutyryl cAMP, and PGE_1 but not by caffeine, AMP or ATP. Orders of magnitude for equieffective concentrations were TSH 0.25 mU/ml, cAMP 3.5 to 10 mM, and dibutyryl cAMP 0.1 mM. The concentration effect curve of PGE_1 was biphasic. PGE_1 enhanced ^{121}I secretion (the amount of ^{121}I radioactivity extracted by butanol) in 6 out of 7 experiments in which the effect of TSH was clearly demonstrated. Generally, the effect was detected at 0.1 µg/ml and decreased for higher concentrations. In one experiment, the effect of PGE_1 was not detectable below 10 µg/ml. In 3 out of 6 experiments, no effect was observed at 10 µg/ml. PGE_1 elicited the formation of intracellular colloid droplets: sometimes at 0.1 µg/ml, most of the times at 1 and 3.5 µg/ml, in 3 out of 6 experiments not at 10 µg/ml, and never at 25 µg/ml. (ART) 22470

1664

WILLIAMS-ASHMAN, H.G. and D.H. LOCKWOOD
Role of polyamines in reproductive physiology and sex hormone action.
Annals of New York Acadmey of Sciences. 171: 882-894. 1970.

Prostaglandins are very briefly mentioned as increasing the transport of sperm in the female reproductive tract. (JRH) 2454

1665

WILLIS, A.L.
A method for studying release of prostaglandins from superfused strips of isolated spleen.
British Journal of Pharmacology. 38: 470P-471P. 1970.

Prostaglandin release from superfused strips of rabbit spleen was detected by isolated tissues suspended in cascade below the spleen. Adrenaline (20-1,300 ng/ml) contracted the spleen and induced an output of prostaglandin activity which was highest during the early part of the infusion. This output increased from 1-25 ng/ml (PGE_2) with the concentration of adrenaline. Acetylcholine (0.2-20 µg/ml) induced non-sustained contractions of the spleen and an output of prostaglandins. (GW) 2231

1666

WILLIS, A.L.
Simplified thin-layer chromatography of prostaglandins in biological extracts.
British Journal of Pharmacology. 40: 583P-584P. 1970.

This brief technique paper describes methods for using commercially prepared plates for thin layer chromatography of prostaglandins from biological tracts. The separation is made chromatographically followed by a bioassay. (JRH) 2200

1667

WILSON, D.E., C. PHILLIPS, and R.A. LEVINE
Inhibition of gastric secretion in man by prostaglandin A_1 (PGA_1).
Gastroenterology. 58: 1007. 1970.

Abstract only. Gastric samples from 11 healthy volunteers were collected during 15-min intervals and analyzed for volume, pH and titrable acidity. After a steady state of gastric output was obtained by intravenous administration of histamine (0.015 mg/kg/hr), PGA_1 was infused for 30 min in either low dose (0.5 to 0.6 µg/kg/min) or high dose (1.0 to 1.25

μg/kg/min) to six and five subjects, respectively. Post-treatment collections upon analysis for the whole group showed that PGA_1 decreased mean volume and acidity, respectively, by 23 and 15 percent during the first 15 min., and 30 and 18 percent during the second 15 min. When the data were separated into high and low dose groups, a significantly greater reduction in volume and acidity was found in the low dose group. The data may indicate a decreased effectiveness of PGA_1 in the presence of increased gastric secretion. The inhibitory effect of PGA_1 on gastric secretion in man may be caused by changes in the gastric microcirculation, such as have been described for PGE_1 in the dog. (ART) 2132

1668

WILSON, D.E., and R.A. LEVINE
  Inhibition of hepatic glucose utilization by prostaglandin E_1. (PGE_1)
  Clinical Research. 18: 468. 1970.

Abstract only. The effect of PGE_1 (85 μg/hr), labeled dibutyryl cAMP and a control Ringer solution on utilization of glucose-U-^{14}C or glucose-1-^{14}C was studied in vitro in the rat liver. It was found that PGE_1 and dibutyryl cAMP suppressed the oxidization glucose-U-^{14}C. PGE_1 failed to inhibit glucose-1-^{14}C oxidation. This suggests that PGE_1 does not inhibit glucose transport or hexose monophosphate oxidization pathway. PGE_1 may affect some stage of intracellular glucose metabolism within the Embden-Meyerhof pathway or beyond it. (JRH) 2158

1669

WINKELMANN, R.K., W.M. SAMS, and J.H. KING
  Human cutaenous vascular smooth muscle responses to catecholamines, histamine, serotonin,
      bradykinin, angiotensin, and prostaglandins.
  Journal of Clinical Investigation. 49: 103a. 1970.

Abstract only. $PGF_{2\alpha}$, PGE_2 and PGA_2 among other biologically active substances were tested for their effect on human cutaneous vascular smooth muscle response in vitro. All of these prostaglandins produced a contraction in the test strips. It was concluded that prostaglandins are mediators of importance for human skin vascular resistance. (JRH) 2216

1670

WIQVIST, N., U. ROTH-BRANDEL, and M. BYGDEMAN
  The effect of prostaglandin E_1 and $F_{2\alpha}$ on the nonpregnant human uterus in vivo.
  International Journal of Gynecology and Obstetrics. 8: 165. 1970.

Abstract only. Prostaglandin E_1 is present in human seminal fluid in substantial amounts and the concentration of this substance has been shown to be low in certain cases of functional infertility. It has been suggested that this correlation is related to the effect of the prostaglandins on the smooth musculature of the female genital tract. Pure prostaglandin E_1 and $F_{2\alpha}$ were administered as single intravenous injections and uterine contractility recorded by a microballon in 16 volunteers during various phases of the cycle. It was found that: (1) Doses above threshold values always increase tone and motility. A significant relaxation of the uterus was never observed. (2) The degree of response was generally dose-dependent. The threshold dose of prostaglandin E_1 varied between 5 and 200 μg and that of $F_{2\alpha}$ between 20 and 50 μg. (3) The nonpregnant uterus seemed to be more sensitive in both substances than the pregnant uterus. (Authors modified) 2417

1671
WIQVIST, N. and M. BYGDEMAN
Induction of therapeutic abortion with intravenous prostaglandin $F_2[_\alpha]$.
Lancet. 1: 889. 1970.

A letter to the editor discusses $PGF_{2\alpha}$ as an abortifacient during the 6th to the 16th week of pregnancy. Administration was a continuous infusion 13-360 μg/min for 7 hours, repeated on the next day. Dosage was increased stepwise until slight or moderate menstrual pains appeared. In 10 to 12 patients, the conceptus was expelled before the end of the second day. Some side effects are noted, and the possible mechanism of abortion is discussed. (RMS) 2048

1672
WIQVIST, N. and M. BYGDEMAN
Therapeutic abortion by local administration of prostaglandin.
Lancet. 2: 716-717. 1970.

The authors report that therapeutic abortion may be induced by intravenous infusion of prostaglandins E_1, E_2, and $F_{2\alpha}$. Intravenous infusion of $PGF_{2\alpha}$ within the first 8 weeks of pregnancy resulted in the complete or partial expulsion of the fetus in 20 out of 22 women, while between the 13th and 20th week only 4 out of 28 pregnancies were successfully terminated. The frequency of side effects as related to dose of $PGF_{2\alpha}$ was analyzed. Intrauterine administration was found to be clinically comparable to intravenous infusion, and also completely eliminated generalized side effects. (GW) 2002

1673
WOLFE, L.S.
Biologically active lipids.
In: Lajtha, A. ed., "Handbook of Neurochemistry IV: Control Mechanisms," p. 149-164. 1970.

This chapter deals with the central theme of prostaglandins although other active acidic lipids or extracts are also mentioned. The complex pharmacological studies of the prostaglandins, the physiological studies in connection with human reproduction, metabolism of adipose tissue, regulation of intrarenal blood flow and platelet adhesion are not considered. (GW) 2333

1674
WOLFE, S.M. and N.R. SHULMAN
Inhibition of platelet energy production and release reaction by PGE_1, theophylline and cAMP.
Biochemical and Biophysical Research Communications. 41: 128-134. 1970.

Human adult male platelets were appropriately suspended and tested for the release of adenine nucleotides and calcium following the addition of thrombin. Thrombin induced a rapid, simultaneous release from the platelets of adenine nucleotides and calcium in a constant molar ratio. PGE_1, theophylline, and dibutyryl cyclic AMP inhibit this release reaction, the attendant lactate production, and the subsequent aggregation of platelets. PGE_1 causes an 18-fold increase in platelet adenyl cyclase activity and it is suggested that inhibition of aggregation is related to increased intracellular levels of cAMP. Compounds which result in increased intracellular levels of cAMP appear to block aggregation by inhibiting the release reaction and energy production. Since platelet aggregation is thought to depend on calcium and ADP, the inhibition by PGE_1 of their release may be the basis for the inhibitory effect of PGE_1 on aggregation. (CWS) 2209

1675
WOLFE, S.M., J. MUENZER, and N.R. SHULMAN
The role of cyclic adenosine-3',5'-monophosphate and prostaglandin in platelet aggregation.
Journal of Clinical Investigation. 49: 104a. 1970.

Abstract only. PGE_1 and theophylline inhibited thrombin-induced platelet aggregation and lactate production. Inhibition decreased with increasing amounts of thrombin. The effects of PGE_1 and theophylline were additive. The inhibitors had to be present before addition of the thrombin was effective. It was concluded that the key to platelet aggregation may be cAMP. Prostaglandin which is known to effect cAMP levels in other tissues may play a role in platelet aggregation. (JRH) 2215

1676
WOLFF, J. and A.B. JONES
Inhibition of hormone-sensitive adenyl cyclase by phenothiazines.
Proceedings of the National Academy of Sciences. 65: 454-459. 1970.

Chlorpromazine $(3 \times 10^{-4} M)$ prevents the stimulation of adenyl cyclase activity in thyroid membranes produced by thyrotropin and prostaglandin, ACTH stimulation of adenyl cyclase in adrenal tissue, and glucagon and epinephrine stimulations of adenyl cyclase activity in the liver. Baseline activity is unaffected. The inhibitory effect of chlorpromazine was not restricted to the polypeptide hormones as shown by the reduction of the PGE_1 response in thyroid membranes and by the abolition of epinephrine stimulation in liver preparations. (GW) 2254

1677
WRIGHT, P.
A new approach to contraception - the abortion pill.
Times (London). Oct. 3: 12-13. 1970.

This newspaper article discusses the role of prostaglandins as contraceptives. Special stress is placed on the "morning after" aspect of contraception. (JRH) 2539

1678
YAMAGATA, S., H. MASUDA, A. ISHIMORI, M. MITA, S. INOUE, H. ARAKAWA, H. SAKURADA, K. NEMOTO and M. SHIMOYAMA
Study of the inhibitory action of prostaglandin on gastric secretion in rat and man.
Gastroenterologia Japonica. 5: 321. 1970.

Abstract only. In the present investigation, the inhibitory effect of Prostaglandin E_1 and E_2 on gastric secretion was compared clinically and in the animal experiment. It was reported already from us that Prostaglandin decreases histamine or AOC tetrapeptide stimulated gastric secretion, but not insulin stimulated one. Using rat it was seen that Prostaglandin E_2 in dose of 30r decreased steady state of gastric secretion due to continuous intravenous infusion of gastrin like AOC tetrapeptide less markedly in comparison with 50r of it or 30 and 50r of E_1. The same study was done clinically to see the inhibitory effect of Prostaglandin E_2 in the patients of gastrointestinal disease under steady state of gastric secretion due to continuous infusion of AOC tetrapeptide and slight reduction of acidity output was observed 20 - 40 minutes after intramuscular injection of Prostaglandin E_2. The reduction of acid output were found to be related to the dose of the administration of Prostaglandin. Except temporal and slight depression of maximal blood pressure, no adverse effects of Prostaglandin were observed. (Authors) 2521

1679
YAMASHITA, K., G. BLOOM, B. RAINARD, U. ZOR, and J.B. FIELD
Effects of chlorpromazine, propranolol, and phospholipase C on thyrotropin and prosta-
glandin stimulation of adenyl cyclase-cyclic AMP system in dog thyroid slices.
Metabolism. 19: 1109-1118. 1970.

The effects of chlorpromazine, propranolol, and phospholipase C on the stimulation, by
thyroid stimulating hormone (TSH) and PGE_1, of the adenyl cyclase-cyclic AMP (AC-cAMP)
system and ^{14}C-1-glucose oxidation were studied in dog thyroid slices. Insofar as
prostaglandins are concerned, chlorpromazine reduced the stimulation of tritiated cAMP
formation and cAMP concentrations induced by PGE_1 (1 μg/ml). PGE_1 and $PGF_{1\alpha}$ (100
mg/ml) increased radioactive phosphorus (^{32}P) incorporation into phospholipids while an
equivalent amount of PGB_1 had no effect and PGA_1 inhibited ^{32}P incorporation. The results
indicate that stimulation of ^{32}P incorporation into phospholipids may be independent of
changes in cAMP concentrations. (Authors modified) 2239

1680
YOSHIMOTO, A., H. ITO, and K. TOMITA
Cofactor requirements of the enzyme synthesizing prostaglandin in bovine seminal vesicles.
Journal of Biochemistry. 68: 487-499. 1970.

The effects of various cofactors on prostaglandin formation from arachidonic acid by the
microsomal enzyme of bovine seminal vesicles were studied. In addition to the known
activators, reduced glutathione and hydroquinone, the enzyme system was found to be
markedly stimulated by hemoglobin, myoglobin or hemin. From the specific effects of these
cofactors on prostaglandin formation or oxygen uptake, or both, it seemed that heme
compounds and hydroquinone were involved in the step in which molecular oxygen became
attached to the unsaturated fatty acids, while reduced glutathione was involved in the
subsequent step of reduction of the peroxide-type intermediate. The facts that hydroquinone
could be replaced by ascorbate, and that the order or addition of reactants to the system
markedly affected both the yield of prostaglandin and the rate of oxygen consumption
suggested that a free radical was formed during prostaglandin biosynthesis. (Authors) 2345

1681
ZAMPAGLIONE, N.G., J.J. LECH and D.N. CALVERT
Diethyl chelidonate, a specific inhibitor of hormone-stimulated lipolysis.
Biochemical Pharmacology. 19: 2157-2164. 1970.

The authors mention several inhibitors of lipolysis, including prostaglandins, without
discussion; 2 references, Butcher and Baird (1968) and Stock et al (1968) are cited with
reference to prostaglandins. (RMS) 2490

1682
ZEITLIN, I.J. and A.N. SMITH
Kinin assays in clinical conditions.
Rendiconti Romani di Gastroentrologia. 2: 176-183. 1970.

The authors found high plasma free kinin levels in patients with carcinoid tumors or
dumping syndrome during periods of flushing and/or diarrhea. Kinin levels returned to
normal when the symptoms abated. However, in other patients who showed these same
symptoms no correlation could be demonstrated between kinin levels and the onset of
symptoms. In light of the work of others, the authors feel that the release of prostaglandins
rather than kinins might be the unifying factor in these patients. (JRH) 2444

438

1683

ZIMMERMAN, B.G. and E.W. DUNHAM
Release of a prostaglandin of the E series from the perfused kidney during nerve stimulation.
University of Michigan Medical Center Journal. 36: 236. 1970.

Abstract only. Prostaglandins (PG) occur in relatively high concentrations in the rabbit renal medulla and in smaller amounts in the cat and dog kidney. Release of PG from the kidney has been demonstrated, but the mechanism of their release is not well understood. The purpose of the present work was to study the influence of sympathetic nerve stimulation and intra-arterially infused norepinephrine on release of PG from the canine kidney.
While the kidney was perfused in situ at constant blood flow, samples of renal venous blood were collected both before and during nerve stimulation or infusion of norepinephrine. Following extraction of the blood, solvent partitioning, and thin-layer chromatography, the samples were bioassayed on the rat fundus and gerbil colon. In 10 animals a low basal release of 29 mμg/min of PG into the renal venous effluent was found, and this was considerably enhanced to 191 mμg/min during nerve stimulation at 10 cps. These values are expressed as PGE_2 equivalents and are not corrected for recovery of 30 percent. Release was also augmented during infusion of norepinephrine.
In 18 other experiments in which the blood-bathed organ technique of Vane was employed, the release of prostaglandin-like material during nerve stimulation and norepinephrine administration was confirmed. These results indicate renal vasoconstriction elicited by endogenously released or exogenously administered norepinephrine evokes the release of a prostaglandin E. The mechanism of this effect is not attributable to a decrement in total renal blood flow, since flow was maintained constant in these experiments. The quantity of PG released from the kidney which reaches the systemic circulation is probably not sufficient to produce a hypotensive effect; however, release of renal prostaglandins may be involved in intrarenal vascular adjustments. (Authors) 2442

1684

ZOR, U., J.B. FIELD and T. KANEKO
Effect of thyrotropin, long-acting thyroid stimulator and prostaglandin E_1 on dog thyroid adenyl cyclase activity and cyclic 3',5' adenosine monophosphate.
Israel Journal of Medical Sciences. 6: 471. 1970.

Abstract only. Thyrotropin (TSH), 10 m units, rapidly increased adenyl cyclase activity (ACA) in thyroid homogenates and slices by 50 to 150% and raised cyclic 3',5'-AMP two to 30 fold. Prostaglandin E_1 increased cAMP and glucose oxidation as much as TSH. (GW) 2326

1685

ZOR, U., T. KANEKO, H.P.G. SCHNEIDER, S.M. McCANN, and J.B. FIELD
Further studies of stimulation of anterior pituitary cyclic adenosine 3',5'-monophosphate formation by hypothalamic extract and prostaglandins.
Journal of Biological Chemistry. 245: 2883-2888. 1970.

An important role has been suggested for adenyl cyclase and cyclic adenosine 3',5'-monophosphate (cAMP) in the control of release of hormones from the anterior pituitary. The study reports on the regulation of cAMP concentrations in rat anterior pituitary gland and its relationship to luteinizing hormone (LH) release. In the course of the study it was found that prostaglandins added in vitro increased formation of cyclic AMP by anterior pituitary, but there was a marked variation in their potency. PGE_1 was the most effective and $PGF_{1\alpha}$ the least. However, none of the prostaglandins increased luteinizing hormone release in vitro.

Hypothalamic extract and prostaglandin E_1 stimulated ^{3}H-adenine incorporation into ^{3}H-labeled cAMP in intact anterior pituitary glands. In this study 0.1 mg per ml of PGE_1 produced significant stimulation. Although PGA_1, PGB_1, and $PGF_{1\alpha}$ also increased cAMP levels in the anterior pituitary, marked differences in effective doses were apparent. (CWS) 2272

1686
ZOR, U.,T. KANEKO, H.P.G. SCHNEIDER, S.M. McCANN, and J.B. FIELD
   Stimulation of anterior pituitary adenyl cyclase activity, cyclic 3',5'-adenosine-monophosphate
      and luteinizing hormone release by hypothalamic extract and prostaglandin E_1.
   Israel Journal of Medical Sciences. 6: 318. 1970.

Abstract only. The role of adenyl cyclase activity and cyclic 3',5'-adenosine-monophosphate in release of anterior pituitary hormones mediated by hypothalmic extract was studied in rat in vitro. Prostaglandin E_1 (0.6 $\mu g/ml$) augmented adenyl cyclase activity (150%) and increased cAMP from 15.5 to 169 nmole/g in anterior pituitary. (GW) 2325

SUBJECT INDEX

0618 0620 0622 0625 0627 0630 0631 0634
0635 0639 0640 0643 0649 0653 0656 0658
0659 0663 0664 0665 0667 0670 0671 0680
0682 0683 0684 0686 0687 0688 0689 0691
0695 0696 0698 0699 0700 0701 0702 0704
0712 0713 0717 0718 0719 0721 0722 0724
0726 0728 0729 0732 0733 0736 0738 0743
0745 0746 0747 0749 0752 0755 0756 0757
0758 0759 0760 0761 0762 0763 0764 0766
0767 0769 0777 0785 0788 0793 0794 0796
0797 0798 0799 0800 0802 0818 0819 0825
0827 0829 0830 0831 0832 0833 0835 0842
0844 0845 0849 0850 0855 0856 0858 0860
0865 0896 0901 0902 0907 0909 0914 0915
0916 0920 0921 0922 0924 0925 0927 0930
0931 0934 0936 0940 0942 0947 0948 0950
0955 0956 0957 0963 0966 0967 0968 0969
0970 0975 0976 0977 0978 0980 0985 0986
0987 0989 0990 0993 0994 0995 0997 0999
1000 1002 1003 1004 1007 1008 1009 1011
1013 1016 1017 1018 1019 1020 1023 1024
1025 1027 1033 1034 1035 1037 1038 1039
1044 1048 1052 1054 1056 1057 1058 1059
1063 1073 1074 1075 1077 1078 1079 1080
1084 1085 1086 1087 1088 1090 1093 1094
1095 1097 1099 1102 1103 1106 1107 1108
1112 1113 1114 1115 1116 1121 1122 1123
1128 1132 1133 1135 1138 1140 1141 1142
1144 1145 1146 1153 1155 1156 1157 1158
1161 1164 1165 1166 1167 1169 1170 1196
1200 1208 1209 1214 1215 1216 1224 1225
1226 1228 1229 1236 1237 1238 1239 1243
1244 1247 1249 1250 1252 1256 1258 1259
1260 1262 1267 1277 1278 1279 1281 1282
1293 1295 1296 1297 1299 1300 1301 1303
1304 1306 1308 1309 1310 1321 1322 1324
1325 1326 1327 1328 1329 1331 1333 1335
1336 1338 1348 1353 1355 1356 1357 1358
1359 1360 1368 1369 1371 1372 1373 1374
1376 1377 1380 1381 1382 1384 1386 1387
1398 1399 1402 1404 1409 1410 1411 1412
1414 1418 1419 1421 1422 1423 1424 1425
1426 1429 1430 1432 1435 1436 1439 1440
1442 1447 1448 1449 1450 1453 1454 1455
1460 1469 1471 1472 1473 1475 1476 1477
1478 1484 1488 1489 1490 1492 1493 1494
1496 1497 1499 1500 1501 1502 1503 1504
1505 1506 1507 1511 1513 1515 1516 1517
1518 1519 1521 1523 1524 1536 1537 1538
1548 1550 1559 1560 1562 1563 1564 1566
1567 1568 1569 1570 1572 1573 1580 1584
1585 1588 1589 1591 1592 1593 1595 1596
1598 1601 1603 1604 1606 1607 1614 1615

1621 1624 1629 1630 1632 1636 1640 1641
1642 1643 1644 1647 1652 1653 1654 1658
1660 1663 1668 1672 1674 1675 1676 1678
1679 1685

PGE$_2$ 0049 0050 0052 0056 0060
0062 0064 0066 0074 0076 0081 0087 0089
0092 0093 0101 0105 0109 0111 0114 0115
0117 0148 0162 0170 0171 0181 0182 0188
0189 0202 0204 0207 0231 0275 0277 0317
0319 0322 0327 0342 0362 0371 0378 0379
0381 0388 0391 0395 0402 0404 0429 0449
0450 0451 0456 0461 0472 0489 0493 0494
0495 0496 0517 0528 0530 0535 0537 0538
0539 0547 0550 0563 0574 0594 0595 0598
0603 0604 0605 0606 0622 0627 0630 0635
0637 0638 0640 0660 0661 0678 0696 0701
0707 0708 0709 0710 0713 0714 0715 0718
0724 0733 0738 0743 0749 0755 0762 0767
0777 0786 0796 0797 0800 0819 0835 0837
0842 0856 0858 0862 0863 0864 0865 0893
0903 0906 0913 0921 0922 0934 0955 0956
0957 0963 0964 0969 0975 0981 0993 0994
1020 1021 1024 1026 1027 1031 1032 1046
1047 1056 1063 1068 1069 1072 1080 1083
1102 1113 1122 1132 1140 1141 1146 1154
1162 1172 1174 1177 1211 1213 1229 1237
1248 1252 1256 1258 1264 1265 1266 1268
1272 1273 1274 1279 1293 1294 1295 1296
1297 1304 1308 1309 1310 1311 1315 1316
1329 1349 1351 1352 1354 1356 1364 1368
1369 1372 1376 1380 1386 1387 1398 1400
1401 1402 1403 1404 1405 1407 1408 1412
1416 1418 1419 1429 1431 1434 1446 1448
1449 1456 1457 1462 1463 1465 1466 1467
1468 1495 1513 1526 1529 1530 1548 1551
1560 1564 1565 1566 1569 1591 1592 1593
1594 1601 1605 1614 1615 1621 1627 1629
1630 1631 1633 1638 1641 1642 1643 1648
1654 1665 1669 1672 1678

PGE$_3$ 0050 0052 0056 0066 0074 0075
0076 0081 0087 0088 0093 0101 0105 0109
0111 0116 0148 0181 0182 0202 0204 0438
0858 0865 1024 1031

PGF, PGF$_1$, PGF$_2$, PGF$_\alpha$ 0018 0028
0033 0041 0042 0043 0107 0172 0176 0179
0223 0533 0713 0726 0815 0823 0858 0967
0979 1027 1157

PGF$_{1\alpha}$ 0051 0055 0056 0066 0074
0076 0092 0093 0105 0111 0137 0148 0152
0169 0174 0202 0204 0208 0253 0254 0256
0275 0277 0302 0331 0346 0354 0371 0374
0379 0381 0404 0421 0423 0431 0435 0436
0438 0449 0450 0451 0456 0461 0513 0528

AUTHOR INDEX

466

JOURNAL INDEX

APPENDIX

I. DIRECTORY OF PROSTAGLANDIN RESEARCH

ARGENTINA

Buenos Aires Medical School
Ramos Mejia Hospital
Buenos Aires

Universidad Nacional de Cordoba
Instituto de Fisiologia
Facultad de Ciencias Medicas
Santa Rosa 1085
Cordoba

Instituto de Investigacion Medica
Mercedes y Martin Ferreyra
Casilla de Correro 389
Cordoba

Instituto de Fisiolgia
Catedra de Bioquimica
Facultad de Ciencias Medicas
Calle 60 y 120
La Plata

AUSTRALIA

S. S. Cameron Laboratory
Reproduction Research Section
Department of Physiology
Werribee 3030

Commonwealth Scientific and Industrial
Research Organization
Division of Animal Physiology
P. O. Box 239
Prospect, New South Wales 2148

University of Melbourne
Department of Physiology
Parkville, Victoria

Monash University
Department of Physiology
Clayton, Victoria 3168

University of Sydney
Department of Obstetrics and Gynecology
Sydney, New South Wales 2006

BRAZIL

Federal University of Bahia
Department of Obstetrics
Maternidade Climerio de Oliveria
Bahia

CANADA

University of Manitoba
Department of Physiology
Winnipeg, Manitoba R3E 073

McGill University
Donner Laboratory of Experimental
Neurochemistry
Montreal Neurological Institute
Montreal, Quebec

Montreal General Hospital
Department of Obstetrics and Gynecology
Montreal, Quebec

University of Western Ontario
Department of Obstetrics and Gynecology,
and
Department of Physiology
London 72, Ontario

FINLAND

University Central Hospital
Department II of Obstetrics and
Gynecology
SF-00290 Helsinki 29

Turku University School of Medicine
Turku

FRANCE

Maternité de Port-Royal
Laboratoire de Chimie Hormonale
23, Boulevard de Port-Royal
75014 Paris

Université Paris — Sud
Institut de Biochemie
Centre d'Orsay
19 Orsay

Université René-Descartes
Laboratoire d'Histologie-Embryologie
45 Rue des Sants-Pères
75006 Paris

GERMANY

Universität Bochum
Bundeswehrzentralkrankenhauses
Essen der Ruhr
Bochum

Max-Planck Institute for Experimental
Medicine
Department of Biochemical Pharmacology
Göttingen

Universität Hamburg
Frauenklinic, Abteilung für klinische
und experimentelle Endokrinologie
Hamburg

HUNGARY

Semmelweiss University
Women's Clinic No. II
Budapest

INDIA

All India Institute of Medical Sciences
Department of Obstetrics and Gynecology
New Delhi 16

Baja Peary Mohan College
Department of Physiology
Uttarpara
Hooghly, West Bengal

Postgraduate Institute of Medical
Education and Research
Department of Pharmacology
Chandigarh

ISRAEL

Weizmann Institute of Science
Department of Biodynamics
Rehovot

ITALY

University of Milan
Institute of Pharmacology
Milan

University of Naples
Institute of Pharmacology
Naples

JAPAN

Aiiku Hospital
Department of Gynecology
Tokyo

Chiba University Medical School
Department of Urology
Chiba-shi, Chiba-ken

Gunma University
Department of Obstetrics and Gynecology
School of Medicine
Maebashi City 371

Hokkaido University Medical School
Department of Obstetrics and Gynecology
Sapporo, Hokkaido 060

Kyoto University
Department of Physiology
Primate Research Institute, and
Faculty of Pharmaceutical Sciences
Inuyama, Aichi

Nihon University
Department of Obstetrics and Gynecology
Tokyo

Niigata University School of Medicine
Asahi Machi Dori 1
Niigata

Sapporo Medical College
Department of Surgery
Sapporo, Hokkaido

University of Tokyo
Faculty of Medicine
Department of Obstetrics and Gynecology
Tokyo 113

MEXICO

Hospital de Gineo-Obstetricia
  Centro Medico Nacional
  Instituto Mexicano del Seguro Social

NETHERLANDS

Unilever Research
  Vlaardingen

University Hospital
  Department of Obstetrics and Gynecology
  Catharijnesingel 101
  Utrecht

NEW ZEALAND

University of Auckland
  Postgraduate School of Obstetrics and
  Gynecology
  Auckland

Otago University Medical School
  Department of Anatomy
  Dunedin

POLAND

ul M. Curie-Sklodowskej 24-A
  Bialystok

SWEDEN

Karolinska Hospital
  Department of Obstetrics and Gynecology
  S 104 01 Stockholm 60

Karolinska Institute
  Department of Medical Chemistry,
  Department of Pharmacology, and
  Department of Physiology
  S 104 01 Stockholm 60

AB Leo
  Research Laboratories
  Helsingborg

University of Lund
  Department of Obstetrics and Gynecology
  Lund

Royal Veterinary College
  Department of Medical Chemistry
  S-104 05 Stockholm 50

Sabbatsberg Hospital
  Department of Obstetrics and Gynaecology
  Kungl

University of Uppsala
  Department of Pathology
  Uppsala

SWITZERLAND

Universität Basel
  Frauenklinick
  Basel

University of Zurich
  Department of Gynecology
  Zurich

THAILAND

Chulalongkorn University
  Faculty of Science
  Department of Biology
  Bangkok

UGANDA

Makerere University Medical School
  Department of Pharmacology
  P. O. Box 7022
  Kampala

UNITED KINGDOM

Cardiff Maternity Hospital
  Department of Gynecology
  Cardiff

A. R. C. Institute of Animal Physiology
  Baraham
  Cambridge

A. R. C. Unit of Reproductive Physiology
and Biochemistry
  Animal Research Station
  30 Huntingdon Road
  Cambridge CB3 OJQ

University of Bristol
  Department of Obstetrics and Gynecology
  Bristol

University of Cambridge
  Physiology Laboratory
  Cambridge CB2 3EG

Chelsea Hospital for Women
  Institute of Obstetrics and Gynecology
  London

University of Edinburgh
  Department of Pharmacology
  1 George Square
  Edinburgh E118 9JZ

University of Glasgow
  Royal Infirmary
  Glasgow

Imperial Chemical Industries, Ltd.
  Pharmaceuticals Division
  Alderley Park
  Macclesfield, Cheshire

Institute of Psychiatry
  Department of Physiology
  De Crespigny Park
  London SE 5

King's College Hospital
  Department of Obstetrics and Gynecology
  London SE 5

Kingston Hospital
  Department of Obstetrics and Gynecology
  Wolverton Avenue
  Kingston-upon-Thames
  Surrey

The London Hospital Medical College
  Department of Pharmacology
  London E1

University of London
  Department of Pharmacology
  School of Pharmacy
  29/39 Brunswick Square
  London WCIN IAX

Miles-Ames Research Laboratories
  Stoke Poges
  Buckinghamshire

Mill Road Maternity Hospital
  Liverpool

National Institute for Research in Dairying
  Shinfield
  Reading RG2 9AT

Oxford University
  Nuffield Department of Obstetrics and
  Gynecology
  The Radcliffe Infirmary
  Oxford QX2 GHE

Queen Charlotte's Maternity Hospital
  Institute of Obstetrics and Gynecology
  Goldhawk Road
  London W6

Queen Mother's Hospital
  Glasgow G3 8SH

Royal College of Surgeons of England
  Institute of Basic Medical Sciences
  Department of Pharmacology, and
  Medical Research Council
  Department of Physiology
  Lincoln's Inn Field
  London WC2A 3PN

Royal (Dick) School of Veterinary Studies
  Department of Veterinary Physiology
  Summerhall
  Edinburgh EH9 1QH

Sheffield University
  Physiology Department
  Sheffield

United Oxford Hospitals
  Division of Obstetrics and Gynecology
  Oxford

University College
  Department of Physiology
  Cardiff

Welsh National School of Medicine
  Department of Obstetrics and Gyaecology
  Cardiff CF2 1XF

Women's Hospital
  Catharine Street
  Liverpool

UNITED STATES

Albany Medical College
  Department of Obstetrics and Gynecology
  Albany, New York

Alza Corporation
  Institute of Biological Sciences
  Palo Alto, California 94304

Batelle Memorial Institute
  Pacific Northwest Laboratory
  Richland, Washington 99352

California State College
  Department of Biological Sciences
  Hayward, California

University of California at Los Angeles
  School of Medicine
  Los Angeles, California 90024

University of California at San Diego
  Department of Obstetrics and Gynecology
  School of Medicine
  San Diego, California

University of Colorado Medical Center
  Department of Obstetrics and Gynecology
  Denver, Colorado 80220

College of Physicians and Surgeons of
Columbia University
  Department of Obstetrics and Gynecology,
  and
  International Institute for the Study of
  Human Reproduction
  New York, New York 10032

Cornell University Medical College
  Department of Obstetrics and Gynecology
  New York, New York

University of Cincinnati College of Medicine
  Cincinnati, Ohio 45229

The George Washington University Medical
Center
  Department of Anatomy
  Washington, D. C. 20037

Harvard Medical School
  Department of Obstetrics and Gynecology,
  Department of Physiology, *and*
  Laboratory of Human Reproduction and
  Reproductive Biology
  Boston, Massachusetts 02115

The Johns Hopkins University
  School of Hygiene and Public Health
  Department of Population Dynamics
  Baltimore, Maryland 21205

The Johns Hopkins University School of
Medicine
  Department of Obstetrics and Gynecology
  Baltimore, Maryland 21205

Eli Lilly and Company
  Division of Pharmaceutical Research
  The Lilly Research Laboratories
  Indianapolis, Indiana 46206

Merck Institute for Therapeutic Research
  Rahway, New Jersey 07065

University of Miami School of Medicine
  Department of Biochemistry,
  Department of Obstetrics and Gynecology,
    and
  Endocrine Laboratory
  Miami, Florida 33152

Michigan State University
  Department of Physiology, *and*
  Department of Medicine
  East Lansing, Michigan 48823

New York Medical College
  Flower Hospital
  New York, New York 10029

University of North Carolina
  Department of Obstetrics and Gynecology
  Chapel Hill, North Carolina 27514

Oklahoma Medical Research Fund
  Oklahoma City, Oklahoma

University of Oklahoma School of Medicine
  Department of Pharmacology, *and*
  Department of Medicine
  800 Northeast 13th Street
  Oklahoma City, Oklahoma 73104

Oregon Regional Primate Research Center
  Biochemistry Department
  Beaverton, Oregon 97005

University of Oregon
  Department of Biology
  Eugene, Oregon

National Institutes of Health
  National Institute of Arthritis and
  Metabolic Diseases
  Building 2 B1-24
  Bethesda, Maryland 20014

Pennsylvania State University
  Department of Biology
  University Park, Pennsylvania 19104

University of Pennsylvania School of
Medicine
  Division of Reproductive Biology
  Philadelphia, Pennsylvania

University of Pittsburgh School of Medicine
  Department of Physiology
  Pittsburgh, Pennsylvania 15213

The Population Council
  Bio-Medical Division
  Rockefeller University
  New York, New York 10021

St. Louis University School of Medicine
  Department of Internal Medicine
  St. Louis, Missouri 63104

G. D. Searle & Company
  Department of Endocrinology
  Chicago, Illinois 60680

Sinai Hospital
  Department of Obstetrics and Gynecology
  Detroit, Michigan

Medical University of South Carolina
  Department of Biochemistry
  Charleston, South Carolina

University of Southern California School
of Medicine
  Section of Reproductive Biology
  Department of Obstetrics and Gynecology
  Los Angeles, California 90033

Southern Illinois University
  Carbondale, Illinois

Space Sciences Research Center
  Columbia, Missouri 65201

Sterling-Winthrop Research Institute
  Rensselaer, New York 12144

Temple University School of Medicine
  Philadelphia, Pennsylvania

University of Texas Medical School at
San Antonio
  San Antonio, Texas 78229

Tulane University
  Department of Obstetrics and Gynecology
  School of Medicine
  New Orleans, Louisiana 70112

United States Naval Hospital
  Department of Obstetrics and Gynecology
  San Diego, California

The Upjohn Company
  Fertility Research
  Kalamazoo, Michigan 49001

Utah State University
  Zoology Department
  Logan, Utah 84321

Vanderbilt University
  Department of Physiology, *and*
  Department of Pharmacology
  Nashville, Tennessee 37203

Washington University School of Medicine
  Department of Obstetrics and Gynecology
  St. Louis, Missouri 63110

University of Washington
  Department of Pharmacology
  School of Medicine
  Seattle, Washington 98105

Wayne State University School of Medicine
  Department of Obstetrics and Gynecology
  Detroit, Michigan 48201

West Virginia University
  Division of Animal and Veterinary Sciences
  Morgantown, West Virginia 26506

Western Michigan University
  Biology Department
  Kalamazoo, Michigan 49001

The Western Pennsylvania Hospital
  Division of Obstetrics and Gynecology
  Pittsburgh, Pennsylvania

University of Wisconsin
  Department of Veterinary Science, *and*
  Institute for Enzyme Research
  Madison, Wisconsin 53706

University of Wisconsin
  Department of Zoology
  Milwaukee, Wisconsin

Yale University School of Medicine
  Department of Obstetrics and Gynecology
  333 Cedar Street
  New Haven, Connecticut 06510

Worcester Foundation for Experimental
  Biology
  Shrewsbury, Massachusetts 01545

YUGOSLAVIA

University of Ljubljana
  Ginecol, Porodniska Klin.
  Slajmarjeva 3
  Ljubljana

Use of Prostaglandins in Fertility Research
1965

O Clinical
● Nonclinical

1965	NUMBER OF PUBLISHED ARTICLES		
COUNTRY	Clinical	Nonclinical	Total
Italy		2	2
Japan		1	1
Sweden	5	2	7
United Kingdom		5	5
TOTALS	5	10	15

Use of Prostaglandins in Fertility Research
1970

Clinical ○
Nonclinical ●

1970	NUMBER OF PUBLISHED ARTICLES		
COUNTRY	Clinical	Nonclinical	Total
Argentina		2	2
Australia		1	1
Finland	1		1
Japan	1	2	3
New Zealand		1	1
Sweden	9	4	13
Uganda	6	2	8
United Kingdom	7	3	10
United States	2	22	24
Unknown		1	1
TOTALS	25	37	64

II. SECONDARY SOURCES OF INFORMATION IN PROSTAGLANDIN RESEARCH

The scientific literature on prostaglandins has grown rapidly, especially since the mid-1960s when the structure of these fatty acid derivatives was determined, thus making synthesis possible. By the late 1960s this growth had begun to overwhelm researchers trying to keep abreast of prostaglandin discoveries in general, or even of prostaglandin research within their own specialty. The proliferation of prostaglandin literature is evident in Fig. 1, which shows in graph form the cumulative number of documents held by the Population Information Program of the George Washington University Medical Center, and identified by the Upjohn bibliography, which was the source material for this study. A count of the prostaglandin titles on fertility research alone showed an increase from 15 documents in 1965 to 237 in 1972 (Table 1). In 1965 published material on fertility research came from only four countries, but in 1972 researchers or writers from 23 countries produced papers. Including all fields of prostaglandin research, authors from over 40 countries are now represented in the literature.

An indication of the breadth of research in prostaglandins can be gained from the list of the journals indexed in this volume. The 20 journals with the largest number of prostaglandin articles are shown in Table 2. The topics covered by these journals include chemistry, biochemistry, pharmacology, physiology, biology and clinical medicine. An individual scientist would work with a subtopic such as the role of prostaglandins in the neuroendocrinology of the eye, or perhaps the effect of prostaglandins on capacitation of sperm, under the broad topic of physiology. The journal list also suggests the countries where the major research is performed. As can be seen, most research originated in the United States, Sweden, and Britain, plus some from other European countries and Japan. A majority of these top 20 journals carry full-length research reports, although several, such as *Nature*, *Federation Proceedings*, *Biochimica et Biophysica Acta*, *Lancet* and *Science*, also publish a significant number of letters, abstracts and brief communications.

To handle this increasing number of publications, librarians and researchers depend on secondary sources of information on prostaglandins such as bibliographies, indexing, alerting, and abstracting services, of which 12 were available

in English in 1970 to 1971. Although none of these services retrieved exclusively in prostaglandins, all provided some citations for prostaglandin research. A subscriber must select a service by the cost, size of data base, currency, availability and suitability of the index. To aid a prostaglandin scientist in his choice, an analysis of secondary sources of information on prostaglandin research was conducted, based on citations provided by the 12 sources up to 1971.

A list of the sources is provided in Table 3, together with their availability, address and telephone, a cost estimate for 1973, and a brief description of their output.

In addition to the services listed in Table 3, the monthly journal *Prostaglandins*, first published in 1972, (333 Cedar Street, New Haven Connecticut 06510) provides a subject-indexed listing of key and current prostaglandin citations prepared by Dr. James R. Weeks of the Upjohn Company. Another comprehensive and timely, but not specifically subject-indexed bibliography is available from the Institute for Scientific Information (325 Chestnut Street, Philadelphia, Pennsylvania 19106) as a weekly printout, ASCAtopics®, DA25, PROSTAGLANDINS.

This analysis used as a reference data base the comprehensive bibliography of the prostaglandin literature, *The Prostaglandins*, issued by the Upjohn Company, pioneers in prostaglandin research and leading U.S. producers. This list is updated and issued periodically, but is indexed only by author and journal. It is not therefore convenient for most researchers who want more selective information.

Each of the 1500 Upjohn citations for the 1906-1970 literature was mounted on an edge-notched index card, and coded for first author, journal, year of publication, type of article (such as journal or abstract), subject category, and year of retrieval. After the output of the other services was obtained, *unique* citations from their lists, that were not included in the Upjohn bibliography, were furnished to Upjohn.

Then each of the secondary services was examined and each citation entered by service name into the reference base of

483

coded Upjohn citations. Biological Abstracts Previews, Chemical Abstracts, Excerpta Medica, MEDLARS, and Science Citation Index (Permuterm Index under the term "Prostaglandin" only) were examined as computer print-outs. The remaining six were searched manually.

A summary of the yearly retrieval of Upjohn citations by the 12 other services is shown in Table 4. MEDLARS and Biological Abstracts Previews consistently retrieved from 35 percent to 60 percent of the Upjohn list; Rindoc, S.C.I. Permuterm, and Excerpta Medica improved their retrieval each year, finally retrieving 43 to 54 percent of the Upjohn data base. These services have a more general orientation than, for example, Index Chemicus and Bibliography of Reproduction, whose narrow domains limited the number of citations.

The data were analyzed in greater depth for the years 1969 and 1970. This analysis showed that Ringdoc, S.C.I. Permuterm, Excerpta Medica and MEDLARS published the largest number of unique citations, that is, citations found by none of the other 11 services. As shown in Table 5 these large index services produced from 12 to 29 unique citations.

Further analysis revealed that three-fourths or more of the documents cited by all but one of the services were full articles rather than abstracts, editorials, reviews, etc. (Table 6). The exception was BioResearch Index, which listed only 35 percent journal articles. Biological Abstracts and Index Chemicus offered citations of complete journal articles only.

To judge the time lag between the publication of an article and its retrieval, the 1970 articles retrieved in 1970 were expressed as a percentage of all 1970 articles retrieved from January 1970 through December 1971. These currency rates are shown in Table 7. The most rapid services were Index Chemicus (93 percent) and Chemical Titles (89 percent), probably because of their limited subject matter and journal variety. Next were S.C.I. Permuterm, Bibliography of Reproduction and Ringdoc, all above 72 percent. The slowest services were the Biological Abstracts group, 20 percent currency at one year later, but including abstracts as well as bibliographic citations.

The 1969-1970 citations were individually coded according to subject matter—biological, chemical, clinical or general—as an aid in choosing a service by its coverage. Table 8 shows the number of retrievals in each subject category for each service. Percentages, calculated as the proportion of retrieval by each service as compared with the total included by Upjohn in that category, give an indication of each service's performance.

Most of Upjohn's documents were biological, and this trend is apparent in all the other services, except Index Chemicus. By percent of Upjohn's data base, B.A. Previews, Chemical Biological Activities and Drug Literature Index found the

greatest percentage of biological items. The comprehensive services, Ringdoc, MEDLARS, S.C.I. Permuterm, Drugdoc and Excerpta Medica, found the greatest overall number of biological titles, however, in addition to their high retrieval in other categories.

Chemical retrievals were near complete in S.C.I. Permuterm, Chemical Abstract Condensates, Ringdoc and Chemical Titles. Only about half of Upjohn's chemically oriented articles were found by Pandex, B.A. Previews, MEDLARS, Excerpta Medica, Drugdoc and Index Chemicus.

Clinical titles were well represented, not in a specialized service, but typically by the comprehensive services: MEDLARS, S.C.I. Permuterm, Ringdoc, Excerpta Medica/Drugdoc, and B.A. Previews which secured a high proportion of Upjohn's citations in all categories.

General articles, often reviews covering information included in the other three categories, were recovered to various extents by the 12 secondary services. The highest proportion were retrieved by Chemical Abstracts Condensates, MEDLARS and Drug Literature Index. The lowest proportion of Upjohn's general articles were cited by Index Chemicus and BioResearch Index, which had the highest proportion of scientific research journal articles of all the services.

Practical considerations in choosing more than one secondary service, beyond format, content and cost, involve primarily the combined number of non-identical citations that two or more services can supply. These figures were computed, for 1969-1970, with the help of the coded index cards. The total number of citations from any two services together are shown in Table 9. This number of identical citations retrieved by any two services in common is shown in Table 10. Several services show high numbers of combined citations; Excerpta Medica and Ringdoc (522), MEDLARS and Ringdoc (472), B.A. Previews and S.C.I. Permuterm (468), Ringdoc and S.C.I. Permuterm (462), and MEDLARS and B.A. Previews (458) have the largest number of citations, but all of these combinations also have many duplications ranging from 212 (Excerpta Medica and Ringdoc) to 256 (Ringdoc and S.C.I. Permuterm) as shown on Table 10.

The advantage gained by adding a second service to an initial service is demonstrated in Table 11. Two sets of figures, the total for a given service and the combined total for two services minus duplicates, were computed as percentages to generate this information. This table indicates the percentage of service II covered by service I. All services are listed both as service I and service II. Table 11 indicates, for instance, that although Ringdoc covers 90 percent of what is included in Chemical Titles, Chemical Titles only covers 14 percent of what is included in Ringdoc. Therefore if a researcher is subscribing to Chemical Titles, it could be useful also to order Ringdoc, but not the reverse.

Conclusions

The major secondary sources of information for researchers in the field of prostaglandin research have been examined and compared to determine the number of documents cited in a given period; the type of subject matter emphasized; the currency of service; the number of citations retrieved; and the degree of duplication found with any possible pairing of services.

All services except for one reflect the heavy emphasis on biological research. Chemically oriented papers are emphasized in six services, but only one service focusses primarily on clinical research. Reviews on the overall status of prostaglandin research were commonly cited in only four services. The ability to index current material rapidly is evident in nine services which show better than 50 percent currency rates. One service showed an 89 percent currency rate, (Chemical Titles) and another 93 percent currency rate (Index Chemicus).

The researcher obtaining citations from one service may want to increase his information flow by adding a second service. But the data presented shows that the selection of a second service should be made with care in that some services may duplicate others by more than 90 percent.

Since different readers or researchers may have different information requirements, it is not possible to evaluate which service is best overall. The preceding analysis is intended only to highlight certain salient characteristics of each service so that researchers may better choose which service of services best meets the individual need.

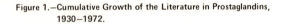

Figure 1.—Cumulative Growth of the Literature in Prostaglandins,
1930—1972.

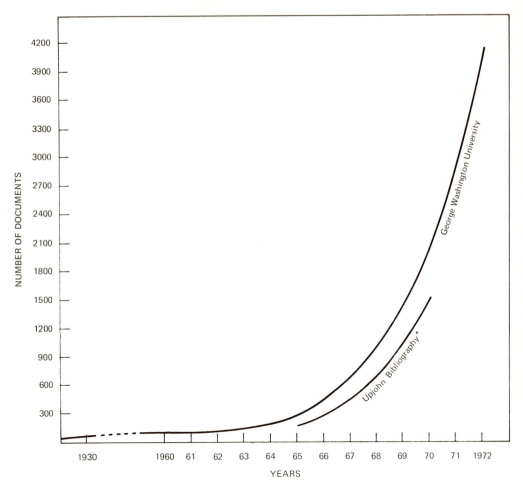

*This curve represents the Upjohn *journal* prostaglandin citations for the years covered by this study. The George Washington University curve is cumulative for *all* prostaglandin citations including popular articles.

Table 1.—Growth of the Prostaglandin Literature on
Fertility Research.

Date	COUNTRY	NUMBER OF PUBLISHED ARTICLES		
		Clinical	Nonclinical	Total
1965	Italy		2	2
	Japan		1	1
	Sweden	5	2	7
	United Kingdom		5	5
	TOTALS	5	10	15
1970	Argentina		2	2
	Australia		1	1
	Finland	1		1
	Japan	1	2	3
	New Zealand		1	1
	Sweden	9	4	13
	Uganda	6	2	8
	United Kingdom	7	3	10
	United States	2	22	24
	Unknown		1	1
	TOTALS	25	37	64
1971	Argentina		5	5
	Australia		2	2
	Brazil	2		2
	Finland	1		1
	France		1	1
	Germany		3	3
	Israel		1	1
	Japan	6	2	8
	Netherlands	1		1
	New Zealand		3	3
	Sweden	9	4	13
	Uganda	10		10
	United Kingdom	10	12	22
	United States	8	35	43
	Unknown		2	2
	TOTALS	48	69	117
1972	Argentina		2	2
	Australia	1	11	12
	Canada	5	7	12
	Finland	3		3
	France		3	3
	Germany	4		4
	Hungary		1	1
	India	1	3	4
	Israel		3	3
	Italy		1	1
	Japan	7	3	10
	Mexico	1	1	2
	Netherlands	1		1
	New Zealand		1	1
	Poland		1	1
	Sweden	12	2	14
	Switzerland	2		2
	Thailand		1	1
	Uganda	4		4
	United Kingdom	22	19	41
	United States	32	80	112
	Unknown	2		2
	Yugoslavia	1		1
	TOTALS	98	139	237

Table 2.—Journals in the Prostaglandin Bibliography In
Order of Number of Prostaglandin Articles
Published 1906-1970.

1. Clinical Research (71)

2. British Journal of Pharmacology (63)

3. Acta Physiologica Scandinavica (56)

4. Nature (48)

5. Journal of Physiology (47)

6. Federation Proceedings (45)

7. Biochimica et Biophysica Acta (33)

8. Lancet (31)

9. Journal of Biological Chemistry (28)

10. Life Sciences (28)

11. Pharmacologist (24)

12. Circulation (23)

13. American Journal of Physiology (23)

14. Journal of the American Chemical Society (18)

15. Journal of Clinical Investigation (18)

16. Science (17)

17. British Medical Journal (14)

18. Naunyn-Schmiedebergs Archiv fur Pharmakologie und
Experimentelle Pathologie (14)

19. Proceedings of the Society for Experimental Biology
and Medicine (14)

20. Shinryo (10)

Table 3.—Secondary Information Services Covering Prostaglandin Research, Including Selective Dissemination of Information (SDI) Costs.

Service*	Most commonly available nonspecialized information package	Yearly cost in dollars	Computer based from year	Approximate cost of prostaglandin SDI in dollars	Remarks	Address of Marketing Offices
1. Bibliography of Reproduction	Monthly, printed. Bibliographic citations. Subject indexed	73.00		2.50 for 10 citations. 0.25 each additional citation.	Manually operated data base.	141 New Market Road Cambridge, CB58HA England Phone: 0223-66675
2. Biological Abstracts	Biweekly, printed. Bibliographic citations, abstracts. Keyword, author, subject indexes.	1,000.00	1960		SDI: Standard profile #831, Prostaglandins. Monthly printout of citations from Biological Abstracts and BioResearch Index.	BioSciences Information Service of Biological Abstracts 2100 Arch St. Philadelphia, Pa. 19103 Phone: 215-568-4016
BioResearch Index	Monthly, printed. Bibliographic citations. Keyword, author indexes.	500.00	1966	50.00		
B.A. Previews	Biweekly, tape. Bibliographic citations. Keyword, subject indexing.	Contact Marketing.	1966		Biological Abstracts and BioResearch Index references available on tape before printed version.	
3. Chemical Biological Activities	Biweekly, printed. Equivalent to sections 1-5 of Chemical Abstracts. Bibliographic citations, research digests. Keyword, author, molecular formula, registration number indexes.	70.00	1965		SDI: Available through individual information centers.	Chemical Abstracts Service Division of the American Chemical Society The Ohio State University Columbus, Ohio 43210 Phone: 614-421-6940
4. Chemical Abstracts Condensates	Tape. (Only printed version is Chemical Abstracts @ $2,900 per year).	4,000.00	1968	Depends on individual centers. Contact Marketing.	Tape version of Chemical Abstracts Index.	
5. Chemical Titles	Biweekly, printed. Bibliographic citations, phrases. Keyword, journal, subject indexes.	60.00	1962		Current awareness service.	
	Tape.	1,700.00				
6. Drug Literature Index	Monthly, printed. Bibliographic citations. Keyword, subject, author indexes.	Contact Marketing.	1969	Depends on individual centers. Contact Marketing.	Printed version of Drugdoc.	Excerpta Medica Foundation P.O. Box 1126 Amsterdam, The Netherlands Phone: 020-222138
7. Drugdoc/Excerpta Medica	Weekly tape service. Tradename, chemical name, biomedical indexing.	15,000.00	1969		A subset of Excerpta Medica tapes.	

Table 3.—Continued

Service*	Most commonly available nonspecialized information package	Yearly cost in dollars	Computer based from year	Approximate cost of prostaglandin SDI in dollars	Remarks	Address of Marketing Offices
8. Index Chemicus	Weekly, printed. Bibliographic citations, abstracts, Keyword, molecular formula, author indexes.	950.00	1967	100.00	SDI: Called ANSA. Issued 12 per year. Index Chemicus now called Current Abstracts of Chemistry.	Institute for Scientific Information 325 Chestnut St. Philadelphia, Pa. 19106 Phone: 215-923-3300
9. MEDLARS	Monthly, printed version is Index Medicus @ $113.00 a year. Bibliographic citations. Keyword, author journal indexing.	SDI only.	1969	Depends on individual Medline center.	Searches available through individual Medline centers.	Mid Atlantic Regional Medical Library P.O. Box 30260 Bethesda, Md. 20014 Phone: 301-496-5116
10. Pandex	Biweekly, printed. Bibliographic citations, abstracts. Keyword, subject, author indexes.	360.00	1970	No direct search service provided.	SDI search programs provided with tape service subscription.	CCM Information 866 Third Ave. New York, N.Y. 10022 Phone: 212-935-4296
	Weekly tape service.	4,000.00				
11. Ringdoc	Tape. Phrases, keyword, subject indexing.	11,750	1964	600.00	SDI: Not available for general use. Only in house, for subscribers.	Derwent Publications Ltd. Rochdale House 128 Theobald Rd. London WC1XRP Phone: 01-242-5823
12. SCI Permuterm	Quarterly, printed. Bibliographic citations. Keyword, journal, organization, author, cited author, cited patent indexes.	1,500.00 to 2,550.00 depending on indexes selected.	1961	95.00 or 135.00	ASCAtopics weekly computer printout of standard profile, $95.00. ASCA weekly printout of individual profile, $135.00.	Institute for Scientific Information 325 Chestnut St. Philadelphia, Pa. 19106 Phone: 215-923-3300

*The author thanks the representatives of all the services listed who provided materials for the performance of this study. Particular gratitude is expressed to Donn Casey of the Bibliography of Reproduction; BIOSIS of Biological Abstracts; Knowledge Availability Systems Center, Philadelphia, Pennsylvania; Excerpta Medica Foundation; The National Lending Library, Boston Spa, England; Smith Kline and French Ltd., Welyn Gardens City, England; and to the Institute for Scientific Information, Philadelphia, Pennsylvania.

Table 4.—Number and Percent of Upjohn Citations Retrieved by Other Services, by Date, pre-1965 to 1970.

Service	Pre 1965		1966		1967		1968		1969		1970		Total	
	No.	%	No.	%	No.	%	No.	%	No.	%	No.	%	No.	%
*Upjohn citations	182	100	98	100	191	100	289	100	355	100	400	100	1515	100
Upjohn not retrieved by any service	56	31	20	20	28	15	15	5.2	91	26	46	12	256	17
1. Bibliography of Reproduction	71	39	31	32	32	17	50	27	46	13	72	18	302	20
Biological Abstracts (BA)	85	47	36	37	85	45	80	42	78	22	112	28	476	31
2. Bioresearch Index (BIOR)	6	3.3	8	8	30	16	78	27	73	21	82	21	277	18
BA Previews (BA + BIOR)[+]	91	50	44	45	115	60	158	55	131	37	194	49	733	48
3. Chemical Abstracts Condensates[+]	—	—	—	—	47	25	109	38	141	40	152	38	449	30
4. Chemical Biological Activities	21	12	10	10	36	19	59	20	78	22	121	30	325	22
5. Chemical Titles	73	40	31	32	38	20	61	21	94	27	102	26	399	26
6. Drug Literature Index	—	—	—	—	—	—	—	—	93	26	138	35	231	15
7. Excerpta Medica/Drugdoc[+]	—	—	—	—	—	—	28	9.7	146	41	169	43	343	23
8. Index Chemicus	9	5	5	5	2	1	13	4.5	25	7.0	28	7	82	5.4
9. MEDLARS[+]	74	41	37	38	66	35	135	47	145	41	206	52	663	42
10. Pandex	—	—	—	—	25	13	35	12	79	22	126	32	265	24
11. Ringdoc	37	20	38	39	79	41	131	45	156	44	217	54	658	43
12. SCI Permuterm[+]	**	—	36	37	65	34	114	39	149	42	206	52	570	38

*Journal + non journal citations
+Computer search
**Citation searching but Permuterm not available 1961-1965

Table 5.—Number of Unique Citations Retrieved by the
Secondary Sources in 1969-1970.

Service	Number of unique citations
1. Bibliography of Reproduction	4
2. ⎧ Biological Abstracts (B.A.)	1
⎨ BioResearch Index (BioR.)	9
⎩ B.A. Previews (B.A. + BioR.)	10
3. Chemical Abstracts Condensates	4
4. Chemical Biological Activities	5
5. Chemical Titles	0
6. Drug Literature Index	9*
7. Excerpta Medica/Drugdoc	13
8. Index Chemicus	0
9. MEDLARS	12
10. Pandex	0
11. Ringdoc	29
12. S.C.I. Permuterm	16
Total Unique	103

*Unique citations of the Drug Literature Index also retrieved by
Excerpta Medica

Table 6.—Number of Full Articles Retrieved in 1969+1970
by the Secondary Sources as a Percentage of Total
Journal Retrievals, Including Articles, Abstracts,
Reviews, Editorials.

Service	Total journal retrievals	Number of full articles retrieved	Percent
1. Bibliography of Reproduction	110	93	84
2. Biological Abstracts (B.A.)	188	188	100
BioResearch Index (BioR.)	151	51	35
B.A. Previews (B.A. + BioR.)	339	239	71
3. Chemical Abstracts Condensates	279	272	98
4. Chemical Biological Activities	199	195	97
5. Chemical Titles	196	189	96
6. Drug Literature Index	230	185	80
7. Excerpta Medica/Drugdoc	303	259	82
8. Index Chemicus	52	52	100
9. MEDLARS	351	313	90
10. Pandex	205	190	93
11. Ringdoc	367	266	73
12. S.C.I. Permuterm	351	262	77
Upjohn	641	430	73

494

Table 7.—Currency Rate of Services by Number of 1970 Citations Retrieved in 1970 as a Percent of All 1970 Citations Retrieved in 1970 and 1971.

Service	Number of 1970 citations retrieved in 1970	Number of 1970 citations retrieved 1970-1971	Percent of 1970 currency
1. Bibliography of Reproduction	42	68	74
2. ⎰ Biological Abstracts (B.A.)	22	111	20
⎱ BioResearch Index (BioR.)	15	73	20
B.A. Previews (B.A. + BioR.)	37	184	20
3. Chemical Abstracts Condensates	80	147	54
4. Chemical Biological Activities	81	121	66
5. Chemical Titles	91	102	89
6. Drug Literature Index	44	137	32
7. Excerpta Medica/Drugdoc	83	168	49
8. Index Chemicus	26	28	93
9. MEDLARS	105	205	51
10. Pandex	80	126	64
11. Ringdoc	155	215	72
12. S.C.I. Permuterm	152	205	74

Table 8.—Retrieval of 1969+1970 Upjohn Journal Prostaglandin Citations by the Secondary Services by Category, Number and Percent.

Service	Biological		Chemical		Clinical		General		Total	%
	No.	%	No.	%	No.	%	No.	%		
Total Upjohn articles	427		88		107		19		641	100
Upjohn articles retrieved by other services.	403	94	79	90	91	85	15	79	588	92
1. Bibliography of Reproduction	42	10	30	23	32	30	6	32	110	17
Biological Abstracts	136	32	25	28	25	23	2	11	188	29
2. BioResearch Index	104	24	22	25	24	22	1	5	151	24
BA Previews	240	56	47	53	49	46	3	16	339	53
3. Chemical Abstracts Condensates	183	43	65	74	22	21	9	47	279	44
4. Chemical Biological Activities	175	41	14	16	9	8	1	5	199	31
5. Chemical Titles	130	30	60	68	5	5	1	5	196	31
6. Drug Literature Index	170	40	18	20	35	33	7	37	230	36
7. Excerpta Medica/Drugdoc	204	48	43	49	51	48	6	31	304	47
8. Index Chemicus	11	26	41	47	0	0	0	0	52	8
9. MEDLARS	238	56	45	57	60	56	8	42	351	55
10. Pandex	122	29	47	53	34	32	2	10	205	32
11. Ringdoc	250	58	62	70	52	49	3	16	367	57
12. SCI Permuterm	228	53	71	80	48	53	4	21	351	55

496

Table 9.—Number of 1969+1970 Upjohn Journal Citations (N=630) Retrieved by Two Secondary Information Services.

B.A. Previews

401	Chemical Biological Activities
427	312	Chemical Condensates
434	402	438	Ringdoc
441	376	419	522	Excerpta Medica/Drugdoc
399	310	337	434	338	Drug Literature Index
458	378	408	472	436	400	MEDLARS
373	278	324	425	350	283	373	Bibliography of Reproduction
391	258	319	407	365	306	379	263	Pandex
468	404	407	462	440	393	460	375	351	SCI Permuterm
396	248	321	386	374	320	384	260	264	364	Chemical Titles
357	231	282	371	336	261	366	141	222	349	199	Index Chemicus

Table 10.—Number of 1969+1970 Upjohn Prostaglandin Journal Citations Retrieved by Two Secondary Services In Common.

B.A. Previews

131	Chemical Biological Activities
185	166	Chemical Condensates
216	164	208	Ringdoc
205	136	173	212	Excerpta Medica/Drugdoc
155	110	163	154	196	Drug Literature Index
226	172	222	245	228	172	MEDLARS
70	31	65	52	83	48	88	Bibliography of Reproduction
147	98	165	165	153	120	177	52	Pandex
216	146	223	256	224	179	242	86	195	SCI Permuterm
133	147	194	177	135	97	163	46	137	183	Chemical Titles
29	21	50	49	30	13	38	22	36	45	50	Index Chemicus

Table 11.—Percent Overlap in 1969+1970 Upjohn Prostaglandin Citations by Two Secondary Services.

Service I*		Biological Abstracts	BioResearch Index	Biological Abstracts Previews	Chemical Biological Activities	Chemical Abstracts Condensates	Ringdoc	Excerpta Medica/Drugdoc	Drug Literature Index	MEDLARS	Bibliography of Reproduction	Pandex	SCI Permuterm	Chemical Titles	Index Chemicus
	Biological Abstracts	●	●	●	54	52	37	41	48	50	44	54	39	52	34
	BioResearch Index	●	●	●	12	15	22	24	23	14	19	21	28	16	34
	Biological Abstracts Previews	●	●	●	66	66	59	65	70	65	64	72	62	68	55
	Chemical Biological Activities	57	16	39	●	59	45	43	50	49	28	48	42	75	40
	Chemical Abstracts Condensates	77	29	56	83	●	57	55	74	63	59	80	63	99	94
	Ringdoc	72	56	65	82	75	●	58	70	70	47	80	73	90	96
	Excerpta Medica/Drugdoc	70	51	62	68	62	58	●	89	65	75	75	64	69	57
	Drug Literature Index	57	34	46	55	58	42	63	●	49	44	59	51	49	25
	MEDLARS	94	34	68	86	79	67	73	77	●	80	86	69	82	72
	Bibliography of Reproduction	26	15	21	16	23	14	27	22	25	●	25	25	23	41
	Pandex	59	30	44	49	59	45	48	54	50	47	●	56	70	68
	SCI Permuterm	73	68	65	73	80	70	72	81	69	78	95	●	93	85
	Chemical Titles	54	22	40	74	70	48	43	44	46	42	67	52	●	94
	Index Chemicus	10	12	9	11	18	14	10	6	11	20	18	13	26	●

Service II (column group header)

*Note: This table should be read as follows: Service I (title) retrieved — percent of the articles retrieved by Service II (title).